Energy, Environment, and Sustainability

Series Editor

Avinash Kumar Agarwal, Department of Mechanical Engineering, Indian Institute of Technology Kanpur, Kanpur, Uttar Pradesh, India

AIMS AND SCOPE

This books series publishes cutting edge monographs and professional books focused on all aspects of energy and environmental sustainability, especially as it relates to energy concerns. The Series is published in partnership with the International Society for Energy, Environment, and Sustainability. The books in these series are edited or authored by top researchers and professional across the globe. The series aims at publishing state-of-the-art research and development in areas including, but not limited to:

- Renewable Energy
- Alternative Fuels
- Engines and Locomotives
- Combustion and Propulsion
- Fossil Fuels
- Carbon Capture
- Control and Automation for Energy
- Environmental Pollution
- Waste Management
- Transportation Sustainability

Review Process

The proposal for each volume is reviewed by the main editor and/or the advisory board. The chapters in each volume are individually reviewed single blind by expert reviewers (at least four reviews per chapter) and the main editor.

Ethics Statement for this series can be found in the Springer standard guidelines here https://www.springer.com/us/authors-editors/journal-author/journal-author-helpdesk/before-you-start/before-you-start/1330#c14214

Sudipta De · Avinash Kumar Agarwal ·
Pankaj Kalita
Editors

Challenges and Opportunities of Distributed Renewable Power

Springer

Editors
Sudipta De
Department of Mechanical Engineering
Jadavpur University
Kolkata, West Bengal, India

Avinash Kumar Agarwal
Department of Mechanical Engineering
Indian Institute of Technology Kanpur
Kanpur, Uttar Pradesh, India

Pankaj Kalita
School of Energy Science and Engineering
Indian Institute of Technology Guwahati
Guwahati, Assam, India

ISSN 2522-8366 ISSN 2522-8374 (electronic)
Energy, Environment, and Sustainability
ISBN 978-981-97-1405-6 ISBN 978-981-97-1406-3 (eBook)
https://doi.org/10.1007/978-981-97-1406-3

This Springer imprint is published by the registered company Springer Nature Singapore Pte Ltd.
The registered company address is: 152 Beach Road, #21-01/04 Gateway East, Singapore 189721, Singapore

Paper in this product is recyclable.

Preface

Currently, the world is moving through a global energy transition. Fossil fuels have supported improved living standards enormously since the industrial revolution. However, this has not only depleted the limited resources of fossil fuels rapidly but also caused a severe threat to future human survival on Earth, i.e., climate change. Emission from the use of fossil fuels is largely responsible for this possible impending disaster. Substituting fossil fuel-based power by renewable options and energy efficiency of power generation systems is a global need now. Distributed renewable power is considered to be one of the potential options to support this transition.

The International Society for Energy, Environment, and Sustainability (ISEES) was founded at the Indian Institute of Technology Kanpur (IIT Kanpur), India, in January 2014 to spread knowledge and awareness and to catalyse research activities in the fields of Energy, Environment, Sustainability, and Combustion. Society's goal is to contribute to the development of clean, affordable, and secure energy resources and a sustainable environment for society; spread knowledge in the areas mentioned above; and create awareness about the environmental challenges the world is facing today. The unique way adopted by ISEES was to break the conventional silos of specialisations (engineering, science, environment, agriculture, biotechnology, materials, and fuels, etc.) to tackle the problems related to energy, environment, and sustainability holistically. This is evident by the participation of experts from all fields to resolve these issues. The ISEES is involved in various activities, such as conducting workshops, seminars, and conferences, etc., in the domains of its interests. The society also recognises the outstanding works of young scientists, professionals, and engineers for their contributions by conferring awards in various categories.

The Seventh International Conference on "Sustainable Energy and Environmental Challenges" (VII-SEEC) was organised under the auspices of ISEES from December 16–18, 2022, at the Indian Institute of Technology, Banaras Hindu University, Varanasi (IIT-BHU), India. This conference provided a platform for discussions between eminent scientists and engineers from several countries, including India, the USA, Spain, Poland, Austria, the Czech Republic, and Korea. At this conference, eminent international speakers presented their views on energy, combustion, emissions, and alternative energy resources for sustainable development and a cleaner environment. The conference presented two high-voltage plenary talks by Dr. Ajay Kumar, Former Union Defense Secretary, GoI, and Dr. S. S. V. Ramakumar, Director (R&D), Indian Oil.

The conference included 29 technical sessions on energy and environmental sustainability topics, including two plenary talks, 15 keynote talks, 80 invited talks from prominent scientists, and 118 contributed talks by students and researchers. The conference included technical sessions on advanced engine technologies, air pollution monitoring, anaerobic digestion, combustion and flames, air pollution control, biodegradation of toxic chemicals, energy and exergy, desalination and wastewater treatment, environmental bioengineering, pollution and climate change: challenges and priorities, alternative transportation fuels and materials, emerging environmental contaminants, sustainable processing of biomass, human health and environmental sustainability, sprays and atomisation, solid waste: challenges and mitigation, solid waste management, sustainable food and agri biotechnology, modelling and simulations, renewable energy technologies, bioremediation, biofuels and biorefineries, engine emissions and control, cleaner technologies for pollution mitigation, microbial processes, energy and environment, coal and biomass gasification, and environmental challenges mitigation.

About 250+ participants and speakers attended the conference, where 14 ISEES books published by Springer, Singapore, under a dedicated series, "*Energy, Environment and Sustainability*", were released. This conference laid the roadmap for technology development, opportunities, and challenges in the Energy, Environment, and Sustainability domains. These topics are relevant for the country and the world in the present context. We acknowledge the support from various funding agencies and organisations for the successful conduct of the VII-SEEC, where the idea of these books germinated. We would therefore gratefully like to acknowledge IIT BHU (Special thanks to Prof. Akhilendra Pratap Singh); SERB, Government of India; Department of Scientific and Industrial Research (DSIR) (Special thanks to Dr. Vipin Shukla); and our publishing partner Springer (Special thanks to Swati Mehershi).

The editors would like to express their sincere gratitude to a large number of authors from all over the world for submitting their high-quality work on time and revising it appropriately at short notice. We want to express our special gratitude to our prolific reviewers: Dr. Sudip Simlandi, Dr. Prakash Ch. Roy, Dr. Santanu P. Dutta, Prof. V. S. Moholkar, Dr. Dudul Das, Dr. Hirakh Jyoti Das, Dr. Farrukh Khalid, Dr. Sandip Sarkar, Dr. Amitava Dutta, Dr. Avishek Ray, Dr. Susanta Roy, Dr. Debangshu Dey, Dr. Suddhasatwa Chakraborty, Dr. Nirmalendu Biswas, Dr. Biswajit Thakur, Dr. Arindam Dutta, Dr. Soupayan Mitra, Dr. Rajiv Ganguly, Dr. Partha Pratim Ray, Dr. Ratan Mondal, Dr. H. Lalhmingsanga, Dr. Pradip Mondal, Dr. Sudip Ghosh, Dr. Nilkanta Barman, Dr. Sudhir Murmu, and Dr. Kaustuv Pradhan who reviewed chapters of this monograph and provided their valuable suggestions to improve them.

This book explores some potential options for a possible successful energy transition using distributed renewable power. However, this transition is not expected to be very smooth specifically with respect to meeting the ever-increasing global energy demand without any compromise for affordability for all. To address this issue, a multi-disciplinary approach involving both technology and policies is required. In this book, several aspects of current technologies with possible retrofits, a few new technologies with good potentials as well as a few aspects of policies and their impacts involving distributed renewable power have been included.

Contributions to this book are on sustainability assessment tools, overview of some emerging technologies, potential new technologies, simulation and modelling of new energy systems, application case studies as well as socio-economic impacts of distributed renewable power. Chapters include recent results and are focussed on current trends in the distributed renewable energy sector. In this book, readers will get an idea about promising as well as future useful technologies of distributed renewable power, their performance assessment, modelling and simulation, application, and socio-economic assessment. A few chapters of this book are based on the overview of state-of-the-art technologies, with a special focus on the theory, development, and applications of these technologies. We hope the book greatly interests the professionals and post-graduate students involved in renewable energy in general and distributed renewable power in particular.

Jadavpur, India — Sudipta De
Kanpur, India — Avinash Kumar Agarwal
Guwahati, India — Pankaj Kalita

Contents

1 **Challenges and Opportunities of Distributed Renewable Power** 1
Sudipta De, Avinash Kumar Agarwal, and Pankaj Kalita

Part I Tool for Sustainability Assessment

2 **Application of MCDM Tool Integrated with SWOT Analysis for Prioritization of Strategies to Optimize Distributed Hybrid Energy Systems for Better Sustainability** 9
Sayan Das, Souvanik De, and Sudipta De

Part II Technology Overviews

3 **Renewable Energy Integration to Electric Power Grid: Opportunities, Challenges, and Solutions** 37
Prithwiraj Purkait, Mousumi Basu, and Sujoy Ranjan Nath

4 **The Promising Role of Thermochemical Conversion in Sustainable Power Generation** 101
Rabindra Kangsha Banik, Samar Das, Urbashi Bordoloi, Hirakh Jyoti Das, Biraj Das, Shayaram Basumatary, Bitupan Das, and Pankaj Kalita

5 **Integration of Anaerobic Digestion and Algal Cultivation—Prospects and Challenges** 141
Kaustav Nath, Rupak Jana, and Ranjana Chowdhury

6 **Renewable Thermal Energy Storage in Polymer Encapsulated Phase-Change Materials: A Comprehensive Overview** 181
Sumit Kumar, Chandrani Sarkar, and Sampa Saha

7 **Challenges and Opportunities of Hybrid Photovoltaic Thermal Collectors and Their Integration into Small-Scale Energy Systems for Prosumers** 215
Madalina Barbu, Monica Siroux, and George Darie

Part III Potential Technologies

8 Older Oilfields as Distributed Geothermal Energy Resources 261
Subha Mondal and Chitta Sahana

9 Distributed Renewables, Smart Solutions: A Blueprint for Sustainable Buildings 279
Kundayi Muchafangeyi and Harpreet Kaur Channi

10 Role of Pumped Hydro Storage to Mitigate Intermittency in Renewable Energy Systems 305
Satyajit Das Karmakar and Himadri Chattopadhyay

11 Harnessing Ocean Wave Energy—Flow Rectifications 323
Doddamani Hithaish and Abdus Samad

Part IV Modeling and Simulations

12 Mathematical Modelling of Reactors Used for Syngas Fermentation—Contemporary Practices and Challenges 353
Dinabandhu Manna, Soumitra Pati, Sudipta De, and Ranjana Chowdhury

13 Hardware-In-Loop Approach of ELC with Battery Storage as Dump Load in Off-Grid SEIG 385
Shalini Sinha and Rajen Pudur

14 Hardware-In-The-Loop Simulation Approach for Grid Integration of a 2.2 kW Self-Excited Induction Generator Using Synchronous Reference Frame Theory 401
Mrinal Kanti Rajak and Rajen Pudur

15 Performance Enhancement of Three-Phase SEIG to Feed Single-Phase Load in Micro-hydro Systems Using a Novel Capacitor Excitation Topology 427
Samrat Chakraborty and Rajen Pudur

16 Modified P-Q Control Algorithm of Shunt Active Power Filter to Reduce Harmonics, Voltage Regulation, Power Factor Correction and Unbalanced Load for Hydro Electric Power Plant 445
Jotirmoy Samanta, Ralli Sangno, and Rajen Pudur

Part V Applications

17 Distributed Power Generation Using a Pilot-Scale Downdraft Gasifier for Rural Application 469
Sweta Sharma, Ayush Garg, Ishita Sarkar, and Chanchal Loha

Part VI Socio-economic Impact Studies

18 Role of Solar Energy in the Development of the Indian Economy 489
Akash Dilip Kamble, Samar Das, Vijaya, Biraj Das, Urbashi Bordoloi, Puja Hazarika, and Pankaj Kalita

About the Editors

Dr. Sudipta De is currently Professor, Mechanical Engineering Department, Jadavpur University, India. His area of interest is sustainable energy. He received his research funding from UGC, DST (India); EU; Swedish Research Council; DFG-Germany; SIU and DIKU-Norway etc. He is an editorial board member of the Springer Nature journal 'Clean Technologies and Environmental Policy'. He received the Indian National Science Academy (INSA) Teacher Award, a lifetime award for distinguished contribution. He was the leader of the Indian Delegation in the 13th Plenary Meeting of ISO/TC 238: Solid Biofuels. He is Fellow of West Bengal Academy of Science and Technology and International Society for Energy, Environment and Sustainability.

Prof. Avinash Kumar Agarwal joined IIT Kanpur in 2001. He worked as Post-Doctoral Fellow at the Engine Research Center, UW@Madison, USA (1999–2001). His interests are IC engines, combustion, alternate and conventional fuels, lubricating oil tribology, optical diagnostics, laser ignition, HCCI, emissions and particulate control, 1D and 3D simulations of engine processes, and large-bore engines. He has published 525+ peer-reviewed international journal and conference papers, 63 edited books, 129 chapters, 16200+ Scopus, and 24300+ Google Scholar citations. He is Editor of FUEL, Associated Editor of the SAE Journal of Engines, and the ASME Open Journal of Engineering. He is India's number one Energy researcher in the recently declared Stanford University listing of the top 2% of researchers globally.

Dr. Pankaj Kalita received his bachelor's degree in mechanical engineering from the Jorhat Engineering College, (Under Dibrugarh University, Assam) and subsequently M.Tech. in Fluids and Thermal Engineering and Ph.D. degrees in clean coal technologies from the Indian Institute of Technology Guwahati. Currently, he is Associate Professor in the School of Energy Science and Engineering, Indian Institute of Technology Guwahati. He worked as Assistant Professor in the Department of Energy at Tezpur University, Assam, before joining at IIT Guwahati in 2015. He has more than 10 years of teaching experience and 16 years of research experience in the field of clean energy technologies. He has published more than 100 research articles in various reputed journals and conferences. So far, he has published 3 edited books and contributed 16 book chapters. He has completed several research and consultancy projects in the areas of clean energy technologies.

Chapter 1
Challenges and Opportunities of Distributed Renewable Power

Sudipta De, Avinash Kumar Agarwal, and Pankaj Kalita

Abstract The global energy sector is now going through a transition. Energy is a basic need of modern civilization and often per capita energy consumption is considered as an index of the living standard of people of a country. Starting from the industrial revolution till date, the majority of the energy supply has been from fossil fuel resources. As a result, sixty per cent of the global fossil fuel reserves have been consumed in the last two hundred years. Moreover, the emissions, mostly carbon dioxide, have caused a severe threat to the future human survival, called climate change. There is a desperate need to replace the fossil fuel-based power sector with suitable renewable power. However, renewable power options are widely varying with several limitations like intermittency, low density, not cost-effective, etc. In addition to large-scale renewable options, smaller distributed power generation with or without grid may be a future suitable option for a smoother energy transition. For this, opportunities are to be explored and simultaneously addressing the challenges involved for distributed renewable power. In this book, this issue is addressed from several viewpoints including sustainability. New technology assessment, overview of potential technologies, reporting of promising future technologies, modelling and simulations as well as applications and socio-economic impacts are reported in different chapters of this book. The book is a state of the art but brief compilation of the field for both researchers/senior students as well as industry professionals.

Keywords Distributed renewable power · Potential technologies · Addressing challenges · Applications · Socio-economic impacts

---

S. De (✉)
Department of Mechanical Engineering, Jadavpur University, Kolkata, India
e-mail: sudipta.de@jadavpuruniversity.in

A. K. Agarwal
Department of Mechanical Engineering, Indian Institute of Technology Kanpur, Kanpur, India

P. Kalita
School of Energy Science and Engineering, Indian Institute of Technology Guwahati, Guwahati, India

S. De et al. (eds.), *Challenges and Opportunities of Distributed Renewable Power*,
Energy, Environment, and Sustainability, https://doi.org/10.1007/978-981-97-1406-3_1

Energy sector is currently going through a global transition. Both the fast depletion of fossil fuels and threat from climate change have forced us to explore a new regime of renewable power with a paradigm shift. However, there are several obstacles to overcome during this transition. Developing suitable technologies may be the major challenge. However, retrofitting new technologies with existing infrastructure may add benefits to this process. With the global goal of sustainable development, this transition path should also comply with sustainability. In addition to large-scale renewable power, distributed renewable generation will add more flexibility to follow the path of sustainable development. This book addresses this aspect of distributed renewable power in a brief but effective compilation. Overall content of this book is distributed in six sections.

The first section includes a chapter on the demonstration of a tool integrating SWOT analysis to assess the overall sustainability of any distributed renewable power generation. As the several criteria of sustainability may not be the best for any single solution, a multi-criteria decision-making (MCDM) has to be done with the decided priority of these criteria. This method of SWOT integrated MCDM tool may be very effective for assessing the overall sustainability of any new future renewable power generation option or the existing ones. Such tools may be very useful for policymakers as well as researchers to decide the acceptable optimum solutions for renewable power.

The second section of the book includes five chapters, each one to discuss an overview of a potential emerging technology of distributed renewable power generation and utilization. Grid integration of distributed renewable power generation has several challenges. Intermittency as well as wide variability over a short period, no control on the resource input according to the demand pattern, etc. are to be addressed properly before distributed renewable generation can be integrated with the main grid. In the first chapter of this section, a good overview is presented on this issue with possible solutions to overcome the challenges. A similar comprehensive overview on the thermochemical conversion of biomass is presented in the second chapter of this section. Though the technology of thermochemical conversion of biomass is old, it has regained its importance again after a long time not only as a renewable power option but also as a way of waste recycling and circular economy towards better sustainability. Different technology options of thermochemical conversion as well as integration of those processes for efficient power generation and assessment of those processes are presented in this chapter. Biochemical process of conversion of biomass is another alternative path for biomass conversion for renewable power. It happens without any heat requirement and by an anaerobic digestion process. In addition, several algae have drawn special attention for good reasons. These can absorb carbon dioxide at a much faster rate and deliver useful value-added chemicals including biofuels. Thus, the integration of biochemical conversion processes with the cultivation of suitable algae may be a good option for distributed power generation. An extensive overview on this integrated process has been presented in a chapter of this section. Energy storage is a critical component of reliable renewable power. With variable demand and more uncertain resource availability in nature, energy storage plays a critical role in reliable renewable power supply. Heat is one of the possible options of energy that can be stored effectively. Phase change materials along with micro/nano encapsulation have been discussed as an effective way of energy storage for renewable power generation. Integration of these energy storage processes with suitable efficient renewable power generation options has been discussed in this chapter. Solar photovoltaic and solar thermal are generally considered as two different technology options for solar power. Also though the higher temperature improves the performance of the solar thermal system, solar photovoltaic operates at a lower temperature. In a new technology of Photovoltaic-thermal collectors, both effects can be combined for a micro-cogeneration system. Such technology not only improves the overall utilization of solar radiation, but also saves the required area for solar installation. A comprehensive overview of such systems including both opportunities and challenges is reported in another chapter of this section.

New technology developments and exploring new options of energy resources are two imperative needs of the global energy transition. In the next section, either such resources or such technologies are explored in four chapters. Geothermal energy is a significant renewable resource, though available in limited areas. The largest share of the cost for power from geothermal energy is due to drilling to reach the source of hot water/steam. However, similar drilling is done during the extraction of oil from oil mines. A significant amount of hot water is produced from many new oil mines at the stage of high water cut. On the other hand, hot water is also available in depleted oil fields which do not require any additional cost of drilling. Utilizing geothermal heat both from the new oil fields as well as from the depleted oil fields can be a good option for future energy sustainability. A chapter in this section includes a discussion on both potential and associated challenges for the utilization of such geothermal energy resources. Buildings consume a significant amount of energy. With the emerging concept of sustainable buildings, buildings are not only made more ‘smart’ with energy-efficient options but also can be used for distributed renewable power generation. Thus, the future energy-efficient buildings are not only energy consumers but also producers of renewable power in a distributed generation process. Real-time monitoring of such ‘smart buildings’ needs automation supported by data analytics integrated with IoT with proper decision logic. Such an efficient building energy management system with distributed renewable power generation is discussed in one chapter. The gap between demand and supply happens in any power generation and consumption. This gap becomes more prominent when uncertainty in the generation is more, specifically for renewable power generation with solar or wind energy. Conventional electrochemical storage devices cannot compensate for this gap with larger values. Pumped hydro storage is an efficient option for such periodic large gaps between generation and consumption. Several aspects of pumped hydro storage systems are discussed as an efficient solution for integrating more distributed renewable power into the main grid. Ocean is a vast resource of renewable energy. Extracting power from the waves is a possible option for distributed renewable power. Though oscillating water column (OWC) in a duct with water waves on one side and open atmosphere on the other side is a common option for wave power, it has poorer efficiency with such periodic change of flow velocity including its direction. A fluidic diode (FD), similar to an electrical diode, is proposed to convert this bidirectional flow into a unidirectional flow with a variable resistance. A detailed discussion on the prospects and challenges of this technology along with numerical study results are reported in another chapter of this section.

For any new technology implementation, modelling and simulation is a standard process as it helps in estimating the expected performance of the developed system at a much lower cost without fabricating it. Subsequently, on the basis of simulation results, the optimum system may be developed as a prototype. Thus, modelling and simulation is an integrated part of engineering system development, specifically new and emerging systems. A few modelling and simulation results are demonstrated in the next section of this book. Biofuels from ligno-cellulosic biomass (LCB) rich in lignin is a prospective option to replace fossil fuels. However, to make this process as a future more sustainable option, detailed process modelling is required. Modelling of different reactors used for syngas fermentation for bioethanol formation has been reported in a chapter of this section. Different challenges along with possible remedies are also recommended in this chapter. Power supply using renewable resources at remote isolated locations faces difficulties associated with fluctuations both on the demand and the supply side. Maintaining constant voltage and frequency with these fluctuations is a critical challenge for such systems. Self-excited Induction Generator (SEIG) is an effective solution for such systems to sustain consistent voltage and frequency at the bus bar. Real-time simulation results are reported for such a system for a better understanding of the operation and performance of such systems in a chapter of this section. Micro hydro-based power generation is more environment friendly as it does not affect local inhabitants through the creation of any water storage upstream of the dam. However, continuous fluctuations both at the generation and demand side may be a difficult challenge to address for such systems. A self-excited induction generator (SEIG) with the power grid for micro hydro small-scale distributed renewable power systems is thoroughly simulated in the MATLAB/Simulink environment with reported results in another chapter. Results are useful for grid integration of such micro hydro systems. In a further extension of this work, simulation results of a new capacitor excitation topology integrated with an induction generator (CET-IG) for single-phase power from a three-phase SEIG driven by the micro hydro system are reported. The results are satisfactory for single-phase power supply with acceptable voltage regulations. In a further extension of studies on the efficient operation of micro hydro power plants, another chapter is included. Harmonics, poor power factor, unbalanced load as well as voltage regulations for standalone power systems are difficulties to address. In this chapter, a shunt active power filter with a modified P-Q controller is reported through simulation in the MATLAB environment.

Successful application of new technologies and commercialization depends on several other factors. It involves socio-economic issues in addition to efficient technology development. A downdraft gasifier-based distributed power generation system in the Indian context is reported in a chapter in the next section.

Technologies are for social development. Proper policy support is also required for promoting technologies suitable for any society. Also review of social impacts after introducing/increasing any technology is a required study to decide the future path of it. Thus, the socio-economic impact studies of major introduction of technologies are required. India has a mission of enhancing solar power at the national level. Several policies are implemented both at national and sub-national levels in India. In the last chapter of the last section of this book, a comprehensive assessment of the impact of Indian solar energy missions/implementations till date is reported. Issues like green job creation and rural development are also reported in this study.

The book consists of seventeen chapters arranged according to the needs of its readers. Starting from the sustainability assessment, the book addresses an overview of the status of a few important technologies, a few promising future technologies, modelling and simulation results, possible applications as well as policy and implementation impacts of different distributed renewable power options. The book may be useful for students at the postgraduate levels of sustainable/renewable energy, researchers as well as practising professionals in this field of distributed renewable power. However, to help any reader to explore articles of specific interests, the whole content is divided into seven different sections: Part I Tool for Sustainability Assessment; Part II Technology Overviews; Part III Potential Technologies; Part IV Modelling and Simulations; Part V Applications; and Part VI Socio-economic Impact Studies. The editors believe that this book will be useful for the current researchers and professionals working in this field.

Part I
Tool for Sustainability Assessment

Chapter 2
Application of MCDM Tool Integrated with SWOT Analysis for Prioritization of Strategies to Optimize Distributed Hybrid Energy Systems for Better Sustainability

Sayan Das, Souvanik De, and Sudipta De

Abstract Energy transition is a current global challenge. Deciding a right path to meet the ever-increasing global energy demand at an affordable cost with low-carbon options is a critical challenge. Deciding appropriate energy strategies and determining their priorities are necessary for proper energy transition. Energy strategy determination is decided on the basis of the country's strengths, weaknesses, opportunities, and threats (SWOT) related to the energy sector. Evaluating proper energy strategies is a perpetual process to prioritize issues in energy planning. Different authorities are responsible for deciding the proper priorities of energy strategies on the basis of new and improved alternative solutions to achieve sustainable development. As energy strategy prioritization correlates several conflicting and interlinked issues and the SWOT method is unable to decide the priority, integrating a multi-criteria decision-making (MCDM) approach is critical. Integrating SWOT with an MCDM method is generic yet novel and one of the best possible options to formulate and prioritize energy strategies for a better energy transition towards sustainability. The study comprehensively reviewed the SWOT-MCDM methods along with the future direction of the research applied in different fields specially focused on energy transition.

Keywords Energy strategy prioritization · SWOT · MCDM

S. Das · S. De (✉)
Mechanical Engineering Department, Jadavpur University, Kolkata 700032, India
e-mail: sudipta.de@jadavpuruniversity.in

S. De
Electrical and Electronics Department, Bits Pilani Hyderabad Campus, Secunderabad 500078, Telangana, India

S. De et al. (eds.), *Challenges and Opportunities of Distributed Renewable Power*, Energy, Environment, and Sustainability, https://doi.org/10.1007/978-981-97-1406-3_2

2.1 Introduction

For the overall sustainable development of a country, energy is an indispensable requirement. The economic and societal development of any country is closely linked to appropriate energy planning [1]. Ever-increasing population, rapid industrialization, and urbanization increase the global energy transition. Till date 2.7 billion people throughout the world are using conventional resources to meet their daily energy needs. Approximately 95% of this is from remote villages [2]. The high dependency on conventional resources not only causes environmental degradation but also has an adverse effect on human health [3]. Considering environmental degradation, countries under the UN signed an agreement in Paris in 2016 [4]. According to the agreement, restricting the global temperature by 1.5 °C from pre-industrial levels is a global target. To achieve carbon–neutral economic development, shifting towards renewable energy resources is a critical need [5]. Additionally 17% of the total world population, especially from remote villages, are living without reliable electricity. For sustainable development of society, providing energy security is the key to success. The United Nations (UNs) decided 2014–2024 as the "Decade of Sustainable Energy for all" by supplying sustainable and affordable energy through non-conventional resources. It is the key factor for achieving the Sustainable Development Goal 7 (i.e., SDG-7) [6]. To address the issues that come in the path of sustainable development, non-conventional resources can play a critical role. Utilizing non-conventional resources without affecting the environment is a crucial goal. Energy management with proper strategy determination and prioritization are essential for this [7].

Due to ever-increasing energy demand and environmental issues, energy policy-making is a widely discussed issue [8]. During this energy transition phase, proper energy conservation and sustainable development have recently become an important aspect of energy planning. Deciding on a proper energy strategy at all levels is a critical need for a better energy transition towards sustainability [9].

Determining appropriate energy strategy and deciding the priority of implementation increases the reliability of the power supply by simultaneously minimizing the cost. It also improves resource utilization [10]. Ideal strategy determination targets several short as well as long-term goals to improve resource distribution, depending on the capabilities of the potential of various regions under different constraints such as technical restrictions, geo-political constraints, and budgetary limitations [11].

The Strengths, Weaknesses, Opportunities, and Threats (SWOT) method is used to develop the appropriate energy strategy [9]. This method is widely used and accepted by all in energy planning and strategic management studies [12]. It is a strategies analysis tool that combines the study of the strengths and weaknesses of the strategies, organizations, or sectors with the study of opportunities and threats in its territory [13]. Besides developing policies and strategies, it must be prioritized for successful implementation [14]. However, the SWOT method has few limitations. It is unable to decide the priority of implementation for determined strategies [9].

Multi-criteria decision-making (MCDM) is a method to decide the most acceptable solution among the alternatives by evaluating them under various criteria [15].

This technique is classified as classical and fuzzy-MCDM, which are used to decide the rank of the alternatives [16]. This MCDM technique is essential to decide the proper ranking of implementing the decided policies. During energy planning, several criteria such as technical, economic, environmental, and social must need to be considered. In energy decision-making problems, alternatives must be evaluated in terms of these criteria. Due to this, it becomes a multi-dimensional composite decision-making process [1]. MCDM method is also effectively used to determine the order of the criteria during decision-making. Various conditions add different aspects to energy planning and make them more complex and uncertain to decide the priority of implementation [9].

Siksnelyte et al. [17] reviewed different MCDM methods for dealing with sustainable energy development issues. Ayan et al. [18] comprehensively reviewed different MCDM methods that were used for determining the weights of the criteria. Espino et al. [19] studied detailed MCDM methods used in the construction industry. Sousa et al. [20] studied different MCDM methods used to achieve SDGs. Abanda et al. [21] studied different MCDM methods used to evaluate National Determination Contribution projects. Benzaghta et al. [22] studied the application of the SWOT method in different sectors. Kabir et al. [23] reviewed the applications of SWOT-Analytic hierarchy method (AHP)-MCDM method in smart power management problems.

2.2 Research Topic

Most of the reported literature focused on analyzing the use of MCDM methods in different sectors. Recent studies reviewed the MCDM and SWOT methods application in energy strategy determination for achieving sustainable development goals. These studies mostly studied these methods separately. No previous analysis comprehensively reviewed the applications of hybrid SWOT-MCDM methods in energy strategy determination and prioritization. Due to the lack of substantial review of this hybrid method, it becomes difficult to select the appropriate MCDM method suitable with SWOT techniques for the specific problems.

Considering the identified research gaps, this study aims to comprehensively review the hybrid MCDM-SWOT method used in energy strategy determination and prioritization for achieving sustainable development. The SWOT method is used to determine the strengths, weaknesses, opportunities, and threats of the energy policies for better energy management. Subsequently, this research also aimed to report the application of different MCDM methods to decide a priority of implementing the decided strategies developed through SWOT analysis.

2.3 Strengths, Weaknesses, Opportunities, and Threats (SWOT) Method

American businessman and management consultant Albert Humphrey developed the SWOT analysis method in the 1960s to 1970s [24]. Evaluating complex strategies, developing a strategy plan, and organizing them as external and internal factors are the major tasks of this method [25]. The objective of the SWOT method is to evaluate the internal and external factors to detect the internal strengths to benefit from external opportunities by highlighting the internal weaknesses and external threats [26]. The methodological steps are described in Fig. 2.1.

SWOT factors are defined as relevant information or data. The structure of this method depends on this information or data which is gathered as input. Internal factors are the factors that can be controllable whereas the external factors are uncontrollable. Strengths and opportunities are the helpful factors that aid success, while the threats and the weaknesses are the harmful factors that prevent the success [26,

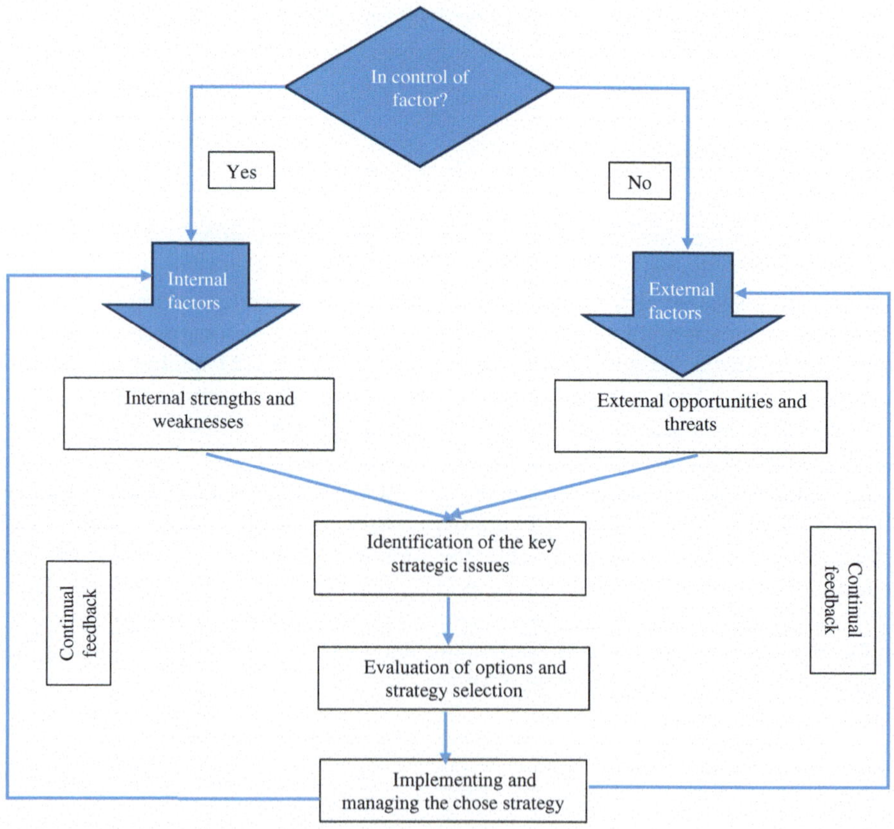

Fig. 2.1 Schematic arrangement of SWOT method

27]. Strength factor is considered as the internal enhancer of the ability, rewarding resource. Internal hindrances to the skills, resources, or qualities required for success are the weaknesses. Opportunities improve performance and can be utilized to one's advantage. The threats are an external factor that reduces the prospective achievement. In a 2 × 2 matrix, strengths are labeled as internal performance enhancers and weaknesses are labeled as internal permeance inhibitors. Opportunities are external inhibitors, while threats impose performance restrictions [28]. This SWOT matrix involves the following:

- Strength-opportunity (SO): It focuses on how to use the system's strength to seize chances on the basis of opportunity.
- Strength-Threats (ST): It ensures to stay safe so that the advantages can be used.
- Weakness-Opportunities (WO): It seeks to reduce gaps to generate new opportunities.
- Weakness-Threats (WT): These are mostly defensive and work to reduce vulnerabilities to neutralize threats.

Based on various considered criteria, i.e., technical, economic, environmental, social, policy, and socio-political issues, the sub-factors of each SWOT factor are formed [29]. To make the questionnaire or to calculate the collected data, this approach does not require any software or tool [30]. The important aspect of this method is open interviews and open groups to determine the strengths, weaknesses, opportunities, and threats [24].

2.3.1 Energy Planning and SWOT Method

To determine the energy strategies, SWOT methods are used. It helps to determine the strengths, weaknesses, opportunities, and threats of the decided energy strategies and policies.

By assessing the internal state of renewable energy of each country, resources, and capabilities of the energy strategies, the sub-factors of the internal factors (Strengths and weaknesses) are developed. Whereas the external factors, i.e., opportunities and threats are decided by examining the environment of the strategies.

Resource potential, financial strength, and technical advancement are the strengths of the developing renewable energy sector. The weaknesses rely on infrastructure, business cost, financial liquidity of the company, tariff, cost and availability of debt, etc. Factors like government policies, green financing, and international threats are the major concerns of the opportunities, and threats are related to the political environment, energy supply, international scenario, availability of fossil fuels, and economic stability.

Initially, a literature review is conducted to design the current energy situation of the country. Then, core decision groups are formed. Subsequently, SWOT factors and sub-factors are decided. Finally, Alternative strategies are identified.

Qaiser [31] compared various energy strategies developed for the renewable and sustainable energy sector of different Asian countries. According to the study of the poor financial situation of the distribution companies, lack of credit opportunities is the major weakness for the growth of the renewable energy sector of this region. There is a conflict of interest between manufacturers of renewable energy equipment and the developers of renewable energy projects in terms of government policies which become the major concern for imposing the import duties. Kamran et al. [32] analyzed the current status and future roadmap for nurturing the renewable energy sector of Pakistan by using the SWOT method. The study reported that the validated resource maps, abundant renewable energy resources, environment-friendly conditions, and growing private sectors are the major strengths, while inefficient technologies, immature institutional framework, hazardous technology-related environment, and huge capital investment are the substantial internal weaknesses. The study also identified significant opportunities as well as threats for the potential renewable energy sector of Pakistan. Elavarasan et al. [33] developed a framework to comprehensively evaluate the drivers and barriers for renewable energy development in significant countries by using the SWOT method. This study majorly focused on India, China, US, Sweden, and Iceland. Agyekum [34] analyzed the strengths, weaknesses, opportunities, and threats of the renewable energy sector of Ghana by using the SWOT method. The study reported that political stability, abundant renewable energy resources, and geographical location are the major strengths, whereas bureaucratic procedure and high capital cost are the key internal weaknesses. The study also identified the opportunities and threats for developing the future roadmap of the renewable energy sector of Ghana. Shi [35] assessed the energy mix situation of ASEAN countries by using the SWOT method. This study reported that the region must provide an effort for establishing cleaner energy by removal of subsidies of fossil fuels and needs to promote renewable and energy efficiency. Fertel et al. [12] analyzed a Canadian energy and climate policy in terms of the coherence between federal and provincial strategies by using the SWOT method. The study showed that the lack of consistency is the major weakness of the Canadian climate and policy. The study also reported different strengths and opportunities of the sector for developing renewable energy in Canada. Terrados et al. [36] developed renewable energy planning through the SWOT analysis method. The study also analyzed the Impact of this method on renewable energy development. Jing and Tao [37] reported a SWOT framework for developing clean energy strategies in China on the basis of environment, social constraints, economic, etc. Zhou et al. [38] designed a model on the basis of 12 SWOT sub-factors for sustainable energy strategies in rural areas.

Most of these studies mainly focused on finding the strengths, weaknesses, opportunities, and threats for the renewable energy sector in different countries. These studies also developed the future pathways for increasing renewable energy share in the power mix for better energy transition towards sustainability. However, due to the above-mentioned limitation, it becomes difficult for this SWOT method to find the proper priority of implementation of these determined strategies. Therefore, implementing strategies becomes critical. It demands to integrate a different MCDM approach with the SWOT method to decide the proper ranking of implementing these determined strategies.

2.4 Multi-criteria Decision-Making (MCDM)

Decision analysis is an important tool to solve a problem with multiple factors, criteria, and objectives. MCDM problems come with five steps: goal, decision-maker's preference, criteria, alternatives, and results [39]. Figure 2.2 classifies the MCDM methods [40]. Depending on the number of alternatives, MCDM methods are classified into two parts: multi-objective decision-making (MODM) and multi-attribute decision-making (MADM); otherwise, the characteristics of both these MCDM are similar [41]. To evaluate continuous alternatives for which we predefined constraints in the form of vectors of decision variables, an objective function set is optimized depending on the constraints when the performance of one or more objectives is degrading [42].

In MADM, the inherent characteristics that are covered lead to considering a lesser number of alternatives and it makes the evaluation more complex because prioritizing the alternatives becomes difficult [43]. The final outcome is decided by pairwise comparing the alternatives with respect to the considered attributes. MCDM is a broader classification and different MCDMs are used in energy strategy prioritization [44]. As per the designer perspective, these models are developed. It may be a direct or an indirect approach. Figure 2.3 shows the classification of such models [45]. Different factors such as technical, institutional, social, economic, environmental, stakeholders, and standards are included in this process which makes MCDM more complex. It involves both managerial and engineering analysis. Based on the priority of the objectives, this process can lead to different outcomes at different times.

Moreover, a particular problem can be solved by different methods on the basis of the decided objective functions. Each method has its own advantages and limitations. Figure 2.4 illustrates the general procedure of the working principle of MCDM.

In energy strategy prioritization problems, different MCDM approaches are used. A few important MCDM models are as follows [46–51].

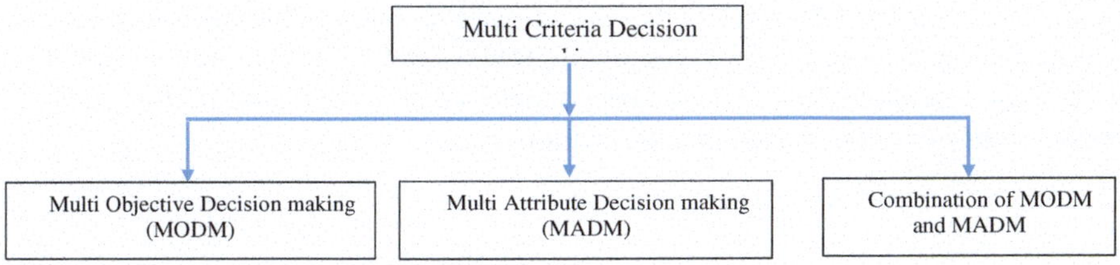

Fig. 2.2 MCDM methods classification [40]

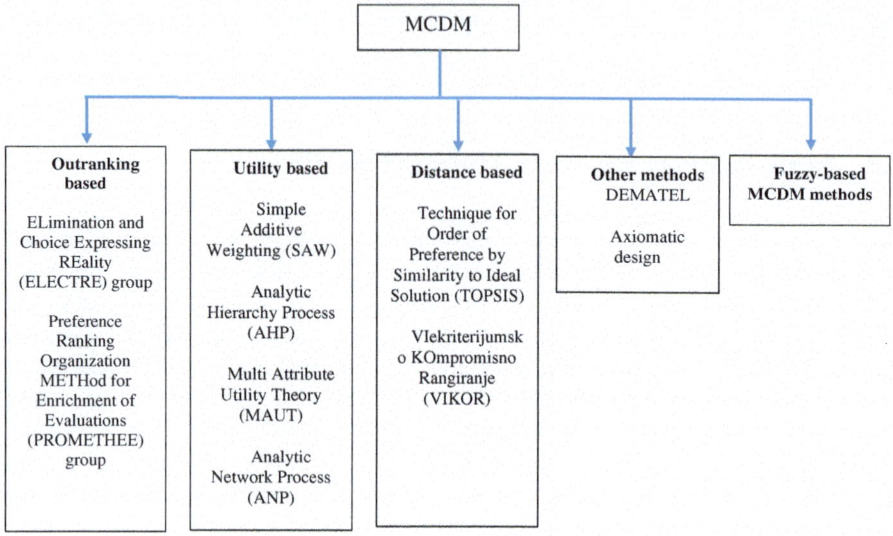

Fig. 2.3 MCDM models [45]

(i) AHP
(ii) ANP
(iii) Weighted Sum Model [47]
(iv) Weighted Product Model
(v) ELECTRE [51]
(vi) PROMETHEE [60]
(vii) TOPSIS
(viii) VIKOR [49]
(ix) Fuzzy-based MCDM.

2.5 Energy Planning and MCDM

In energy planning, different MCDM models are successfully utilized. These methods are considered as most suitable options for solving energy planning-related issues. Different MCDM models that are utilized in energy planning are reviewed in this section. Broadly three types of MCDM models, i.e., outranking methods, utility-based methods, and distance-based methods are considered in this analysis. These models have been used in combination also.

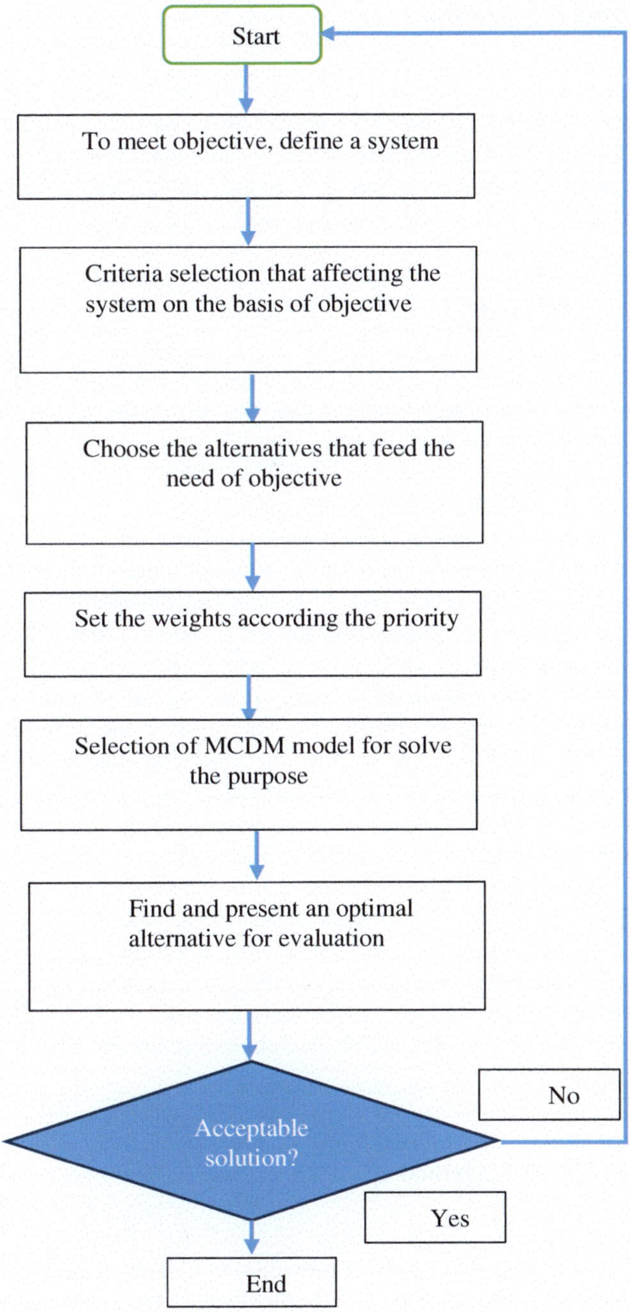

Fig. 2.4 Common methodology of MCDM

2.5.1 *Outranking Methods*

In this group, ELECTRE and PROMETHEE methods are the most important models. Out of these models, ELECTRE methods are the most common ones that are considered for energy planning. This method has the broader perception to give a practical view for the problem by inculcating all the queries and suspicions. This ELECTRE method is mostly preferred in energy allocation at the demand side.

2.5.1.1 ELECTRE Method

This MCDM method established strong preferences on the basis of given information. Different mathematical functions are used to indicate the dominance degree of one alternative over another. It is the operation research method and uses pseudo-criteria and binary outranking relations [52]. This technique uses a pseudo-criteria for comparison to evaluate a number of strategy options.

Lin et al. [53] reported that ELECTRE method were used to prioritize the alternatives for sustainable hydrogen production pathways determination. Beccali et al. [54] proposed to use ELECTRE method for selecting renewable energy technologies for different locations. Hwang and Frank [55] compared five different MCDM methods and presented the use of this outranking method in a distributed energy system. In ref. [56], authors studied the energy planning situations of Greek Island. The study identified the energy strategy factors, selected the criteria, and formulated the alternative strategies. In this study, the researchers use the ELECTREE III method and highlight those aspects which are crucial in regional energy planning problems. In ref. [57], ELECTRE III methods were used in the power distribution method followed by a simple ranking method. The study showed that ELECTRE III was a healthy and reliable method to support power distribution support planning for Cynthia. Flourentzou et al. [58] considered the ELECTRE method for renovating and rating the present building scenario considering different criteria such as energy required for heating, cooling, and impact of other appliances on foreign and indoor environment quality and cost. In ref. [59], researchers used the ELECTRE method to determine the strategies for improving the economic investment in solar photovoltaic (PV) systems.

2.5.1.2 PROMETHEE Method

The Preference Ranking Organization Method for Enrichment Evaluation (PROMETHEE) is an outranking MCDM method that decides the rank of the finite set of alternatives on the basis of the conflicting criteria [60]. Brans developed this method in 1982 which was used for partial ranking of alternatives. Vincke and Brans further expanded this method in 1985 to PROMETHEE II for complete ranking determination of the alternatives [60]. Recently PROMETHEE III and IV were also used. It is the preference function-based MCDM and it is the most accepted one for determining ranking. It is applicable in many fields for ranking purposes [61].

Bortoluzzi et al. [62] proposed to use PROMETHEE MCDM methods for selecting energy technologies to develop an economic distributed energy system. Demirel et al. [63] reported the PROMETHEE method for deciding energy-efficient steam boilers for industrial applications. Chisale et al. [64] reported that PROMETHEE method was suitable to decide a techno-economically optimal solution for Malawi. Georgopoulou et al. [65] used the PROMETHEE II method in a group decision support system to advise the decision experts in promoting renewable energy resources usage. Tsoutsos et al. [66] exploited the decision-making methodology to plan the sustainable energy for the island of Crete in Greece. For this study, authors used the PROMETHEE I and PROMETHEE II methods for finding the alternative energy strategy to develop techno-economic, environmentally friendly, and socially acceptable renewable energy systems.

2.5.2 Utility-Based Method

Different utility-based MCDM methods, i.e., Analytic Hierarchy Process (AHP), Weighted Sum Model (WSM), MAUT, SAW, and ANP are extensively used in energy strategy planning. To rank the decided strategies, these methods are commonly used.

2.5.2.1 AHP, ANP, Delphi, and MAUT Methods

Saaty in 1980 developed the AHP decision-making method. This method decomposes a complex MCDM method into a hierarchical order. Hierarchical structuring of the model, pairwise comparison of alternatives, and the criteria and priorities synthesis are the three steps of the AHP MCDM method [67].

ANP considers the dependencies between elements in internal and external sets. Inner dependency shows the interaction between criteria in the same cluster and external dependency represents the interaction of criteria in a different set. The dependencies are shown in arrows. This method is a control hierarchy used to evaluate the different dimensions of criteria. Defining clusters and deciding the dimension of the criteria, designing the model, creating interdependencies, and pairwise comparison are the four steps performed under this method, respectively [68].

Delphi method systematically collects the data from the anonymous experts and connects it in written, discussion and provides feedback on a specific topic. This method enables a chance for a completely diverse expert's opinion [67].

Multiple Attribute Utility Theory (MAUT) method is based on Vonn Neumann and Morgenstern's utility theory foundation. It was developed by Keeney and Raiffa. It simplifies the multi-attribute evaluation process. It gives a logical and reasonable process to decide between conflicting criteria. In this method, quantitative and qualitative both the criteria are used [69].

According to the literature for energy planning, the AHP method is the most common method over MAUT. However, this AHP method has some limitations such as the mutual outcome environment required for defining utility functions and the complexity in process to calculate the scaling constraints over MAUT methods [70]. Recent literature used both the AHP and MAUT methods for energy planning [45].

Alizadeh et al. [71] showed that the ANP method with Benefit, Opportunity, Cost, and Risk (BOCR) is suitable to determine energy strategy planning and decide the rank of the strategies to reduce the barrier for renewable energy resources utilization in Iran. Meyar- Naimi et al. [72] used AHP to assess sustainability factors for the energy generation system on an upgraded policy structure. Luthra et al. [73] ranked the barriers that are categorized under seven dimensions by the AHP method. Pathak et al. [74] studied the modified DELPHI and AHP methods to identify and rank the different barriers like socio-economic, environmental, and political for advancing different renewable energy technologies.

2.5.3 Distance-Based Method

The factors considered under the decision-making process have their own internal impact. It is reclassified into a common scale. So, it is critical to determine the relative impact of each criterion for obtaining a sustainable energy solution. To indicate the relative importance, weights need to be assigned for the criteria. The equal weights and the rank-order weights methods are the two types used to estimate the weights of the criteria. The details of these types are described in Fig. 2.5 [75].

Assigning weights for the criteria in MCDM evaluation is for better and accurate decision-making analysis as both qualitative and quantitative data are associated with it. Weight assigned for the criteria using qualitative data is influenced by decision-maker's preferences. Due to this limitation, Saaty proposed a 1–9 linguistic scale for changing qualitative data into quantitative one. Either equal weights can be assigned to the criteria or can be assigned on the basis of the rank-order. This rank-order method is of three types, i.e., subjective, objective, and combined.

Determining subjective weights is based on an expert's opinion and to obtain the judgment decision-makers must need to present a set of questions in the process to the expert. AHP is the subjective weight determination process.

Objective weight determination is derived from the gathered information in each criterion with the help of mathematical calculation without any intervention of decision-makers. The decision-makers should be entrusted with the final decision. This process does not consider all the factors and involves some inaccuracy.

In the combined method, both the principles of the above-mentioned process are considered [75].

The energy system involves different conflicting criteria that need to be assigned proper weights for decision-making evaluation. This requires a more accurate weight assigned process. In previous literature [10, 76–78], different weighting methods are used to assign the weights of the criteria. The weight determination processes are

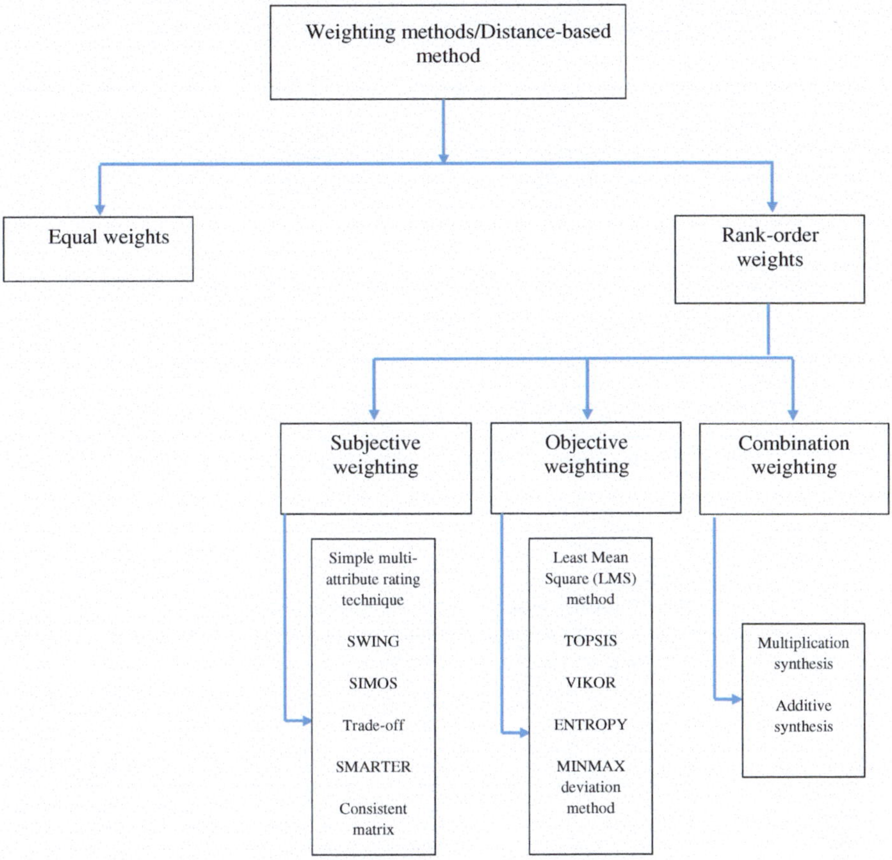

Fig. 2.5 Different weight ranking methods [75]

applicable in different fields of energy systems. Most of the studies reported that objective weighting methods are used to decide the weights of the criteria and the rank of the alternatives.

Based on these assigned weights alternatives are ranked. Distance-based methods, i.e., TOPSIS and VIKOR are mostly used for ranking the alternatives. The working principles of these two methods are described in Figs. 2.6 [4] and 2.7 [7], respectively.

The TOPSIS method is considered as a weighting method as well as a MCDA method. The Working principle is clearly discussed in Fig. 2.6. In the TOPSIS method, the selected best alternative has the shortest distance from the positive ideal solution. This method is used in prioritizing energy strategies. In ref. [79], solar PV cleaning technologies are ranked by TOPSIS method. In ref. [80], building renovation solution is obtained through TOPSIS MCDM approach. In ref. [4, 7], references authors used the TOPSIS MCDM method to decide the overall sustainable decentralized hybrid energy system for remote villages of India.

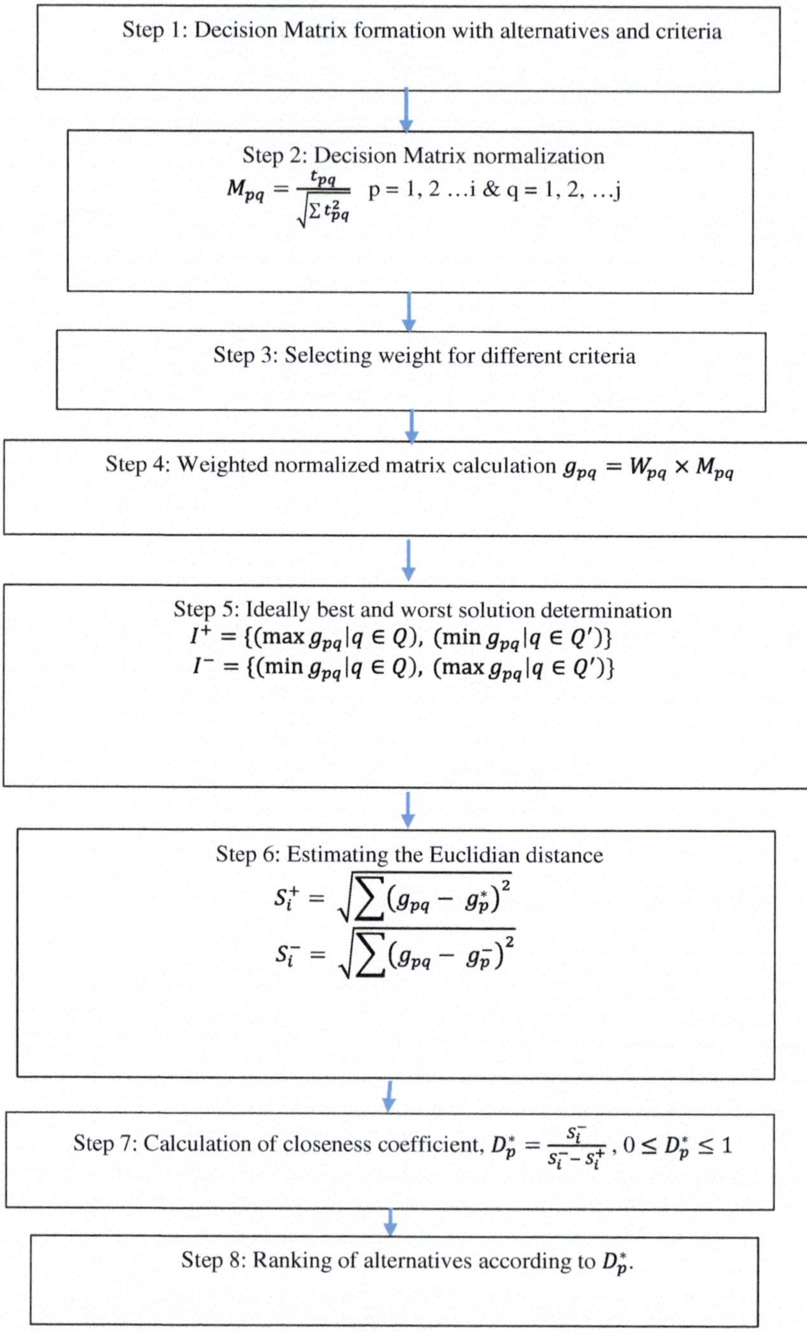

Fig. 2.6 TOPSIS method working principle [4]

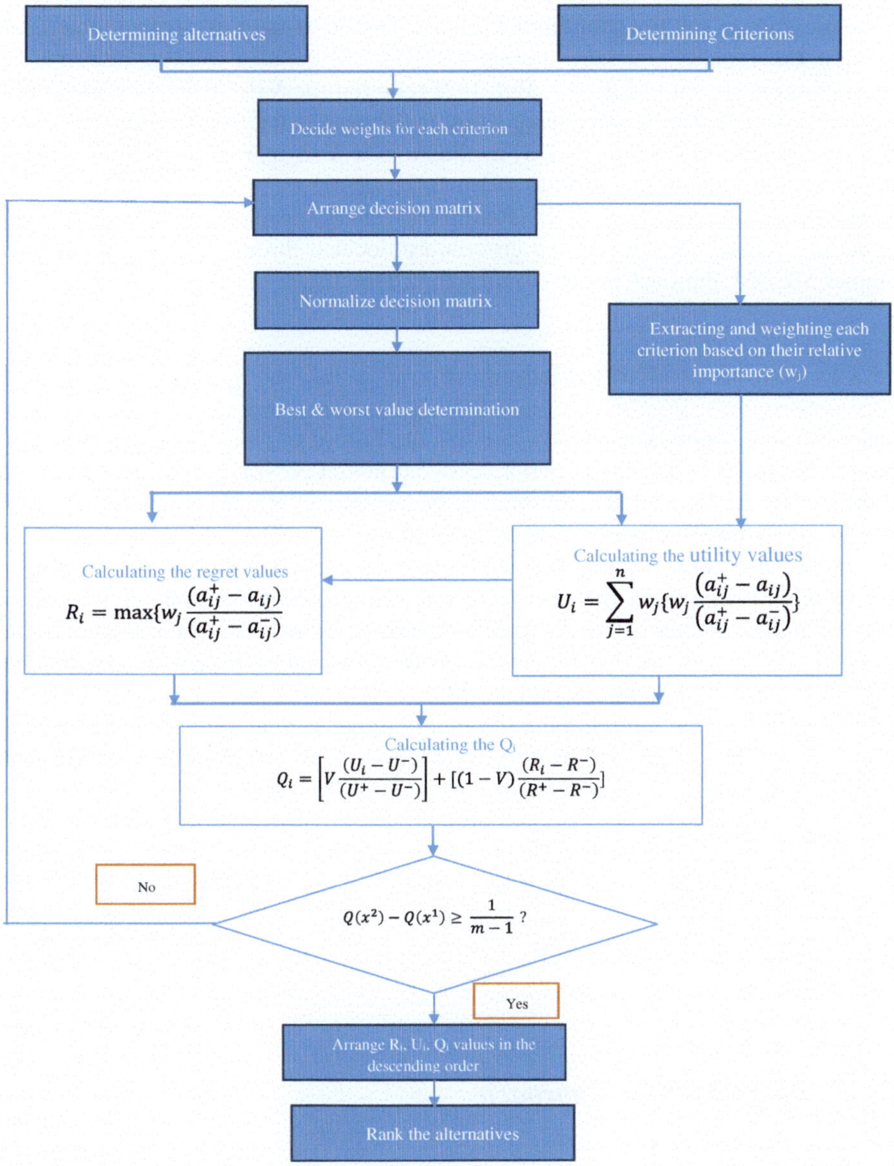

Fig. 2.7 VIKOR method working principle [7]

For the optimization of complex systems on the basis of multiple criteria, the VIKOR method was developed. The compromise ranking list is determined by this process which leads to a compromised solution. The stable weights interval for the preferred stability of the compromised solution is mostly obtained from the given weights. The VIKOR method ranks and selects a solution from the set of alternatives in the presence of multiple conflicting criteria. On the basis of closeness to the ideal solution, this process ranks the alternatives [81]. The detailed working principle of this method is shown in Fig. 2.7. Ref. [82] reported the VIKOR method to decide the overall sustainable hybrid energy solution for rural India. Rojas-Zerpa and Yusta [83] considered the VIKOR MCDM method to select an appropriate electric supply planning for rural and remote areas.

Most of the literature showed that for different energy problems combined MCDM methods are used. Weights of the criteria are decided through one technique and another MCDM approach is used to rank the alternatives.

2.5.3.1 Combined MCDM Methods

Balezentis and Streimikiene [84] prioritize the European Union's energy development scenarios by a two-integrated assessment model. WSM, ARAS, and TOPSIS models were used to rank the alternatives. Villacreses et al. [85] aimed to find suitable locations for developing wind firms in Ecuador based on different criteria. In this study, AHP, VIKOR, and TOPSIS were used in combination to decide a suitable solution. Billig and Thraen [86] aimed to evaluate 99 different thermochemical and bio-chemical alternatives in techno-economic aspects of renewable methane production. The study used AHP and the utility value analysis process for alternative utilization. Buyukozkan and Karabulut [87] reported that integrated AHP-VIKOR MCDM methods were used to rank the energy projects. Sanchez-Lozano et al. [88] proposed that integrated AHP-TOPSIS-ELECTRE methods with GIS were suitable to decide the best location for solar firm development in Spain.

With the growing energy demand, proper energy policy determination becomes essential. Different aspects are added to energy making problems and that makes the process more complex. There are uncertainties in linguistic terms associated with criteria weights determination and alternative ranking. Hence, it is proposed to use Fuzzy-MCDM approaches to cope with these linguistic uncertainties in decision-making.

2.5.4 Fuzzy-MCDM Approach

Zadeh introduced the Fuzzy set theory (FST) concept that was designed to consider the vagueness of imprecision of cognitive processes. The concept is that the elements have a degree of membership in a set theory. The membership function maps the elements in the set to the degree of membership within a certain interval, i.e., 0–1. If the membership function value is zero then it is not in the set and if it is one then it is completely in the set. In fuzzy logic, different types of membership functions are used. Monotonic, triangular, and trapezoidal are the three most common methods. Triangular fuzzy number (TFN) is the commonly used method due to its

Table 2.1 Fuzzy linguistic scale [89]

Linguistic terminology	Linguistic terminology for performance	TFN	Reciprocal value
Equally important	Very poor (VP)	1, 1, 1	1, 1, 1
Weakly important	Poor	2, 3, 4	1/4, 1/3, 1/2
Strongly important	Fair	4, 5, 6	1/6, 1/5, 1/4
Demonstrated important	Good	6, 7, 8	1/8, 1/7, 1/6
Absolutely important	Very good	8, 9, 9	1/9, 1/9, 1/8

computational simplicity and the nature of evaluating expert's linguistic opinion [89]. The considered fuzzy linguistic scale is shown in Table 2.1.

Abdullah and Najib [90] reported that an intuitionistic fuzzy-AHP (IF-AHP) was used for sustainable energy planning in Malaysia. Kulkarni et al. [91] considered fuzzy-AHP (F-AHP) to decide the rank of the different energy alternatives. Kabak et al. [92] evaluated different buildings under overall energy performances by F-ANP method. Papapostolou et al. [93] considered a F-TOPSIS approach for prioritizing alternative energy policy scenarios to increase renewable energy share in the power mix by 2030. Guo and Zhao [94] reported that the F-TOPSIS method was used to decide suitable sites for designing EV charging stations. Cavallaro and Ciraolo [95] compared different solar energy technologies by the F-PROMETHEE method. Lee et al. [96] integrated F-AHP and DEA methods for promoting the national agenda of Korea for Low carbon and green growth. Kaya and kahraman [97] integrated VIKOR and F-AHP methods to determine the best energy policy for Istanbul.

The MCDM approaches prioritize the energy policies without determining the appropriate strengths, weaknesses, opportunities, and threats of the policies. Only the SWOT method is not suitable to prioritize the strategies and also MCDM methods are unable to decide the internal and external strategies. Hence, integration of both these methods may be useful to develop strategies and prioritize them. Firstly, this SWOT method determines the strengths, weaknesses, opportunities, and threats of the energy policies and secondly the MCDM methods are used to decide the suitable rank of the strategies. In recent studies, SWOT-MCDM integrated methods are used for deciding the appropriate energy strategies and the suitable rank of implementing these strategies.

2.6 SWOT-MCDM Methods for Energy Planning

The SWOT-MCDM methods are applied in several fields. A detailed review of the literature is discussed in Table 2.2.

Table 2.2 SWOT-MCDM applications

Sl. no	Authors and year	Methods used	Application area and study objective	Outcomes
1	Rahimirad and Sadabadi [98]	SWOT-grey MCDM	Prioritize strategies for waste to energy	Abundant municipal solid waste generated in Pakistan can be green and useful to generate green electricity
2	Almutairi et al. [99]	SWOT-best worst method–grey aras	Renewable energy strategy development for Saudi Arabia	Low-cost renewable energy generation is the highest priority for the country
3	Tavakoli et al. [100]	SWOT-fuzzy-MCDM	Prioritize strategies for water consumption management	Advertising the correct way of the water consumption gets the highest priority
4	Longsheng et al. [101]	SWOT-grey MCDM	Strategy prioritization for waste to energy management	Abundant municipal solid waste can be useful to generate green and sustainable energy
5	Almutairi et al. [102]	SWOT-SWARA-grey ARAS	Best renewable energy plans for Iran	Efficient wind-solar technology with storage loss and output energy fluctuation is the best strategy for the considered study area
6	Akcaba and Eminer [10]	SWOT-ANP-TOPSIS	Strategy prioritization for energy management in Cyprus	Establishing an interconnected connection to the mainland is obtaining the highest priority
7	Yonter and Derse [103]	SWOT-AHP-ANP	Action plan determination and prioritization for energy management	Increasing renewable share in buildings is the key priority for the study area
8	Buyukozkan et al. [104]	SWOT-AHP-MARCOS	Digital transformation strategy prioritization for Air Industry	Focusing of differentiated digital customer is the most accepted strategy

(continued)

Table 2.2 (continued)

Sl. no	Authors and year	Methods used	Application area and study objective	Outcomes
9	Mukeshimana et al. [105]	SWOT-MICMAC	Renewable energy strategy determination and prioritization for Rwanda	Raising investment, providing incentives, policy support, and creating favorable conditions for renewable energy are the four top strategies
10	Wang et al. [106]	SWOT-F-AHP	Strategy prioritization for selecting renewable energy resources in Pakistan	Economic and socio-political criteria are the two most important factors
11	Solangi et al. [11]	SWOT-AHP-F-TOPSIS	Evaluating strategies for sustainable energy planning in Pakistan	Low-cost and sustainable electricity supply is obtained as the highest priority
12	Stoilov and Martinov [107]	SWOT-AHP-PROMETHEE	Selecting the location for rail-road intermodal terminal	Quantity of the manipulated containers is getting the highest priority
12	Ervural et al. [108]	SWOT-ANP-TOPSIS	Analyzed the energy strategy alternatives and prioritized them	Geo-strategic position within the framework of the local region is the most important priority to make the country a renewable energy hub

2.7 Case Study

Das et al. [7] performed an techno-economic with environmental impact assessment and risk in return on investment (ROI) analysis for obtaining an overall sustainable decentralized energy system for a remote village of Gujrat, India. Considering available local renewable resources supported with storage and diesel generator (DG) backup, the hybrid energy systems are developed. The techno-economic factors such as cost of electricity (COE), net present cost (NPC), and unmet load (UL) are assessed with a detailed life cycle assessment (LCA) of the combinations. The risk in ROI is also evaluated using the Monte Carlo Simulation (MCS) method. The integrated analysis may not have a single optimal solution. A MCDM method is used to decide an optimally sustainable energy solution for a remote village.

2.8 Conclusions

In recent times, energy strategy planning is an emerging topic in the literature. Deciding suitable energy strategies is an important aspect for governments because of their potential impact. Due to strategic importance, governments are trying to accurately and effectively manage the country's energy policy. Deciding the proper priority of implementing energy strategies includes several conflicting criteria and it creates complexities. Prioritizing the various internal and external energy planning strategies obtained through the SWOT analysis method is a very critical yet necessary decision-making process. To achieve sustainable development goals, proper strategy development and prioritization is required.

Firstly, different MCDM methods were used to prioritize these strategies. However, finding the strengths and opportunities to minimize the threats and weaknesses is not possible for this method. Few studies only found the internal and external factors using the SWOT method. The limitation of this method is that it is unable to prioritize the strategies. This limitation is discussed in this study. Therefore, performing SOT and MCDM methods separately creates difficulties to find the positive and negative factors as well as difficulty to prioritize the strategies, respectively. Hence, performing both the studies simultaneously may solve the issues. In the first phase, SWOT analysis decides the positive and negative factors of the SWOT factors, and then by using MCDM, it is prioritized. This study shows that previously single MCDM method was used to decide the rank. After that, a combination of MCDM methods was used to find the priority of implementing the strategies.

As the multiple factors are included in energy strategy design, the complexity is increasing. The literature review shows that the AHP method is mostly used with the SWOT method due to its simplicity. The study shows that a few outranking methods say ELECTRE III, TOPSIS, and PROMETHEE methods are also gaining popularity. However, no single method is able to rank the alternatives, as every method has its own merits and demerits. Hybrid MCDM methods are thereby developed for addressing these issues. In a little literature hybrid MCDM methods are integrated with the SWOT method.

With the increase of different aspects, the linguistic complexities and uncertainties are ever-increasing. Hence, in a recent analysis, Fuzzy-MCDM methods were used to cope with these challenges. However, still more effort is required to obtain a better decision-making solution to the energy strategy prioritization problems. It requires more sophisticated and efficient decision-making methods to deal with more complexities and uncertainties.

Considering the challenges, the researchers have the opportunities to work on developing new MCDM methods. It may solve the challenge of overcoming these complexities and uncertainties associated with energy strategies prioritization problems.

Acknowledgements Mr. Sayan Das gratefully acknowledges the West Bengal State Government for the SRF fellowship for this research work.

References

1. Kaya İ, Çolak M, Terzi F (2019) A comprehensive review of fuzzy multi criteria decision making methodologies for energy policy making. Energy Strateg Rev 24:207–228. https://doi.org/10.1016/j.esr.2019.03.003
2. Valickova P, Elms N (2021) The costs of providing access to electricity in selected countries in Sub-Saharan Africa and policy implications. Energy Policy 148.https://doi.org/10.1016/j.enpol.2020.111935
3. Sovacool BK (2012) Design principles for renewable energy programs in developing countries. Energy Environ Sci 5:9157–9162. https://doi.org/10.1039/c2ee22468b
4. Das S, Ray A, De S (2020) Optimum combination of renewable resources to meet local power demand in distributed generation: a case study for a remote place of India. Energy 209:118473. https://doi.org/10.1016/j.energy.2020.118473
5. Das S, De S (2023)Technically efficient, economic and environmentally benign hybrid decentralized energy solution for an Indian village: multi criteria decision making approach.J Clean Prod 388.https://doi.org/10.1016/j.jclepro.2022.135717
6. Yilan G, Kadirgan MAN, Çiftçioğlu GA (2020) Analysis of electricity generation options for sustainable energy decision making: the case of Turkey. Renew Energy 146:519–529. https://doi.org/10.1016/j.renene.2019.06.164
7. Das S, De S (2023) MCDM for simultaneous optimum economy, investment risk and environmental impact for distributed renewable power: demonstration with an Indian village data. Energy Convers Manag 277.https://doi.org/10.1016/j.enconman.2022.116631
8. Energy Security (2021) IEA, 2021a. In: IEA. https://www.iea.org/topics/energy%0A-security. Accessed 27 Jun 2022
9. Das S, De S (2023) Strengths, weaknesses, opportunities and threats determination and strategy prioritization using hesitant fuzzy decision-making approach for better energy sustainability: demonstration with Indian data. Energy Convers Manag 281. https://doi.org/10.1016/j.enconman.2023.116847
10. Akçaba S, Eminer F (2022) Evaluation of strategic energy alternatives determined for Northern Cyprus with SWOT based MCDM integrated approach. Energy Rep 8:11022–11038. https://doi.org/10.1016/j.egyr.2022.08.227
11. Solangi YA, Tan Q, Mirjat NH, Ali S (2019) Evaluating the strategies for sustainable energy planning in Pakistan: an integrated SWOT-AHP and Fuzzy-TOPSIS approach. J Clean Prod 236. https://doi.org/10.1016/j.jclepro.2019.117655
12. Fertel C, Bahn O, Vaillancourt K, Waaub JP (2013) Canadian energy and climate policies: a SWOT analysis in search of federal/provincial coherence. Energy Policy 63:1139–1150. https://doi.org/10.1016/j.enpol.2013.09.057
13. Zhang L, Li S, Hu Y, Nie Q (2022) Economic optimization of a bioenergy-based hybrid renewable energy system under carbon policies—from the life-cycle perspective. Appl Energy 310.https://doi.org/10.1016/j.apenergy.2022.118599
14. Ioannou A, Angus A, Brennan F (2017) Risk-based methods for sustainable energy system planning: a review. Renew Sustain Energy Rev 74:602–615. https://doi.org/10.1016/j.rser.2017.02.082
15. Kiliç M, Kaya I (2015) Investment project evaluation by a decision making methodology based on type-2 fuzzy sets. Appl Soft Comput 27:399–410. https://doi.org/10.1016/j.asoc.2014.11.028
16. Kahraman C, Öztayşi B, Uçal Sari I, Turanoğlu E (2014) Fuzzy analytic hierarchy process with interval type-2 fuzzy sets. Knowl Based Syst 59:48–57. https://doi.org/10.1016/j.knosys.2014.02.001
17. Siksnelyte I, Zavadskas EK, Streimikiene D, Sharma D (2018) An overview of multi-criteria decision-making methods in dealing with sustainable energy development issues. Energies 11.https://doi.org/10.3390/en11102754
18. Ayan B, Abacıoğlu S, Basilio MP (2023) A comprehensive review of the novel weighting methods for multi-criteria decision-making. Inf 14.https://doi.org/10.3390/info14050285

19. Jato-Espino D, Castillo-Lopez E, Rodriguez-Hernandez J, Canteras-Jordana JC (2014) A review of application of multi-criteria decision making methods in construction. Autom Constr 45:151–162. https://doi.org/10.1016/j.autcon.2014.05.013
20. Sousa M, Almeida MF, Calili R (2021) Multiple criteria decision making for the achievement of the un sustainable development goals: a systematic literature review and a research agenda. Sustain 13.https://doi.org/10.3390/su13084129
21. Abanda FH, Chia EL, Enongene KE et al (2022) A systematic review of the application of multi-criteria decision-making in evaluating nationally determined contribution projects. Decis Anal J 5.https://doi.org/10.1016/j.dajour.2022.100140
22. Benzaghta MA, Elwalda A, Mousa M et al (2021) SWOT analysis applications: an integrative literature review. J Glob Bus Insights 6:55–73. https://doi.org/10.5038/2640-6489.6.1.1148
23. Habibul Kabir K, Aurko SY, Saifur Rahman M (2021) Smart power management in OIC countries: a critical overview using SWOT-AHP and hybrid MCDM analysis. Energies 14:1–50. https://doi.org/10.3390/en14206480
24. Bayraktar M, Yüksel O (2023) Analysis of the nuclear energy systems as an alternative propulsion system option on commercial marine vessels by utilizing the SWOT-AHP method. Nucl Eng Des 407.https://doi.org/10.1016/j.nucengdes.2023.112265
25. Bas E (2013) The integrated framework for analysis of electricity supply chain using an integrated SWOT-fuzzy TOPSIS methodology combined with AHP: the case of Turkey. Int J Electr Power Energy Syst 44:897–907. https://doi.org/10.1016/j.ijepes.2012.08.045
26. Wang K (2007) A process view of SWOT analysis. In: Proceedings of the 51st annual meeting of the ISSS-2007, Tokyo, Japan
27. Sarsby A (2016) SWOT analysis. Lulu. com
28. Watkins R, Leigh D (2009) Handbook of improving performance in the workplace. John Wiley & Sons, The handbook of selecting and implementing performance interventions
29. Kazemi F, Abolhassani L, Rahmati EA, Sayyad-Amin P (2018) Strategic planning for cultivation of fruit trees and shrubs in urban landscapes using the SWOT method: a case study for the city of Mashhad, Iran. Land use policy 70:1–9. https://doi.org/10.1016/j.landusepol.2017.10.006
30. Kumar S, Srikanth R, Mahalingam S (2023) Regulatory framework for India ' s energy security and sustainability. Energy Policy 162:112815. https://doi.org/10.1016/j.enpol.2022.112815
31. Qaiser I (2022) A comparison of renewable and sustainable energy sector of the South Asian countries: an application of SWOT methodology. Renew Energy 181:417–425. https://doi.org/10.1016/j.renene.2021.09.066
32. Kamran M, Fazal MR, Mudassar M (2020) Towards empowerment of the renewable energy sector in Pakistan for sustainable energy evolution: SWOT analysis. Renew Energy 146:543–558. https://doi.org/10.1016/j.renene.2019.06.165
33. Madurai Elavarasan R, Afridhis S, Vijayaraghavan RR et al (2020) SWOT analysis: a framework for comprehensive evaluation of drivers and barriers for renewable energy development in significant countries. Energy Rep 6:1838–1864. https://doi.org/10.1016/j.egyr.2020.07.007
34. Agyekum EB (2020) Energy poverty in energy rich Ghana: a SWOT analytical approach for the development of Ghana's renewable energy. Sustain Energy Technol Assess 40:100760. https://doi.org/10.1016/j.seta.2020.100760
35. Shi X (2016) The future of ASEAN energy mix: a SWOT analysis. Renew Sustain Energy Rev 53:672–680. https://doi.org/10.1016/j.rser.2015.09.010
36. Terrados J, Almonacid G, Hontoria L (2007) Regional energy planning through SWOT analysis and strategic planning tools. Impact on renewables development. Renew Sustain Energy Rev 11:1275–1287. https://doi.org/10.1016/j.rser.2005.08.003
37. Jing W, Tao M (2021) Research on clean energy development strategy of China three gorges corporation based on SWOT framework. Sustain Energy Technol Assess 47:9–13. https://doi.org/10.1016/j.seta.2021.101335
38. Zhou J, He P, Qin Y, Ren D (2019) A selection model based on SWOT analysis for determining a suitable strategy of prefabrication implementation in rural areas. Sustain Cities Soc 50.https://doi.org/10.1016/j.scs.2019.101715

39. Mateo JRSC (2012) Multi criteria analysis in the renewable energy industry. Springer Science & Business Media
40. Wang JJ, Jing YY, Zhang CF, Zhao JH (2009) Review on multi-criteria decision analysis aid in sustainable energy decision-making. Renew Sustain Energy Rev 13:2263–2278. https://doi.org/10.1016/j.rser.2009.06.021
41. Hayashi K (2000) Multicriteria analysis for agricultural resource management: a critical survey and future perspectives. Eur J Oper Res 122:486–500. https://doi.org/10.1016/S0377-2217(99)00249-0
42. Korhonen P, Moskowitz H, Wallenius J (1992) Multiple criteria decision support-a review. Eur J Oper Res 63:361–375
43. Hwang C-L, Masud ASM (2012) Multiple objective decision making—methods and applications: a state-of-the-art survey. Springer Science & Business Media
44. Belton V, Stewart T (2002) Multiple criteria decision analysis: an integrated approach. Springer Science & Business Media
45. Kumar A, Sah B, Singh AR et al (2017) A review of multi criteria decision making (MCDM) towards sustainable renewable energy development. Renew Sustain Energy Rev 69:596–609. https://doi.org/10.1016/j.rser.2016.11.191
46. Miller DW, Starr MK (1969) Executive decisions and operations research. (No Title)
47. Fishburn PC (1967) Additive utilities with incomplete product sets: application to priorities and assignments. Oper Res 15:537–542
48. Saaty TL (1980) The analytic hierarchy process. McGrawhill, Juc New York
49. Opricovic S (1998) Multicriteria optimization of civil engineering systems. Fac Civ Eng Belgrade 2:5–21
50. Edwards W (1977) How to use multiattribute utility measurement for social decisionmaking. IEEE Trans Syst Man Cybern 7:326–340
51. Benayoun R, Roy B, Sussman B (1966) ELECTRE: Une méthode pour guider le choix en présence de points de vue multiples. Note Trav 49:2–120
52. Ghobadipour B, Mojarradi B (2015) Comment on: M. Abedi, S. A. Torabi, Gh. Norouzi and M. Hamzeh, "ELECTRE III: A knowledge-driven method for integration of geophysical data with geological and geochemical data in mineral prospectivity mapping." J Appl Geophys 117:136–137. https://doi.org/10.1016/j.jappgeo.2015.03.005
53. Lin R, Lu S, Yang A et al (2021) Multi-criteria sustainability assessment and decision-making framework for hydrogen pathways prioritization: an extended ELECTRE method under hybrid information. Int J Hydrogen Energy 46:13430–13445. https://doi.org/10.1016/j.ijhydene.2021.01.018
54. Beccali M, Cellura M, Mistretta M (2003) Decision-making in energy planning. Application of the electre method at regional level for the diffusion of renewable energy technology. Renew Energy 28:2063–2087. https://doi.org/10.1016/S0960-1481(03)00102-2
55. McCahon CS, Hwang CL, Tillman FA (1983) A multiple attribute evaluation of Bayesian availability estimators. IEEE Trans Reliab 32:496–503
56. Georgopoulou E, Lalas D, Papagiannakis L (1997) A Multicriteria decision aid approach for energy planning problems: the case of renewable energy option. Eur J Oper Res 103:38–54
57. Zhang T, Yuan J (2005) Decision-aid for power distribution system planning problems using ELECTRE III. In: 2005 International power engineering conference. IEEE, pp 1–317
58. Roulet C-A, Flourentzou F, Labben HH et al (2002) ORME: a multicriteria rating methodology for buildings. Build Environ 37:579–586
59. Azzopardi B, Martínez-Ceseña EA, Mutale J (2013) Decision support system for ranking photovoltaic technologies. IET Renew Power Gener 7:669–679
60. Brans J-P, Vincke P (1985) Note—a preference ranking organisation method: (the promethee method for multiple criteria decision-making). Manage Sci 31:647–656
61. Debbarma B, Chakraborti P, Bose PK et al (2017) Exploration of PROMETHEE II and VIKOR methodology in a MCDM approach for ascertaining the optimal performance-emission trade-off vantage in a hydrogen-biohol dual fuel endeavour. Fuel 210:922–935. https://doi.org/10.1016/j.fuel.2017.08.016

62. Bortoluzzi M, Furlan M, Colombo SG et al (2021) Combining value-focused thinking and promethee techniques for selecting a portfolio of distributed energy generation projects in the brazilian electricity sector. Sustain 13.https://doi.org/10.3390/su131911091
63. Demirel YE, Simsek E, Ozturk E, Kitis M (2021) Selection of priority energy efficiency practices for industrial steam boilers by PROMETHEE decision model. Energy Effic 14:1–20. https://doi.org/10.1007/s12053-021-10007-8
64. Chisale SW, Eliya S, Taulo J (2023) Optimization and design of hybrid power system using HOMER pro and integrated CRITIC-PROMETHEE II approaches. Green Technol Sustain 1:100005. https://doi.org/10.1016/j.grets.2022.100005
65. Georgopoulou E, Sarafidis Y, Diakoulaki D (1998) Design and implementation of a group DSS for sustaining renewable energies exploitation. Eur J Oper Res 109:483–500
66. Tsoutsos T, Drandaki M, Frantzeskaki N et al (2009) Sustainable energy planning by using multi-criteria analysis application in the island of Crete. Energy Policy 37:1587–1600. https://doi.org/10.1016/j.enpol.2008.12.011
67. Nassereddine M, Eskandari H (2017) An integrated MCDM approach to evaluate public transportation systems in Tehran. Transp Res Part A Policy Pract 106:427–439. https://doi.org/10.1016/j.tra.2017.10.013
68. Ozkaya G, Erdin C (2020) Evaluation of smart and sustainable cities through a hybrid MCDM approach based on ANP and TOPSIS technique. Heliyon 6.https://doi.org/10.1016/j.heliyon.2020.e05052
69. Çetinkaya C, Erbaş M, Kabak M, Özceylan E (2023) A mass vaccination site selection problem: an application of GIS and entropy-based MAUT approach. Socioecon Plann Sci 85.https://doi.org/10.1016/j.seps.2022.101376
70. Edwards TG, Chelst KR (2004) The HSOR project: insinuating OR into high school mathematics classrooms. INFORMS Trans Educ 4:1–8
71. Alizadeh R, Soltanisehat L, Lund PD, Zamanisabzi H (2020) Improving renewable energy policy planning and decision-making through a hybrid MCDM method. Energy Policy 137.https://doi.org/10.1016/j.enpol.2019.111174
72. Meyar-Naimi H, Vaez-Zadeh S (2013) Sustainability assessment of a power generation system using dsr-hns framework. IEEE Trans Energy Convers 28:327–334
73. Luthra S, Kumar S, Garg D, Haleem A (2015) Barriers to renewable/sustainable energy technologies adoption: Indian perspective. Renew Sustain Energy Rev 41:762–776. https://doi.org/10.1016/j.rser.2014.08.077
74. Pathak SK, Sharma V, Chougule SS, Goel V (2022) Prioritization of barriers to the development of renewable energy technologies in India using integrated Modified Delphi and AHP method. Sustain Energy Technol Assess 50:101818. https://doi.org/10.1016/j.seta.2021.101818
75. Odu GO (2019) Weighting methods for multi-criteria decision making technique. J Appl Sci Environ Manag 23:1449. https://doi.org/10.4314/jasem.v23i8.7
76. Nasrollahi S, Kazemi A, Jahangir MH, Aryaee S (2023) Selecting suitable wave energy technology for sustainable development, an MCDM approach. Renew Energy 202:756–772. https://doi.org/10.1016/j.renene.2022.11.005
77. Tazzit S, Jing L, Ma J, Jaradat R (2023) Systems-thinking skills preferences evaluation model of practitioners using hybrid weight determination and extended VIKOR model under COVID-19. Adv Eng Inf 57.https://doi.org/10.1016/j.aei.2023.102107
78. Brodny J, Tutak M (2023) Assessing the energy security of European Union countries from two perspectives–a new integrated approach based on MCDM methods. Appl Energy 347:121443. https://doi.org/10.1016/j.apenergy.2023.121443
79. Aljaghoub H, Abumadi F, AlMallahi MN et al (2022) Solar PV cleaning techniques contribute to sustainable development goals (SDGs) using multi-criteria decision-making (MCDM): assessment and review. Int J Thermofluids 16.https://doi.org/10.1016/j.ijft.2022.100233
80. Pinzon Amorocho JA, Hartmann T (2022) A multi-criteria decision-making framework for residential building renovation using pairwise comparison and TOPSIS methods. J Build Eng 53.https://doi.org/10.1016/j.jobe.2022.104596

81. Opricovic S, Tzeng GH (2004) Compromise solution by MCDM methods: a comparative analysis of VIKOR and TOPSIS. Eur J Oper Res 156:445–455. https://doi.org/10.1016/S0377-2217(03)00020-1
82. Dutta R, Das S, De S (2023) Multi criteria decision making with machine-learning based load forecasting methods for techno-economic and environmentally sustainable distributed hybrid energy solution. Energy Convers Manag 291:117316. https://doi.org/10.1016/j.enconman.2023.117316
83. Rojas-Zerpa JC, Yusta JM (2015) Application of multicriteria decision methods for electric supply planning in rural and remote areas. Renew Sustain Energy Rev 52:557–571. https://doi.org/10.1016/j.rser.2015.07.139
84. Baležentis T, Streimikiene D (2017) Multi-criteria ranking of energy generation scenarios with Monte Carlo simulation. Appl Energy 185:862–871. https://doi.org/10.1016/j.apenergy.2016.10.085
85. Villacreses G, Gaona G, Martínez-Gómez J, Jijón DJ (2017) Wind farms suitability location using geographical information system (GIS), based on multi-criteria decision making (MCDM) methods: the case of continental Ecuador. Renew Energy 109:275–286. https://doi.org/10.1016/j.renene.2017.03.041
86. Billig E, Thraen D (2017) Renewable methane–a technology evaluation by multi-criteria decision making from a European perspective. Energy 139:468–484. https://doi.org/10.1016/j.energy.2017.07.164
87. Büyüközkan G, Karabulut Y (2017) Energy project performance evaluation with sustainability perspective. Energy 119:549–560. https://doi.org/10.1016/j.energy.2016.12.087
88. Sánchez-Lozano JM, García-Cascales MS, Lamata MT (2016) Comparative TOPSIS-ELECTRE TRI methods for optimal sites for photovoltaic solar farms. case study in Spain. J Clean Prod 127:387–398. https://doi.org/10.1016/j.jclepro.2016.04.005
89. Tadić S, Zečević S, Krstić M (2014) A novel hybrid MCDM model based on fuzzy DEMATEL, fuzzy ANP and fuzzy VIKOR for city logistics concept selection. Expert Syst Appl 41:8112–8128. https://doi.org/10.1016/j.eswa.2014.07.021
90. Abdullah L, Najib L (2016) Sustainable energy planning decision using the intuitionistic fuzzy analytic hierarchy process: choosing energy technology in Malaysia. Int J Sustain Energy 35:360–377. https://doi.org/10.1080/14786451.2014.907292
91. Kulkarni SH, Jirage BJ, Anil TR (2017) Alternative energy options for India—a multi-criteria decision analysis to rank energy alternatives using analytic hierarchy process and fuzzy logic with an emphasis to distributed generation. Distrib Gener Altern Energy J 32:29–55
92. Kabak M, Köse E, Kirilmaz O, Burmaoğlu S (2014) A fuzzy multi-criteria decision making approach to assess building energy performance. Energy Build 72:382–389. https://doi.org/10.1016/j.enbuild.2013.12.059
93. Papapostolou A, Karakosta C, Doukas H (2017) Analysis of policy scenarios for achieving renewable energy sources targets: a fuzzy TOPSIS approach. Energy Environ 28:88–109
94. Guo S, Zhao H (2015) Optimal site selection of electric vehicle charging station by using fuzzy TOPSIS based on sustainability perspective. Appl Energy 158:390–402
95. Cavallaro F, Ciraolo L (2013) Sustainability assessment of solar technologies based on linguistic information. In: Assessment and simulation tools for sustainable energy systems: theory and applications. Springer, pp 3–25
96. Lee SK, Mogi G, Hui KS (2013) A fuzzy analytic hierarchy process (AHP)/data envelopment analysis (DEA) hybrid model for efficiently allocating energy R&D resources: in the case of energy technologies against high oil prices. Renew Sustain Energy Rev 21:347–355
97. Kaya T, Kahraman C (2010) Multicriteria renewable energy planning using an integrated fuzzy VIKOR & AHP methodology: the case of Istanbul. Energy 35:2517–2527
98. Rahimirad Z, Sadabadi AA (2023) Green hydrogen technology development and usage policymaking in Iran using SWOT analysis and MCDM methods. Int J Hydrogen Energy 48:15179–15194. https://doi.org/10.1016/j.ijhydene.2023.01.035
99. Almutairi K, Almutairi MS, Harb KM, Marey O (2023) A thorough investigation of renewable energy development strategies through integrated approach: a case study. Energy Sour Part A Recover Util Environ Eff 45:708–726. https://doi.org/10.1080/15567036.2023.2169786

100. Tavakoli M, Torabi SA, GhanavatiNejad M, Nayeri S (2023) An integrated decision-making framework for selecting the best strategies of water resources management in pandemic emergencies. Sci Iran
101. Longsheng C, Ali Shah SA, Solangi YA et al (2022) An integrated SWOT-multi-criteria analysis of implementing sustainable waste-to-energy in Pakistan. Renew Energy 195:1438–1453. https://doi.org/10.1016/j.renene.2022.06.112
102. Almutairi K, Hosseini Dehshiri SJ, Hosseini Dehshiri SS et al (2022) Determination of optimal renewable energy growth strategies using SWOT analysis, hybrid MCDM methods, and game theory: a case study. Int J Energy Res 46:6766–6789. https://doi.org/10.1002/er.7620
103. Yontar E, Derse O (2022) Evaluation of sustainable energy action plan strategies with a SWOT/TWOS-based AHP/ANP approach: a case study. Environ Dev Sustain. https://doi.org/10.1007/s10668-022-02804-7
104. Büyüközkan G, Havle CA, Feyzioğlu O (2021) An integrated SWOT based fuzzy AHP and fuzzy MARCOS methodology for digital transformation strategy analysis in airline industry. J Air Transp Manag 97.https://doi.org/10.1016/j.jairtraman.2021.102142
105. Mukeshimana MC, Zhao ZY, Nshimiyimana JP (2021) Evaluating strategies for renewable energy development in Rwanda: an integrated SWOT–ISM analysis. Renew Energy 176:402–414. https://doi.org/10.1016/j.renene.2021.05.104
106. Wang Y, Xu L, Solangi YA (2020) Strategic renewable energy resources selection for Pakistan: based on SWOT-Fuzzy AHP approach. Sustain Cities Soc 52.https://doi.org/10.1016/j.scs.2019.101861
107. Stoilova SD, Martinov S V. (2019) Selecting a location for establishing a rail-road intermodal terminal by using a hybrid SWOT/MCDM model. IOP Conf Ser Mater Sci Eng 618.https://doi.org/10.1088/1757-899X/618/1/012060
108. Cayir Ervural B, Zaim S, Demirel OF et al (2018) An ANP and fuzzy TOPSIS-based SWOT analysis for Turkey's energy planning. Renew Sustain Energy Rev 82:1538–1550. https://doi.org/10.1016/j.rser.2017.06.095

Part II
Technology Overviews

Chapter 3
Renewable Energy Integration to Electric Power Grid: Opportunities, Challenges, and Solutions

Prithwiraj Purkait, Mousumi Basu, and Sujoy Ranjan Nath

Abstract With rapid urbanization, the need for energy usage has become more dominant than any other demand in modern society. With rapidly depleting fossil fuel sources, seconded by the need for environmental protection, the search for new and renewable energy (RE) sources has gained immense importance as sustainable alternatives. However, because of the intermittent and unpredictable nature of many of these RE sources, it has become imperative that their greatest benefit can only be harnessed when they are integrated into existing electric power networks. In this context, the greatest impetus toward comprehending a sustainable future is evolution of technology for integration of RE sources in electric power network and its control. There are various issues to be addressed while integrating RE sources to the power grid, including technical, economic, societal, and other challenges. This chapter begins by introducing the opportunities of integrating RES into the power grid, followed by their scopes and benefits. The various challenges faced while integrating RE into the power grid are further explained with specific case studies. Possible solutions to these issues are discussed in the next sections. Progress in the area of development of mathematical models of RE systems for studies related to grid integration is highlighted. Simulation tools that are being used by the research community for RE integration and control applications are introduced. Finally, some recent research works on RE integration-related activities are presented. This book chapter thus aims to enrich the knowledge base in green energy systems and opens up new horizons for simulation and experimental studies to address the crucial need of power industry in exploring power electronics, control technology, information and communication technology (ICT), and smart grid technology for supporting RE integration to grid.

Keywords Renewable energy · Grid integration · Distributed energy resources

---

P. Purkait (✉) · M. Basu · S. R. Nath
Power Engineering Department, Jadavpur University, Kolkata, India
e-mail: prithwirajp.pe@jadavpuruniversity.in

S. De et al. (eds.), *Challenges and Opportunities of Distributed Renewable Power*,
Energy, Environment, and Sustainability, https://doi.org/10.1007/978-981-97-1406-3_3

Abbreviations

AI	Artificial intelligence
ANFIS	Adaptive neuro-fuzzy inference system
ANN	Artificial neural network
BPP	Bioenergy power plant
BESS	Battery energy storage system
BMS	Battery management system
CHP	Combined heat and power
DER	Distributed energy resources
DFIG	Doubly fed induction generator
DG	Distributed generation
DR	Demand response
DSM	Demand side management
DISCOMs	Distribution companies
DSTATCOM	Distribution static synchronous compensator
ES	Energy storage
ESS	Energy storage system
EMS	Energy management system
FRT	Fault Ride-Through
EV	Electric Vehicle
FACTS	Flexible AC transmission systems
FESS	Flywheel energy storage system
FSP	Forecast Service provider
GA	Genetic algorithm
GENCOs	Generation companies
GHG	Greenhouse gas
GPS	Global positioning system
G2V	Grid to vehicle
HPSO-TVAC	Self-organizing hierarchical particle swarm optimizer with time-varying acceleration coefficients
HVRT	High-Voltage Ride-through
IEA	International energy agency
ICT	Information and communication technology
IGBT	Insulated gate bipolar transistor
IoT	Internet of Things
LVRT	Low-Voltage Ride-through
MG	Microgrid
Max	Maximum
Min	Minimum
MPPT	Maximum power point tracking
NZE	Net zero emission
OLTC	On-load tap-changers
PCA	Principal component analysis

PCC	Point of common coupling
PD	Protective device
PEV	Plug-in electric vehicle
PHS	Pumped hydro storage
PMSG	Permanent magnet synchronous generator
PSO	Particle swarm optimization
PMU	Phasor measurement unit
PV	Photovoltaic
PWM	Pulse width modulation
RE	Renewable energy
RES	Renewable energy sources
SCADA	Supervisory control and data acquisition
SCED	Security constrained economic dispatch
SHPP	Small-hydro power plant
SC	Supercapacitor
SOC	State of charge
STATCOM	Static synchronous compensator
SSSC	Static synchronous series compensator
SVC	Static VAR compensator
T&D	Transmission and Distribution
TCSC	Thyristor-controlled series capacitor
UC	Ultra-capacitor
UPFC	Unified power flower controller
UPQC	Unified power quality conditioner
VPP	Virtual power plant
VRE	Variable renewable energy
VRT	Voltage ride-through
V2G	Vehicle-to-grid
WTG	Wind turbine generator
WSP	Weather service provider

Nomenclature

$Apvt$	Total area of a solar PVT panel
a_d, b_d, c_d	Fuel price coefficients of Diesel generator
C_T	Total price
C	Capacitance
$C_{UC-total}$	Total UC system capacitance
E_{UC}	Amount of energy taken from UC bank
$fw(y)$	Weibull probability distribution function of wind power y
$f_{PV}(y)$	Lognormal probability distribution function of solar power y
G_r	Solar irradiation forecast

G_{rT}	Solar irradiation on unit area of a PVT panel at time t
H_l	Height of the waterfall of mini-hydro plant
$H_{PVT,t}$	Heat output of solar PVT panel at time
Kpv	Direct cost coefficient of solar PV plant
Kw	Direct price coefficient of WTG
LHV_{MSW}	Lower heat value of the material used in combustion (kJ/kg)
N_d	Number of diesel generators
N_{PV}	Number of solar PV units
N_{EV}	Number of PEVs
N_h	Number of small-hydro power plants
N_w	Number of WTGs
n_P	Number of series strings in parallel
n_s	Number of capacitors linked in series
$Oe_{pvt}(P_{pvt})$	Reserve cost function owing to overestimation of solar PV plant at time t
$Ow_t(Pw_t)$	Reserve price function owing to overestimation of WTG at time t
P_{Bt}	Power generation of BPP at hour t
Pb_t^c, Pb_t^d	Charging and discharging power of BESS at time t
Pdg_t	Power generated by the diesel generator at time t
Pdg^{min}, Pdg^{max}	Minimum and maximum limits of output power for diesel generator
$P_{PVT,t}$	Output power of solar PVT panel at time t
P_{PVt}	Power generation of solar PV plant at hour t
P_{PVr}	Rated power output of solar PV plant
Pw_t	Available power generation of WTG at time t
Pwr	Rated power generation of WTG
Pw_t^{min}, Pw_t^{max}	Minimum limit and maximum limit of power generation for WTG at time t
P_{ht}	Power output of mini-hydro plant at time t
Psp_{gt}	Power generation of PHS at time t
Psp_{pt}	Pumping power of PHS at hour t
$Ps_{gh}^{min}, Ps_{gh}^{max}$	Minimum and maximum power production limits of PHS
$Ps_{ph}^{min}, Ps_{ph}^{max}$	Minimum and maximum pumping power limits of PHS
P_{EVt}	Charging power of PEV at time t
$P_{EV}^{min}, P_{EV}^{max}$	Minimum and maximum limits of charging power of PEV
$PESSC_t$	Power charge to BESS at time t
$PESSD_t$	Power discharge from BESS at time t
$Ppev_t^{tr}$	Power consumption throughout traveling time
P_{Dt}	Predicted power demand at time t
Q_t	Inflow rate of mini-hydro plant at time t
$Q_{gt}(Psp_{gt})$	Release rate of PHS at hour t
$Q_{Pt}(Psp_{pt})$	Pumping rate of PHS at hour t

R	Internal resistance
$R_{UC-total}$	Total UC system resistance
$(SOC)_t^{pev}$	SOC of PEV at time t
$(SOC)_{max}^{pev}$ and $(SOC)_{min}^{pev}$	Maximum and minimum stored energy of PEV
t, T	Time index and scheduling eon
UR_{dg}, DR_{dg}	Ramp-up and ramp-down rate limits of diesel generator
$Ue_{PVt}(P_{PVt})$	Penalty cost function owing to underestimation of solar PV plant at time t
$Uw_t(Pw_t)$	Penalty price function owing to underestimation of WTG at time t
ue_{PV}, oe_{PV}	Penalty cost and reserve cost for solar PV plant
ue_w, oe_w	Penalty price and reserve price of WTG
Ub_t^c, Ub_t^d	Binary variables demonstrating status of BESS at time t
$U_t^{pev,c}, U_t^{pev,d}$	Binary variables demonstrating charging mode and discharging mode of PEV at time t
$uloss$	Heat loss coefficient
V_{in}	Cut-in wind velocity
V_{out}	Cut-out wind velocity
Vr_t	Rated wind velocity at time t
Vf_t	Forecasted wind velocity at time t
V_{MSW}	Quantity of material used in combustion (kg) in bio-energy plant
$V_{res,t}$	Water capacity in upper reservoir of PHS at hour t
$V_{res}^{min}, V_{res}^{max}$	Minimum and maximum upper reservoir stowage limits of PHS
$V_{res}^{start}, V_{res}^{end}$	Opening and finale stowed water capacity in upper reservoir of PHS
$V(t)$	Voltage after time t
V_i	Initial voltage of UC bank before the discharging starts
V_f	Final voltage of UC bank after which discharging starts
T_{ref}, T_{ambt}	Reference and ambient temperature
$T_{cell,PVT,t}$	Solar PV cell temperature at time t
$T_{amb,t}$	Ambient temperature at time t
V_{MSW}	Quantity of material used in combustion of BPP
ΔD_t^{pev}	Traveling distance of PEV at time t
$\eta_{ePVT,t}$	Electrical efficiency of the solar PVT panel at time t
$\eta_{ref,PVT}$	Reference electrical efficiency of solar PVT panel
$\beta_{ref,PVT}$	Power temperature coefficient of solar PVT panel
$\tau_{\alpha PVT}$	Transmission and absorption coefficient of solar PVT panel
ε_r	Temperature coefficient
ρ	Specific weight of water
η_G	Efficiency of hydro-generator
η_T	Efficiency of hydraulic turbine

η_g	Generator efficiency of BPP
η_p	Conversion efficiency of BPP
η_b^c, η_b^d	Charging and discharging efficacy of BESS
$\eta^{pev,c}$, $\eta^{pev,d}$	Charging and discharging efficacy of PEV
η^{pev}	Total efficacy of PEV

3.1 Introduction

Following the Paris Agreement on climate change (November 2016), international efforts to deploy RES, especially wind and solar technologies have reached record levels in recent years. In the 450 Scenario, nearly 60% of the power generated globally in 2040 is projected to come from renewables [1]; wind and solar PV make up almost half of it (Fig. 3.1) [2]. Wide-scale deployment expects solar PV cost to be reduced by a further 40–70% and onshore wind by 10–25% by 2040 [1]. However, reduced cost of renewables, on their own, will not be able to meet the NZE target by 2050 as set by the International Energy Agency (IEA). Structural changes are necessary for the planning, design, and operation of the power system so as to encourage investments in integrating RES into the existing power network.

RES share in total electricity generation is projected to rise from around 25% in 2020 to over 40% in 2030, and nearly 80% in 2050 [2]. The IEA's 450 scenario predicts an energy scenario by 2030 with a substantial degree of renewable energy penetration, remarkable improvement in EV technology as well as process efficiency.

The power transmission and distribution infrastructure need to expand to meet the growing electrification, integration of innumerable new renewable energy projects, and reinforce systems that have to adapt to the system dynamics and flexibility. Power transmission and distribution networks worldwide currently include around 80 million km of lines. It is projected that 13 million km of distribution lines and about 1.6 million km of transmission lines are going to be added by 2030 [2]. By 2050, a further 45 million km of distribution line and 4 million km of transmission line are going

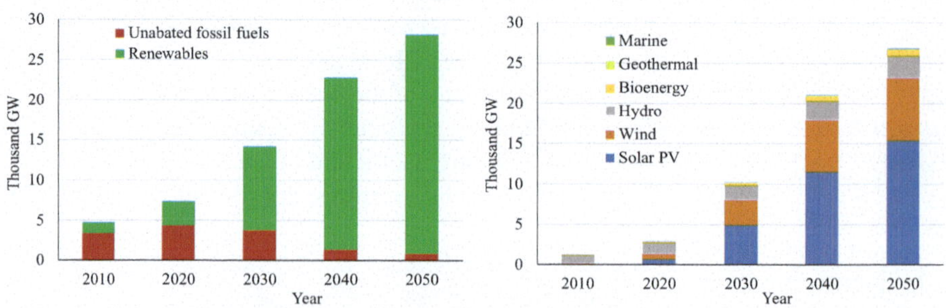

Fig. 3.1 Total installed capacity from unabated fossil fuels drops substantially, whereas there is marked growth in low emission renewable sources in the NZE scenario during 2010–2050 [2]

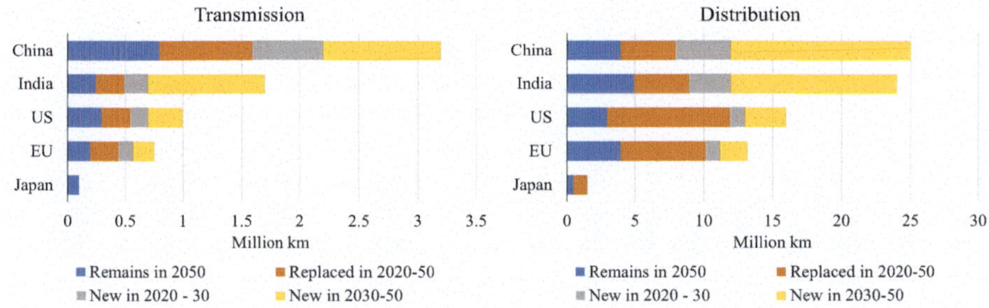

Fig. 3.2 Power network expansion in the NZE scenario during 2020–2050 [2]

to be added [3] (Fig. 3.2) along with various power transfer equipment including primary equipment, transformers, sensing, protection, and control equipment.

Such overwhelming growth in electric power infrastructure is aimed at evacuating the enhanced renewable energy generation. Integration of solar PV and wind with a penetration share of around 25% in the power mix is affordable in terms of system design and operation. The variability and unpredictability of wind and solar output can be accommodated by a suitably implemented grid that has backup support from power plants that are ready to dispatch at short notice. At higher levels of renewable penetration, the options of energy storage and demand response become crucial to cope with the variability of solar PV and wind energy outputs. In the NZE scenario, grid-scale battery storage deployment is looking at an above 4000% growth rate from 2022 to 2030 [4]. Batteries not only store the excess energy output of wind and solar when they are abundant, but they also support the grid at times of increased demand, or unavailability of solar/wind, and for enhancing grid stability and reliability. Other forms of storage, such as gravity-based, heat-based, or flywheels, are under development and are soon expected to emerge as a supplement to battery storage.

Integration of variable and intermitted RE generation sources harshly affects the predictability of power flow through the system. Such variabilities lead to system conditions that are more dynamic and complex to manage. Digital and smart technologies play a major part in overcoming the challenges arising out of these changes. Smart sensors and control devices with associated software can provide dynamic system information in real time. Such real-time assessment of system health enables efficient utilization of existing resources and allows the network to be fully utilized without sacrificing stability, reliability, and resilience. The involvement of digital technology and high-speed communication between the stakeholders ensures coordinated planning, operation, information, and knowledge sharing for enhanced grid security.

Paris Agreement 2016 has thus promoted unprecedented government investments in the modernization of the power system, the driving factor of which is RE integration to the grid. Overcoming the jolt of the Covid-19 pandemic, the energy sector is set to see an overall increase in investment [3, 5]. The main boom in recent years has been in the electric power sector—mostly in renewables and grids. A growing share

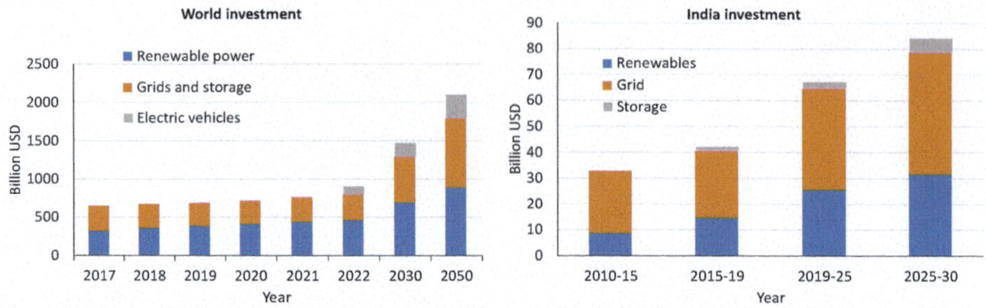

Fig. 3.3 Average annual investment in the power sector, worldwide, and in India [3, 5]

of investments in the power sector is captured by RE and gird (Fig. 3.3)—which is suggestive of the significance of grid integration of RES for supporting clean energy transitions. Abiding by the pledge of NZE 2050, power grids need to be expanded, modernized, and digitized to allow deep penetration of the large share of renewables that are set to come up. Storage solutions and electric vehicles gain importance to add flexibility to the future power system.

It is thus obvious that the world outlook for future power systems is large-scale integration of RES into the power grid. This chapter thus outlies the opportunities for grid integration of RES, the challenges encountered, and also possible ways to overcome those. The later part of the chapter summarizes the modern research avenues for modeling and simulation analysis of RES and their grid integration performance with the help of an example.

3.2 Grid Integration of RES

In the 450 Scenario, RE is projected to account for 60% of electricity generation in 2040 [1]. VRE is less likely to impose significant challenges on the power network at low penetration levels. However, when VRE is deployed to the extent of the 450 Scenario, it will require significant enhancement in the grid integration measures. These measures include technical, commercial, operational, institutional, policy, societal, and market design aspects to permit efficient and cost-effective amalgamation of RE in the power network.

3.2.1 Benefits and Opportunities

In order to harness full benefit from the large amount of DERs, grid integration is essential. DERs include renewable sources of generation, various forms of energy storage devices, electric vehicles, microgrids, and various forms of loads. The legacy grid that was planned, designed, constructed, and operated for evacuation of large amount of conventional power, has to be upgraded to accommodate such DER too. This has led to the development of smart or intelligent grids with advanced sensing, communication, control, and computation technology.

A distributed generation (DG) system may include renewable as well as non-renewable sources of energy. The integration of DG into the existing distribution system can substantially influence the power flow along the network. Such influences are beneficial at times, while challenging at other perspectives. The positive benefits that act as "system support benefits", include:

- **Reduction in Transmission and Distribution (T&D) losses**—DG systems being geographically distributed along the entire network, the length of the power line required to evacuate such power to the nearest load centers is shorter, resulting in reduced T&D losses. A cost saving of about 10–15% can be achieved through 6–7% reduction in grid losses.
- **Improved reliability**—With market liberalization, customers are driven to become more aware of their rights to uninterrupted quality power supply. These include not only domestic customers, but also industrial and large commercial installations. Because of this, many government policies now focus on energy security and sustainability. Improving reliability of existing power system comprising primarily of conventional generating sources would require to high investment and maintenance costs. Grid-supplied electricity is blamed for being unreliable in such cases, which has led to investments in DG being integrated into the grid. With greater flexibility, affordability, cost-effectiveness, reduced lead-time, and provisions for storage, grid-connected DG is proving to be the key provision for future reliable power.
  On the other hand, PV and wind being intermittent and unpredictable in nature, the reliability and availability of electricity from these sources are questionable when operating in stand-alone mode. In this respect, a grid connection is a much better option that has the feature of bi-directional power flow.
- **Voltage support and improved power quality**—Geographical proximity of DG to consumers makes the line length shorter leading to reduced voltage drops. Long lines are also prone to external disturbances causing flickers and other power quality issues. DG units, because of their fairly quick response time, and with smart control of interconnecting converters, can be deployed for both active and reactive power supports as and when necessary. This ensures better voltage regulation, frequency profile, and power quality improvements.
- **Deferments of new or upgraded transmission and distribution infrastructure**—Electricity demand is poised to increase by 1.5 times in the coming two decades. Relying on centralized conventional generating stations will mean huge investments in upscaling transmission and distribution infrastructure. When alternate fuels are locally available, customers can be encouraged to invest in DG thereby bypassing transmission and distribution upgradation costs.
- **Easy and quicker installation on account of prefabricated standardized components**—Adding capacity to the system with conventional sources and associated installation of T&D lines are often time-consuming. In comparison,

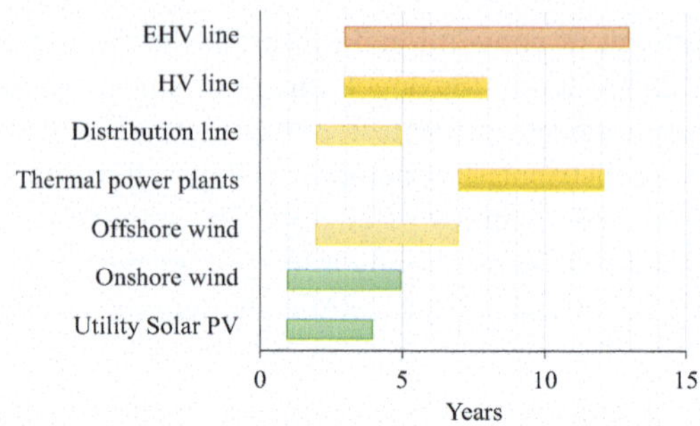

Fig. 3.4 Typical lead times for electricity grids, solar PV, wind, and coal-fired power plant [3]

deployment times for wind and PV are much faster because of the availability of the prefabricated standardized components that need to be added in modules (Fig. 3.4). Moreover, adding DG to the grid reduces the need for the installation of long-distance power lines.

- **Environment friendliness**—Major drivers for opting distributed generations are the uses of RE and combined heat power (CHP) that restrict GHG emissions. Environmental policies are the major driving force for the demand for DG integration to power grid. Environmental regulations compel electricity market players to explore and invest in cleaner and more efficient energy solutions.
- **Response flexibility**—The activation time for most DERs is low, meaning that such resources have much faster response than other sources in adjusting their power output. Such units can provide electric flexibility by prompt adjustments of power outputs. Those adjustments are crucial for system operation, system balancing, trading in spot markets, network congestion management, price flexibility, and demand response management.
- **Utilization of local resources**—Renewable energy sources are generally small scale and are geographically distributed over the grid. Installing DG along the grid thus provides the opportunity to explore cheap fuels along the locality, such as run-of-the-stream micro hydro, burning landfill gases, or other locally available biomass products.
- **Avoidance of the construction of large conventional generating plants and new transmission lines**—Use of RES and mini-CHP as part of DG avoids building new conventional power plants to support the enhanced load demand. DGs being geographically distributed, do not require immediate revamp in the transmission infrastructure when their capacity can be accommodated within existing power transfer limits of the lines.
- **Limited financial risks**—Competitiveness in the electricity market has encouraged the players to invest in small capacity DG plants rather than large conventional generation stations. Financial risks associated with market uncertainties are proportionately smaller with DG sources.
- **Improved supply quality and customer support**—The availability of DG close to load centers is beneficial to the customers in terms of supply quality and reliability. Locally available DGs can not only provide better voltage profiles at

customer premises, but also reduce the number of customers' minutes lost during grid outages when the DG can be made to operate in islanded mode.

- **Reduced electricity price**—Distribution companies (DISCOMs) and bulk consumers can install their own DG systems to reduce power purchase from the grid. This will reduce their power purchase costs, especially during peak hours when, typically, the power is costlier. With customers having such an option of choosing between grid power and power from their own DG units, the market power price is set to be reduced by generation companies (GENCOs).

3.2.2 Obstacles and Challenges

In the 450 Scenario, it is predicted that solar PV and wind are going to cover almost 30% of the world's electricity supply in 2040. Around 2030–2035 itself, VRE is projected to have a leading share in electricity supply in the United States, the European Union, China, and India (Fig. 3.5). Such high penetration level of DGs into the grid poses several challenges that include technical, institutional, policy, operational, financial, market design, and societal issues. Failing to address these system integration issues could mean under-utilization of the available renewable-based generation leading to a shortfall in decarbonization options.

Technical challenges Following are the specific technical issues to be addressed for integration of DG to the grid.

Interfacing: Most RE sources such as wind and solar energy cannot be directly connected to the grid. Various interfacing topologies and power electronic components are required to maintain constant voltage, frequency, and other parameters to ensure a stable connection with the grid following specified grid codes. A typical solar PV cell produces DC voltage of varying amplitude. Grid connection in such cases requires a closed-loop controlled boost chopper and inverter with an in-built maximum power point tracker (MPPT) that converts the solar PV output to 3-phase

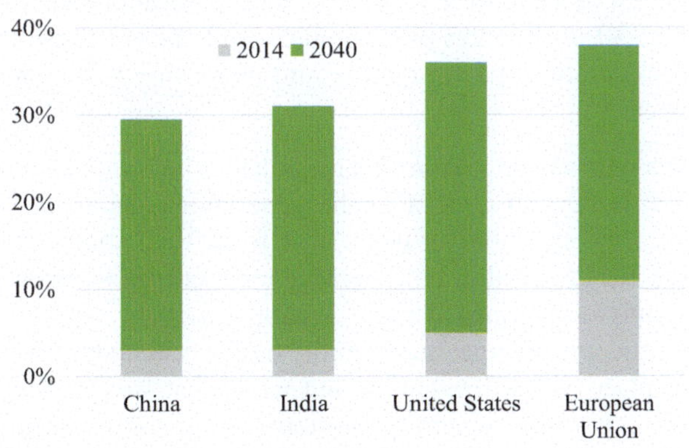

Fig. 3.5 Proportion of solar PV and wind in total electricity generation in the 450 Scenarios [1]

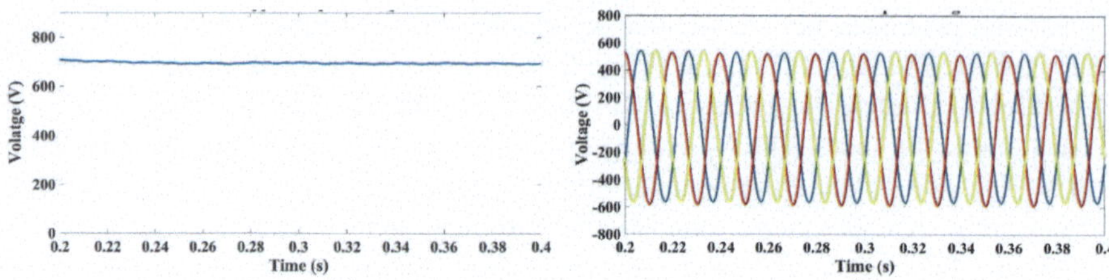

(a) (b)

Fig. 3.6 Solar PV generator output. A boost chopper is first used to step up the solar PV output DC voltage (**a**), and then the inverter is used to generate 3-phase AC voltage (**b**) to be fed to the grid

AC of voltage and frequency matching to those of the grid. Wind generator systems also require power electronic converters to match their output voltage and frequency to that of the grid before being synchronized. Suitably designed phase locked loops (PLL) need to be included in the converter circuit that makes the RES output voltage and frequency to follow those of the grid for maintaining synchronization under varying operating conditions. RESs such as small hydro, bio-fueled generators, and diesel generators that do not have power electronic converters, depend on closed-loop controlled automatic regulators for grid-synchronization. Figure 3.6a shows the output DC voltage from a typical boost chopper connected to a solar PV array. A properly designed inverter is essential to produce a 3-phase sinusoidal AC signal as shown in Fig. 3.6b, from the chopper output with proper amplitude, frequency, and phase angle so that it can be synchronized to the grid.

Planning and risk management: Introducing RE brings a host of new devices, often non-linear devices, interfacing and control modules, and topologies into the system, which might affect the power quality, system stability, and efficiency unless carefully planned and compensated for, wherever needed. Power electronic converters that are integral parts of all solar PV, wind, and other RE generators, include non-linear semiconductor devices. These non-linear devices are one major source of harmonics in the power network. Figure 3.7 shows the voltage and current profiles of the inverter output of a solar PV generator. Non-sinusoidal nature of the signals as evident from Fig. 3.7 are sources of harmonics in the power network. Harmonics are responsible for degrading power quality and increasing losses in the transmission/distribution devices.

Uncontrollable variability: Solar and wind output vary throughout the day in a way that is beyond control of the generation operators. Figure 3.8 shows a pattern of hourly dispersions of generation from VRE sources in India [5]. Such fluctuations ask for the need to employ additional energy sources/sinks to balance demand and supply on the grid at every instant. Ancillary support systems for maintaining grid frequency, voltage, and stability as per grid codes for DG integration need to be put in place. Solar irradiation and wind gust vary almost instantaneously making the corresponding generating systems of very low inertia. Figure 3.9 shows variations in solar PV output with varying solar irradiation and Fig. 3.10 shows PMSG-based

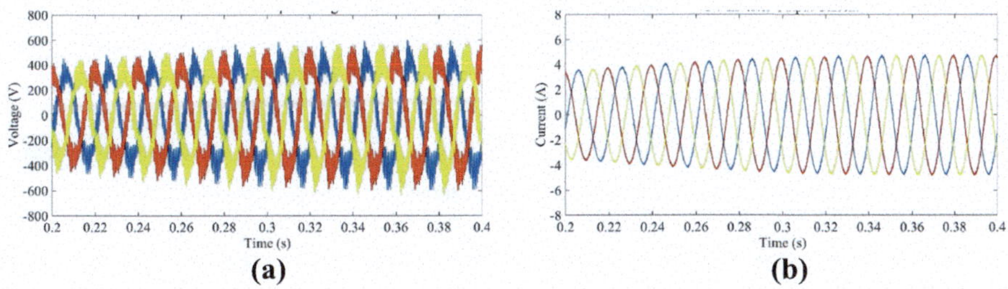

Fig. 3.7 Solar PV generator inverter output (**a**) voltage and (**b**) current containing harmonics

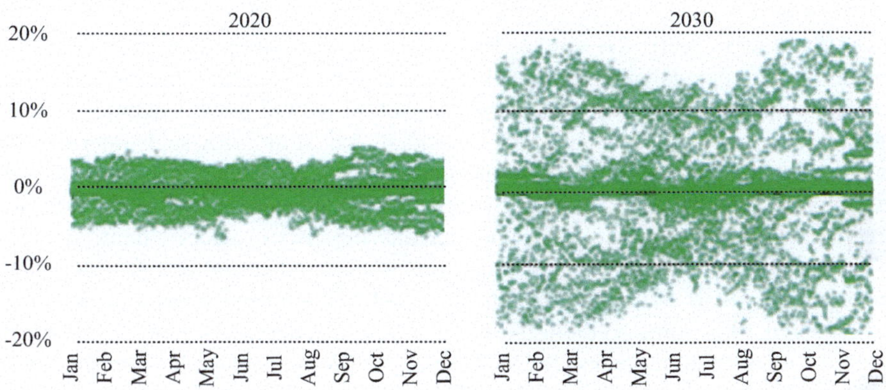

Fig. 3.8 Hourly deviations in VRE generation in India as percentage of average annual demand [5]

wind turbine generator output variations at varying wind speeds. Conventional base generators can never have enough ramping rates to match such sudden variations so as to compensate for solar PV and wind generation variability. Fast acting ESSs with adequate sensing tools need to be made available to balance out such subtle and unpredictable variations in RE generation.

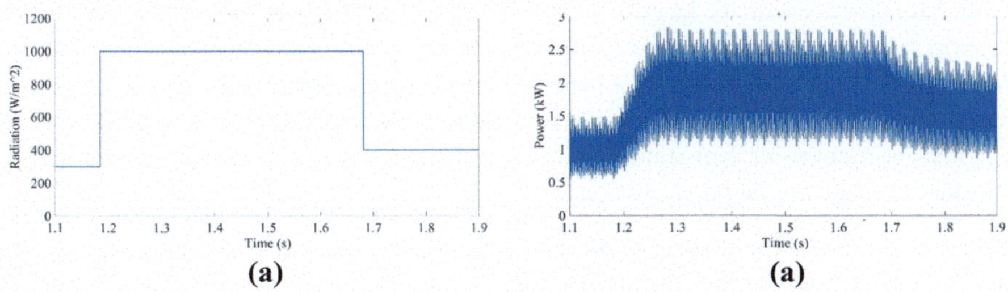

Fig. 3.9 Variation in solar PV output power (**b**) with varying solar irradiation (**a**)

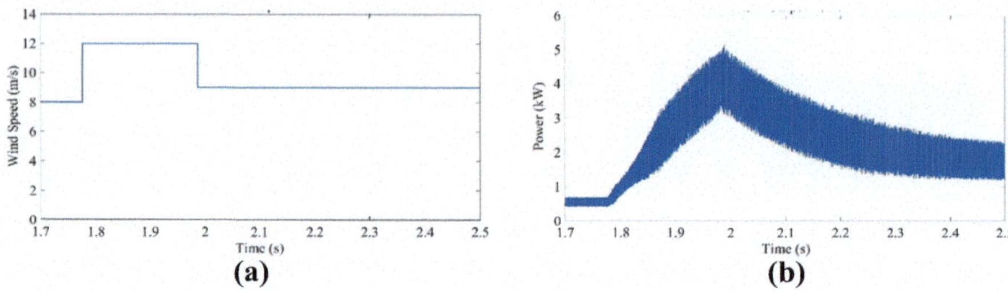

Fig. 3.10 Variation in PMSG-based wind turbine generator output power (**b**) with varying wind speed (**a**)

Reliability and Security: RE sources being variable and intermittent, their constant availability and dependency are questionable. In such a situation, keeping the grid balanced and ensuring a continuous and adequate supply of power becomes a challenge. Matching the aggregate load demand in real time with such intermittent sources makes power system operation much more challenging. The inclusion of fast acting ESS such as batteries, flywheels, super-capacitors, and also EVs in bi-directional mode into the grid can tackle such intermittency to a good extent. Modern power systems are moving toward ***smart grid*** wherein advanced ICT features with smart sensors, smart controllers, IoT devices, and AI-based analyses tools are being implemented. Short-time prediction (of both load and generation), analytics of AMI, advanced monitoring tools and SCADA, optimal grid management tools, dynamic load flow management system, and dynamic compensators are some of the features of smart grid that attempt to cope with grid reliability and security under RES variability and intermittency.

Congestion management: Adding DG at the distribution level has a major impact on the amount of power to be handled by power system equipment such as cables, lines, and transformers. This problem is observed during lean load periods, as well as during some specific peak load periods when base generators and peaking DGs both tend to feed power to cash in premium price resulting in exceeded grid capacity. High incentives have attracted substantial investments in RE generation both from the public and private sectors, whereas augmentation in the capacity of power evacuation and distribution is largely dependent on state-owned incorporations. This scenario has put immense pressure on the planning and design aspect of modern power systems for matching generation with evacuation capacity. Figure 3.11 shows the investments in RES generation in India vis-à-vis transmission and distribution capacity augmentation. Substantial growth in investments in network infrastructure is the key to meeting the gap between the two growth rates for alleviating congestion.

Voltage Regulation: Inclusion of DG in power network generally reduces the system inertia. With unpredictable variations in DG generation, voltage regulation is critically affected. At times when generation is more than load demand, there are possibilities of voltage escalation at the customer end. At other times, there may be a voltage drop. Such frequent voltage variations render traditional voltage regulating

Fig. 3.11 Average investment in India's renewable energy growth and power network augmentation [5]

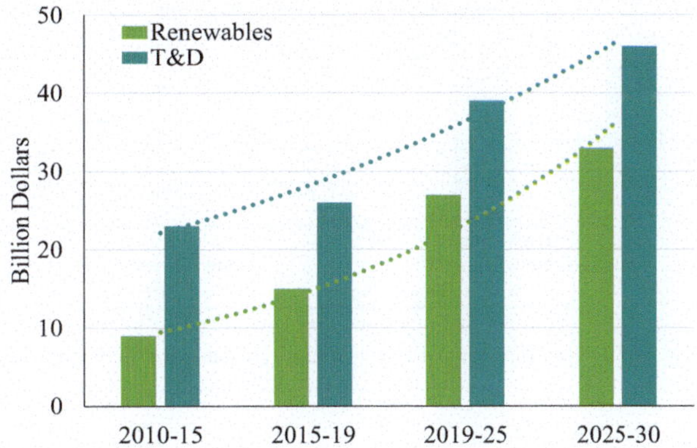

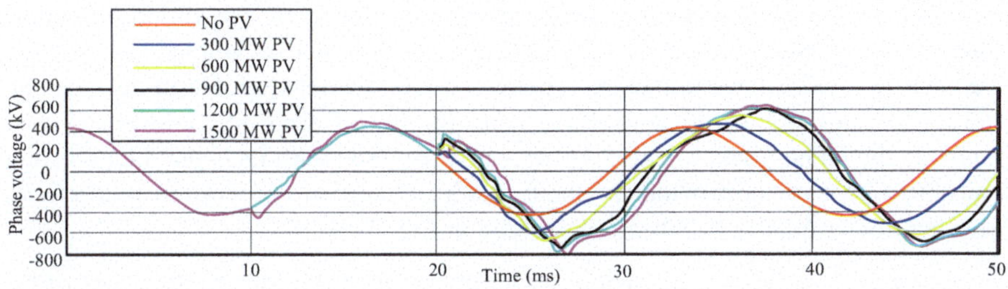

Fig. 3.12 Bus voltage variations due to different degrees of PV penetration in the grid [6]

devices such as step voltage regulators, transformer on-load tap-changers (OLTC), and switched capacitor banks to respond inappropriately. Figure 3.12 (reproduced from [6]) shows simulated phase A voltage profiles in bus number 29 in the IEEE 39 bus system with different degrees of PV penetration in the grid.

Frequency Regulation: Operating frequency of the power network can be maintained at its nominal value only when the active power generation is balanced against active load demand. Any imbalance in generation and load will make the frequency to deviate from its nominated value. The inertia of rotating machines present in the power system can be useful in reducing such frequency deviations. With increased RES penetration, maintaining frequency becomes a major issue due to unpredictable and abrupt variations in solar and wind power availability. Moreover, due to the much-reduced inertia of the power electronic converters used for connecting the RES to the grid, frequency stability becomes a big challenge. Figure 3.13 (reproduced from [7]) shows the effect of change in frequency upon the occurrence of an abrupt system fault. It is seen that due to low system inertia induced by penetration of wind and PV power (lower the inertia constant H), the amount of frequency deviation is larger, even crossing the critical limit of 49.5 Hz. It clearly shows that the integration of RES affects the frequency stability unless fast-acting controllers are employed.

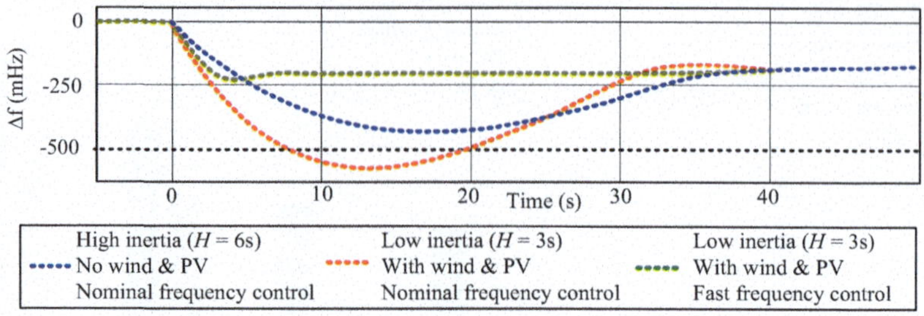

Fig. 3.13 System dynamic frequency response under influence of RES [7]

Power Quality: Under certain scenarios, DG can either contribute to improve or sometimes deteriorate power quality. Installation and integration of DG will definitely affect the system frequency unless properly controlled. In addition, the intermittent and unpredictable nature of RES causes issues such as voltage sag, voltage swell, harmonics, voltage unbalance, flicker, etc., in the network. Harmonics can be produced from the generation itself, or from the power electronic converters used alongside the DGs. Flicker is also a power quality issue with DGs that lacks strong control systems, and occurs mostly due to sudden changes in generation and/or load. Different power quality events are described in Fig. 3.14 [6].

Protection System Design: Inclusion of DG in power grid brings in the scope of bi-directional power flow that renders the hitherto radial network to the possibility of unbalancing. Inclusion of DG makes coordination of protective devices complicated and calls for dynamic setting. The size, type, and location of the DG influences the extent to which relay coordination is affected. Different aspects that complicate the system protection schemes with integration of DG include: (1) protection of DG

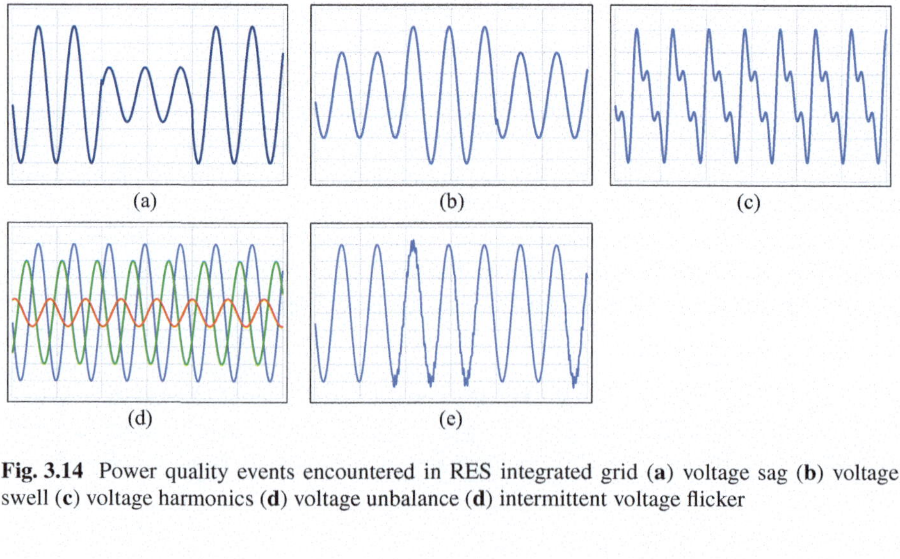

Fig. 3.14 Power quality events encountered in RES integrated grid (**a**) voltage sag (**b**) voltage swell (**c**) voltage harmonics (**d**) voltage unbalance (**d**) intermittent voltage flicker

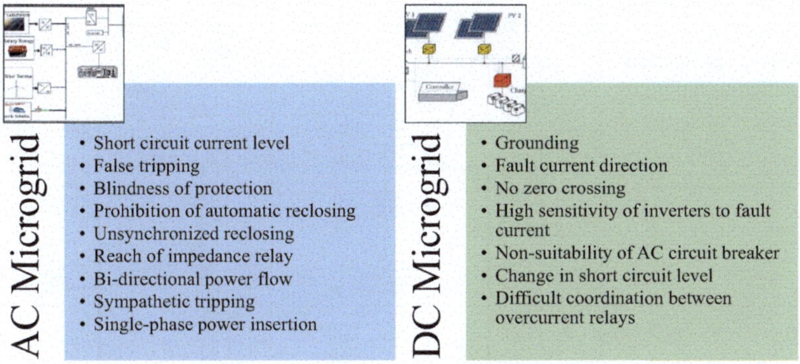

Fig. 3.15 Challenges in microgrid protection system design due to integration of RES

from internal faults, (2) fault current fed by DG to the distribution network, (3) loss-of-mains or anti-islanding protection, and (4) impact of DG on existing network protection equipment settings. Synchronous generators in wind generators influence the protection scheme and fault level due to their inherent large short-circuit currents. Other DG units that are connected to grid through power electronic converters (solar PV for instance), however, have negligible effect on the existing protection system. Various challenges encountered during protection scheme design of AC and DC microgrids with RES are summarized in Fig. 3.15.

A typical case of "blindness to protection" can be demonstrated when a fault occurs in a downstream feeder shown in Fig. 3.16a [8]. Both the utility grid and the RES unit contribute to the total fault current. Often because of its higher voltage at PCC, current contribution from the RES is large enough to reduce current contribution from the utility grid. Thus, PD1 may become "blind" to the fault and may not trip even though feeder 1 carries large fault current. In Fig. 3.16b, it is sufficient that PD3 trips to clear the fault at feeder 1. However, if the RES unit contributes a large share of the fault current, PD2 may experience "sympathetic tripping" thereby unnecessarily disconnecting feeder 2 from utility grid supply.

System Stability: Traditionally, distribution network stability considerations took a back seat because of its apparent passive behavior. With increased penetration of RES at distribution level, their contribution to system stability can no longer be overlooked. Variability, unpredictability, and starting time of DGs all affect the distribution system stability. Issues that need to be carefully designed include both transient as well as long-term dynamic stability and voltage collapse. Authors in [9] have presented report on frequency stability of inter-area oscillation in a grid subjected to disturbance. Upon occurrence of a 3-phase fault, deviations in oscillation frequency and system damping ratios with varying level of PV penetration is shown in Fig. 3.17 (reproduced from [9]). With increase in PV penetration, the oscillation frequency is seen to increase almost linearly, while damping ratio decreases. Both these two scenario points to degraded stability index with increasing PV penetration level.

Fig. 3.16 Impact of RES on protection system (**a**) blindness to protection (**b**) sympathetic tripping of PD

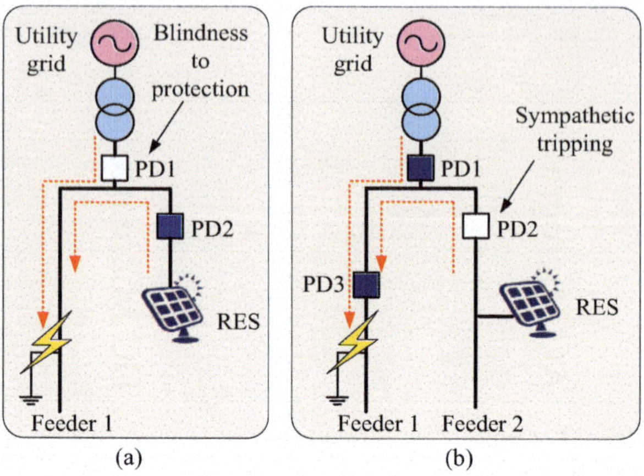

Fig. 3.17 Effect of PV penetration on frequency stability of inter-area oscillation due to sudden disturbance [9]

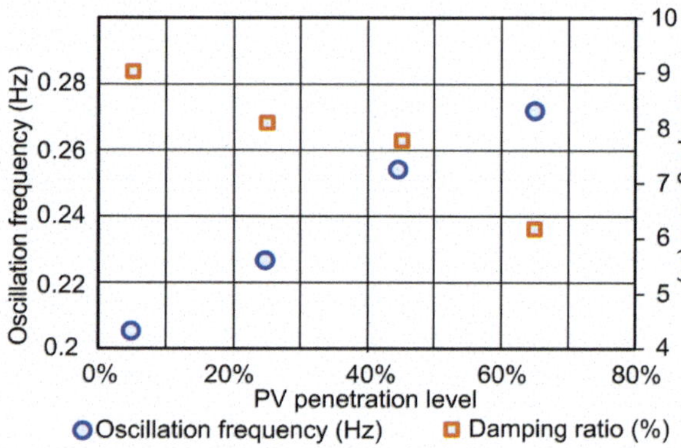

Power Conditioning: While solar PV produce DC power at their output, wind generator frequency often deviates from that of the grid. Thus, in order to integrate these DG sources to the grid, it is essential to introduce power converters such as chopper and inverters between the DG and grid. Controlling these converters requires knowledge and skill to have seamless synchronization between DG and grid. Additionally, these converters inject harmonics into the system unless properly designed. Figure 3.18 shows solar PV inverter output voltage without any filter and then with a properly designed filter. The plots demonstrate the utility of proper filter design for generating good quality power from RES for integration to grid.

Power Loss: Location of the DG along the line, its size and degree of penetration in the grid determines the reliability and losses. When DG power is injected into the distribution feeder close to load clusters, then line power losses are less. In addition, the flow of utility power from main grid to the load being reduced, the risks of grid overloading are reduced. On the other hand, if the DGs are connected to the HV/MV grid, away from the load clusters, or when the injected DG power is more than

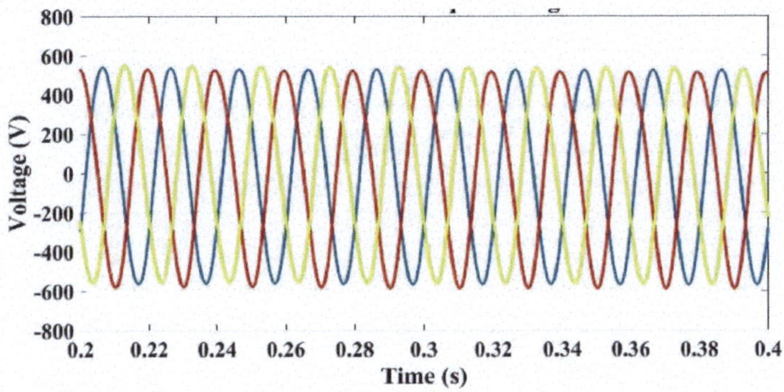

(a)

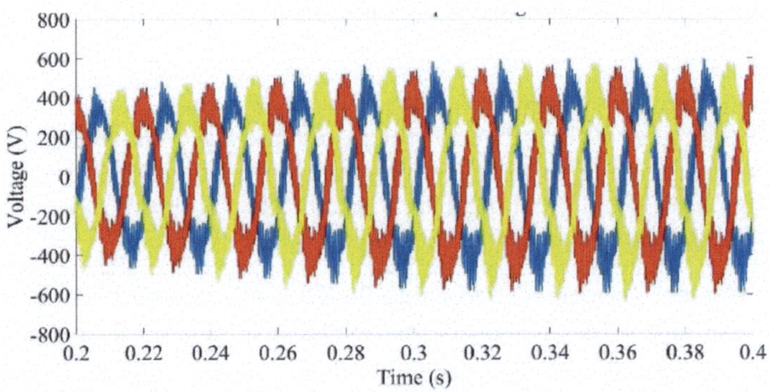

(b)

Fig. 3.18 Solar PV inverter output voltage profile (**a**) with and (**b**) without filters to demonstrate the effect of power conditioning during RES integration to grid

feeder capacity, line losses will increase. Increased line losses will ultimately boil down to increased tariff imposed on the consumers.

Islanding: Unintentional/accidental islanding of DG from main grid may occur due to localized faults or large generation/load mismatch causing protecting devices to trip. Such islanding is harmful to the field workers who may be unaware of the situation, and also will result in loss of phase and loss of voltage synchronization. Transitions from islanding to synchronized mode and vice versa cause transients, recloser maloperation, and false tripping. Common problems with unintentional islanding of DGs include voltage and frequency instability, safety, power quality, and interference. It is specified in IEEE 1547–2003 standard to detect an unintentional islanding within 2 s, and IEEE 929–1988 standard recommends disconnection of the DG upon unintentional islanding detection. Authors in [10] have demonstrated the frequency instability effect at the instance of an unintentional islanding operation of a DG microgrid. Figure 3.19 shows the droop in DG microgrid frequency when islanded, with all DGs running at their maximum capacity, and the MG distribution

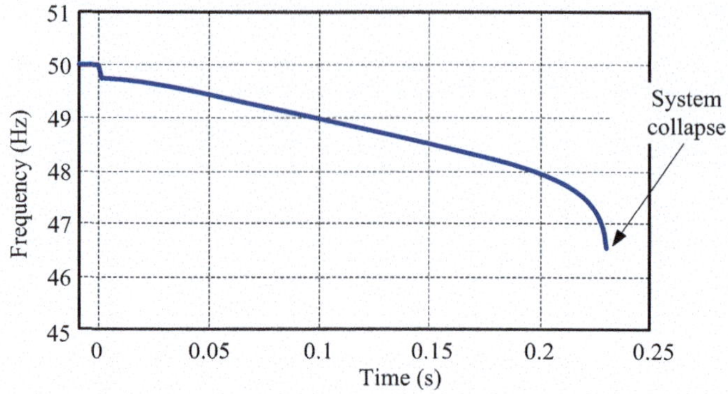

Fig. 3.19 DG microgrid distribution system frequency collapse due to unintentional islanding [10]

system encountering sudden power deficiency due to islanding. It is seen that within 0.24 s of islanding, the MG distribution system collapses. In isolated DG microgrids, frequency stabilization can be achieved by using selective load shedding since most of the DGs are already running at their full capacity and no further increase in RES generation is achievable.

Voltage and Current Transients: Switching the DGs ON and OFF can cause short-term voltage and current transient oscillations that can destabilize the power network. Figure 3.20 (reproduced from [6]) shows nature of transient voltage at bus number 29 in IEEE 39 bus system upon connection of a 300 MW wind farm at 20 ms to the grid.

Short Circuit Levels: Inclusion of DG increases the short-circuit fault level of a network thereby affecting safety and reliability. The fault level depends on type and size of DG and its distance from fault location. Accurate estimation of short-circuit level is important for deciding protective device ratings and relay coordination in a power network. With inclusion of DGs in a grid, the equivalent network impedance reduces leading to increase in the fault level. Increased fault levels may exceed CB breaking capacity, push CTs to saturation, and disturb relay coordination resulting

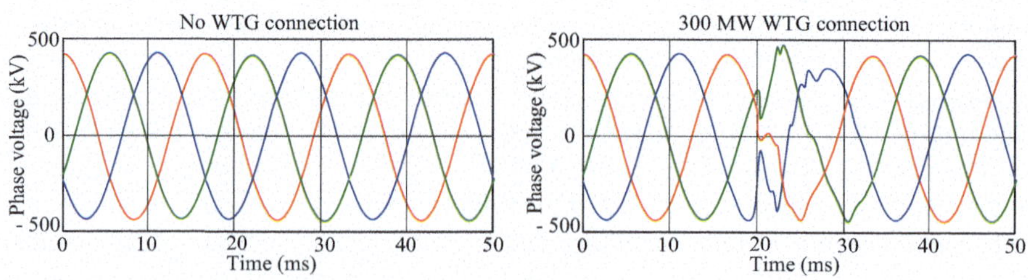

Fig. 3.20 Voltage transients due to WT connection [7]

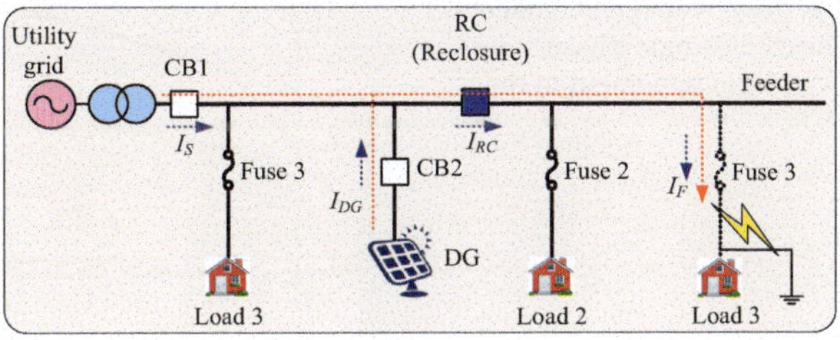

Fig. 3.21 Influence of DG integration on fault current level

in protection system failure. Figure 3.21 can be used to demonstrate the effect of DG on fault current level of a network.

Without the DG, the fault current nearly equals the circuit breaker 1 (CB1) current and the reclosure (RC) current ($I_F \approx I_S \approx I_{RC}$). Rupturing current setting of the Fuse 3 and operating current value of PDs (CB1 and RC) are coordinated accordingly so that Fuse 3 isolates the fault first without tripping RC and CB1 so that power can be continued to remaining unaffected parts of the network. However, when DG is connected to the grid, it also feeds the fault along with the main utility source, thereby $I_F \approx I_{RC} = I_S + I_{DG}$. RC current level is thus higher than the main CB1 level. The RC may trip in that case thereby disconnecting a greater section of the network, many of whom are not in the faulted line.

Voltage Unbalance: Single phase DGs (such a solar PV) can cause voltage unbalance when integrated to a 3-phase power network. Authors in [11] used real-time simulation model of a medium voltage grid with different degrees of PV penetration to study the effect of voltage unbalance due to single-phase PV power injection. Figure 3.22 shows percentage voltage unbalance injected into the grid at various buses in a typical 12-bus distribution network [11] due to different degree of penetration of solar PV generation. It is observed that voltage unbalance effects are lower on buses 2, 7, and 12 that are closer to the PCC. Whereas voltage unbalance go up at buses 5 and 11 that are further away from the PCC. Its is thus common that the problem of voltage unbalance is more dominant on buses that are further away from the utility power source. ANSI C84.1–1995 standard limits such unbalances to below 3%.

Transmission/Distribution, communication, and security: In countries where solar PV and wind sites tend to be concentrated at certain regions far away from load centers, evacuating such generated energy faces the challenge of transmission bottleneck for inter-state and intra-state transfers. Their locations also require to meet challenges of remote monitoring, secure communication and control for seamless grid integration and efficient operation. High concentration of rooftop PV systems integrated to distribution grids can lead to network congestions as well. Different bottlenecks encountered during evacuation of highly penetrated renewable energies to the grid are highlighted in Fig. 3.23.

Accurate forecasting: Solar and wind power availability are characterized by randomness, intermittency, and volatility that poses operational challenges for their integration into the grid. Therefore, accurate prediction of PV and wind power have become

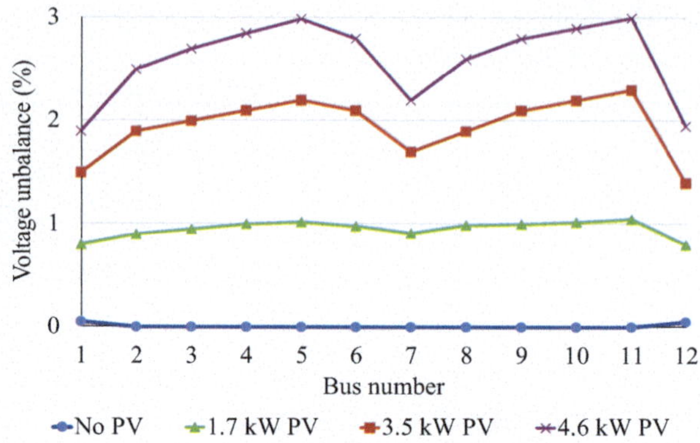

Fig. 3.22 Voltage unbalance caused by single-phase solar PV system integrated to the grid

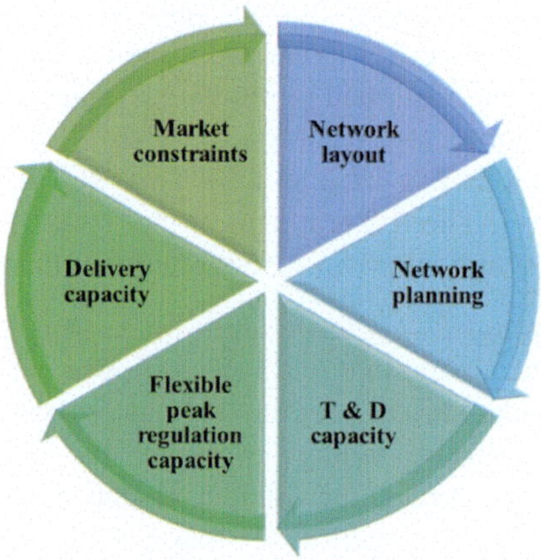

Fig. 3.23 Bottlenecks encountered during high penetration of renewables in grid

essential for safe and stable power system operation. The forecasts need to be accurate in both short term and long term. Without adequate forecasting information, it is difficult for RES management centers to schedule and manage RES with grid. Even with improved weather and RES generation forecasting, there will still be instances of unpredictability when RES generation may either exceed or fall short of the predicted energy generation. Standby reserves (storage devices) and dispatchable loads need to be properly designed and integrated to balance such instances of unpredictability.

Schmitt et al. in [12] have presented results of actual solar irradiance versus predicted pattern, as reproduced in Fig. 3.24. It is observed that an accurately designed forecasting algorithm makes prediction fairly accurate, except for a small delay on sunny days (Apr. 27). On a cloudy day (Apr. 28), short duration forecasts can better follow the random variations of actual irradiance. However, prediction errors are still present even for a 2 h ahead forecast. Errors in forecasting will largely affect the momentary peak-pricing costs in addition to creating generation-load imbalance.

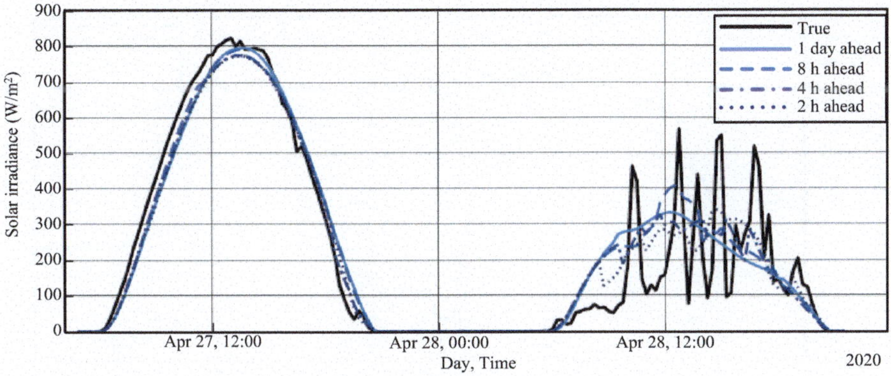

Fig. 3.24 Typical solar irradiation pattern forecasting

Storage Needs: Imprecise forecasting of RES will lead to mismatch between actual generation and prediction. To compensate such imbalances, energy storage systems are essential as a part of DG. Currently, available energy storage options are either low on efficiency or too expensive. Improved storage solutions will facilitate the integration of RE into the grid. Storage systems may include Batteries, Supercapacitors, Flywheels, Super Conducting Magnetic Energy Storage Systems, Pumped Hydro Storage Systems, Compressed Air Storage Systems, etc. In modern power grids, intelligent energy management systems (EMS) are put in place that monitors actual RES power output in real time and accordingly sends charging or discharging commands to the ESS so as to meet total system output power requirements of the utility. Figure 3.25 demonstrates the smoothing effect achieved by a suitably controlled BESS of 1.5 MW capacity that compensates for variations in a 1 MW solar PV module [13].

Active power control: Unless total load demand is supported by the amount of active power supply, be it from DG or utility, there are chances of frequency fluctuations and instability. A 2% deviation in frequency could mean 40–60% deviation in active power. It is thus necessary to have prompt and precise control of RES (some of which are unpredictable) and storage components of the DG to match grid power and load demand.

Reactive power control: Integration of DGs in grid can cause both over-voltage as well as undervoltage issues at load bus depending on the share of DG power output, utility power availability, and load demand. When voltage variation goes beyond the limit prescribed by corresponding grid code, the DG should be configured to provide reactive power support to the grid to restore the voltage. Grid code also specifies power factor range of the DGs so as to maintain desired level of reactive power support to the grid.

Fault Ride-Through (FRT) and Voltage Ride-Through (VRT): FRT/VRT is a situation where the DG is required to maintain synchronism with the grid even when a fault

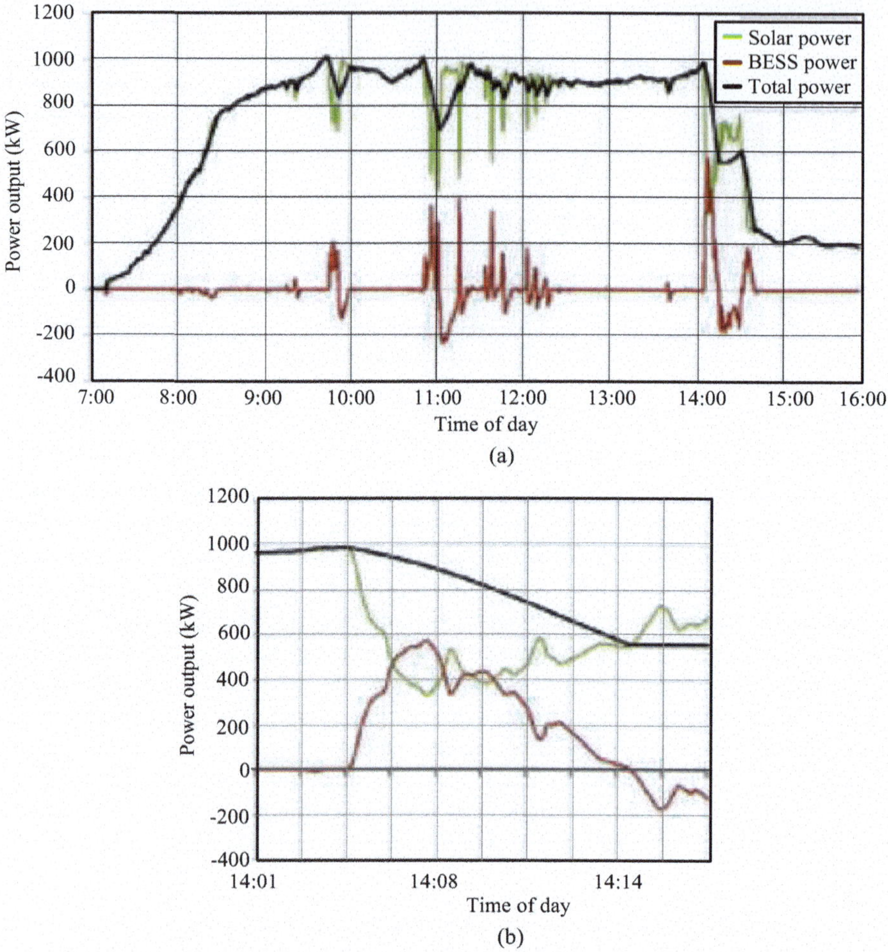

Fig. 3.25 BESS for smoothening variability in solar PV output (**a**) full day (**b**) detailed view of maximum variation event

occurs in the system. VRT controllers need to ensure grid system stability when large scale RES is integrated with the grid. Grid codes prescribe very stringent FRT/VRT capability to be fulfilled by the DG system owners. Figure 3.26 [14] shows the voltage VRT limiting zones that generating units in a microgrid need to sustain to conform to various grid codes. The plot also contains VRT requirements as specified by several RES rich countries including India.

Accurate metering: With DG coming into the picture, it is necessary that production and consumption are properly monitored and billed fairly. This mandates the need of smart meters which can ensure accurate billing and constantly monitor the production and consumption in real-time to ensure grid balancing and flexibility.

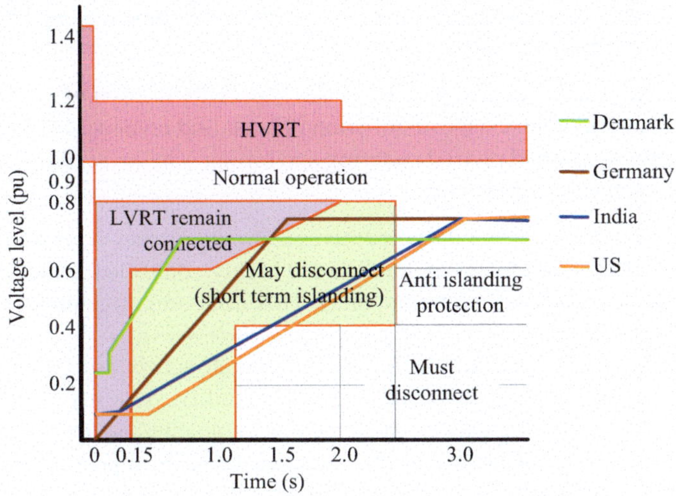

Fig. 3.26 VRT requirements specified by standards and various country's requirements

Economic challenges Type of DG and technology used influences the financial aspects of grid integration. Some critical financial issues that need to be addressed for large scale grid integration of DG are as follows:

Investment cost: With the present-day available technology, initial investments for DG installation is still high, often comparable to conventional centralized stations (Fig. 3.27). Unless there is a big market for electricity generated from DGs or there are subsidies from the government, the high initial investments for DG installation as a part of the grid is still expensive till date.

Integration costs: This includes the installation, operation, and maintenance costs for electricity generation from RES. RESs, though cost-effective in the long run, require

Fig. 3.27 Comparison of unsubsidized levelized cost of energy [15]

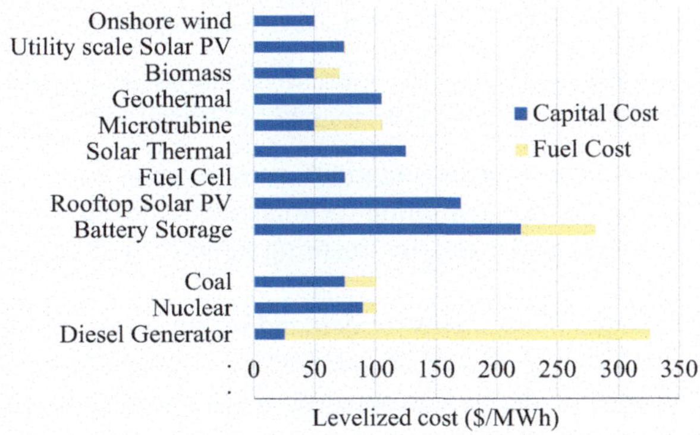

higher initial setup costs due to the energy conversion equipment along with control, interfacing and storage modules.

Cost Allocation: A difficult question that arises especially in cases of grid-connected private RE generation units is the distribution of costs and revenues between utilities and consumers. A new and revised metering and billing scheme needs to be put forth to overcome such challenges.

Effect on power market: A liberal, competitive electricity market may promote investments in RES. But most electricity markets are highly controlled, either by governmental bodies or by a few major companies This discourages competition and the integration costs rise.

Trading prices of bulk electricity depend on accurate scheduling. With variability and unpredictability of solar and wind power availability, volatility of market price is an issue that need to be addressed by power companies having large share of RES integrated to grid. Generation uncertainties of RES impedes their effective participation in short-term power markets. Absence of innovative incentive packages for RES trading as a part of appropriate market framework have retarded DG investors initiatives.

Financial implications to T&D companies: High level of DG penetration to grid requires high level of active management of the power transfer network, both at transmission and distribution level. T&D companies need to be provided with business incentives for permitting third party DG operators to use their existing infrastructure. New commercial arrangements need to put into place in order to support the development of active distribution network with increased share of RES.

Taxes/Trade Barriers: Various taxes and government policies on the generation and distribution of electrical energy from natural resources, or the taxes and trade barriers on important technologies for these processes create barriers in the penetration of DG into the energy landscape.

Operational challenges The system operators face challenges of varying nature and complexity while managing the grid integrated with DG. Those operational challenges that are predominant, are discussed below:

Regulatory Policies: In order to promote integration of DG to grid, clear and articulated regulatory policies need to be put in place:

- Ancillary service regulations, products and power markets
- Adequate compensation for solar and wind curtailment to lift investor confidence
- Prudent grid codes for ensuring network stability with high penetration level of DG
- Adequate incentives for demand response and storage systems
- Liquidity of short-term wholesale power markets
- Augmented policy to support DISCOMs who now have to pay fixed charges for the conventional power plants in addition to meeting the obligation of purchasing national RE power

Grid Operation: Some key requirements for smooth RE integration from the grid operators' perspective include swift response, flexibility in supply/demand, accuracy in forecasting, optimized scheduling, reliable SCADA system, and robust ancillary services.

Operational Mode: Some of the DG systems are flexible and controllable. They can be brought into/out of service in short notice so that they can be operated at either standby or peak shaving mode. Such flexibility in DG operation makes them well suited to respond to either load variations or electricity price fluctuations. It is thus necessary to monitor and control the operating mode of DGs continuously so that it can be changed on demand.

Dispatchability: Unlike conventional thermal or hydro power units whose output power are well controlled, most RE generator power output solely reply upon availability of the input driving energy (e.g., solar irradiation or wind flow). Such RE sources thus become non-dispatchable. Output power of such RESs cannot be varied as per network demand when forecasting goes wrong or the storage device malfunctions.

Active Distribution Network: Traditionally distribution networks are passive while serving the sole purpose of transferring the energy from transmission networks to the customers. However, with large scale integration of DG, the distribution network operators need to actively take part in the energy management process. This requires distribution network operators to actively participate in forecasting, data collection and analysis, communication, device control, demand side management, system integration, automation, etc. Advanced management and control actions need to be part of the active distribution network to ensure prompt, safe, and secure operation.

Localized RES: Operating localized DGs such as rooftop solar and EVs pose great challenge to the system operators since they induce concerns such as local voltage issues, bi-directional power flow, limited visibility of new installations, accurate forecasting, etc.

RE management: RES has typical features of intermittency, variability, unpredictability, and wide geographical dispersion. For seamless connectivity of RES to grid, it requires real-time monitoring and analysis of its dynamic impact on the system. Dedicated RES management and control centers are necessary with the following functionalities:

- Addressing variability, intermittency, and ramping aspects of RE integration
- Effectively managing power system operations while maintaining security and stability of grid
- Integrating information from Forecast Service provider (FSP) and Weather Service provider (WSP) to provide forecasting of RE generation
- Coordinating with respective load dispatch centers for scheduling of RE generation

- Integration with existing SCADA technology for real-time tracking and visualization of RE generation.

Environmental and Societal Challenges In spite of the fact that RES are considered as clean energy, there are some negative impacts on the environment. In addition, there are certain societal issues that need to be addressed while harnessing the benefits of RES integrated to grid.

Environmental effects of solar PV: Installation of solar PV farms and parks require large project areas that may affect the natural habitat. Manufacturing of solar cells use up large quantities of treated water and the cleaning process involves materials that are often hazardous. Wastes from PV industries are toxic that can contaminate soil, water, and air unless stringent waste management and disposal procedures are not followed.

Environmental effects of wind power: Even though wind power is one of the cleanest sources of energy, there are certain environmental concerns that need to recognized and mitigated. These include land use issues, visual and sound pollution, threat to wildlife and habitat.

Apart from these environmental issues, citizen skepticism in accepting change in general, poses another barrier to widespread RES adoption throughout the world.

Consumer involvement: Lack of awareness among consumers keeps a large share of them from getting involved in the process of harnessing potential benefits of RES integration to grid. Here, institutions and organizations need to play central role through awareness programs, community education, and social networks.

Control of resources: Another possible issue that might arise is the question of who owns or rather has the control of the RE generation equipment. A domestic consumer who has installed a solar PV system, for example, might want to control the generation but the grid operators might want to control the generation as per their need depending on load demand and power availability.

Demand side management (DSM): Integrating RE in the form of distributed generation will need better management of load as well because demand and generation must be equal to ensure smooth functioning of the grid. DSM at its core needs to control the consumer load in response to availability and price of electricity. It involves motivating and incentivizing customers to take active part in the DSM process.

Myths and misconceptions: A considerable chunk of population today still remain uneducated. Several misconceptions about these relatively unknown sources of energy continue to plague parts of the society.

3.2.3 Roadmap for Addressing RES Integration Challenges

Extensive work has been carried out in the recent past, both in research and in field, for devising possible solutions for DG-grid integration challenges. Several statistical, numerical, and computational algorithms are being deployed for overcoming the issues associated with the uncertainty and variability of RE power generation. Cost-effectiveness, grid infrastructure, operational practices, type of RES generation, policies and regulations, and network topology are some of the main criteria for selection of the most suitable solution technology. Increased flexibility requirements in view of accommodating the additional variability of renewables are to be addressed by robust forecasting tools, real-time SCADA, energy storage, DSM, dispatchable generation, and other monitoring and control mechanisms.

Grid balancing: Deep penetration of large-capacity RE into the grid results in uncertainty and variability. Meanwhile, unexpected disturbances will continue to originate from load variations, network faults, and conventional generator outages. New solutions to address this conjunction of difficulties include state-of-the-art technologies and operational practices to give space for more flexibility while managing constant efficient load-generation balancing (Fig. 3.28). Then only it will be possible to run a reliable system while maintaining supply security avoiding interruptions of power supply. On the generator side, conventional sources can provide more controlled flexibility to the power system than RES. On the load side, flexibility can be achieved through demand response, and also by the use of storage systems that can act as both load and source. Recent researches on grid integrated PV power fluctuation smoothing have investigated several advanced control techniques including principal component analysis (PCA), deep neural network, Elman neural network, wavelet transform-based ANN, wavelet-fuzzy filtering, ramp-rate control, and GA-based feedback control.

Interfacing solutions: At the point of common coupling (PCC) between DG and the grid, their voltage, frequency, and phase sequence (in case of 3-phase grid) must be matched at all instants. Power electronic converters with feedback have seen remarkable development in the recent years to meet these requirements (Fig. 3.29, 3.30, 3.31 and 3.32). In case the RES output power is DC, such as in Solar PV, inverters are used to convert it to AC to be coupled to the grid. For wind generators, the output power at variable voltage and frequency (due to wind speed variations) needs to be converted to power at specified voltage and frequency by multi-stage and often multi-level converters.

Maintaining frequency stability: Integration solar PV and wind reduces total inertia of the system. Wind generators being connected to grid through power converters, and solar PV having no rotating mass, they cannot contribute to system inertia. Reduced inertia reduces frequency rigidity thereby degrading the frequency response. New controllers emulating inertial behavior of a conventional rotating generator are being designed for increasing system inertia in order to enhance frequency stability of RES

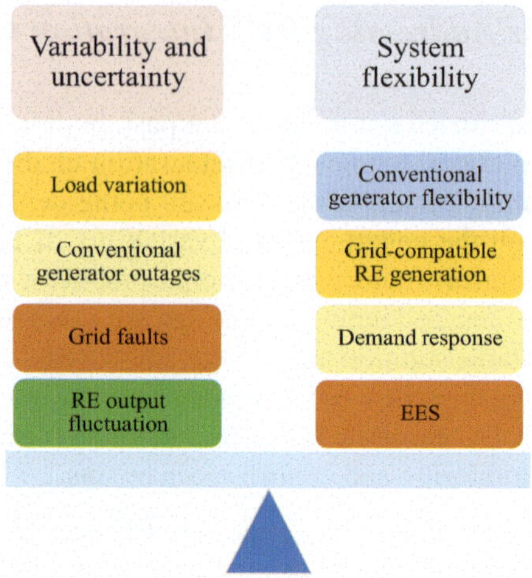

Fig. 3.28 System flexibility balancing variability and uncertainty

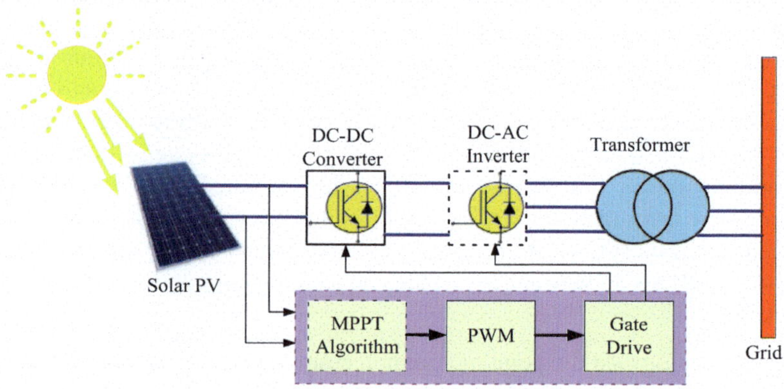

Fig. 3.29 Interfacing system for solar PV power integration to grid

integrated grids. Frequency in RES generators can also be regulated by conserving its active power using de-loading operation or by including ESS (Figs. 3.33 and 3.34).

Wind energy systems can take part in grid frequency stabilization through innovative techniques such as de-loading, inertial response, and droop-control. Suitably controlled power electronic converters that release the stored kinetic energy of wind turbine can also help in frequency stabilization of wind integrated systems (Fig. 3.33).

Solar PV systems that are made to work at maximum power point, do not have any reserve power margin as such for frequency support, especially during negative frequency excursion. One possible way how Solar PV systems can help frequency stabilization is by de-loading technique, where it is operated at reduced output power mode thereby keeping some reserve. Using ESS and inertial response technique

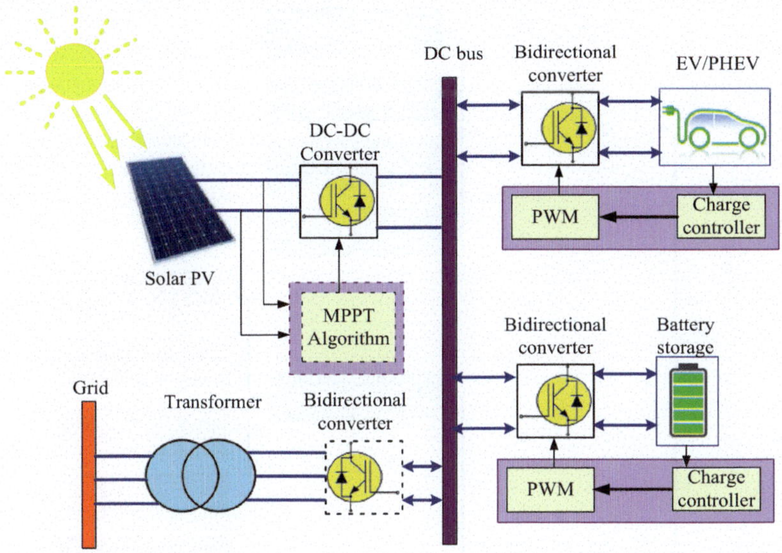

Fig. 3.30 Interfacing system for solar PV power integration to grid with provision for connecting EV and ESS

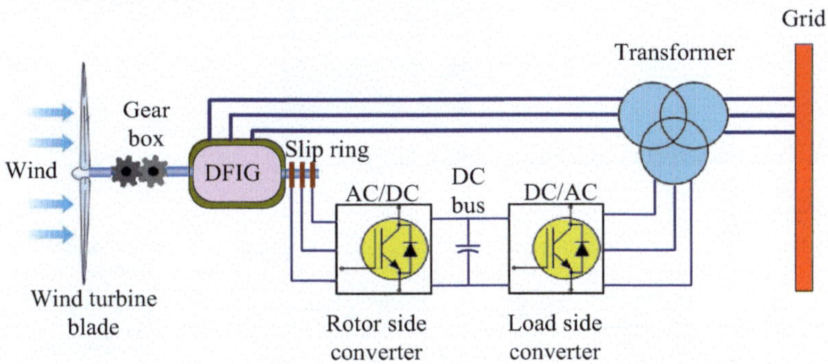

Fig. 3.31 Interfacing system for DFIG-based wind generator integration to grid

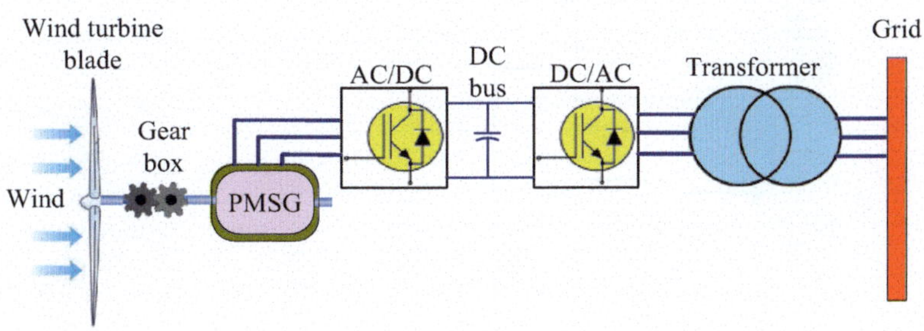

Fig. 3.32 Interfacing system for PMSG-based wind generator integration to grid

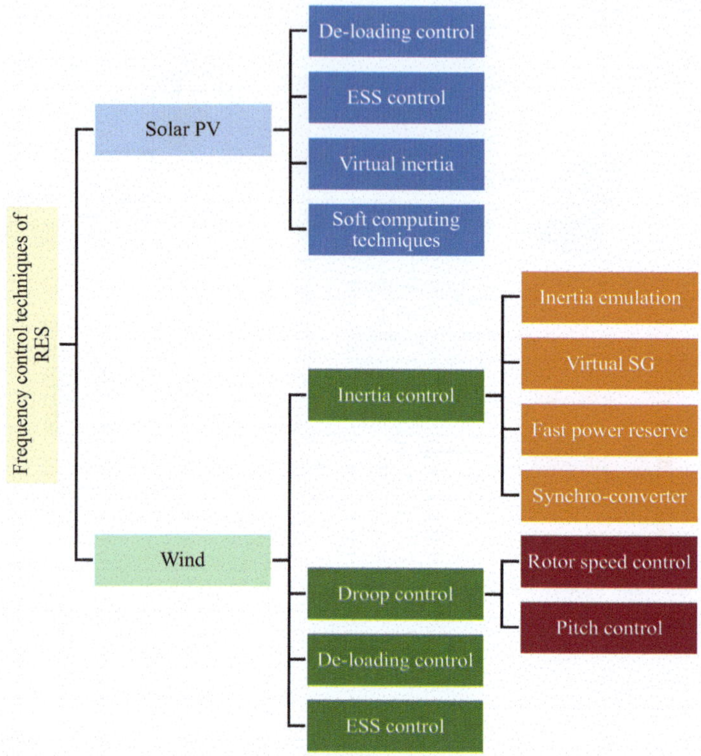

Fig. 3.33 Frequency control techniques for RES [16, 17]

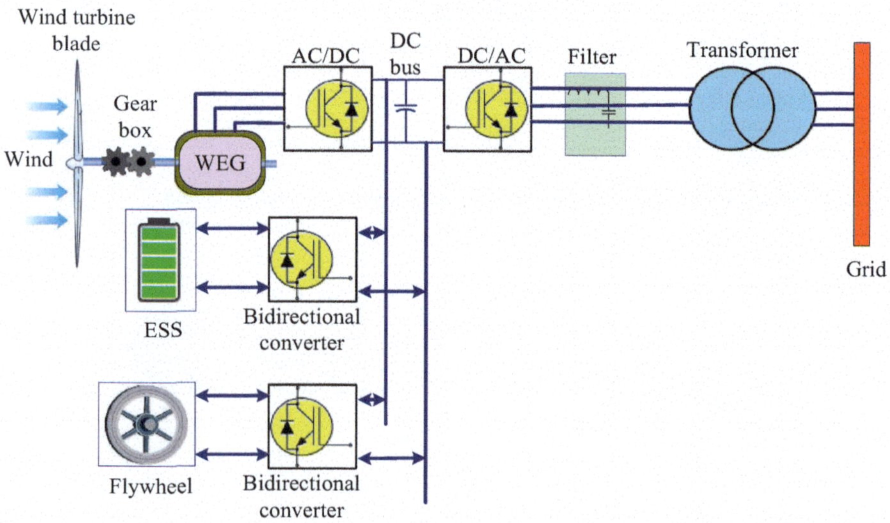

Fig. 3.34 Grid connected wind energy system with ESS and flywheel for frequency support

can also help in frequency regulation of solar PV integrated grids. Soft computing techniques such as AI, fuzzy, ANN, and optimization techniques have also been employed to regulate solar PV power output pattern in order to improve frequency response.

Voltage control: Overvoltage at consumer end is a common issue with DG connected grids, especially at instants of excess RE generation. One of the simpler ways to keep the voltage within allowable limits is to use OLTC to reduce the primary substation voltage that has feeders with DG. Another solution, though expensive to some extent, is to use increased diameter of feeders that have DG connected to it. This option can be avoided if the EMS is designed to curtail RE generation at times of low load. Modern techniques however, rely on reactive power flow control in grid. DG units with controlled converters as well as conventional synchronous generators can be made to absorb excess reactive power to limit voltage escalation. FACTS devices such as SVC or STATCOM and ESS devices are used for dynamic reactive power control in distribution feeders. Other centralized and decentralized strategies for voltage regulation that have recently been explored include stochastic and heuristic algorithms like GA, ANN, Fuzzy control, multi-agent control, seeker optimization algorithm, gravitational search algorithm, etc. Managing reactive power has the added benefits of enhancing system stability, and reducing power quality issues.

Forecasting of RES generation: Accurate forecasting of wind and solar generation is the key to the associated uncertainties. It helps grid operators to put in or take out units more effectively to absorb variations in solar PV and wind generation without having to invest in the requirement of operating reserves. Accurate forecasting in collaboration with meteorological departments for power specific requirements are being explored. Forecasting technologies are employed to predict weather and hence output from wind and solar generations at different timescales (hour ahead, day ahead, week ahead, and month ahead) more accurately (Fig. 3.35). These predictions are further communicated to grid operators that enable them to more effectively schedule and dispatch resources so as to optimally utilize all assets under grid operation. One major function of all RE management centers is therefore, forecasting of RE generation in their jurisdiction area.

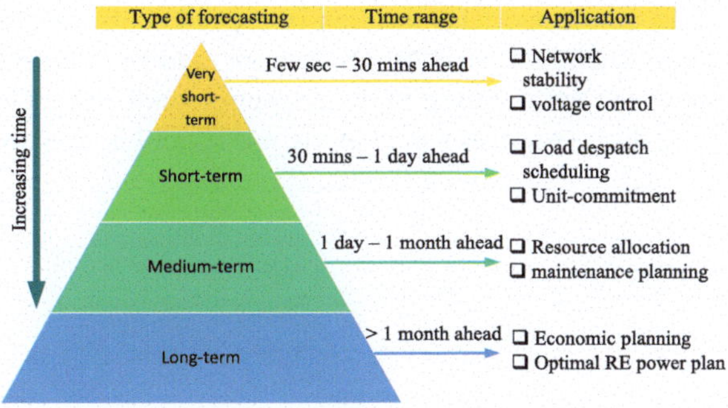

Fig. 3.35 Forecasting time horizons [18]

The various forecasting models that have been explored for this purpose include:

- Time series
- Regression models
- Genetic algorithm/fuzzy logic
- Econometric
- Gray prediction
- ANN and expert systems.

Power Quality Improvement: Integrating RES to grid invariably requires power electronic converters. These non-linear devices are often accountable for injecting harmonics into the system thereby degrading power quality. Modern day inverters designed with IGBT switches using PWM technique produces a cleaner output with fewer harmonics. Use of improved control strategies and appropriate auxiliary devices can further improve the power quality. Installing various line-conditioning systems such as harmonic filters, FACTS devices, energy storage systems, constant voltage transformers, surge suppressors, isolation transformers, and converter control techniques can further suppress PQ related disturbances.

Active power filters are employed for improving power quality, reactive power control, and mitigating current harmonics.

Recently, many FACTS devices such as DVR, STATCOM, DSTATCOM, SSSC, SVC, TCSC, UPFC, etc., are also being employed in high penetration RES networks for improving power quality aspects including harmonics, power factor, electrical quantity oscillations, and voltage dip.

Improvement of power quality of RES, especially power smoothing, have been effectively achieved through uses of ESS such as battery, supercapacitor, and flywheel energy storage.

Various techniques that are being investigated in recent times for improvement in power quality in RES integrated networks are summarized in Fig. 3.36.

Incorporating Fault Ride-Through (FRT) and LVRT Capabilities: As discussed earlier, even minor faults in the interconnected network can make the wind or solar PV to isolate themselves from the grid leading to system instability. In such situations, FRT capability makes the wind/PV plants to stay connected to the grid for a specified period during faults. During contingencies, it is important that the RESs continue to supply power and maintain reliability of the system. FRT capabilities are incorporated to grid-connected RES through improved control techniques and different auxiliary devices such as ESS, FACTS devices, and fault current limiters (FCL) (Fig. 3.37). Two of the mostly used FCLs are superconducting and non-superconducting types, due to their low losses in standby mode, and high insulation strength.

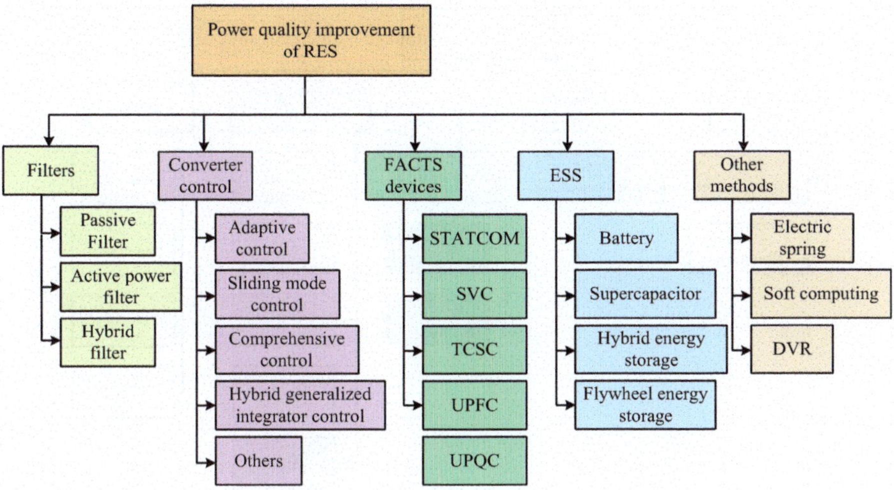

Fig. 3.36 Power quality improvement techniques in RES integrated grid [16]

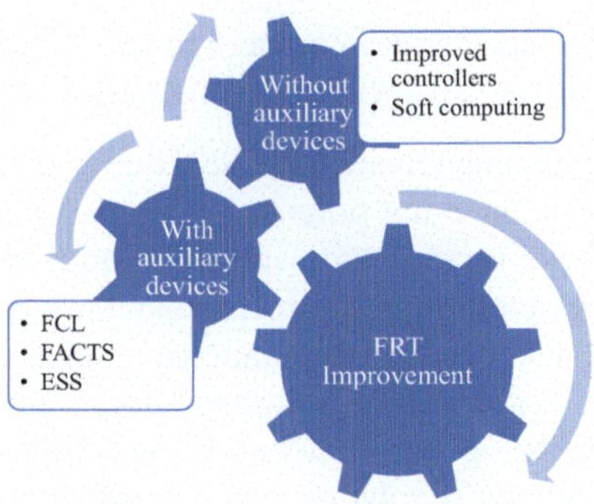

Fig. 3.37 FRT improvement techniques of RES

Most commonly used methods for enhancing LVRT capability of wind turbines during faults are:

- blade pitch control for adjusting wind turbine input
- switched capacitors, SVC, STATCOM, UPQC, DVR, etc., for adjusting WTG output
- active crowbar and series antiparallel thyristors for adjusting DFIG based WTG output.

Handling Uncertainty Issues: Intermittent nature of RES generation brings in a great deal of uncertainty in DG integrated power networks. Full or partial mitigation of

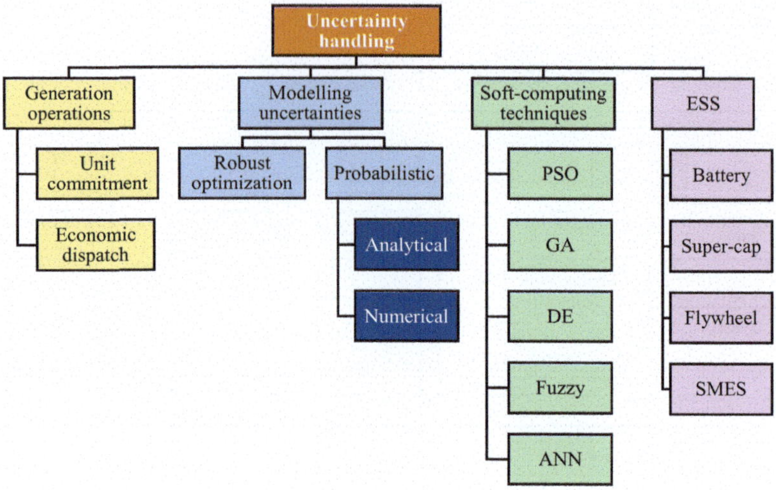

Fig. 3.38 Uncertainty mitigation techniques in grid-connected RES systems

these uncertainty issues are dealt with in modern times using various soft computing techniques, optimization tools, advanced control algorithms, and ESS (Fig. 3.38).

Storage Solutions: Energy Storage Systems (ESS), due to their wide range of configurations and characteristics, have the potential of assisting RES integration to grid in a number of ways. These uses include storing surplus RES generation thereby minimizing the need for generation curtailment, matching generation with loads by time-shifting, absorbing RE generation fluctuations, grid balancing through ancillary services, load-following, load-leveling, ramping, improving power quality, acting as reserves for managing RE generation uncertainty, and smoothening RE plant outputs. ESS also aids frequency regulation, serve as spinning reserves, supplemental reserves, and emergency power supply/back start support. ESSs for grid-support purposes project enormous growth in coming two decades both worldwide as well as in India (Fig. 3.39).

EVs are fast emerging as an important component of ESS in DG integrated grid. While a single EV can only store a relatively small quantity of energy in its battery, an aggregation of EVs all plugged into the grid at the same time can act as a single large energy storage device. Such an electric vehicle virtual power plant (EV-VPP) can serve as time-shifting storage device and can be used to store RE at times of low demand, and release it back to grid to meet peak demand. EVs working in vehicle-to-grid (V2G) mode can also be employed as frequency regulating device for enhancing reliability and stability of DG connected grids.

Demand Response (DR) Opportunities: DR is a vital feature of modern power grid where load management is performed in the demand side rather than the generation side so as to reduce the need for reserves, especially when RE resources are integrated to the grid. DR can provide additional flexibility in the system to compensate the

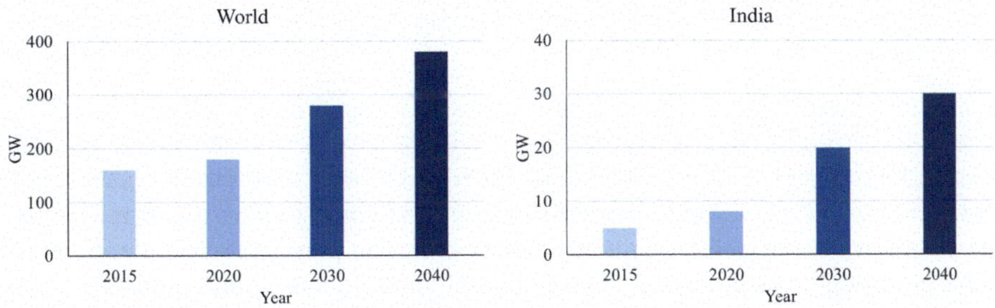

Fig. 3.39 Installed capacity of ESS in 450 scenario [1]

effects of fast ramps, supply reserves, ancillary services, and also manage peak demand. DR is used to balance the system in the event of RE fluctuations thereby enacting cost saving from maintaining reserves. Advancements in ICT makes it possible to monitor and control demand in near real time and in large scale. The entire balancing paradigm is posited to shift from generation matching the load to load dynamically matching the generation. Long-term DR looks at proliferation of energy efficient devices and practices toward reducing peak demand. On the other hand, short-term DR involves energy price as a driver in addition to generation-demand imbalance for dynamic variation of load profile. Load shifting, peak shaving, valley filling, and incentive-based DR are some of the approaches being approached today to enable smooth participation of RES in the grid. Scope of DR implementation in various consumption sectors in India is highlighted in Fig. 3.40.

Interconnecting more distributed resources: When smaller RE units are concentrated in one aera then its intermittent nature affects availability and reliability of power supply. Instead, if large number of such small RESs distributed over a wider geographical spread be interconnected, then the impacts of intermittency can be minimized. In that case, fluctuations in overall output power availability will be reduced

Fig. 3.40 Technical potential of DR in India in the 450 scenario [1]

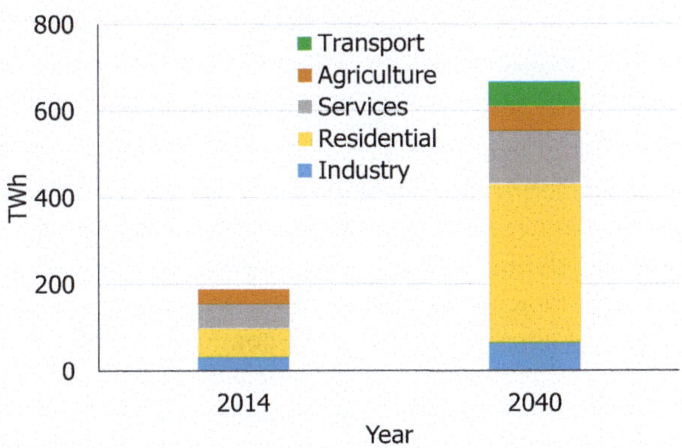

since local variations will affect the smaller units only without much impact on the overall output.

Transition toward Smart Grid: Enhancing the existing grid capabilities with ICT, thereby transforming it to a smarter grid would largely benefit the RE integration process. Digital and smart technologies—such as real-time monitoring systems, IoT applications, and AI-powered predictive analysis is going to play major role in the transformative process. Consequently, smart grids can play a crucial role in facilitating the smooth integration of high shares of variable renewables and supporting distributed generation. Smart grid offers the following enabling technologies for supporting RES integration to the grid:

- *Better forecasting*: Advanced and smart sensors, integration with GPS, satellite imagery system, and widespread communication with active support from advanced computer models allow fast and accurate prediction and management of RE variability and uncertainty.
- *Smart converters*: Smart and intelligent power electronic converters provide robust control opportunities to system operators for RE integration management and grid support.
- *Demand response*: Smart meters, coupled with smart appliances and secure communication network allows demand response management for grid balancing in the presence of RES.
- *Integrated storage*: ESS integrated to the grid with EMS and BMS provides opportunities to smoothen short-term variations in RE output so as to meet the gap between generation and demand. Real-time and distributed network state monitoring using GPS enabled PMU triggers the integrated ESS into operation so as to ensure grid balancing during peak demand periods.
- *Real-time system evaluation and management*: Smart instrumentation and control equipment integrated to smart grid T & D networks provides system operators with real-time information on system conditions, thereby enabling better management of RE integration process and efficient RES power evacuation.
- *PEV*: Smart grids promotes the use and management of PEVs. PEVs can act as one major "behind-the-meter-storage" component that can aid storing excess RE generation, thereby reducing the need for distribution upgrade. PEVs in V2G mode can act as fast acting ESS for grid support in presence of fluctuating RE.

Operational Practices: Power system operation with DG integrated grid is focused on maneuvering all system resources to overcome the issues arising out of variability and intermittency of large-capacity RES integration. These processes involve planning, scheduling, dispatch and control of power (Fig. 3.41). Measurements on power system parameters as obtained from SCADA are analyzed by the EMS in operations center and appropriate corrective commands related to secure and economic dispatch and control actions are communicated back to field devices. The EMS need to be robust and up-to-date to cope with the new and evolved challenges that modern grid has to face due to large scale integration of RES. Major enhancements related to EMS applications can be summarized as:

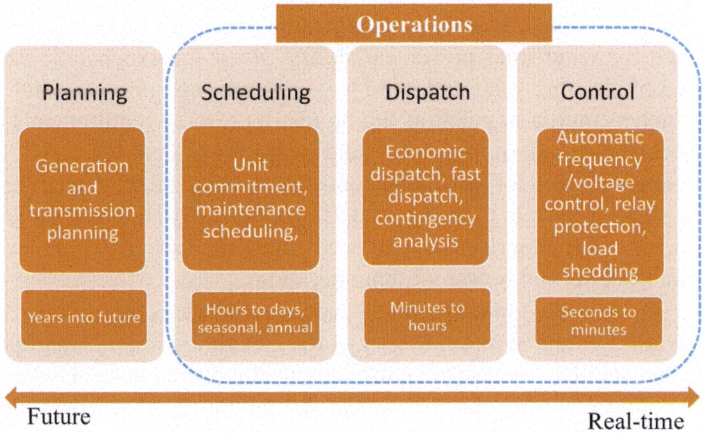

Fig. 3.41 Power system planning and operation processes with timelines [19]

- Unit commitment (UC)—Enhancements in timeline, reducing from traditional day-ahead UC to 4–6 h in response to updates in RE generation forecast.
- Contingency analysis (CA)—Contingency situations in RE plants need to be added to the traditional grid-contingency measures. RE power forecast plays an important role in this case also.
- Online dynamic security analysis (DSA)—Purpose of DSA in power system operation centers is to evaluate system stability and security limits based on network topology and operating state. With RE power modeling and forecast, it is possible to augment the DSA to an online DSA that can handle the operational challenges arising out of increased level of RE penetration.
- Security constrained economic dispatch (SCED)—SCED is aimed at specifying how to dispatch generators so that they produce electricity at the most economic way under the constraints of reliability and operational limits of generation and transmission capacities. With RE being integrated to grid, the SCED now has to include the modified scenarios of RE generation, demand response, smart grid applications, and associated operational costs.
- Fast dispatch—Fast dispatch is one of the main operational practices that helps seamless integration RES to grid. Faster dispatch can provide efficient operation and management of RE power variability by reducing the need for additional regulating resources. It also ensures better grid balancing through closer matching of load and generation level via access to larger set of resources.
- Automatic generation control (AGC)—Traditional AGCs, whose function were limited to generation-load balance in real time for ensuring frequency stability, are now being upgraded to address the variability and uncertainty of RE generation based on forecasts. Advanced sensing, communication, and control capabilities are now being incorporated in the AGC system to regulate RE power plants in conjunctional with the conventional plants and tie-lines.

- Reserves management—Reserves management operations need to be updated to compensate for wind/solar power variability. Need for expensive reserves can be reduced by limiting solar and wind power ramps, and provide other ancillary services such as inertia, regulation, etc. to restrict reserve requirements.
- Cyber Security—Modern power system with integrated RE need elaborate communication system for fast acting control systems to be effective. This renders the cyber-physical system to be exposed to cyber security challenges. Cyber security measures in RE integrated system needs integrated security policies between forecasting systems and power system operations.
- Ancillary services (AS) management—Ancillary services have immense contribution towards managing intermittency of RE generation in grid integrated mode. For instance, ancillary dispatch is extremely useful during natural calamities and sudden weather disturbances. To expand the gamut of AS for RE integration, many ancillary markets are now being opened up to include more RE generators and promote ASs to be benefited by market driven price.
- Tariff restructuring—Recent financial amendments like ABT, DSM (Deviation Settlement Mechanism) and frequency-linked DSM have contributed toward a healthy power sector thereby promoting RE integration. These mechanisms motivate stakeholders to take active part in load-generation balance, thereby limiting injections and withdrawals to specified boundaries. Such a disciplined grid is more welcoming to RE power integration.
- Power system studies—Studies on load flow, fault simulation, and fault analysis by national and regional load dispatch centers lead to scopes for improvements in grid parameters. These include consolidation of transmission corridor, upgradation of substations, installing voltage and reactive power compensating reactors, installing FACTS devices, SVCs, STATCOMs, etc. These studies have promoted improvement in grid characteristics and contributed to seamless integration of RE.

Renewable Energy Policies: In is an established notion now that RES provide a prime opportunity that can fulfill the growing global energy demand by ensuring energy security and climate change issues. Large scale deployment of RES and their grid integration calls for favorable and supportive government policies. The different policies being implemented worldwide for supporting RES proliferation include:

- feed-in-tariffs (FiT)
- investment tax credit (ITC)
- renewable portfolio standards (RPS)
- net metering
- green certificates
- capital subsidy
- low-interest bank loans
- national renewable energy targets.

Having said so, these regulatory policies are strongly influenced by availability of the RES and projected investment costs involvement. It is thus necessary to encompass all stakeholders to develop appropriate regulatory policies for promotion of RES integration to grid. There can be several cycles of regulatory changes before a country can adopt the most suitable combination.

Power market framework: Ensuring security and reliability of electricity is the prime driver for all policies, rules, and regulations governing power market. Traditional market designs for vertically integrated power system are influenced by economic efficiency, environmental policies, emission restrictions, and investments in capacity additions. With increasing share of variable RE in the power network, the power market design in poised to change. Investment range need to be broadened to encompass flexible power plants (both RES and conventional), assets, ESS, DR, T&D networks, and ancillary services. Modern power markets have dynamic price signals as their primary stimuli. To meet this objective, it is necessary that more flexibility be unlocked in the existing power system, prioritize high value RES projects for grid integration to scale down integration challenges, and attract new investments in the market. Competitive wholesale markets encourage with short-term electricity pricing encourage more RE generators to take part in the power market. Flexible power plants can be rewarded through market prices for their promptness in responding to system needs. Market for RES integrated power system makes use of short-term generation forecasting to estimate electricity price with respect to time and geographical location.

With increased share of RES, the market design needs to reflect short-term system operation costs for ensuring power system balance near to real time. Therefore, the market design should look toward:

- Removing electricity price barriers at times of scarcity
- Ensuring that the market design gives equal opportunity to all players
- Encouraging flexibility in all forms (generation, storage, DSM, etc.)
- Long-term coordination between geographically spread players
- Developing new risk mitigating strategies to reassure RES players to invest
- Developing prosumer market consistent with wholesale and retail prices.

3.3 Modeling and Simulation Studies

Extensive studeis are being carried out throughout the world in recent times, both in acadmic and industrial research facilities for accurate modeling of RES and their Grid-integration. All aspects of grid-integration challenges as detailed in the previous section are being studied with the help of such models through both offline and real-time simulation. This section describes the mathematical models used for simulation of different RES performance, followed by case-study of a realistic RES-connected micro-grid operation.

3.3.1 Modelling of RES

Electricity produced from burning of fossil fuel is a main contributor to climate change. Moreover, through decades of extensive utilization of fossil fuel, energy crisis is evident in the near future. In spite of that, energy consumption is continuing to upsurge due to the worldwide growth of economy. Hence, it is imperious to explore and develop inexpensive as well as clean energy alternatives to facilitate sustainable global society and to reduce climate change as much as possible. To attain this, several efforts are made to enhance the utilization of RESs. Among these RESs, wind and solar are recently most favorable, which jointly produce more than 50% of the worldwide installed renewable energy capacity. This section summarizes mathematical tools for modeling of various RESs and demonstrates the results of integrating such RESs into the grid.

Wind turbine generator (WTG): Wind power has emerged as a very important RES over the last few decades. Due to pollution free and eco-friendly aspect, wind energy is one of the most apposite selections of renewable group.

Natural wind flow being unpredictable, probability distribution of wind speed is frequently used to evaluate wind energy potential. Most commonly used probability distribution function for wind is the *Weibull distribution.* If V is the wind speed in m/s, *Weibull* probability distribution function $f(V)$ is articulated as [20]:

$$f(V) = \frac{k}{c}\left(\frac{V}{c}\right)^{(k-1)} \exp\left[-\left(\frac{V}{c}\right)^{(k)}\right] \tag{3.1}$$

Where k is the dimensionless *Weibull* shape parameter and c is *Weibull* scale parameter (m/s).

It is a known fact that output power of a WTG is largely dependent on wind velocity [21]. The non-linear relationship between output power and wind velocity is explained by means of operating constraints of the WTG. It operates in four phases:

- First phase: in this standby phase, wind speed is lower than the cut-in wind speed (V_{in}) and no power is produced
- Second stage: power production upsurges non-linearly with wind velocity ranging from cut-in wind speed (V_{in}) to rated wind speed (V_r)
- Third stage: maximum power i.e., the rated power is produced, when wind velocity ranges from rated wind velocity (V_r) to cut-out wind velocity (V_{out})
- Fourth stage: while wind velocity is more than cut-out wind velocity (V_{out}), WTG is disconnected for safety reasons and power generation stops (cut-off stage).

Output power of a WTG at any time t is expressed as

$$\mathrm{P}w_t = 0, \;\; \text{for } Vf_t < V_{in} \text{ and } Vf_t > V_{out}$$

$$\mathrm{P}w_t = \left(\mathrm{A}_w + \mathrm{B}_w \times Vf_t + C_w \times Vf_t^{\,2}\right) \times \mathrm{P}wr, \;\; \text{for } V_{in} \le Vf_t \le Vr, \;\; t \in \mathrm{T} \tag{3.2}$$

$$\mathrm{P}w_t = \mathrm{P}wr,\ \text{for}\ Vr \le Vf_t \le V_{out}$$

A_w, B_w and C_w are calculated as:

$$\mathrm{A}_w = \frac{1}{(V_{in} - Vr)^2}\left[V_{in}(V_{in} + Vr) - 4 \times V_{in} \times Vr \times \left(\frac{V_{in} + Vr}{2 \times Vr}\right)^3\right] \tag{3.3}$$

$$\mathrm{B}_w = \frac{1}{(V_{in} - Vr_t)^2}\left[4 \times (V_{in} + Vr) \times \left(\frac{V_{in} + Vr}{2 \times Vr}\right)^3 - (3 \times V_{in} + Vr)\right] \tag{3.4}$$

$$C_w = \frac{1}{(V_{in} - Vr_t)^2}\left[2 - 4 \times \left(\frac{V_{in} + Vr}{2 \times Vr}\right)^3\right] \tag{3.5}$$

Figure 3.42 depicts a typical power output curve of WTG whose rated power is 600 kW, rated wind velocity is 15 m/s, cut-in wind velocity is 5 m/s and cut out velocity speed is 25 m/s.

Typical cost function of an WTG comprises of direct cost, penalty cost (Ue_w) and reserve cost (Oe_w).

$$fw_t(\mathrm{P}w_t) = \mathrm{K}w \times \mathrm{P}w_t + Oe_{wt}(\mathrm{P}w_t) + Ue_{wt}(\mathrm{P}w_t) \tag{3.6}$$

Reserve cost owing to overestimation and penalty cost owing to underestimation on available wind power is expressed as:

$$Oe_{wt}(\mathrm{P}w_t) = oe_w \times \int_{\mathrm{P}w_t^{\min}}^{\mathrm{P}w_t} (\mathrm{P}w_t - y) \times fw(y)dy \tag{3.7}$$

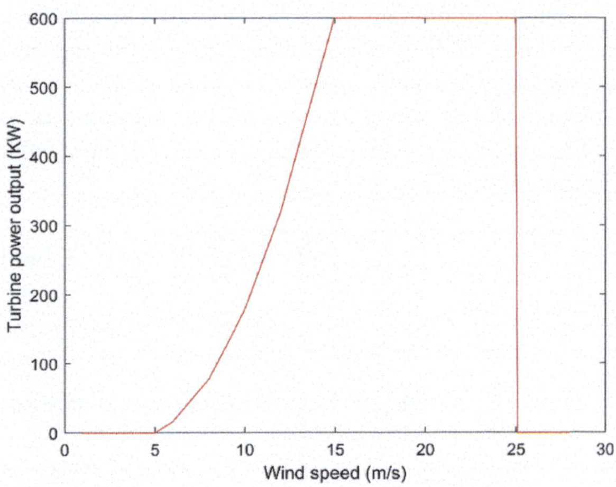

Fig. 3.42 Typical power output curve of WTG

$$Ue_{wt}(\mathrm{Pw}_t) = ue_w \times \int_{\mathrm{Pw}_t}^{\mathrm{Pw}_t^{\max}} (y - \mathrm{Pw}_t) \times fw(y)dy \tag{3.8}$$

Solar PV plants: The power produced by solar PV plant is mainly dependent on solar irradiation, plus deviation among reference temperature and ambient temperature [22]. Power output from solar PV plant at hour t is expressed as:

$$\mathrm{P}_{PVt} = \mathrm{P}_{PVr}\left[1 + \alpha_r \times \left(\mathrm{T}_{ref} - \mathrm{T}_{ambt}\right)\right] \times \frac{G_r}{1000} \tag{3.9}$$

Cost function of solar PV plant consists of direct cost, penalty cost (U_w) and reserve cost (O_w).

$$fpv_t(\mathrm{P}_{PVt}) = \mathrm{K}pv \times \mathrm{P}_{PVt} + Oe_{PVt}(\mathrm{P}_{PVt}) + Ue_{PVt}(\mathrm{P}_{PVt}) \tag{3.10}$$

Reserve cost owing to overestimation and penalty cost owing to underestimation on available solar power is expressed as:

$$Oe_{PVt}(\mathrm{P}_{PVt}) = oe_{PV} \times \int_{\mathrm{P}_{pvt}^{\min}}^{\mathrm{P}_{pvt}} (\mathrm{P}_{PVt} - y) \times f_{PV}(y)dy \tag{3.11}$$

$$Ue_{PVt}(\mathrm{P}_{PVt}) = ue_{PV} \times \int_{\mathrm{P}_{pvt}}^{\mathrm{P}_{pvtt}^{\max}} (y - \mathrm{P}_{Pvt}) \times f_{PV}(y)dy \tag{3.12}$$

Solar PVT panel: Photovoltaic-thermal panel (PVT) combines photovoltaic as well as solar thermal technology, to maximize their benefits in a single piece of equipment. Solar PVT panel generates electricity and heat concurrently for consumer, to maximize available space and obtain a better return on investment compared to isolated solar photovoltaic or solar thermal systems.

Electrical power and efficacy obtained from a solar PVT panel at time t may be computed as [23]:

$$\mathrm{P}pvt_t = \mathrm{A}pvt \times Gr_t \times \eta_{ePVT,t} \tag{3.13}$$

$$\eta_{ePVT,t} = \eta_{ref,PVT}\left[1 - \beta_{ref,PVT}\left(\mathrm{T}_{cell,PVT,t} - \mathrm{T}_{ref}\right)\right] \tag{3.14}$$

$\mathrm{T}_{cell,PVT,t}$ is dependent on environmental conditions and may be assessed as:

$$\mathrm{T}_{cell,PVT,t} = 30 + 0.0175(Gr_t - 300) + 1.14\left(\mathrm{T}_{amb,t} - 25\right) \tag{3.15}$$

Thermal power as well as thermal efficacy obtained from a solar PVT panel may be calculated as:

$$\mathrm{H}_{PVT,t} = Apvt \times Gr_t \times \eta_{th,i} \tag{3.16}$$

$$\eta_{th,t} = \tau_{\alpha PVT}\left(1 - \eta_{ePVT,t}\right) - \frac{uloss\left(\mathrm{T}_{cell,PVT,t} - \mathrm{T}_{amb,t}\right)}{Gr_t} \tag{3.17}$$

Hydro power plants: Hydro power is one of the oldest RESs in use, with size ranging from few kW to few GW. Hydro power plants can be categorized on the basis of system size as:

>30 MW	Large
1 MW–30 MW	Small
100 kW–1 MW	Mini
<100 kW	Micro

Mini-hydro plant: Mini-hydro power plants are mostly run-of-river type which has no accretion of water in any form of reservoir. Affluent river inflow rotates a turbine connected to an electrical generator. Generated power is proportional to flow of water that rotates the turbine and also to the height of waterfall. Power generated from mini-hydro plant [24] at time *t* is calculated as:

$$\mathrm{P}_{ht} = \rho Q_t \mathrm{H} \eta_G \eta_\mathrm{T} \tag{3.18}$$

Bioenergy (Waste-to-energy): Bioenergy could be referred to as heat or electricity derived from organic materials. The approaches to produce such energy relies on the availability of specific material at the location of the project and then the process to generate electricity, either by direct combustion approach to generate heat and then electricity or converting raw material to biofuel (liquid or gas). Municipal solid waste (MSW) one of the main biomass source, and heat is generated following the relevant combustion processes. In order to calculate the amount of electricity generated from the bio-energy power plant (BPP) [25], (3.19) is developed:

$$\mathrm{P}_{Bt} = \eta_g \times \eta_p \times LH\, V_{MSW} \times V_{MSW} \times \frac{1}{3600} \tag{3.19}$$

η_g is the generator efficiency that is around 70% and η_p is the conversion efficiency which is of the order of 20%.

3.3.2 Modeling of ESS

Electric grid should have sufficient power generation capacity for meeting demand of the consumer. But electric power demand varies significantly both daily as well as seasonally, and committing generators for matching the extensive peak-to-base electric demand is a major challenge. Electricity providers should have ample installed capacity for matching peak load demand and should always operate at sufficient capacity for meeting real-time load demand. To meet these necessities, whereas electricity providers are made to operate their generators at 20% above the predictable demand, but an overall average of only 55% of installed power generation capacity is utilized over the course of one year.

These inefficiencies are mostly instigated by perishable nature of energy in electric grid. Owing to lack of energy storing devices in grid system, energy should be immediately delivered to and used by consumers. Further, energy storage in grid would permit many power plants for running nearer to full capacity and decrease energy losses in the course of electricity transmission. Energy storage is a chief element in branching out energy sources as well as adding more RESs into energy market. By using energy storing devices, generation sources need not be ramped up or down, instead can be run at optimum efficacy while energy storing devices make up for variations in load demand.

Applications that might benefit from energy storing in electric grid have an extensive range of requirements. In isolated areas, seasonal energy storing of MW-hr capacity for months at a time is required. On the other hand, stabilization of transmission and distribution networks need to store energy for simply several minutes before discharging it and be able to have energy capacities on watt-hour scale. Different types of energy storing devices are now available to work across all of these time and energy scales.

With rising concern over reliable and secure energy, energy cost, and environmental concern over fossil fuel-based sources, RESs have experienced large deployment in recent years. Moreover, RESs are more available worldwide but electrical shortages are not that common due to poor substructure. Majority of growth in renewable energy system has been happening in solar photovoltaic and wind energy. When permeation of these sources rises in grid, it becomes necessary for them to participate in grid support functions. But wind and solar power are sporadic in nature which averts them from acting as independent and reliable energy sources. Sporadic nature of these sources requires support from ESSs. ESS plays a crucial role in supplying continuous and high-quality power to grids that have solar PV and wind integrated to it. Existing ESS uses several technologies, including hydroelectricity, batteries, super-capacitors, flywheels and others as detailed in Fig. 3.43.

ESSs have been applied in a wide range of electrical systems, from consumer electronics to automotive industry, and to utility-level transmission as well as generation.

Support for renewable energy systems is needed at several time lengths ranging from short term (less than 5 s) to long term (several hours). Short-term support

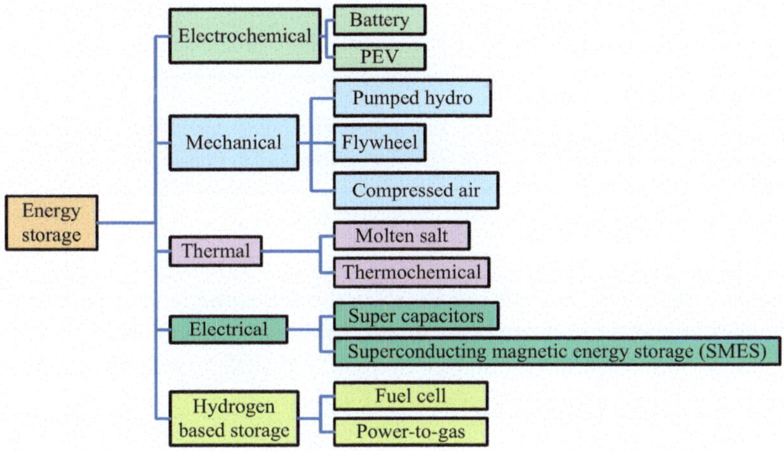

Fig. 3.43 Types of energy storage systems

is required for the period of solar PV clouding, wind power gusting, and power ramp-rate limitations. Appropriate energy storing for such applications should have large power capacity. Long-term support may consist of power shifting and spinning reserve functions.

Electrical utility systems may take advantage of ESS at distribution, transmission, and generation levels. ESS can stabilize electric power system by supplying voltage as well as frequency support, load smoothing plus peak power shearing, spinning reserve, and additional auxiliary services.

Battery Energy Storage System (BESS): BESSs are used by three principal groups: power producing utilities, power allocating utilities, and major power consumers. Principal advantage of BESSs include: load smoothing, enhancing grid stability as well as power quality, frequency balancing, spinning reserve, decreasing source irregularity, zero emission, and deferred investment for new generation as well as transmission equipment. Power allocating utilities as well as major power consumers may evade pricey demand changes by discharging their battery at peak demand period and then recharging with lower price during off-peak demand period.

Because of technological advancements and substantial development in battery chemistries, a varied range of battery options for storage applications have materialized in recent times. BESS offers flexibility in capacity, siting, and quick response needed for meeting application demand over an extensive range of functions than some of the other types of storing devices. Different types of batteries used for energy storage applications are shown in Fig. 3.44.

BESS in essence is considered to be a limited capacity dispatchable generator. Performance of a battery is represented by its state of charge (SOC) value at each time instant. SOC is an indicator of residual charge in terms of percentage of battery capacity. SOC value increases while charging the battery and decreases at the time of discharging. BESS can supply power at any time when SOC is greater than minimum

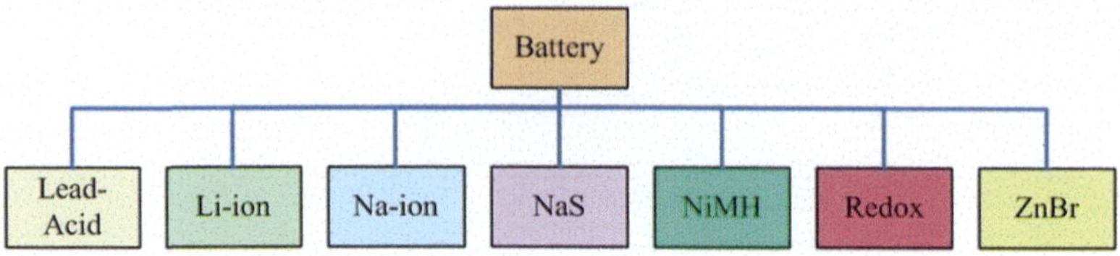

Fig. 3.44 Types of batteries for energy storage applications

permissible value $(SOC)_{min}$. SOC of BESS in the following hour depends on SOC and net charging power in the existing hour [26]. It is articulated as:

$$(SOC)_t = (SOC)_{(t-1)} + \mathrm{P}b_t^c \times \eta_b^c - \frac{\mathrm{P}b_t^d}{\eta_b^d} \tag{3.20}$$

SOC of a BESS must be less than $(SOC)_{\max}$ and more than $(SOC)_{\min}$.

$$(SOC)_{\min} \leq (SOC)_t \leq (SOC)_{\max} \tag{3.21}$$

Charging power of BESS is limited by its maximum charging power limit $\mathrm{P}b^{c,\max}$.

$$0 \leq \mathrm{P}b_t^c \leq \mathrm{P}b^{c,\max} \times U_t^c \tag{3.22}$$

Discharging power of BESS is limited by its maximum discharging power limit $\mathrm{P}b^{d,\max}$.

$$0 \leq \mathrm{P}b_t^d \leq \mathrm{P}b^{d,\max} \times U_t^d \tag{3.23}$$

$$0 \leq Ub_t^c + Ub_t^d \leq 1 \tag{3.24}$$

BESS should not be charged and discharged at the same time, otherwise their lifetime will be shortened.

Plug-in electric vehicle: Compared to fossil-fuel-based cars, main benefits of EVs are their silent, efficient, and emissions-free operation. PEV is fetching a promising approach as it reduces GHG emission as well as fossil-fuel consumption. In addition, PEV power has the potential to be utilized effectively as a portable ESS and offer V2G service.

It is difficult to have a suitable model of PEV in order to evaluate its performance. The PEV model is intricate as it comprises of a wide number of interconnected components. The PEV *connection mode* provides information about whether it is connected in charging or discharging mode. SOC is a percentage assessment of amount of energy available in its battery at a given time. SOC specifies the lasting time of a battery beforehand which determines the requirement for charging or substitution. If PEV's battery SOC value is more than specified value, it links to discharging mode, else it links to charging mode. In discharging mode, PEV links to the power grid in V2G mode. In charging mode, PEV's battery charges from power grid, making it to operate in G2V mode.

Modeling of PEV parking lot and power consumption of the PEVs throughout its traveling period is expressed as follows [26]:

$$Ppev_t^{tr} = \Delta D_t^{pev} \times \eta^{pev} \tag{3.25}$$

Permissible charging and discharging rates of each PEV are expressed as:

$$0 \leq \mathrm{P}pev_t^c \leq \mathrm{P}pev^{c,\max} \times U_t^{pev,c} \tag{3.26}$$

$$0 \leq \mathrm{P}pev_t^d \leq \mathrm{P}pev^{d,\max} \times U_t^{pev,d} \tag{3.27}$$

$$0 \leq U_t^{pev,c} + U_t^{pev,d} \leq 1 \tag{3.28}$$

Charging and discharging modes are segregated at each time. The parking lot can switch between V2G and G2V modes.

Energy content of battery at each time period is equal to energy content in preceding time period, charging energy minus discharging energy and energy consumption of PEV's battery because of traveling at that time period. Energy capacity of battery in each PEV is expressed as:

$$(SOC)_t^{pev} = (SOC)_{(t-1)}^{pev} + \mathrm{P}pev_t^c \times \eta^{pev,c} - \frac{\mathrm{P}pev_t^d}{\eta^{pev,d}} - \mathrm{P}pev_t^{tr} \tag{3.29}$$

Minimum and maximum limits of SOC for PEVs are expressed as:

$$(SOC)_{\min}^{pev} \leq (SOC)_t^{pev} \leq (SOC)_{\max}^{pev} \tag{3.30}$$

Ultra-capacitor energy storage: Super or ultra-capacitor (UC) ranging from 800 to 1500F behaves similar to very high-power and low-capacity battery. However, an UC stores electric energy by collecting and segregating opposite charges physically, as opposed to a battery, which stores energy chemically. One fundamental aspect of supercapacitor is its exceptional life cycle. It is possible to cycle UC very rapidly without much reduction in its lifecycle that most of the chemical battery experiences. UC has a high cycle efficacy comparing with chemical battery. The main obstacle with UC is its less specific energy and large voltage dispersal among individual cells. UC also has distinctive feature that its voltage is directly proportional to its state-of-charge. Hence, either its operational range should be restricted to high SOC regions, or control electronics should be used for compensating extensively varying voltage.

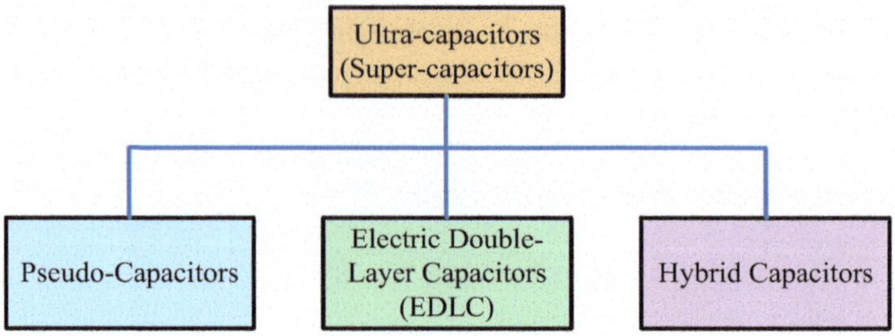

Fig. 3.45 Classification of ultra-capacitors based on the electrode design

UC also can act as energy storage device and has the ability of rapid charge and discharge providing high power for a very short time. Through the use of a considerably bigger surface area for electrodes and thinner dielectrics, UCs are looking at substantial enhancement in capacity and energy density over classical capacitor, while maintaining reasonably high power density values. Comparing with other energy storage devices, UC has a very high power density, while its energy density is considerably lesser than that of electric battery. UC is particularly suited for applications that require high ramp rate with short deep cycles. UCs are being explored in recent times for storing electrical energy during periods of surplus generation and supply high power within a short time during peak load demand. Energy density of UC is 100 times larger than classical electrolytic capacitor. UC has many attractive properties such as rapid charge–discharge ability, longer life, almost maintenance-free and eco-friendly.

On the basis of electrode design, UCs can be categorized into three main groups: electrical double-layer, pseudo-capacitors, and hybrid capacitors as shown in Fig. 3.45. Double-layer capacitor relies on electrostatic field among two plates, while pseudo-capacitor employs electrochemical reactions for storing electric charge. Hybrid capacitor combines both these two phenomena and comprises popular Li-ion UC.

UC is superior to BESS considering life cycling and energy efficacy. In real-world applications, needed amount of terminal voltage and energy or capacitance of UC storing system may be attained by using multiple UCs in series and parallel [27]. Terminal voltage defines number of capacitors which should be linked in series for forming a bank and total capacitance defines number of capacitors which should be linked in parallel in bank. UC units may be arranged for building a UC bank, capable of supplying short-range peak demands. Amount of energy taken from UC bank is directly proportional to capacitance and change in the terminal voltage. Voltage state of an UC with a capacitance C taking charge into a resistance R can be articulated as:

$$V(t) = V_i \exp\left(-\frac{t}{RC}\right) \tag{3.31}$$

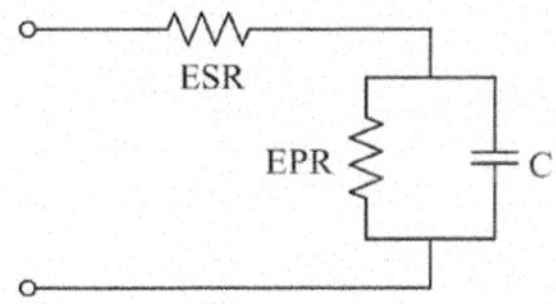

Fig. 3.46 Equivalent circuit of a UC unit

RC time constant defines real period of charging and discharging procedures for some initial voltage on capacitor. SOC of an UC bank is available energy capacity expressed as a percentage of its rated energy capacity.

Amount of energy taken from UC bank is directly proportional to capacitance and change in terminal voltage, and is formulated as:

$$\mathrm{E}_{UC} = \frac{1}{2}C\left(V_i^2 - V_f^2\right) \tag{3.32}$$

where $V(t)$ is voltage after time t, V_i is initial voltage of UC bank before discharging starts, V_f is final voltage of UC bank after which discharging, C is a capacitance, R is an internal resistance, E_{UC} is amount of energy taken from UC bank.

Figure 3.46 shows the classical equivalent circuit of an UC unit. The model includes a capacitance (C), an equivalent series resistance (ESR) signifying charging and discharging resistance, and an equivalent parallel resistance (EPR) signifying self-discharging losses. EPR models leakage effect and only impacts long-term energy storage performance of UC.

To attain desired amount of energy taken from UC bank, and a reasonable charge and discharge efficacy, total resistance and capacitance of an UC bank depicted in Fig. 3.47 may be computed as:

$$R_{UC-total} = n_s \frac{ESR}{n_p} \tag{3.33}$$

$$C_{UC-total} = n_p \frac{C}{n_s} \tag{3.34}$$

where n_s is number of capacitors linked in series and n_P is the number of series strings in parallel. $C_{UC-total}$ is total UC system capacitance [F] and $R_{UC-total}$ is total UC system resistance [Ω].

Pumped hydro storage: PHS is an environmentally friendly and cost-effective process for storing energy in a wide range of capacities. PHS has very long life as well as low maintenance cost. Installing a big PHS system is highly dependent on geographical characteristics of a particular region and requires huge investments. PHS is mainly dependent on prospective site and system size. PHS consists of a pump, a turbine, as well as reservoirs.

PHS unit operates as a classical hydro plant throughout peak demand period [28]. In PHS unit, water is pumped from a lower reservoir to an upper reservoir during

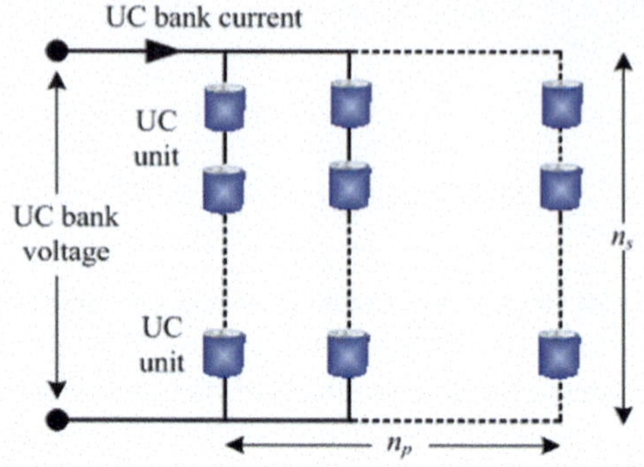

Fig. 3.47 Arrangement of capacitors in a UC bank

off-peak demand period, while the stored water in upper reservoir is released for generating power during peak demand period. PHS unit thus acts in generating mode during peak-load period, and in pumping mode during lean load periods. However, a PHS unit should be rested for an hour during the changeover time owing to physical limitations. Pumping rate is controlled by variable-speed drive which controls pump speed to regulate voltage as well as frequency. Turbine flow rate is controlled using solenoid valves placed at output of reservoir. Figure 3.48 depicts schematic diagram of PHS unit. Water capacity of the reservoirs, and related pumping and generation expressions of a typical PHS unit are modelled using equations (3.35–3.41):

$$V_{res,(t+1)} = V_{res,t} + Q_{Pt}(\mathrm{P}sp_{Pt}),\ t \in \mathrm{T}_{pump} \quad (3.35)$$

$$V_{res,(t+1)} = V_{res,t} - Q_{gt}(\mathrm{P}sp_{gt}),\ t \in \mathrm{T}_{gen} \quad (3.36)$$

$$V_{res,(t+1)} = V_{res,t},\ t \in \mathrm{T}_{change\_over} \quad (3.37)$$

$$\mathrm{P}sp_g^{\min} \le \mathrm{P}sp_{gt} \le \mathrm{P}sp_g^{\max},\ t \in \mathrm{T}_{gen} \quad (3.38)$$

$$\mathrm{P}sp_P^{\min} \le \mathrm{P}sp_{Pt} \le \mathrm{P}sp_P^{\max},\ t \in \mathrm{T}_{pump} \quad (3.39)$$

$$V_{res}^{\min} \le V_{res,t} \le V_{res,}^{\max},\ t \in \mathrm{T} \quad (3.40)$$

$$V_{res,0} = V_{res,\mathrm{T}} = V_{res}^{start} = V_{res}^{end} \quad (3.41)$$

Flywheel Energy Storage System (FESS): Frequency regulation and voltage leveling are two fundamental aspects for safe and reliable operation of any electricity network. Although BESS may be used for short-duration service but it suffers due to its short lifespan and need for disposal of toxic material. FESS [29] is an encouraging substitute to BESS for providing short-duration backup to grid. FESS may play a key

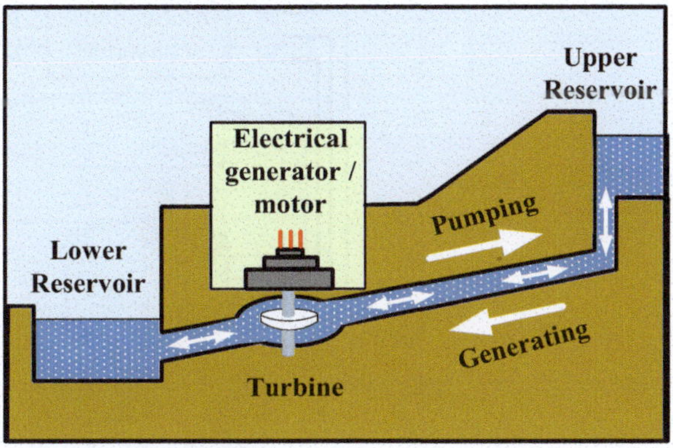

Fig. 3.48 Pumped hydro storage system

role in grid for maintaining its frequency by toning electric demand and supply for a small period. FESS is a mechanical storing system which transforms electricity into kinetic energy and this kinetic energy is transformed to electricity when required. FESS has a long lifespan, usually 15–20 years, high efficacy as well as high power density. FESS mostly comprises of rotor, motor-generator set, enclosure, mechanical bearing, vacuum pump, and power conversion system. A bi-directional converter is used to supply electricity to the FESS while it is storing kinetic energy, whereas it transfers the generated electrical power to grid when utilizing its stored kinetic energy.

FESS uses an electro-mechanical device which stores rotating kinetic energy (E),which is a function of rotating speed (ω) and rotor's prime moment of inertia (I_P):

$$E = \frac{1}{2} I_P \omega^2 \tag{3.42}$$

Vital components of FESS shown in Fig. 3.49 are: (1) rotor/flywheel to store kinetic energy, (2) bearing system for supporting rotor/flywheel, (3) power converter system to charge and discharge, electric machine and power electronics, (4) additional auxiliary components.

3.3.3 Sample Simulation Case Study for RESconnected Microgrid (MG)

Day-ahead scheduling of a RES-connected isolated MG comprising of diesel generators, small-hydro power plants, solar PV plants, WTGs, BESS, PEVs, and load has been studied. Figure 3.50 shows schematic diagram of the proposed MG structure.

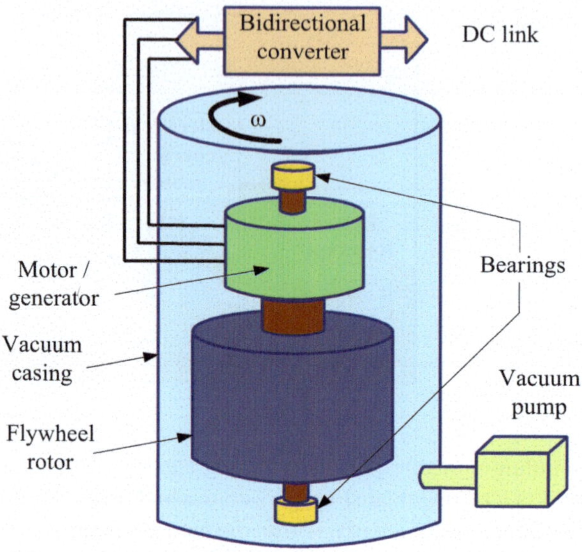

Fig. 3.49 A typical flywheel energy storage system

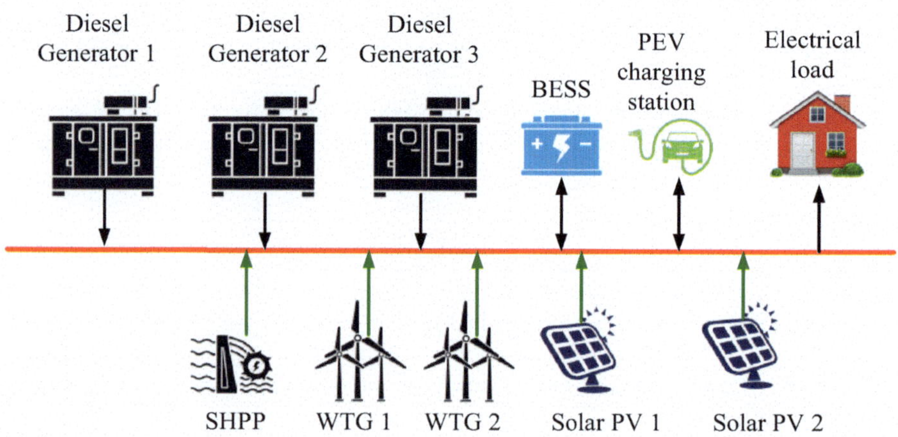

Fig. 3.50 Configuration of proposed MG

Self-organizing hierarchical particle swarm optimizer with time-varying acceleration coefficients (HPSO-TVAC) is used to analyze power scheduling of the MG.

Problem Formulation

Mathematical formulation of day-ahead scheduling of the isolated MG are stated as:

1. Objective function
   Total cost function is articulated as

$$C_T = \sum_{t=1}^{T}\left[\sum_{i=1}^{N_d} fdg_{it}(Pdg_{it}) + \sum_{j=1}^{N_w} fw_{jt}(Pw_{jt}) + \sum_{k=1}^{N_{PV}} fpv_{kt}(Ppv_{kt}, Hcs_{lt})\right] \quad (3.43)$$

Fuel cost function of ith diesel generator at time t is expressed as:

$$fdg_{it}(Pdg_{it}) = a_{di} + b_{di}Pdg_{it} + c_{di}Pdg_{it}^2 \quad (3.44)$$

Cost function of WTG comprises of direct cost, underestimation penalty cost plus overestimation reserve cost.

$$fw_{jt}(Pw_{jt}) = Kw_j \times Pw_{jt} + Ow_{jt}(Pw_{jt}) + Uw_{jt}(Pw_{jt}) \quad (3.45)$$

Overestimation reserve cost and underestimation penalty cost of obtainable power from WTG is formulated as:

$$Ow_{jt}(Pw_{jt}) = ow_j \times \int_{Pw_{jt}^{min}}^{Pw_{jt}} (Pw_{jt} - y) \times fw(y)dy \quad (3.46)$$

$$Uw_{jt}(Pw_{jt}) = uw_j \times \int_{Pw_{jt}}^{Pw_{jt}^{max}} (y - Pw_{jt}) \times fw(y)dy \quad (3.47)$$

The cost function of solar PV unit comprises of direct cost, underestimation penalty cost plus overestimation reserve cost:

$$fpv_{kt}(Ppv_{kt}) = Kpv_k \times Ppv_{kt} + Opv_{kt}(Ppv_{kt}) + Upv_{kt}(Ppv_{kt}) \quad (3.48)$$

Overestimation reserve cost and underestimation penalty cost of obtainable power from solar PV unit is expressed as:

$$Opv_{kt}(Ppv_{kt}) = opv_k \times \int_{Ppv_{kt}^{min}}^{Ppv_{kt}} (Ppv_{kt} - y) \times fw(y)dy \quad (3.49)$$

$$Upv_{kt}(\mathrm{P}pv_{kt}) = upv_k \times \int_{\mathrm{P}pv_{kt}}^{\mathrm{P}pv_{kt}^{\max}} (y - \mathrm{P}pv_{kt}) \times fw(y)dy \tag{3.50}$$

2. Constraints

(i) Power equilibrium constraints

$$\sum_{i=1}^{\mathrm{N}_d} \mathrm{P}dg_{it} + \sum_{m=1}^{\mathrm{N}_h} \mathrm{P}h_{mt} + \sum_{j=1}^{\mathrm{N}_w} \mathrm{P}w_{jt} + \sum_{k=1}^{\mathrm{N}_{PV}} \mathrm{P}pv_{kt} + \mathrm{P}ESSD_t - \sum_{p=1}^{\mathrm{N}_{EV}} \mathrm{P}_{EVpt} - \mathrm{P}_{Dt} = 0,\ t \in \mathrm{T} \tag{3.51}$$

$$\sum_{i=1}^{\mathrm{N}_d} \mathrm{P}dg_{it} + \sum_{m=1}^{\mathrm{N}_h} \mathrm{P}h_{mt} + \sum_{j=1}^{\mathrm{N}_w} \mathrm{P}w_{jt} + \sum_{k=1}^{\mathrm{N}_{PV}} \mathrm{P}pv_{kt} - \mathrm{P}ESSC_t - \sum_{p=1}^{\mathrm{N}_{EV}} \mathrm{P}_{EVpt} - \mathrm{P}_{Dt} = 0,\ t \in \mathrm{T} \tag{3.52}$$

Equation (3.51) is applied while battery is in discharging mode and Eq. (3.52) is applied while battery is in charging mode.

Power output of jth wind turbine generator at any time t is expressed as:

$$\mathrm{P}w_{jt} = 0,\ \text{for } Vf_t < V_{in} \text{ and } Vf_t < V_{out}$$

$$\mathrm{P}w_{jt} = \left(\mathrm{A}_w + \mathrm{B}_w \times Vw_t + C_w \times Vw_t^2\right) \times \mathrm{P}wr_j,\ \text{for}\ V_{in} \le Vf_t < Vr_t,\ j \in \mathrm{N_w},\ t \in T \tag{3.53}$$

$$\mathrm{P}w_{jt} = \mathrm{P}wr_j,\ \text{for} Vr_t \le Vf_t \le V_{out}$$

A_w, B_w, and C_w are computed as:

$$\mathrm{A}_w = \frac{1}{(V_{in} - Vr_t)^2}\left[V_{in}(V_{in} + Vr_t) - 4 \times V_{in} \times Vr_t \times \frac{(V_{in} - Vr_t)^3}{2 \times Vr_t}\right] \tag{3.54}$$

$$\mathrm{B}_w = \frac{1}{(V_{in} - Vr_t)^2}\left[2 - 4\frac{\left(\frac{4}{V_{in}} + Vr_t\right)^3}{2Vr_t}\right] \tag{3.55}$$

$$C_w = \frac{1}{(V_{in} - Vr_t)^2}\left[4 \times (V_{in} + Vr_t) + \frac{(V_{in} + Vr_t)^3 - (3 \times V_{in} + Vr_t)}{2 \times Vr_t}\right] \tag{3.56}$$

BESS has capacity to supply power when SOC is more than its lower limit of allowable value $(SOC)_{\min}$. SOC of a BESS in succeeding hour is dependent on its SOC and net charging power in existing hour.

$$(SOC)_{(t+1)} = (SOC)_t + PESSC_t \times \eta - PESSD_t \tag{3.57}$$

Each time, SOC of a BESS must be within $(SOC)_{\min}$ and $(SOC)_{\max}$.

$$(SOC)_{\min} \leq (SOC)_t \leq (SOC)_{\max} \tag{3.58}$$

Charging power of BESS is limited by its upper charging power limit $\mathrm{P}ch_{\max}$.

$$PESSC_t \leq \mathrm{P}ch_{\max} \tag{3.59}$$

Discharging power of BESS is limited by its maximum discharging power limit $\mathrm{P}dis_{\max}$.

$$PESSD_t \leq \mathrm{P}dis_{\max} \tag{3.60}$$

(ii) PEV charging power limit

Charging/discharging power of PEV at each hour is limited by charging/discharging capacity plus power requirement.

$$\sum_{t=1}^{\mathrm{T}} \mathrm{P}_{EVpt} = \mathrm{P}_{EVp,total} \tag{3.61}$$

Power necessary of pth PEV is total power $\mathrm{P}_{EVp.total}$ that is to be charged in 24 h and calculated by average mileage of personal itinerant vehicles. Charging/discharging capacity is limited by maximum and minimum charging/discharging power of PEV. Obtained charging power of pth PEV from MG is limited to minimum and maximum charging capacity limits signified by (3.62):

$$\mathrm{P}_{EVp}^{\min} \leq \mathrm{P}_{EVpt} \leq \mathrm{P}_{EVp}^{\max} \tag{3.62}$$

(iii) Small-hydro plant

Power generated by i^{th} small-hydro plant at time t is calculated by

$$\mathrm{P}_{hit} = \gamma Q_{it} \mathrm{H}_i \eta_{Gi} \eta_{\mathrm{T}i} \tag{3.63}$$

(iv) Capacity limits of DGs

$$Pdg_i^{\min} \le Pdg_{it} \le Pdg_i^{\max},\ i \in \mathrm{N}_d,\ t \in \mathrm{T} \tag{3.64}$$

(v) Ramp rate limit constraints of DGs

$$Pdg_{it} - Pdg_{i(t-1)} \le UR_{dgi},\ i \in \mathrm{N}_d,\ t \in \mathrm{T} \tag{3.65}$$

$$Pdg_{i(t-1)} - Pdg_{it} \le DR_{dgi},\ i \in \mathrm{N}_d,\ t \in \mathrm{T} \tag{3.66}$$

(vi) Solar power model

Output power attained from ith solar PV unit at hour is stated as

$$\mathrm{P}_{PVit} = \mathrm{P}_{PVri} \times \left[1 + \varepsilon_r \times \left(\mathrm{T}_{ref} - \mathrm{T}_{ambt}\right)\right] \times \frac{G_r}{1000},\ i \in \mathrm{N}_{PV}, t \in \mathrm{T} \tag{3.67}$$

3. Numerical Results

Day-ahead scheduling of the modelled MG is simulated in MATLAB (Version: 8. 1. 0. 604 (R2013a)) environment by using HPSO-TVAC.

The modelled MG comprises of three diesel generators, two solar PV units, two WTGs, one small-hydro power plant, one BESS and 30 PEVs. Energy consumption of every PEV is assumed to be 20 KWh/day. Installed capacity of BESS is 500 KW. Data of diesel generators is provided in Table 3.1. Height of the water fall of small-hydro power plant is 10 m and inflow rate is given in Table 3.2. Efficacy of turbine and generator is taken as 80% and 70% respectively. Rating of WTGs is $\mathrm{P}_{wr1} = 6$ MW and $\mathrm{P}_{wr2} = 4$ MW. Cut in, cut out, and rated wind velocities are $v_{in} = 5$ m/sec, $v_o = 25$ m/s, and $v_r = 15$ m/s, respectively. Direct cost coefficient, reserve cost, and penalty cost for each WTG are taken as 0.07, 0.02, and 0.01, respectively. Rating of solar PV units is $\mathrm{P}_{PV1} = 5$ MW and $\mathrm{P}_{PV2} = 3$ MW. Direct cost coefficient, reserve cost, and penalty cost for every solar PV unit are taken as 0.06, 0.02, and 0.01, respectively. Reference temperature and temperature coefficient are taken as 25° C and −0.25%/K (per Kelvin). Maximum and minimum prediction limits of solar irradiation and wind velocity are given in Figs. 3.53 and 3.54. Hourly power demand and temperature are provided in Table 3.3. PEVs are connected to MG for charging from 18th h to 24th h.

The parameters of HPSO-TVAC are taken as $\mathrm{N_P} = 100$, $w_{\max} = 0.25$, $w_{\min} = 0.05$, $c_{1i} = 2.5$, $c_{1f} = 0.2$, and $c_{2i} = 0.2$, $c_{1f} = 2.5$. Maximum iteration number is chosen as 100.

Power generation obtained from diesel generators, small-hydro plant, WTGs, solar PV plants, PEVs, and BESS corresponding to the minimum cost attained from HPSO-TVAC is portrayed in Fig. 3.51. Cost convergence characteristic attained from this algorithm is depicted in Fig. 3.52.

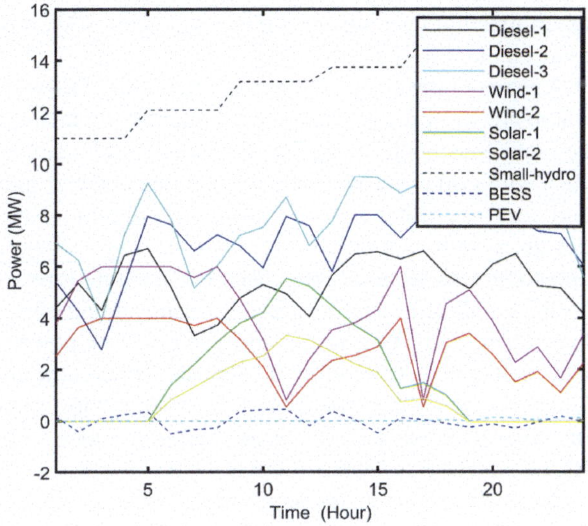

Fig. 3.51 Power generation obtained from diesel generators, small-hydro plants, WTGs, solar PV plants, PEVs, and BESS

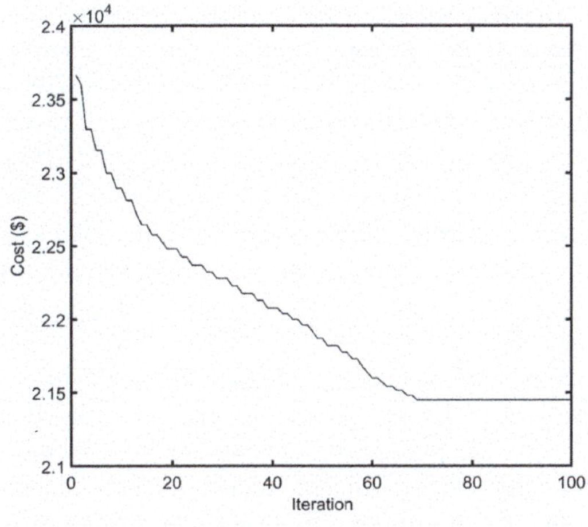

Fig. 3.52 Cost convergence characteristics

3.4 Conclusion

With the world power scenario marching toward achieving the Paris Agreement 2016, renewable energy sources are projected to reach a level of 60% of the total power generated by 2040. The intermittent and unpredictable nature of RES, especially wind and solar can be best dealt with when they are integrated with the main power grid. Integrating renewables to the grid provides abundance of benefits and opportunities in the form of improved reliability, reduction in T&D losses, voltage support, improved power quality, deferment in infrastructure investments, response flexibility, environmental friendliness, limited financial risks, improved customer support, and reduced electricity price.

In spite of the array of benefits, integrating DG with the grid does not come easy. There are several technical, environmental, financial, and operational challenges to be overcome while doing so. Technical challenges involve interfacing of variable RES with grid ensuring stability, planning and risk management, network reliability and security, voltage and frequency regulation, power quality, protection system design, system stability, power conditioning, voltage and current transients, T&D and communication security, need for accurate forecasting, storage needs, FRT and LVRT capabilities, and accurate metering among others. Financial issues range from investment costs, integration costs, to effect on power market, financial institutions, and tax/trade restructuring. The system operators also have to face several challenges to match the characteristics of RES with traditional power network. These include regulatory policies, grid operation, mode of operation, dispatchability, and RE management to name a few.

Extensive research has been carried out and practical implementation is underway in recent times to explore possible solutions to these issues. Advancements in power electronic devices and systems have made it possible to implement sophisticated converters for integration of RES with the grid. Advanced control tools are being explored for accommodating more flexibility in the power system while maintaining grid balance. New tools and techniques have been developed for maintaining frequency stability and achieve voltage control in RE integrated grids. Robust statistical and numerical models have been tested for accurate prediction of RES generation. Power quality issues arising out of RES integration are being addressed through the use of improved converter designs, filters, and FACTS devices. There have been remarkable advancements in storage technologies in recent years that have been extremely useful for RES to grid integration, especially to deal with variability and uncertainty. Several operational practices and RES policies have been undated/amended to facilitate grid integration. Power markets are also being transformed to welcome RES into the power grid and related ancillary services.

This chapter also reported results and findings on one sample study case on a RES-integrated isolated microgrid. Day-ahead scheduling was performed considering uncertainty of renewable energy sources to replicate real life scenario. A novel HPSO-TVAC approach has been used to solve this scheduling problem. The microgrid test system consists of diesel generators, small-hydro power plants, solar PV plants, wind turbine generators, battery energy storage system, and plug-in electric vehicles. It is observed that during excess generation, battery energy storage system is charged by taking power from the grid, and during generation deficit, battery energy storage system is discharged thereby supplying power back to the microgrid. It is also demonstrated that the developed energy management algorithm makes the plug-in electric vehicles to charge during off-peak hours. The presented case study shows the way for future research in RES microgrids, especially to implement demand side management and demand response programs.

It is envisaged that in-depth research and development is essential to ameliorate the grid integration process of RES and its implications on power network. Refined mathematical model of DG components are being used in research laboratories for accurate simulation studies.

Acknowledgement This work has been supported by DST – FIST Project in Power Engineering Department, Jadavpur University [SR/FST/ETII/2018/237].

Appendix

See Figs. 3.53 and 3.54, Tables 3.1, 3.2, and 3.3.

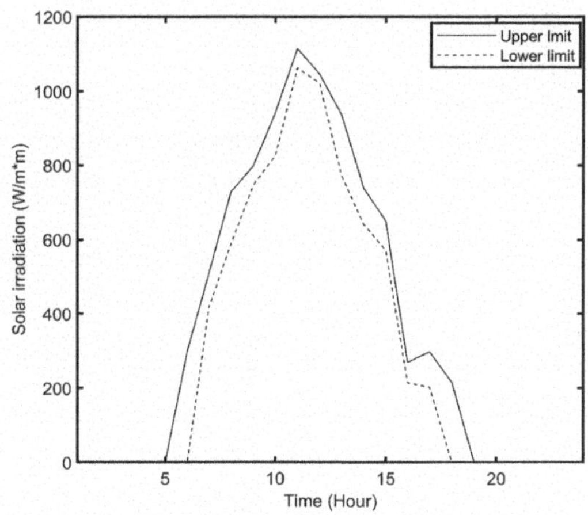

Fig. 3.53 Minimum and maximum prediction limits of solar irradiation

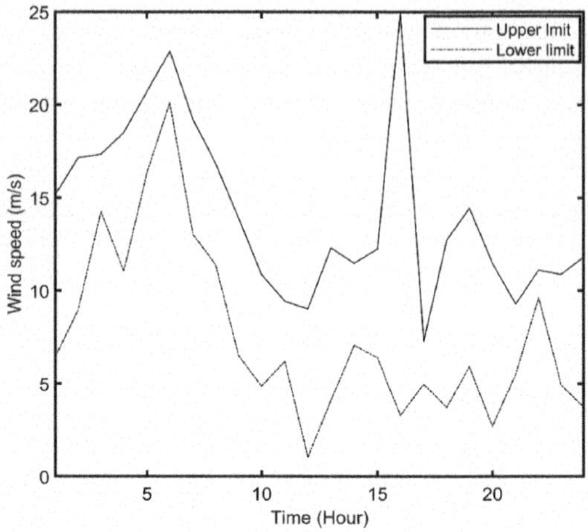

Fig. 3.54 Minimum and maximum prediction limits of wind speed

Table 3.1 Data of diesel generators

Unit	P_d^{min} (MW)	P_d^{max} (MW)	a_d ($/hr)	b_d ($/MWh)	c_d ($/MW^{2}h)	UR_d (MW/hr)	DR_d (MW/hr)
1	0	7	10.193	10.518	6.256	3	3
2	0	8	2.035	6.028	4.400	3	3
3	0	10	1.1525	6.534	4.400	4	4

Table 3.2 Hourly inflow of small-hydro power plant

Hour	Q (m^3)	Hour	Q (m^3)	Hour	Q (m^3)
1	0.2	9	0.24	17	0.27
2	0.2	10	0.24	18	0.27
3	0.2	11	0.24	19	0.27
4	0.2	12	0.24	20	0.27
5	0.22	13	0.25	21	0.23
6	0.22	14	0.25	22	0.23
7	0.22	15	0.25	23	0.23
8	0.22	16	0.25	24	0.23

Table 3.3 Hourly load demand and temperature

Hour	Load demand (MW)	Temperature (°C)	Hour	Load demand (MW)	Temperature (°C)	Hour	(Load demand (MW)	Temperature (°C)
1	34.00	23	9	46.00	27	17	42.50	26
2	36.00	23	10	44.00	28	18	46.00	25
3	32.00	24	11	45.00	29	19	44.00	24
4	40.00	25	12	44.00	29	20	45.00	24
5	46.00	25	13	46.00	29	21	40.00	24
6	45.00	25	14	50.00	28	22	38.00	24
7	40.00	26	15	50.00	28	23	36.00	23
8	44.00	27	16	48.00	27	24	34.00	23

References

1. IEA (2016) World Energy Outlook (2016). IEA, Paris. https://www.iea.org/reports/world-energy-outlook-2016. License: CC BY 4.0
2. IEA (2022) World Energy Outlook (2022). IEA, Paris. https://www.iea.org/reports/world-energy-outlook-2022. License: CC BY 4.0
3. IEA, Annual clean energy investment 2017–2022. IEA, Paris. https://www.iea.org/data-and-statistics/charts/annual-clean-energy-investment-2017-2022
4. IEA, Global installed grid-scale battery storage capacity in the Net Zero Scenario, 2015–2030. IEA, Paris https://www.iea.org/data-and-statistics/charts/global-installed-grid-scale-battery-storage-capacity-in-the-net-zero-scenario-2015-2030, IEA
5. IEA (2021) India Energy Outlook (2021). IEA, Paris. https://www.iea.org/reports/india-energy-outlook-2021.
6. Ghania SM, Hashmi AM (2021) Transient over-voltages simulation due to the integration process of large wind and photovoltaic farms with utility grids. IEEE Access 9:43262–43270. https://doi.org/10.1109/ACCESS.2021.3065874
7. Ulbig A, Borsche TS, Andersson G (2014) Impact of low rotational inertia on power system stability and operation. IFAC Proceed Vol 47(3):7290–7297. https://doi.org/10.3182/20140824-6-ZA-1003.02615
8. Che L, Khodayar ME, Shahidehpour M (2014) Adaptive protection system for microgrids: protection practices of a functional microgrid system. IEEE Electrif Mag 2(1):66–80. https://doi.org/10.1109/MELE.2013.2297031
9. You S et al (2017) Impact of high PV penetration on the inter-area oscillations in the U.S. Eastern interconnection. IEEE Access 5:4361–4369. https://doi.org/10.1109/ACCESS.2017.2682260
10. Mahat P, Chen Z, Bak-Jensen B (2011) Review on islanding operation of distribution system with distributed generation. In: 2011 IEEE power and energy society general meeting. Detroit, MI, USA, pp 1–8.https://doi.org/10.1109/PES.2011.6039299
11. Lucas A (2018) Single-phase pv power injection limit due to voltage unbalances applied to an urban reference network using real-time simulation. Appl Sci 8(8):1333. https://doi.org/10.3390/app8081333
12. Schmitt T, Rodemann T, Adamy J (2021) The cost of photovoltaic forecasting errors in microgrid control with peak pricing. Energies 14:2569. https://doi.org/10.3390/en14092569
13. Rasolomampionona D, Kłos M (2023) Energy storage systems and their role in smart grids', smart grids technology and applications. IntechOpen. https://doi.org/10.5772/intechopen.103945

14. Popavath L, Kaliannan P (2018) Photovoltaic-STATCOM with low voltage ride through strategy and power quality enhancement in a grid integrated wind-PV system. Electron J 7(51). https://doi.org/10.3390/electronics7040051
15. Lazard's Levelized Cost of Energy Analysis–Version 8.0. September 2014. https://www.seia.org/sites/default/files/resources/Levelized%20Cost%20of%20Energy%20-%20Version%208.0.pdf
16. Alam MS, Al-Ismail FS, Salem A, Abido MA (2020) High-level penetration of renewable energy sources into grid utility: challenges and solutions. IEEE Access 8:190277–190299. https://doi.org/10.1109/ACCESS.2020.3031481
17. Kumar GVB, Sarojini RK, Palanisamy K, Padmanaban S, Holm-Nielsen JB (1996) Large scale renewable energy integration: issues and solutions. Energies 12. https://doi.org/10.3390/en12101996
18. Santhosh M, Venkaiah C (2019) Sustainable energy, grids and networks short-term wind speed forecasting approach using ensemble empirical mode decomposition and deep boltzmann machine. Sustain Energy Grids Netw 19:100242. https://doi.org/10.1016/j.segan.2019.100242
19. IEC White paper (2012) Grid integration of large-capacity renewable energy sources and use of large-capacity electrical energy storage. Geneva, Switzerland, ISBN 978–2–8322–0340–8
20. Seguro JV, Lambert TW (2000) Modern estimation of the parameters of the Weibull wind speed distribution for wind energy analysis. J Wind Eng Indust Aerodyn 85(1):75–84
21. Hariria A-M, Hashemi-Dezakia H, Hejazia MA (2020) A novel generalized analytical reliability assessment method of smart grids including renewable and non-renewable distributed generations and plug-in hybrid electric vehicles. Reliab Eng Syst Saf 196:106746
22. Shilaja C, Ravi K (2017) Optimization of emission/economic dispatch using Euclidean affine flower pollination algorithm (eFPA) and binary FPA (BFPA) in solar photo voltaic generation. Renew Energy 107:550–566
23. Yousefi H (2017) Mohammad Hasan Ghodusinejad, Alibakhsh Kasaeian, "Multi-objective optimal component sizing of a hybrid ICE + PV/T driven CCHP microgrid." Appl Therm Eng 122:126–138
24. Basu S, Basu M (2022) Social group entropy optimization for day-ahead heat and power scheduling of an isolated microgrid. Eng Optim 54(7):1193–1216
25. Basu M, Jena C, Khan (2023) Dynamic optimal power flow for multi-operator renewable energy-based virtual power plants. First published: 03 June 2023. https://doi.org/10.1049/rpg2.12775
26. Basu M (2023) Optimal day-ahead scheduling of renewable energy-based virtual power plant considering electrical, thermal and cooling energy. J Energy Storage 65:107363
27. Ahmadi S, Bathaee SMT, Hosseinpour AH (2018) Improving fuel economy and performance of a fuel-cell hybrid electric vehicle (fuel-cell, battery, and ultra-capacitor) using optimized energy management strategy. Energy Conversion Manag 160:74–84
28. Wood AJ, Wollenberg BF (2012) Power generation, operation and control. John Wiley & Sons
29. Xiaojun L, Palazzolo A (2022) A review of flywheel energy storage systems: state of the art and opportunities. J Energy Storage 46:103576

Chapter 4
The Promising Role of Thermochemical Conversion in Sustainable Power Generation

Rabindra Kangsha Banik, Samar Das, Urbashi Bordoloi, Hirakh Jyoti Das, Biraj Das, Shayaram Basumatary, Bitupan Das, and Pankaj Kalita

Abstract The rising demand for petroleum-based fuels, coupled with escalating environmental concerns, has encouraged the scientific community to focus on exploring alternative energy resources for power generation. Biomass, being renewable, carbon–neutral, and abundant, has significant potential for generating power with minimal environmental impact. Biomass energy has an estimated annual production capacity of 181.5 billion metric tonnes of which only 8.2 billion metric tonnes are presently utilized. There is significant potential in terms of technological advancement for the conversion of biomass energy. Among the various biomass conversion routes, the thermochemical conversion process has gained prominence since the late nineteenth century. This well-established technology encompasses various categories such as pyrolysis, torrefaction, carbonization, liquefaction, gasification, and combustion, depending on the feedstock and desired end product. The present chapter provides a comprehensive overview of the various thermochemical conversion processes, their working principles, and operating criteria. Furthermore, this chapter briefly explores various power generation pathways such as steam power plants, gasifier-IC engine combined technology, gasifier gas turbine combined technology, and gasification fuel cell combined technology available for thermo-chemically converted biomass. The various limitations and challenges associated with the thermochemical conversion process have been reported in this chapter. Currently, the global power generated from biomass sources is about 764.2 TWh as of 2021 having an average annual growth rate of about 7%. The government initiatives aimed at promoting biomass-based power generation are also discussed in detail. Moreover,

R. K. Banik · S. Das · H. J. Das · S. Basumatary · B. Das · P. Kalita (✉)
School of Energy Science and Engineering, Indian Institute of Technology Guwahati, Guwahati 781039, India
e-mail: pankajk@iitg.ac.in

U. Bordoloi · B. Das
School of Agro and Rural Technology, Indian Institute of Technology Guwahati, Guwahati 781039, India

S. De et al. (eds.), *Challenges and Opportunities of Distributed Renewable Power*, Energy, Environment, and Sustainability, https://doi.org/10.1007/978-981-97-1406-3_4

techno-economic assessments, recent advancements in this field on the commercial scale, and their potential impact on CO_2 reduction are examined.

Keywords Biomass · CO_2 reduction · Sustainable power generation · Techno-economic analysis · Thermochemical conversion

4.1 Introduction

In the dynamic landscape of global energy transformation, the allure of sustainable alternatives has ushered in an era of profound change. This shift is primarily motivated by the aftermath of the Covid-19 pandemic, coupled with a collective response to the ongoing global energy crisis. As the world navigates this transformation, energy demand is projected to rise by a steady 2.1% annually between 2021 and 2050. This growth is attributed to the nexus of factors including population expansion, rapid economic development, and urbanization. A critical driver, however, is the urgency to address environmental challenges in alignment with the guidelines of the United Nations Framework Convention on Climate Change (UNFCCC), binding nations worldwide to adopt sustainable energy solutions. This dual pursuit of economic prosperity and environmental preservation underscores the imperative of this energy revolution [1].

Amid this transition, renewable power sources have emerged as a pivotal force reshaping the electricity generation landscape. Across the globe, substantial investments are being directed towards these sustainable options. Notably, India has made significant strides in its renewable energy sector, positioning the nation as a global leader in curbing greenhouse gas emissions while enhancing energy security for its burgeoning population. This ascent has been propelled by a confluence of ambitious renewable energy targets, supportive governmental policies, technological breakthroughs, and heightened environmental awareness. The country's swift progress in harnessing its abundant renewable resources is evidenced by its current position as the fourth-largest producer of renewable energy based on installed capacity. Figure 4.1 graphically illustrates this significant achievement. The nation's commitment to sustainability is evident in its pursuit of installing 500 GW of electricity capacity from non-fossil sources by 2030, as announced by the Prime Minister during the Conference of the Parties 26th meeting (COP26) [2]. This ambitious target is being diligently pursued by the Ministry of New and Renewable Energy (MNRE) in India. The proactive approach of the Indian government, characterized by initiatives, incentives, and supportive policies, has effectively catalyzed investments and expedited the implementation of renewable energy projects across the country. India's geographical diversity and varying climatic conditions have endowed it with a rich spectrum of renewable resources, including solar, wind, hydro, biomass, and tidal energy [3]. Of these, solar energy has garnered significant attention and is poised to contribute substantially to global renewable electricity generation, projected to reach a commendable 25% by 2030. Meanwhile, the installed capacity of solar energy in

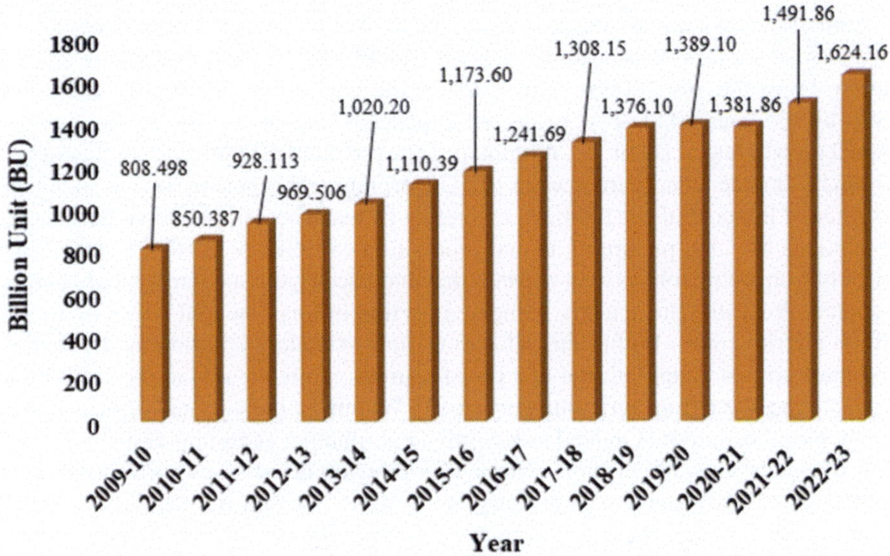

Fig. 4.1 Total generation and growth of RE in India from 2009–2010 to 2022–2023 [4]

India has already reached an impressive 67.07 GW. Equally striking is the remarkable growth of wind power, evidenced by a substantial increase in global electricity generation—reflected in an unprecedented rise of 265 terawatt-hours (TWh) in 2022, signifying a robust 14% surge. As of the latest data, India's installed wind capacity is a formidable 42.86 GW [4].

While solar and wind energy undoubtedly have been frontrunners in the renewable energy arena, the evolution of biomass energy has unfolded uniquely over time, carving a distinctive niche for itself. Initially confined to traditional uses such as cooking and heating, biomass energy's journey took a decisive turn in the late twentieth century with significant technological advancements. The dawn of the twenty-first century witnessed an era of addressing the intermittency issues posed by solar and wind power. It was during this juncture that biomass energy truly came into its own, emerging as a steadfast and dependable baseload solution. Its ability to provide consistent power generation irrespective of climatic conditions positioned it as a game-changer in the quest for reliable and sustainable energy sources. A notable facet of biomass energy is its carbon–neutral nature, aligned seamlessly with the global focus on combatting climate change. Biomass energy leverages organic waste materials, contributing to the circular economy by reducing landfill waste. Moreover, it plays a pivotal role in nurturing local economies, fostering job creation and rural development—a distinguishing feature that sets it apart from solar and wind power. Beyond its stationary applications, biomass energy also plays a pivotal role in the transportation sector. The blending of 10% bioethanol in petrol, mandated by the Indian government, has triggered a reduction in fossil fuel consumption and associated emissions.

Embarking on a remarkable trajectory, the field of bioenergy is poised to witness a substantial surge, with aspirations to nearly double its electricity generation contribution. From the 764.2 TWh mark in 2021, the goal is set ambitiously at 1386.8 TWh by the year 2030. This projected expansion reflects an impressive average annual growth rate of about 7%, illuminating the escalating significance of bioenergy within the broader energy framework [5]. Underpinning this potential is the abundant resource of lignocellulosic biomass, extending its reach across the global landscape. Projections sketch a picture of annual production capacity, projecting a staggering 181.5 billion metric tonnes—an expanse that underscores the sheer magnitude of this resource. Presently, the actively harnessed portion of this potential tallies up to 8.2 billion metric tonnes. Within this allocation, approximately 7 billion metric tonnes find their origins in agricultural and forest sources, while the remaining 1.2 billion metric tonnes stem from agricultural waste [6]. Within the Indian context, the nation's yearly biomass output is indeed noteworthy, culminating at an impressive range of approximately 450–500 million tonnes. This biomass cornerstone corresponds to a substantial 32% of the country's primary energy needs. Demonstrating an unwavering commitment to sustainable energy practices, India is vigorously prioritizing biofuels. India's dedication extends towards biomass power generation, with a capacity of an impressive 10.2 GW, complemented by an equally notable 0.55 GW in the waste-to-energy sector, figures reflecting the prevailing momentum as of 2023. In tandem with these pursuits, the exploration of pioneering technologies, notably energy storage and green hydrogen, unfolds as a strategic response, addressing the unique challenges that accompany intermittent renewable sources. This multifaceted approach is poised to fortify the grid's stability while perpetuating India's commitment to a sustainable and thriving energy landscape.

Within these compelling statistics, a profound realization surfaces: a significant portion of potential remains untapped, highlighting a realm of unexplored opportunities that could play a pivotal role in attaining the energy objectives outlined for 2050. This reservoir of untapped potential not only serves as a compelling impetus for action, but also underscores the immediate opportunity to fully harness the inherent capacities of biomass. This endeavour could potentially make a substantial contribution to the tangible realization of the energy targets set for the landscape of 2050. Within this realm of possibilities, a distinct pathway emerges—one that unequivocally underscores the paramount significance of thermochemical conversion. Thermochemical conversion represents a transformative avenue that capitalizes on biomass and waste feedstock to yield clean and renewable energy. Unlike conventional fossil fuels, which release harmful greenhouse gases and contribute to the depletion of finite resources, thermochemical conversion offers a route to power generation that doesn't exacerbate environmental challenges [2]. This technology encompasses a diverse array of processes, such as pyrolysis, torrefaction, hydrothermal treatment, gasification, and combustion, each presenting unique advantages and applications [3]. This trajectory represents a pivotal juncture—an evolutionary conduit that holds the promise of transforming untapped potential into practical and sustainable outcomes. Through the mechanism of thermochemical conversion, a transformative process unfolds, allowing the intricate architecture of biomass to metamorphose into an array of high-value energy products. This narrative journey traces a path from potential to realization, where foundational scientific principles intersect with pioneering technological innovation. This convergence ultimately culminates in the unlocking of a clean, efficient, and scalable energy solution. As the world advances, this trajectory resounds as a resounding call, to embrace the power of thermochemical conversion, an indispensable route that not only aligns with the energy aspirations of 2050, but also establishes a sustainable legacy for future generations. The fusion of scientific advancement and pragmatic implementation finds synergy within this pathway, illuminating a course that holds the pivotal key to unlocking the full potential of biomass. This trajectory propels humanity towards a more ecologically balanced and prosperous future, guided by the principles of innovation, sustainability, and responsible resource stewardship.

In this context, this chapter aims to explore the compelling role of thermochemical conversion in sustainable power generation, with a specific emphasis on its application to biomass as the primary feedstock. Biomass, due to its abundant and renewable nature, stands out as a promising feedstock for thermochemical conversion technologies. The choice of biomass as the primary feedstock for thermochemical conversion is substantiated due to its abundance, replenishable, and carbon–neutral nature. Furthermore, biomass resources are diverse, encompassing agricultural residues, forestry waste, energy crops, and organic waste materials. These resources, which are often underutilized, hold the potential to not only reduce waste, but also mitigate greenhouse gas emissions while fostering rural development. By examining its versatility, environmental benefits, and potential for widespread implementation, we seek to shed light on how this innovative technology can contribute significantly to global efforts towards a low-carbon and sustainable energy landscape. As the global community strives to address the urgent challenges of climate change, energy security, and sustainable development, understanding and harnessing the potential of thermochemical conversion with biomass becomes imperative. Doing so will be vital in shaping a cleaner, greener, and more sustainable future for generations to come.

4.1.1 Importance of Biomass in RE Power Generation

Biomass holds immense importance in renewable energy (RE) power generation due to its unique advantages and contributions to sustainability. Derived from organic materials and continuous replenishment nature, biomass processes ensure a consistent supply of energy, promoting long-term energy security. Moreover, biomass power generation is considered carbon–neutral, as the carbon dioxide (CO_2) emitted during combustion is balanced by the CO_2 absorbed during plant growth, making it a vital tool in combating climate change and reducing GHG emissions [7]. Additionally, biomass-based power plants can also utilize organic waste as feedstock which helps in mitigating waste management challenges [8]. Biomass power plants provide baseload electricity, operating continuously to meet stable energy demands, thereby maintaining grid stability [9]. Furthermore, in rural and remote areas with limited access to traditional electricity grids, biomass offers decentralized energy solutions, supporting local economic development and improving living standards. Its technological flexibility allows for various conversion processes, catering to diverse energy needs and optimizing energy production based on available biomass feedstock [10].

The adoption of biomass power generation also supports sustainable agriculture and forestry practices. The cultivation of energy crops encourages responsible land use, promoting sustainable agricultural practices like crop rotation and diversification. By integrating biomass with existing fossil fuel-based power infrastructure, the transition to RE becomes smoother and more efficient, maximizing existing resources and minimizing new investments [11].

4.2 Thermochemical Conversion Process

The thermochemical conversion process is a broad process used for conversion of solid biomass in presence of heat by combustion, generation of secondary fuel having higher heating value or valuable chemicals [12]. Thermal conversion involves direct combustion or heating of solid biomass to produce solid char, liquids (bio-oils), and syngas to generate process heat and electricity. Depending upon the type of product obtained, the thermochemical conversion process for solid biomass can be classified as direct combustion or co-firing, torrefaction, carbonization, pyrolysis, liquefaction, and gasification [13]. In this chapter, the authors will provide a detailed overview of the principles, applications, and recent developments in the above-mentioned thermochemical conversion techniques.

4.2.1 Torrefaction

Torrefaction is considered to be a modern process developed for upgrading biomass into a clean and energy-dense solid fuel. It is a slow thermochemical treatment process performed between 200 and 300 °C in an oxygen deficit environment for a suitable residence time [14, 15]. The process alters the chemical structure of a wide variety of lignocellulosic biomass enhancing its carbon content while reducing its hydrogen and oxygen content.

The process of torrefaction can be divided into two categories: dry and wet torrefaction. In case of dry torrefaction, a hot inert gas or indirect heating method is utilized, while in case of wet torrefaction, the high moisture biomass is heated in the presence of subcritical or compressed water.

During the dry torrefaction process, the decomposition of hemicellulose molecules occurs with the dissociation of structural oxygen at around 190 to 270 °C. However, there is slight decomposition of cellulose below 270 °C but lignin hardly dissociates at such lower temperature. As the temperature increases above

270 °C, degradation of cellulose and lignin takes place leading further decrease in the mass yield. The cellulose molecules break down into smaller molecules which are more reactive due to the breakage of hydrogen bonds. Besides, the formation of non-polar unsaturated structures makes the torrefied biomass hydrophobic to a certain extent. The torrefaction process produces a highly reactive feedstock for thermochemical conversion having 60–90% of the energy retained in the original feedstock [16]. Torrefaction of biomass results in the reduction of its oxygen and hydrogen contents while its carbon content increases, which ultimately reduces the H/C and O/C ratio of the torrefied product. Moreover, the net calorific value of the torrefied biomass increases to about 20–30 MJ/kg depending upon the feedstock and torrefaction conditions [16].

Wet torrefaction is an important process to transform biomass with high moisture content and waste streams into valuable solid fuel. In this process, the biomass is added to hot compressed water or subcritical water between 180 and 265 °C and at a pressure of 2–10 MPa for a shorter residence time in an inert environment [17]. The wet biomass undergoes hydrolysis, polymerization, decarboxylation, and dehydration during the process. Among all the lignocellulosic components, hemicellulose is mainly decomposed due to its better reactivity in that temperature range [18]. In general, the wet torrefaction process also reduces the oxygen and hydrogen contents of the biomass increasing its carbon content. Figure 4.2 represents the mass and energy balance for the torrefaction of chopped wood.

During wet torrefaction, the hemicellulose present in the biomass gets depolymerized to generate monomers and oligomers via hydrolysis. This degradation of hemicellulose and cellulose results in the production of solid biochar with reduced moisture content. The high moisture biomass is usually mixed with a sufficient amount of water to fully submerge (dry biomass: water = 1:6) inside the reactor. The main

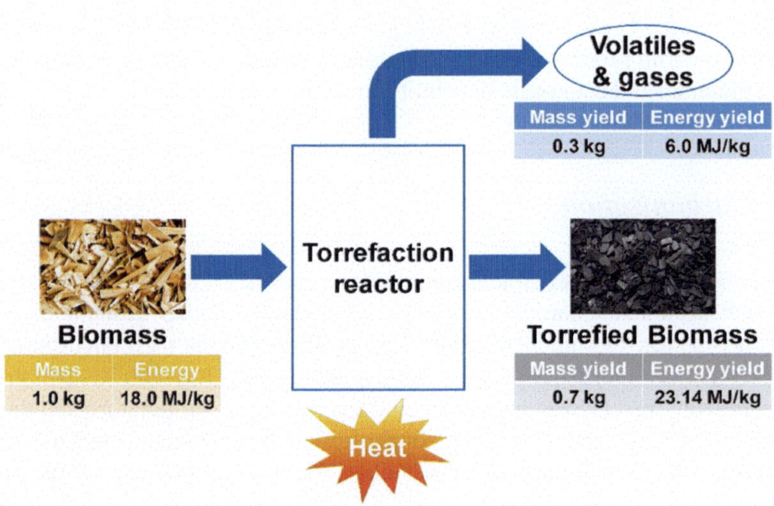

Fig. 4.2 Mass and energy balance for torrefaction of chopped wood [19]

advantage of wet torrefaction is that the heat required for drying the biomass is not needed as the process takes place in wet conditions. The end product of the process contains solid torrefied biomass, volatile acids, water, and gases [20].

The process of torrefaction is influenced by various parameters such as torrefaction temperature, pressure, residence time, type of rector, and the physio-chemical properties of biomass [21]. During the torrefaction process, temperature and residence time control the degree of torrefaction of the biomass material. With the rise in torrefaction temperature and residence time, the mass and energy yield drop, which improves the fuel properties. The mass and energy yield of the torrefied material which are one of the very important parameters of the process are used to assess the extent of biomass torrefaction [15].

The expression for mass yield [22] is given as

$$\text{Mass yield } (Y_{mass}) = \frac{\text{mass of torrefied biomass } (m_p)}{\text{mass of feed biomass } (m_f)} \times 100\% \tag{4.1}$$

The expression for energy yield [23] is given as

$$\text{Energy yield } (Y_{energy}) = \frac{\text{Higher heating value of torrefied biomass } (HHV_p)}{\text{Higher heating value of feed biomass } (HHV_f)} \times Y_{mass} \tag{4.2}$$

Based on the degree of torrefaction, the torrefied biomass has a better combustion characteristic with lesser smoke generation compared to the original feedstock. Improvement in properties like hydrophobicity, calorific value, energy density, and grindability of the biomass material can be observed with the severity of the torrefaction process [20]. The use of torrefied biomass for gasification improves the yield and quality of the syngas along with cold gas efficiency (CGE) [24]. The integration of torrefaction with processes like gasification, co-firing, and pyrolysis enhances the overall efficiency and economic feasibility towards sustainability [25].

4.2.2 Carbonization

Carbonization is a thermochemical conversion process used to produce mainly solid carbon-rich products (charcoal) having a significantly higher amount of carbon content. It is a slow thermochemical process performed in an inert environment for suitable residence time, usually in the temperature range of 300–500 °C. This process is different from torrefaction with respect to the operating temperature and the residence time. The residence time for carbonization can be extended from 2 h to several days. Carbonization of biomass yields in char, which highly resembles the properties of coal used in thermal power plants. The processes involved in carbonization are drying (20–200 °C), torrefaction (200–300 °C), low-temperature carbonization

(300–400 °C), and high-temperature carbonization (400–500 °C) [26]. This process is used to produce both charcoal and biochar depending on the requirement.

4.2.3 *Pyrolysis*

Pyrolysis is an endothermic thermochemical process where the carbonaceous biomass is converted into value-added products in the absence or partial supply of oxygen at a temperature range normally higher than 400 °C under atmospheric pressure [27]. The decomposition of the organic feedstock produces new molecules that possess some superior characteristics than the original biomass. The products of pyrolysis are biochar, bio-oil, and non-condensable gases. Biochar is a solid porous carbonaceous material mostly comprised of carbon with some inorganic compounds like alkali and alkaline earth metals. Bio-oils are dark brown, free-flowing organic liquids that are comprised of highly oxygenated compounds. It is a mixture of many chemicals, including acids, aldehydes, ketones, furfural, anhydro sugars, phenolic compounds, insoluble high molecular weight organics, and water. Bio-oil is a value-added intermediate chemical that may be used to generate energy after purification. The liquid fuel can be supplied to diesel engine for power generation in remote places and in boilers to generate heat in cogeneration plants. The main gases in biomass pyrolysis include carbon dioxide, carbon monoxide, hydrogen, methane, ethane, ethylene, propane, sulphur oxides, nitrogen oxides, and ammonia. The main application of produced syngas is typically the generation of power and heat. This can be realized either in stand-alone combined heat and power (CHP) plants or through co-firing of the product gas in large-scale power plants.

The reaction parameters affecting the process of pyrolysis are feedstock type, temperature, heating rate, retention time, gas flow rate, rector geometry, particle size, and physical structure of the biomass feedstock. The most common feedstock for pyrolysis is lignocellulosic biomass such as wood chips, forest and agricultural residues, solid waste, etc. These biomasses are primarily composed of cellulose 38–50 wt%, hemicellulose 23–32 wt%, and lignin 10–25 wt%. Temperature and heating rate are dominant factors that can determine the characteristics of the products. The primary function of temperature changes in the pyrolysis process (which is dependent on the rate at which the system heats up) is the provision of suitable heat to guarantee the proper production of products. Below 200 °C, dehydration of the biomass occurs, followed by hemicellulose decomposition in the range of 200–280 °C. The cellulose decomposition happens in the range of 240–350 °C, and lastly, lignin decomposes at a temperature above 400 °C. Faster heating rates favour the evolution of fluid (liquid and gaseous) products compared to solid products. To create an inert atmosphere, normally N_2, H_2, CO, CO_2, CH_4, and steam can be used as carrier gas inside the reactor [28].

The mechanism of pyrolysis involves mainly primary and secondary reactions. The primary reactions include charring: condensation of the benzene ring during pyrolysis; depolymerization: cracking of bonds and formation of gas and volatiles;

Table 4.1 Parameters for different pyrolysis processes for liquid and gas production [28]

Conversion process	Temperature (°C)	Residence time	Heating rate (°C/min)	Major products	Product yield		
					Liquid (wt%)	Gas (wt%)	Solid (wt%)
Fast pyrolysis	400–650	0.5–5 s	100	Bio-oil (thinner) Gases Char	65–75	13–25	12–19
Flash pyrolysis	700–1000	Less than 0.5 s	500 and more	Gases Bio-oil	60–70	10–15	15–25

fragmentation, and formation of compounds with short-chain and some incondensable gases. Secondary reactions involve the recombination or further cracking of the unstable compounds produced in the primary reactions [29].

Depending on the process parameters, pyrolysis can be classified into three main categories, namely fast pyrolysis, slow pyrolysis, and flash pyrolysis. Fast pyrolysis refers to the process where the temperature range is given as 800–1300 °C with a heating rate of 10–200 °C/min at a residence time of 1–10 s. Fast pyrolysis is carried out to get oil because biochar production in this method is very less (typically 15–20% of the net output). The particle size of fast pyrolysis should be fine (lower than 1 mm). Fast pyrolysis starts with the decomposition of biomass to produce vapours, aerosol, and some solid char. Then condensation and cooling of these vapour and aerosol form a dark brown liquid. For flash pyrolysis to be effective, temperature, heating rate, and residence time would be within the range of 450–600 °C, 103–104 °C/min, and < 1 s. Due to the reduction in secondary cracking reactions, the yield of bio-oil is maximized to about 60–75 wt% of the total products in flash pyrolysis. The particle size of flash pyrolysis is usually less than 0.2 mm [30]. The important parameters for the production of liquid and gas from the pyrolysis process have been mentioned in Table 4.1.

As bio-oil and pyrolytic gas are the prime interest for power generation, the desired products can be obtained by varying the operational conditions:

i. To obtain a high yield of liquid fuel, low temperature with a high heating rate and short residence time is preferred
ii. To obtain the maximum yield of gaseous fuel, high temperature, long residence time, and low heating rate would be used

4.2.4 Liquefaction

Liquefaction, also known as hydrothermal or aqueous liquefaction, refers to a thermochemical process conducted under high pressure and temperatures, where biomass is decomposed and transformed to liquid fuel under specific conditions involving sub or supercritical water, as well as occasional organic solvents and/or catalysts

[31]. The flow diagram for hydrothermal liquefaction is shown in Fig. 4.3. The standard conditions for hydrothermal liquefaction involve temperatures ranging from 250 to 375 °C and operating pressures between 4 and 22 MPa. A closely related process occurs under supercritical water conditions, where temperatures exceed 374 °C and pressures surpass 22 MPa [32]. At temperatures and pressures near the critical point, water undergoes significant changes in its properties, such as density, dielectric constant, permittivity, and polarity. Despite being highly compressed, water remains in a liquid state, but exhibits a relatively higher ionic product (hydroxyl OH^- and hydronium H^+) and lower permittivity compared to ambient conditions [33]. Here, the strength of hydrogen bonds in water weakens, leading to an increase in the solubility of non-polar organic compounds. At such conditions, various reactions take place, including depolymerization, bond breaking, rearrangement, and decarboxylation. These reactions effectively break down the solid structure of biomass, resulting in the formation of oily components called biocrude. Additionally, it also generates an aqueous phase containing accumulated carbon, nitrogen, and phosphorus (C/N/P) compounds, along with a minor amount of gas [34].

Liquefaction is an efficient thermochemical method for processing biomass with high water content, unlike pyrolysis, which requires costly pre-treatment steps to remove moisture from feedstock. This process remains unaffected by the water content level and the specific type of biomass, resulting in high conversion rates and the production of relatively pure products [35]. A significant advantage of utilizing water as both a reactant and reaction medium in biomass conversion is the elimination of the expensive drying step for wet feedstocks. This characteristic makes water a promising component for biomass liquefaction [36]. Compared to pyrolysis, the liquefaction process produces biofuel of higher quality with a higher heating value due to lower levels of oxygenated compounds. Although water, CO_2, and CO are generated during liquefaction through decarbonization, dehydration, and

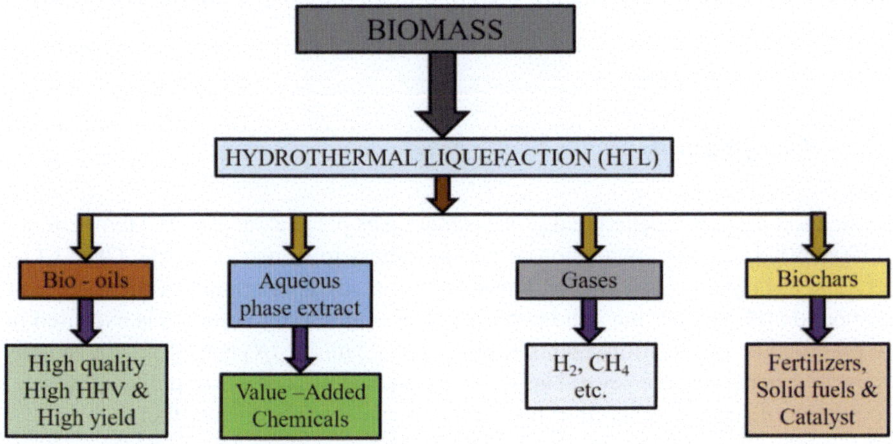

Fig. 4.3 Hydrothermal liquefaction flow diagram [39]

decarboxylation reactions that remove some oxygen, a significant amount of oxygen remains. Consequently, despite its superior quality, the biofuel obtained from liquefaction exhibits instability, low heating value, and high viscosity and acidity [37]. The heavy oil generated from the liquefaction process exhibits insolubility in water and possesses a viscous nature, which can occasionally present challenges during handling. However, this issue can be mitigated by incorporating organic solvents such as propanol, butanol, acetone, etc., into the reaction system. Additionally, the presence of solvent-reducing gases like carbon monoxide (CO) or hydrogen (H_2), along with a catalyst, is necessary in conjunction with the biomass. The inclusion of a catalyst not only lowers the required liquefaction temperature, but also enhances reaction kinetics, ultimately leading to improved oil yield during the liquefaction process [38].

The properties of the biofuel depend on factors such as the biomass type and operating conditions. Distinct varieties of biomass exhibit diverse reaction patterns and display varying responses to process conditions, including heating rate, temperature, pressure, and pH [40]. Typically, biomass with moisture contents below 90 wt% did not show significant changes in biocrude yield. However, when the moisture contents exceeded 90 wt%, variations in biocrude yield exhibited different tendencies, potentially influenced by the specific biomass type and reaction conditions. Moreover, the impact of moisture content was closely linked to the pressure since pressure alterations influenced the effective concentration of biomass by modifying the phase behaviour of water [34]. The heating rate is an important factor to consider, along with the temperature and duration of operation. For lower subcritical temperatures, a slower heating rate is preferable, while higher subcritical and supercritical temperature operations benefit from a faster heating rate. When operating at high temperatures, a slow heating rate can lead to the formation of unwanted by-products like char, as intermediates and bio-oil fractions repolymerize or carbonize. Conversely, processes with high heating rates can cause secondary cracking of long chain hydrocarbons and aromatics, resulting in the production of gaseous products [41].

4.2.5 Gasification

Gasification is a thermochemical process that takes place at an elevated temperature of 550–950 °C for the conversion of solid biomass to syngas (CO, H_2, CH_4, CO_2, H_2O, and light hydrocarbons), condensable tars, nitrogen compounds (NH_3), sulphur compounds (H_2S), and fine solid particles. This endothermic process is executed in the presence of gasification agents such as air, steam, or oxygen [42]. The syngas produced can be used for several applications such as for combustion in gas engines, production of second-generation biofuels and valuable chemicals, fuel cells, direct firing, and co-firing in existing coal-fired boilers.

Gasification is a complex process that can be segregated into four different zones: drying, thermal decomposition or pyrolysis, combustion/partial combustion, and

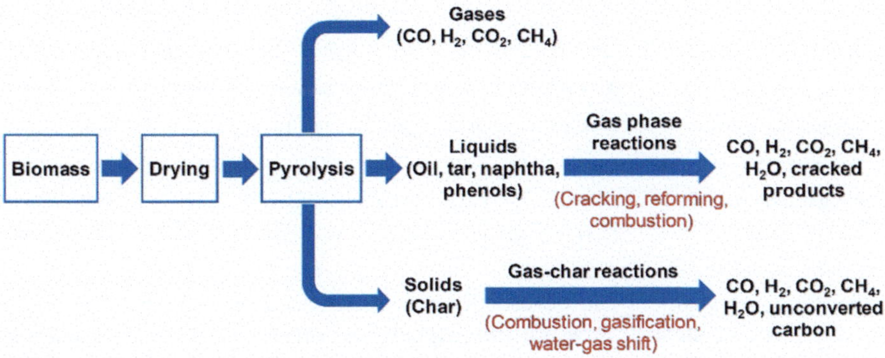

Fig. 4.4 Schematic of the gasification process showing the various zones and products [19]

reduction zone [19] as depicted in Fig. 4.4. In the drying zone, the moisture present in the biomass is removed at about 100–200 °C. Normally, solid biomass with a typical moisture content of 10–20% is recommended for gasification. Higher moisture content results in considerable energy loss and deteriorates the quality and yield of syngas. The dried biomass then undergoes pyrolysis resulting in the formation of solid biochar, bio-oils, and gases. The pyrolysis process initiates with the decomposition of hemicellulose molecules within 150 to 270 °C, forming vapours, char, and tar. Cellulose molecules present in the biomass degrade at 275 to 350 °C. The lignin present in the biomass decomposes between 250 and 500 °C to produce aromatics mainly solid char. However, above 500 °C, the yield of gaseous products accelerates with a negligible yield of tar.

In the combustion zone, exothermic reactions take place in the range of 900–1200 °C. The type and amount of gasifying agent usually control the combustion zone temperature preventing it from attaining the ash melting temperature [43]. The combustion of biomass results in the formation of exhaust gases (CO_2, H_2O, CO, and H_2) along with a sufficient amount of heat which is utilized for partial drying and pyrolysis. In the reduction zone, the unburnt char and pyrolysis gases along with the tar particles are reduced between 600 and 1000 °C. The reduction reactions are endothermic in nature and hence the temperature of this zone consequently decreases. At the end of this zone, the char and the majority of the tar are consumed producing syngas and ash.

The reactions taking place during the gasification process are shown below.

Oxidation reaction:

$$C + O_2 \rightarrow CO_2,\ \Delta H = -406\ \text{kJ/mol} \tag{4.3}$$

$$C + \frac{1}{2}O_2 \rightarrow CO,\ \Delta H = -111\ \text{kJ/mol}$$

$$H_2 + \frac{1}{2}O_2 \rightarrow H_2O,\ \Delta H = -242\ \text{kJ/mol}$$

Steam reforming:

$$CH_4 + H_2O \rightarrow CO + 3H_2,\ \Delta H = +206\ \text{kJ/mol} \tag{4.4}$$

Water gas reaction:	$C + H_2O \rightarrow CO + H_2,\ \Delta H = +131\ kJ/mol$	(4.5)
	$C + 2H_2O \rightarrow CO_2 + 2H_2,\ \Delta H = +97\ kJ/mol$	
Boudouard reaction:	$C + CO_2 \rightarrow 2CO,\ \Delta H = +172\ kJ/mol$	(4.6)
Water gas shift reaction:	$CO + H_2O \rightarrow CO_2 + H_2,\ \Delta H = -41\ kJ/mol$	(4.7)
Methanation reaction:	$C + 2H_2 \rightarrow CH_4,\ \Delta H = -75\ kJ/mol$	(4.8)
Tar reforming:	$C_xH_y + xH_2O \rightarrow xCO + \left(x+\frac{y}{2}\right)H_2$	(4.9)

The types of gasifiers can be broadly classified based on the gasifying agent (air, O_2, CO_2, or steam), pressure (atmospheric or pressurized), fluid dynamics, temperature (slagging or non-slagging), and heat supply (direct or indirect heating using heat exchanger) [44].

On the basis of fluid dynamics, gasifiers are categorized into fixed bed, fluidized bed, entrained bed, and plasma gasifiers. Fixed bed gasifiers are the oldest and most widely used gasifiers on the commercial scale owing to their simple design and ease of operation. The typical syngas composition obtained from a fixed bed gasifier is 10–15% CO, 15–20% H_2, 3–5% CH_4, 10–15% CO_2, and the rest is N_2 [45] with gross calorific value lying within 5–10 MJ/Nm^3. Based on the contact between solid fuel and gasification agent, fixed bed gasifiers can be categorized as downdraft, updraft, and cross-draft gasifiers.

Downdraft gasifier, as shown in Fig. 4.5a, operates with top feed for biomass, and the gasification agent (air) is fed at the middle of the gasifier near the combustion zone. Biomass dries in the drying zone, pyrolysis into char and gases, followed by combustion with the liberation of exothermic heat, and finally enters the reduction zone [19]. The updraft gasifiers have a configuration with downward biomass flow due to gravity and bottom entry for the gasifying agent, as depicted in Fig. 4.5b. These counter-current gasifiers have a combustion zone at the base, heating the gasifying agent to 750 °C. The hot gases flowing upwards are reduced near the reduction zone and then rise for pyrolysis and drying, yielding higher thermal efficiency. However, as the gas passes through the unburnt fuel, it carries minimal ash content but contains tar and water vapour. Cross draft gasifier (Fig. 4.5c) has side air inlet and top biomass feed. Syngas exits through the nozzle exactly opposite the gasifying agent inlet. As the gasifying agent boosts the combustion reactions, it operates at a very high temperature with close combustion and reduction zones, diminishing thermal efficiency and gas quality for commercial use.

Fluidized bed gasifiers provide uniform temperature distribution in gas–solid mixtures, enhancing reactions and minimizing agglomeration risk. The commonly used bed materials are silica sand, dolomite, glass beads, olivine, calcite, etc., for the gasification of a wide range of biomass feedstocks. Usually, fluidized bed reactors are operated below the ash and bed material melting temperature, i.e., around 650–950 °C. The efficiency of carbon conversion is significantly high (up to 95%) in these gasifiers [46]. It can be broadly classified into bubbling fluidized bed (BFB), circulating fluidized bed (CFB), and dual fluidized bed (DFB) based on reactor configuration, superficial velocity, and circulation of bed material. In bubbling fluidized bed (BFB)

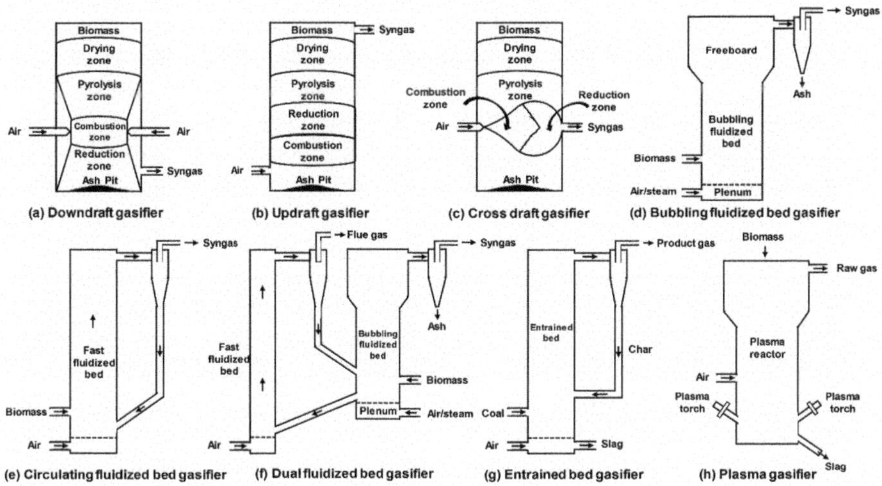

Fig. 4.5 Schematic of different types of gasifiers [19]

gasifiers, as represented in Fig. 4.5d low-velocity gasifying agent is supplied at the bottom of the bed made of mainly silica sand. Fluidization creates rigorous percolating motion of the bed materials at the lower dense region due to the presence of more solids. As the particles move upwards, the drag force and gravity counterbalances, and after attaining a certain height, particles start to fall back to the dense bed [47]. The syngas exits from the top and is separated from ash and bed material via cyclone. Active catalytic bed aids tar conversion, making them efficient for high-ash biomass with good fuel flexibility. In case of a circulating fluidized bed (CFB) gasifier, as shown in Fig. 4.5e, the bed material escapes the combustor due to its high fluidization velocity. Cyclone separates syngas and solids; entrained solid returned to combustor via downcomer for char circulation. The fast fluidization condition increases heat and mass transfer characteristics, circulation of char, and prolongs the residence time, facilitating the required reduction reactions and minimizing char loss [48]. CFB yields better quality syngas having higher calorific value operating at a superficial gas velocity of about 3–7 m/s. However, the higher superficial velocity creates problems such as severe back mixing, solid buildup, or agglomeration if moisture is sufficiently high in the feedstock [49]. DFB gasifier employs two reactors: a bubbling fluidized bed gasifier (700–870 °C) operated with air or steam and a circulating fluidized bed riser (850–1000 °C) operated with air [50]. The feedstock is supplied to the BFB gasifier and reacts with limited air and steam to produce char, tar, and syngas. The solid char generated in the gasifier is transported through the loop seal to the riser along with the bed materials where it is combusted in the presence of excess air. The bed material absorbs the exothermic heat and is circulated to the gasifier via a cyclone. Bed material is continuously circulated between these two zones to provide the endothermic heat required for gasification. The tar formed within the gasifier reacts with the excess steam and decomposes producing lighter hydrocarbons and gas. The calorific value of syngas from steam gasification is normally 12–20 MJ/Nm^3 [51]. The configuration of the DFB gasifier is shown in Fig. 4.5f.

Operating with turbulent flow, high temperature (around 1300 °C), and pressure, with short residence time (~1–2 s), entrained bed gasifiers ensure swift conversion and large-scale suitability. The solid biomass and oxidant are fed simultaneously from the top and achieve a high throughput. Entrained bed gasifiers, as shown in Fig. 4.5g, have the versatility to handle a wide variety of feedstocks and can produce clean tar-free syngas. Plasma gasification employs thermal plasma generated through a plasma torch to decompose waste like municipal solid waste (MSW). The thermal plasma dissociates the electrons from the molecules of the gasification agent resulting in a stream of ionized plasma [52]. The high-temperature plasma disrupts feedstock's chemical bonds, forming active species and converting organic matter to H_2 and CO-rich syngas, as shown in Fig. 4.5h.

In the past 15 to 20 years, gasification technology in terms of Integrated Gasification Combined Cycle (IGCC) plants and Integrated Gasification Fuel Cell Combined Cycle (IGFC) technologies have started to be used by the power industry. However, it should be noted that the highest theoretical thermal efficiency achievable with conventional IGCC/IGFC systems remains below 52% [53].

4.2.6 Combustion and Co-generation

Combustion is a chemical process that involves an exothermic reaction between a fuel and oxygen present in air resulting in the formation of flue gas (CO_2, CO, water vapour, etc.) and liberation of energy in the form of heat. It is one of the ancient processes that is used in many rural areas worldwide for cooking and space heating using fuel wood or any other solid biomass. The heat released during this process is a result of the conversion of the chemical energy stored in the fuel into thermal energy. The combustion process is widely utilized in power plants, industrial processes, and transportation. In power plants, fossil fuels are burned to produce high-pressure steam, which is then used to drive steam turbines and generate electricity. In industrial processes, combustion is employed for heating, melting, or chemical reactions. In transportation, internal combustion engines burn fuel to propel vehicles. The efficiency of combustion processes can vary depending on several factors, including the type of fuel, combustion technology, and system design. Efforts are being made to improve combustion efficiency by adapting the cogeneration of electricity and heat, reducing emissions, and exploring alternative fuels to mitigate the environmental impact of combustion. The general equation for the combustion of a fuel is given as follows:

$$C_xH_y + O_2 \rightarrow CO_2 + H_2O + Heat \tag{4.10}$$

Unlike conventional power generation, which typically wastes a significant amount of heat energy, cogeneration systems capture and utilize the waste heat, thereby increasing overall energy efficiency. The waste heat produced during electricity generation is recovered and used for various heating, cooling, or industrial processes. This heat can be utilized for space heating, domestic hot water production, district heating systems, or industrial applications like drying or steam generation [54].

4.3 Power Generation Routes Through Thermochemical Conversion

The selection of a thermochemical conversion process for power generation depends upon various parameters including the type and availability of the feedstock, geographical location, climatic conditions, and the desired outputs. The desired output from the process may be heat, electricity, value-added products, or secondary fuel. Therefore, based on these criteria, proper selection of the process has been usually performed [55]. For example, if we need to generate electricity primarily using wood or agricultural residue, the better option would be a combination of processes, torrefaction followed by gasification. On the other hand, if bio-oil is the target product along with heat from a relatively high moisture feedstock, pyrolysis would be the proper thermochemical process.

The thermochemical conversion process produces diverse combustible fuels, which can be employed for producing power utilizing a range of conversion technologies. Cogeneration is the simultaneous production of combined heat and power (CHP) using the same fuel source. The energy produced through thermochemical conversion processes like combustion and gasification can be used for cogeneration, as it has immense potential due to its inherent properties. Combustion has become a firmly established and commercially viable method for biomass conversion. On the other hand, biomass gasification, being more complex than combustion, has not yet achieved full commercial status due to the difficulties in handling biomass as a fuel, especially when compared to conventional fossil fuels [56]. Biomass gasification for CHP generation is a viable technique, particularly for small-scale plants having electric power generating capacity below 10 MW, giving an appealing alternative to biomass combustion [57]. The syngas obtained from biomass gasification can be a fuel source for generating electricity and heat through various means, including internal combustion (IC) engines, gas turbines, Stirling engines, and even fuel cells [58].

4.3.1 Steam Cycle

The steam turbine is the main prime mover of vapour-dominated power plants, which converts the thermal energy of steam to mechanical energy in the form of shaft work. Steam generation can be performed by the combustion of fossil fuel or different biomass in the boiler. The steam turbine operation is based on the principle of the Rankine cycle. Steam turbine applications are widely found in combined heat and power systems, where steam at lower pressure is extracted for direct use. Steam turbines are classified broadly into impulse and reaction turbines. In the impulse turbine, the high-velocity steam expands in the nozzle and strikes the blades on the fixed blade of the turbine at a constant pressure. On the contrary, in a reaction turbine, the steam expands as it flows over the blades of the turbine. The configuration of the blades can be made in such a way that the blades act as a nozzle [59]. Advancements in steam turbine technology have been driven by the need for increased efficiency, flexibility, and sustainability in power generation. Reheat and regenerative steam turbines have been developed to improve the overall efficiency of the power cycle. The reheat cycle utilizes multiple stages of steam expansion, whereas regenerative cycles incorporate feedwater heaters to preheat the water prior to entering the boiler [60]. A condensing turbine is also designed which optimizes energy extraction from steam by condensing exhaust, boosting efficiency, and creating a vacuum. In contrast, a non-condensing turbine releases steam into the air without condensing, sacrificing efficiency but being useful when utilizing waste heat is advantageous [59].

The biomass-fired steam generators are the widely used technology, as shown in Fig. 4.6. Over the years, biomass feedstock is fed to the boiler, along with air and then water is circulated through the boiler tubes. Rice husk, groundnut shells, firewood, coconut, and other agro waste and municipal solid waste (MSW) are some of the most used biomass in case of biomass-based steam power plants [61]. Steam turbines are employed mostly for a biomass-based power generation range from 50 kW to 250 MW [62]. The electrical efficiency of biomass steam Rankine cycle depends on the installed capacity of the plant and is usually in the range of 15–35% [63]. The biomass-based steam generation can be also applied to the organic steam Rankine cycle where the operating temperature is less than 300 °C [63].

Another technological advancement to increase the thermal efficiency in the steam cycle is the use of the supercritical Rankine cycle. With this cycle, an enhanced efficiency of 2–3% has been observed along with lower emission compared to that of the subcritical Rankine cycle. To further enhance the efficiencies at lower loads, an ultra-supercritical Rankine cycle has been employed where the temperature and pressure range is usually 600–700 °C and 240–310 bar, respectively [64]. The practical supercritical and ultra-supercritical plants are mostly based on coal thermal power plants. More technological interventions are required for the corrosion and ash handling issues with the use of biomass in supercritical steam cycles.

Steam turbines have been adapted to accommodate biomass co-firing in conventional coal-fired power plants [65]. The potential of hybrid system in power generation is very promising in terms of efficiency, performance, and flexibility in operation.

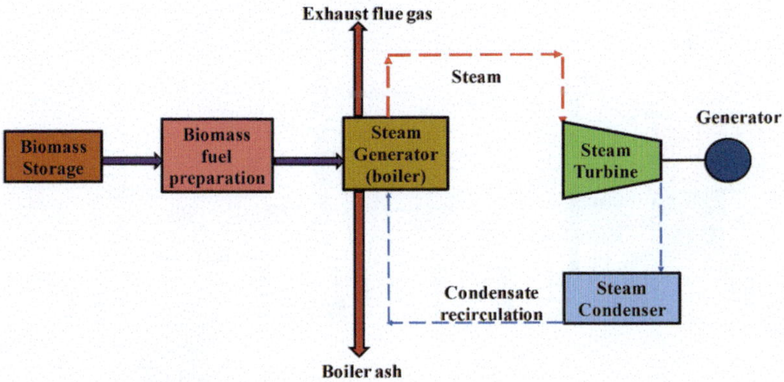

Fig. 4.6 Biomass feed thermal power plant [55]

The hybrid system comprises of conventional regenerative Rankine steam cycle with some renewable energy integration like solar, wind, and biomass. A hybrid system with a solar parabolic collector and biomass combustion system for steam generation results in improved plant fuel efficiency [66]. Overall, advancements in steam turbine technology have significantly improved the efficiency, flexibility, and sustainability of power generation, making steam turbines a vital component in various energy systems worldwide.

4.3.2 Gasifier-IC Engine Combined Technology

Among the different prime movers, Internal combustion (IC) engines are the preferred choice for small-scale biomass gasification CHP plants due to their numerous advantages over alternative power generation units, such as low capital cost, modularity, reliability, good part-load performance, and high operating efficiency [58]. Utilizing syngas obtained from biomass gasification in conjunction with an IC Engine is a feasible technology for generating heat and power on a small scale [67]. Syngas can be utilized in two ways: as a single fuel mode in spark ignition (SI) engines and as a dual fuel mode in compression ignition (CI) engines. When combined with an IC engine, the gasifier system can achieve an electrical energy conversion efficiency ranging from 35 to 45% [68]. A schematic diagram of a gasifier system coupled with the engine is shown in Fig. 4.7.

Gasifier-IC engine power generating system encompasses several components, such as a biomass feeder, gasifier, gas cooling and cleaning system, an engine, gaseous fuel injection system, and generator, to function effectively. In this plant, the gasifier converts the biomass (e.g., wood chips) to syngas. After the gas-conditioning process at the plant, the syngas is employed as fuel for a gas engine coupled with a generator, which generates electricity for the grid and heat. If the engine remains

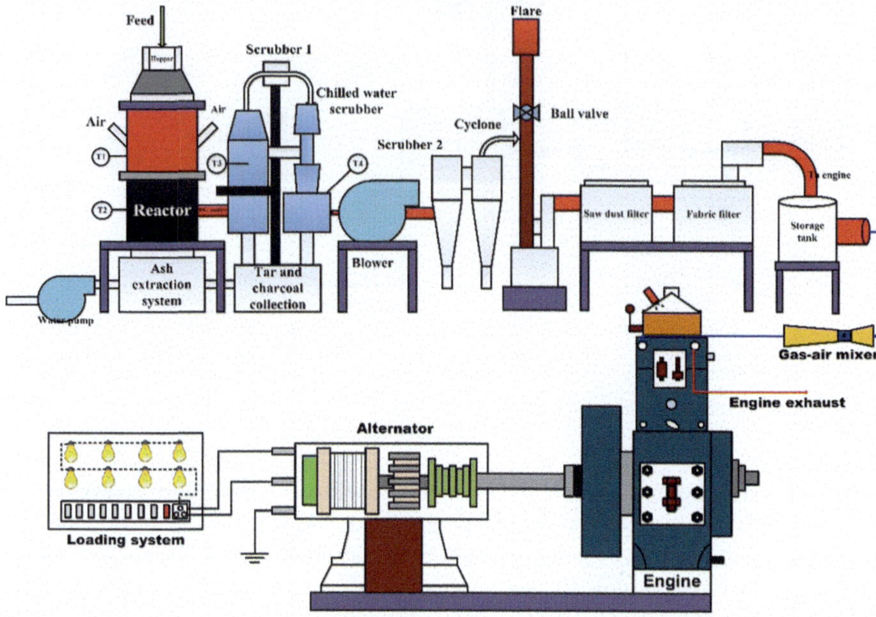

Fig. 4.7 Schematic diagram of gasifier-IC engine power generating system

inactive, the syngas can be directed to a boiler, which is burned solely to produce heat [57]. However, the waste heat (WH) source or utility area may differ based on the requirement. A biomass gasifier-IC engine-generator integrated system for power production and utilization of waste heat to heat biomass feedstocks using the syngas and exhaust gas from the IC engine was studied by Li et al. [69], as depicted in Fig. 4.8. When using redwood pellets and woodchips, the combined electricity and waste heat efficiencies were 43.0–54.6% and 40.0–60.4% of the total energy input [69]. Similar investigations were also carried out by the authors Soares and Oliveira [70] and HajiHashemi et al. [71].

Before utilizing syngas in the engine, it is crucial to thoroughly understand the engine's operation to ensure the successful implementation of the system. To achieve optimal performance of syngas in the IC engine, the biomass should have a density greater than 300 kg/m^3, moisture content of up to 20%, particle size ranging from 20 to 50 mm, and impurity levels in syngas should be reduced to acceptable levels (<50 mg/m^3) [72]. Additionally, an essential step in the process involves cleaning the gaseous fuel before introducing it into the engine. This is effectively achieved by employing the wet scrubbing method, wherein the gas is cooled to temperatures below 150 °C before passing to the scrubbing stage. Wet scrubbing stands as a proven cleaning technology, known for its ability to eliminate alkali metals, ammonia, particulates, and tar from gaseous fuel. To ensure the maximum retention of the gaseous fuel, it is preferable to incorporate either catalytic or thermal cracking of the tar prior to the wet scrubbing stage [73]. The distinct advantage of syngas over other combustible

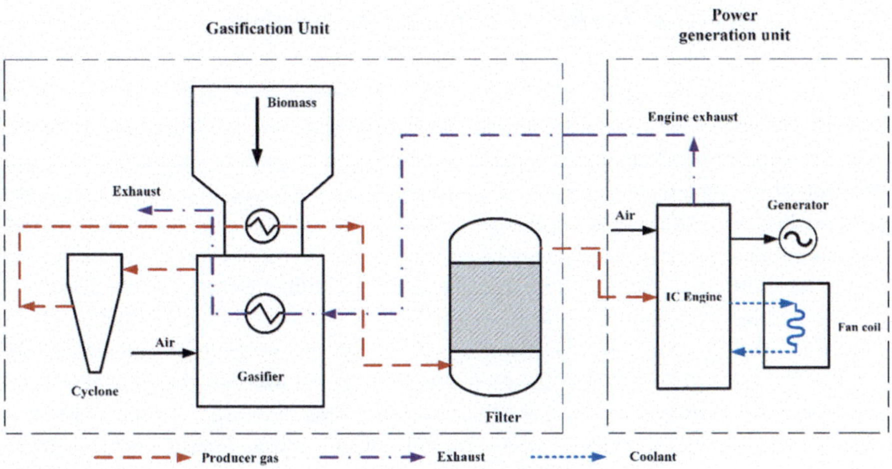

Fig. 4.8 Schematic of the gasifier–engine–generator system [69]

gases lies in its superior knocking behaviour, making it well-suited for IC engine applications. This advantage stems from the presence of hydrogen (H_2) in syngas, resulting in an elevated laminar flame velocity that combats knocking tendencies effectively. Furthermore, the existence of inert gases like CO_2 (12–15%) and N_2 (48–50%) in syngas serves to suppress pre-flame reactions and effectively control engine knock [74, 75]. Additionally, higher concentrations of H_2 in syngas lead to heightened thermal efficiency, while simultaneously curtailing the emission of unburned CO and HC through the substitution of carbon particles within the fuel mixture. Despite its benefit, the higher combustion flame temperature attributed to H_2 content results in amplified heat release rate and cylinder pressure culminating in the emission of higher levels of NO_X. Conversely, the existence of CO in syngas acts to reduce the adiabatic flame temperature, thereby regulating the release of NO_X pollutants. Moreover, a greater fraction of CO_2 heightens the specific heat capacity of the fuel mixture, contributing to a reduced in-cylinder combustion temperature and, consequently, lower NO_X emissions [76, 77].

India has implemented a diverse array of gasifiers in rural areas, spanning from 20 to 500 kW capacities, to meet the energy needs at the village level. In an effort to bridge the gap between electricity demand and supply, the Indian government is adopting new policies and actively promoting the use of renewable energy technologies. This increasing focus is also fostering the adoption of biomass-based technologies for rural electricity generation. One of the significant challenges faced by this technology in the market is the limited availability of engines specifically designed for syngas, leading to system de-rating when employed in this context. Nevertheless, optimizing engine performance can be accomplished through adjustments in the injection system. Furthermore, an effective alternative involves utilizing a mixture of syngas and biogas, obtained from anaerobic digestion, as a fuel source for the engine.

4.3.3 Gasifier-Gas Turbine Combined Technology

The gas turbine is renowned for its exceptional efficiency, minimal emissions, ease of operation, and reliability, all achieved at a relatively low capital cost [78]. One of the recognized approaches to thermochemical power generation is through the utilization of gas turbine technology. Utilizing biofuels to power gas turbines presents a viable solution to address the challenges posed by dwindling fossil fuel reserves and environmental concerns. Direct firing of gas turbines with biomass is not feasible due to potential damage to the turbine blades caused by combustion products. However, the syngas generated from the biomass gasifier can be used in a gas turbine (GT) cycle after the cleaning and conditioning of the syngas to generate electricity as shown in Fig. 4.9. The exhaust gases from the GT are used in the heat recovery steam generator (HRSG) to generate the required thermal energy for the industrial process [79]. Further, the exhaust gases from HRSG are utilized in the biomass dryer to reduce the moisture content of the feedstocks (e.g., sugarcane baggage) [80]. It can be observed that to meet the straightening requirement of a gas turbine, the tar concentration has to be <100 mg/Nm^3. The biomass gasification process followed by utilization of syngas in the GT can provide an electricity generation efficiency of approximately 50%, whereas, for steam turbines, it is below 20% [81]. There are some limitations in this system as the gasification process produces syngas with impurities and generates tar which may damage the blades of the gas turbine and release volatile organic compounds which is harmful to the environment.

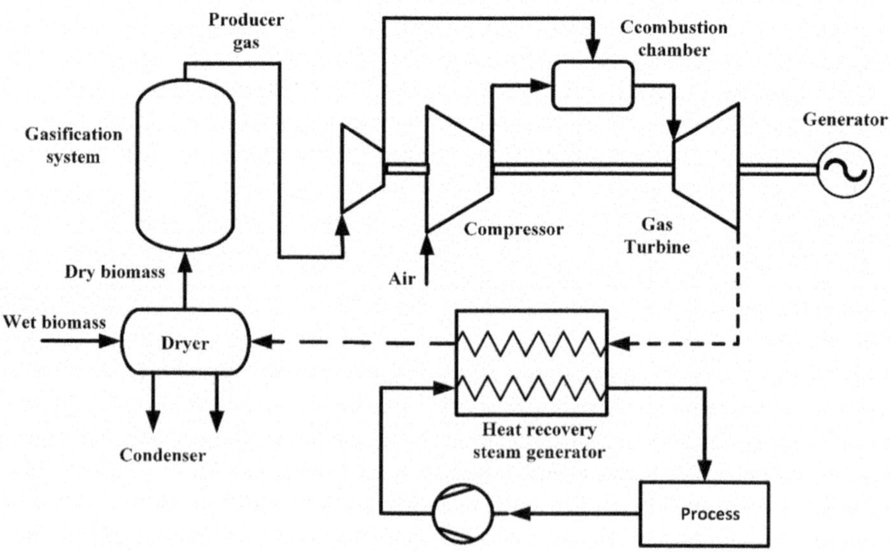

Fig. 4.9 Gasifier-gas turbine CHP system [80]

4.3.4 Gasification-Fuel Cell Combined Technology

A fuel cell is an electrochemical device that consistently converts chemical energy into electric energy, along with some heat, as long as a fuel supply and oxidant is maintained. In this sense, fuel cells share similarities with batteries, as they both generate power through electrochemical processes, and with engines, which continuously consume fuel for operation. However, there are significant differences. Unlike engines or batteries, fuel cells do not require recharging, operate silently and efficiently, and when hydrogen is used as fuel, they only produce electricity and water. As a result, fuel cells are often referred to as zero-emission engines [82].

Since the early 1990s, there has been substantial research interest worldwide in combining gasifiers with high-temperature fuel cell systems like Molten Carbonate Fuel cell (MCFC) or Solid Oxide Fuel Cell (SOFC) to develop power generation systems with exceptionally high efficiencies and minimal emissions. Figure 4.10 depicts the top-level block flow diagram for a large-scale coal-based SOFC power plant, comprising three main sections. The gasification section contains a coal gasifier and supporting sub-systems/components. To facilitate an O_2 blown gasifier, an air separation unit supplies oxygen. The resulting raw syngas undergoes particulate removal and then enters the gas clean up and CO_2 sequestration section. In this phase, sulphur, mercury, and other impurities are eliminated from the fuel gas, preparing it for feeding into the downstream SOFC power module. Crucial to this section is the CO_2 separation sub-system, responsible for removing enough CO_2 to enable the necessary carbon capture from the syngas. The by-products from the gas clean up section are captured CO_2 for sequestration and sulfur removed from the syngas [83].

The purified syngas become the fuel source for power generation within the SOFC power section. The fuel cell operates at atmospheric pressure, while its waste heat is utilized in a combined cycle, potentially featuring a gas turbine, steam turbine, or both. Additionally, supplemental power is generated through the expansion of high-pressure syngas using a fuel gas expander, upstream of the atmospheric pressure

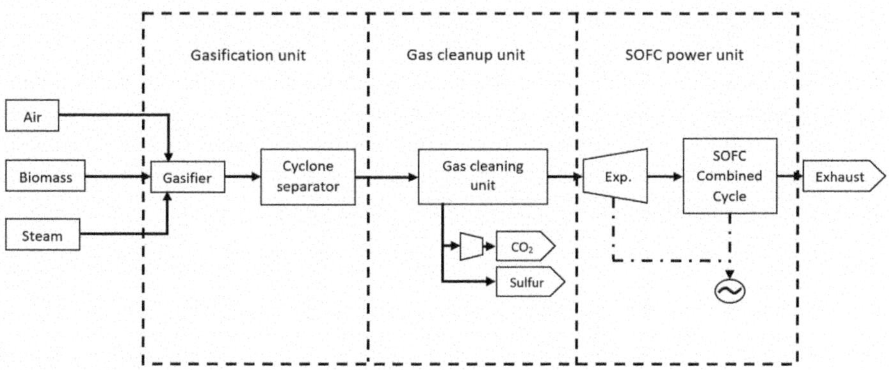

Fig. 4.10 Flow diagram for a large-scale coal-based SOFC power plant [83]

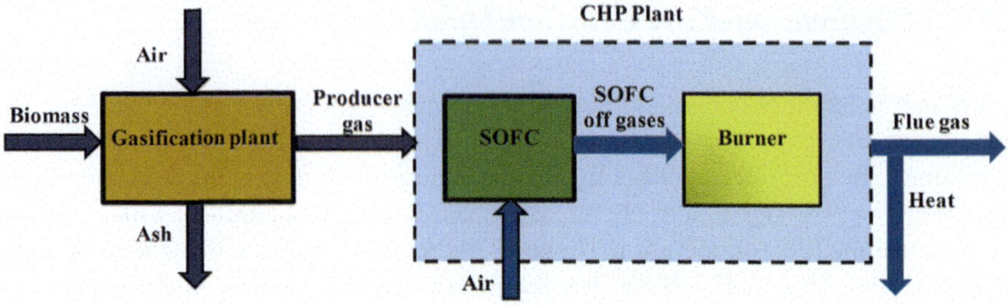

Fig. 4.11 Cogeneration plant using a gasifier and fuel cell [85]

fuel cell stack [84]. The power plant demonstrates excellent integration, ensuring high electrical efficiency. Notably, thermal integration is achieved, allowing waste heat recovery and its utilization for generating the process steam required by the plant. This close coupling among the gasification, clean up, and SOFC Power islands exemplifies the plant's well-integrated design.

Various studies reported power and heat generation potential using biomass gasification systems and fuel cells. A typical plant set-up for CHP using a gasifier and fuel cell is presented in Fig. 4.11. This type of plant is grouped into two sections, viz., biomass gasification and CHP plants. In the gasification process, the biomass is transformed into syngas, subsequently utilized in the CHP plant to generate electric power and heat by the fuel cell (like a SOFC) and a burner [85]. Several researchers consider SOFC the most appropriate for integrating biomass gasifiers. Biomass gasification and fuel cell integration appear to be favourable choices for generating CHP using solid biomass, including high efficiency and minimal environmental impact [86].

Renewables and fuel cell technology are fast-growing sectors, with efforts focused on establishing a sustainable connection between them for power generation. In the coming years, there will be a significant increase in biogas fuel cell installations worldwide, along with active research on wood and biomass gasifiers for H_2 production. Some locations will utilize wind power for H_2 generation, while others will use solar to power small vehicle fleets [86]. Industries and businesses with a need for heat and power, especially those susceptible to power interruptions, will operate additional stationary power plants fuelled by natural gas. In the residential sector, the initial focus will be on powering homes using natural gas-based fuel cell systems. Due to the liberalized power market, power outages have become more frequent, leading to increased demand for fuel cell-based backup systems and combined heat and power systems to provide reliable power supply during such occurrences.

4.3.5 *Power Generation from Torrefaction, Pyrolysis, and Liquefaction Process*

When co-firing raw biomass in coal-fired power plants, significant challenges arise, including the generation of a substantial volume of flue gas, the potential for rapid and uneven combustion of volatiles leading to unstable flame propagation, non-uniform heat distribution, and the need for a larger feeding system due to the comparatively low energy density of the fuel [87]. Torrefaction process value adds to the raw biomass to be used as a replacement for coal in power industries. Torrefied wood is a promising energy source with properties similar to coal in terms of energy content and density, lower ash content, and is more environmentally friendly. Also, due to the higher energy density of torrefied biomass, handling and shipping costs are less as compared to fossil fuels, which makes torrefied biomass more profitable [88]. In the case of combined heat and power (CHP) plants, torrefied biomass is used for co-firing with coal without any change in the plant design [89]. Studies showed that torrefied fuel, when pelletized, can completely replace coal in the power sector. In contrast, untreated chips and pellets can replace only 8–15% of coal in comparable scenarios [90].

Both bio-oil and biochar from pyrolysis and liquefaction processes can be used as a fuel for power production. Bio-oil offers versatile applications, including its use in boilers to generate heat and power, co-firing alongside natural gas or coal in power plants, blending with ethanol or gasoline, or undergoing conversion into fuels such as ethanol and diesel [91]. Biochar can be utilized by co-combusting in existing coal-fired power plants, directly burning in most systems to generate heat as a replacement for coal without modification, or in CHP plants to achieve clean heat and power generation. Biochar co-firing can lead to reduced ecological, climate, and resource-related impacts while simultaneously mitigating carbon emissions from thermal power plants [92].

4.4 Challenges of the Thermochemical Conversion Process

The challenges associated with thermochemical biomass conversion are diverse and demand careful consideration to ensure successful technological implementation. One of the primary hurdles in this process is the variability of biomass feedstock, which significantly impacts conversion efficiency. Various types of biomass require different process conditions for optimal conversion, complicating the process [93].

Maintaining precise reactor temperature and other process conditions during thermochemical conversion poses another significant challenge. Impurities and contaminants, such as ash, alkali, and alkaline earth metals in biomass, can lead to equipment fouling and corrosion, diminishing process efficiency. Additionally, the elevated oxygen content in biomass creates difficulties in its conversion and leads towards higher tar production [94]. Furthermore, thermochemical conversion processes often demand substantial external energy inputs to achieve the necessary temperatures and pressures. This reliance on non-renewable energy sources can reduce the overall plant efficiency and counteract the environmental advantages of biomass conversion.

During biomass pyrolysis and gasification, the formation of tar and particulate matter can occur as by-products. These substances can obstruct equipment, compromise gas quality, and pose environmental and health hazards. Effective management of the resulting ash, a product of mineral content in biomass, is crucial to prevent environmental pollution. The economic viability of thermochemical biomass conversion presents another challenge. Initial capital costs for specialized equipment can be prohibitive, and competition from fossil fuels and other renewable energy sources can affect project feasibility. Despite these challenges, ongoing research and development endeavours are actively addressing these issues, striving to enhance the efficiency, cost-effectiveness, and environmental friendliness of thermochemical biomass conversion technologies. As technology advances and supportive policy frameworks evolve to favour sustainable energy sources, the outlook for biomass conversion as a clean energy solution continues to improve.

4.5 Government Initiatives for Biomass-Based Power Generation

Policy support emerges as the key catalyst behind fostering sustainability and success of the biofuel sector, especially in its early developmental phase. Given its current position, this backing is essential due to the industry's existing challenges, the elevated cost of production, inadequate supporting infrastructure, and immature technology, making it less competitive than conventional energy sources. But, over the past decade, the biofuels sector has witnessed remarkable growth, primarily due to proactive government policy interventions across the globe. Regarding the biofuel policies adopted by different nations, Brazil, the United States, and the European Union were among the pioneers in recognizing the significance of biofuels and the spotlight has now turned to Asian nations [95].

Over the years, India has placed a significant emphasis on energy policy, centred on renewable sources, including biomass and bioenergy. The inception and establishment of renewable energy policies in India, which encompassed biomass energy, can be attributed to the oil crisis that unfolded during the 1970s. In line with this background, the Government of India formed the Fuel Policy Committee (FPC) in 1974, and subsequent advancements were made in 1979 with the inception of the Working Group on Energy Policy (WGEP). The core responsibility of WGEP is to devise a robust plan and propose policy measures addressing both conventional and non-conventional energy aspects [96]. In 1981, the Commission for Additional Sources of Energy (CASE) was established in this regard, exclusively engaged with the implementation and deployment of policies. Then in 1982, CASE was revamped into the Department of Non-Conventional Energy Sources (DNES) under the Ministry of Energy, mandated with the crucial role of propelling the adoption and advancement of non-conventional energy sources throughout India. Later in 1992, DNES was upgraded to a ministry, named Ministry of Non-Conventional Energy Sources (MNES). This progressive action was undertaken with the goal of enlarging and exploring the avenues for promoting renewable energy activities in India, through augmented financial and institutional support [97]. This process gained increased momentum after the establishment of the Ministry of New and Renewable Energy (MNRE) in 2006. MNRE since its inception, dedicated its efforts to the advancement of the bioenergy sector, addressing challenges associated with biomass management, including agricultural and municipal solid waste (MSW), and striving to unlock its full energy utilization potential.

With the year 2003 as a starting point, the Planning Commission of India introduced the National Biofuel Mission (NBM) aiming to confront socio-economic and environmental challenges. The mission was primarily centred on the gradual expansion and enhancement of biofuel feedstock plantations, oil processing, trading, research, and development [98]. Continuing the trajectory of policy evolution, the National Policy on Biofuel was announced in December 2009, specifically targeting the utilization of biomass energy resources, integrating advanced technology deployment, and ensuring sustainability as its primary mission [99]. Dedicated initiatives were undertaken to devise various policies aimed at nurturing biomass to bioenergy technologies. A number of policies in India have been put into practice to prioritize the development of biogas technology and the utilization of organic manure. On the other hand, numerous policies have been directed towards the promotion of biomass technologies through thermochemical conversion routes. In 2003, a programme related to biomass gasification was initiated with the core aim of electrifying isolated hamlets and villages. This programme was later revamped by MNRE in 2009, refocusing its scope on comprehensive biomass gasifier-based power facilities, featuring a 100% syngas engine and a boiler-turbine generator (BTG) [100]. In November 2017, a policy was introduced to emphasize the promotion of biomass pellets by co-firing a minimum of 5–7% biomass pellets, derived primarily from agro residue in coal-based power plants. This policy was revised in October 2021, obligating integration of biomass co-firing as a compulsory element for coal-fired thermal plants in India [101]. In December 2018, a scheme was notified focusing on facilitating the advancement of biomass-based cogeneration in sugar mills and industries. Within the framework of this scheme, central financial assistance (CFA) of INR 25 lakhs per MW and INR 50 lakh per MW was allocated to bagasse and non-bagasse based cogeneration projects [102]. To support the growth of briquette and pellet manufacturing, along with promoting biomass (excluding bagasse) based cogeneration in various industries across the nation, a scheme was released by MNRE, until March 2026. The scheme offers CFA of INR 9 lakh per metric tonne/hour (MTPH) of pellet/briquette manufacturing capacity and INR 40 lakh per MW of installed capacity for biomass (excluding bagasse) cogeneration initiatives [103]. Moreover, to encourage paddy straw palletization and mitigate stubble burning, Union Minister of Environment, Forest & Climate Change in October 2022 released a guidelines regarding a subsidy of INR 50 crore from Central Pollution Control Board (CPCB) to pellet manufacturers in Punjab, Haryana, Rajasthan, Uttar Pradesh, and Delhi [104].

The substantial biomass energy capacity in India serves as a vital catalyst in the journey towards achieving a sustainable energy distribution. However, the existence of issues like sintering, formation of tar, agglomeration, and the significant reliance on biomass contribute to noteworthy challenges to widespread commercial adoption.

4.6 Commercial Biomass-Based Power Generation in India

Commercial biomass-based power generation in India has experienced significant growth in recent years, with the sector witnessing substantial progress as of 2022. Presently, there are over 800 biomass power and bagasse/non-bagasse cogeneration projects operating across the country, aggregating to an impressive capacity of 10,205.61 MW. Among the leading states for biomass power projects are Chhattisgarh, Madhya Pradesh, Gujarat, Rajasthan, and Tamil Nadu. The Biomass IPP segment contributes 1,871.11 MW to the overall capacity and represents the capacity installed by independent power producers using biomass as the primary feedstock. The Bagasse Cogeneration segment, on the other hand, constitutes a significant portion of the installed capacity with 7,562.45 MW. Bagasse, a by-product of the sugarcane industry, is utilized to generate electricity and heat in cogeneration plants, making it a valuable renewable energy source. Finally, the non-bagasse cogeneration segment contributes 772.05 MW to the total capacity. This includes cogeneration plants that use non-bagasse biomass feedstock, such as agricultural residues and other organic materials, for power generation and heat production [105].

Hosahalli village in Karnataka took the lead by implementing the first decentralized power generation system utilizing a biomass gasifier. Another gasifier unit of 20 kW capacity was installed in Hanumanthanagara village. These two villages in Karnataka stand as exemplary models of effective decentralized power plants, particularly in utilizing gasifier-based systems with continuous monitoring at the village level. According to reports, the gasifier plant operating in the dual fuel (DF) mode, with a diesel replacement ranging from 70 to 85%, demonstrates a commendable 21% conversion efficiency in converting wood to electricity [106]. Another biomass gasifier-based power plant with an average capacity of 400 kW was installed on Chhota mollakhali island of Sundarbans, West Bengal, used mainly by domestic, commercial, and industrial users [107]. Moreover, a biomass gasifier plant in Vebra village, Araria, Bihar, with a connected load capacity of 60 kW, serves to power essential operations such as rice mills, wheat mills, and water pumps [108]. Similarly, in Gaiyari village, Araria, Bihar, the power generated from a biomass gasifier plant, with a connected load capacity of 125 kW, is distributed to 15 micro-enterprises, supporting their operations and fostering economic growth. A 9 kW capacity, biomass gasifier plant fully operated on syngas was installed in Pithoragarh, Uttarakhand [108]. Besides generating electricity, the project aims to proactively enhance water recharge, promote biodiversity regeneration, and elevate the overall quality of life.

Table 4.2 State-wise installed biomass-based power plants

State	Co-gasification of coal with biomass (in MW)	Cogeneration with bagasse/non-bagasse (in MW)	Total installed capacity (in MW)
Andhra Pradesh	2	14	64
Chhattisgarh	24	46	70
Gujarat	17	21	38
Karnataka	20	47	67
Madhya Pradesh	22	39	61
Maharashtra	16	36	52
Punjab	11	14	25
Rajasthan	20	28	48
Tamil Nadu	15	27	42
Uttar Pradesh	17	40	57
Others	2	10	12
Total	**140**	**230**	**370**

Again, the power generated from a 500 kW gasifier plant in Gosaba Island, West Bengal is utilized for various purposes such as street lights, school lighting, drinking water supply, irrigation, and powering small-scale industries, benefitting around 1113 households [109]. Table 4.2 shows the state-wise installed capacity of co-gasification plants of biomass with coal and cogeneration plants with bagasse/non-bagasse as biomass feedstock.

4.7 Economic Feasibility

The economic feasibility analysis of biomass power plants is crucial for assessing the financial viability of the project and making informed decisions regarding investments, resource allocation, and potential revenue streams. The economic evaluation of any power plant is determined through total capital investment, operating cost, and revenue generation from the electricity produced by the power plant. Table 4.3 explains the different costs associated with biomass power plants. The effective tools for assessing the economic feasibility of biomass-based power plants are discussed in the next section.

- **Life cycle cost analysis (LCC)**: LCC analysis provides a comprehensive framework for comparing the economic viability of bioenergy technologies against conventional energy technologies. LCC analysis is performed considering all the costs mentioned in Table 4.3, over the lifespan of the project. LCC analysis enables the evaluation of different component reliabilities, facilitates cost comparisons between designs, and helps determine the cost-effectiveness of the plant. The

Table 4.3 Different costs associated with biomass power plants [110]

Total capital investment		Total operating cost
Direct cost	Indirect cost	
A = Plant section + equipment cost + fumes treatment B = Piping section C = Electrical section D = Civil works E = Installation + auxiliary services + instrumentation and controls + site preparation) = 0.65 × A Where, Equipment cost, $Y = aX^b$ where Y = equipment cost in a million USD, a and b are specific coefficients based on the condition of the plant and electrical output X = electrical output in MW	Engineering costs = 0.12 × A Startup costs = 0.1 × A	Labour cost + Fuel cost + Contingency cost + Ash disposal cost + Maintenance cost • Labour number = $13.761 \times (MW_e)^{0.4328} \pm 5\%$, where MW_e refers to the nominal electrical capacity output of the plant • Fuel cost involves biomass collection, handling, transportation, and storage The annual biomass demand tonne/yr, $B_{ad} = \frac{OH \times 3.6 \times P_c}{e_f \times LHV}$ where OH = annual operating hours of the power plant in h/yr P_c = plant capacity in MW_e, e_f = foreseen plant efficiency in percentage LHV = low heating value of the biomass in MJ/kg Biomass transportation cost, $B_T = \frac{B_f}{V_c} \times C_t$ where B_f = biomass flow rate in tonne/yr V_c = vehicle capacity in tonne/vehicle C_t = unit transport cost in EUR/vehicle, considering an average route from the barycenter of the area where the plant is assumed to be located • Maintenance cost is reported to be 1.5–4% of the total capital cost for the combustion plants • 10% of the total non-fuel operating costs were considered as the contingency costs

value of LCC is determined using the following formula as shown in Eq. 4.11 [110].

$$LCC = C + M_{wp} + E_{wp} + R_{wp} - S_{wp} \tag{4.11}$$

where

C Installation cost of the plant
M_{wp} Annual maintenance costs
E_{wp} Annual energy cost, including fuel cost
R_{wp} Annual replacement cost
S_{wp} Salvage value of the plant

- **Net present value (NPV)**: NPV is the sum of the initial investment of the biomass power plant and the present values of future cash flows at a particular discount rate. Though the process of calculation of NPV is complex, this method is effective as it considers the time value of money. A positive NPV is desirable to predict the profitability of the plant. The NPV of a biomass plant is determined as per the following Eq. 4.12 [110].

$$NPV = C_f \times d_f - TIC \tag{4.12}$$

where

TIC Total capital cost
$d_f = \frac{(1+r)^n - 1}{r(1+r)^n}$ Cumulative discount factor
r Discount rate
$C_f = N + D$ Process operation cash flows over the lifetime of the project
$N = G(1 - Q)$ Net profit after tax
$G = S - C - D_i$ Gross benefit, where C is total costs minus annuity cost and D is capital depreciation costs.

- **Internal rate of return (IRR)**: IRR is a very effective tool to check the viability of the biomass-based plant. It represents the discount rate at which the net present value (NPV) of the project's cash flows becomes zero.
- **Discounted payback period (DPB)**: DPB is a measure of the risk involved in the investment of the biomass-based power plant as it indicates the period to recover the investment. This period is the time when the biomass plant reaches the 'break-even point' of the investment. DPB is measured as per Eq. 4.13 [110].

$$\sum\nolimits_{1}^{DPB} \frac{CF_n}{(1+r)^n} = 0 \tag{4.13}$$

where CF_n is the cash flow related to the nth year and r is the discount rate

- **Levelized cost of electricity (LCOE)**: LCOE is a metric used to assess the average cost of producing electricity over the entire lifespan of the power plant. It considers all costs, including initial investment, operation, maintenance, and fuel, and allows for comparisons between different routes of biomass power generation. The LCOE of biomass plants should be comparable to the electricity produced by fossil fuel-based plants. The LCOE is determined by using the below-mentioned Eq. 4.14 [110].

$$LCOE = \frac{\sum_{n=0}^{T} \frac{TIC(n)+OM(n)+FC(n)}{(1+r)^n}}{\sum_{n=0}^{T} \frac{EP(n)}{(1+r)^n}} \quad (4.14)$$

where

TIC	Total investment cost in the year
OM	Annual operation and maintenance cost
FC	Annual fuel cost
T	Life span of the project in years
EP	Estimated annual generation
I_o	Initial investment

- **Profitability index (PI)**: The *PI* is a financial ratio that measures the profitability of an investment by comparing the future value of its expected cash flows to the initial investment cost, as shown in Eq. 4.15 [110]. A biomass plant will be profitable if the *PI* index is greater than 1.

$$PI = \frac{NPV}{I_o} + 1 \quad (4.15)$$

- **Sensitivity analysis**: Sensitivity analysis of a power plant is a method used to assess the impact of changes in key variables or assumptions on the financial, operational, or performance aspects of the power plant. It helps identify how sensitive the project's outcomes are to variations in factors such as fuel costs, electricity prices, technology efficiency, capital costs, or environmental regulations, allowing for a better understanding of the project's risks and potential opportunities.

A case study was conducted to assess the techno-economic feasibility of biomass pellets- based power generation in India. Biomass palletization provides improved volumetric energy density than the original biomass feedstock. The biomass pellets can be used for both individual combustion or co-fire with coal in the boiler of a power plant. It is predicted that, by the year 2030, the potential for annual electricity generation from biomass pellets is projected to reach 244 terawatt-hours (TWh), contributing to a total electricity production of 4000 TWh in India during the same period. Utilizing comprehensive assumptions related to input expenses and technical considerations, the calculation indicates that the expense associated with a biomass pellet would amount to €64 per metric tonne. Consequently, the levelized cost of electricity is anticipated to be €0.12 per kilowatt-hour (kWh), surpassing even the expense of imported coal-based electricity [111].

In assessing the economic viability of biomass Power facilities, the pivotal consideration lies in transportation distance, the feedstock type, the capital investment of the plant, operational and maintenance cost of the plant as per the technology adopted, and revenue generation from the plant. The integration of biomass resources with other renewable sources, such as solar, offers an economically viable method for generating electricity. A microgrid system was designed for Punjab, featuring a peak load demand of 50 kW in the summer. This system was proven to be optimal, with an estimated unit Cost of Electricity of $0.0735/kWh, lower than the grid's cost of $0.1/kWh. The system's payback period is approximately 3.83 years, indicating a strong potential for increasing renewable energy usage. Sensitivity analysis revealed that higher biomass and biogas generator costs lead to increased net present cost and electricity expenses. Conversely, electricity costs decrease as inflation rates rise, while higher inflation rates result in higher net present costs [112]. Local market conditions, government support, and specific project characteristics will also influence the financial outcomes of any biomass-based plant.

4.8 Impact on CO_2 Reduction

The U.S. Energy Information Administration (EIA) predicted that world energy consumption will rise by 56% between 2010 and 2040 [113]. Fossil fuels are extensively used to meet energy demands, releasing carbon dioxide (CO_2) and other greenhouse gases. Therefore, it is crucial to reduce fossil fuel consumption and enhance the utilization of readily accessible renewable energy sources within the energy sector [114]. In the present discussion, the impact of power generation by the thermochemical conversion process of biomass on CO_2 reduction is considered. This process mainly comprises combustion, gasification, pyrolysis, and liquefaction of biomass. CO_2 capture for thermochemical processes can be achieved by utilizing various methods like pre-combustion, post-combustion, and oxy-combustion technologies [115]. Bioenergy with CO_2 capture and storage (BECCS) technique plays an essential role in CO_2 collection from industrial sources and storing them in geological places, thereby achieving negative emissions. Shahbaz et al. [116] extensively studied the BECCS concept. They reported that the combustion and gasification processes have been primarily emphasized within BECCS, while liquefaction and pyrolysis methods have received comparatively less emphasis [116]. Various power plants and industries are using biomass as fuel for the generation of electricity and heat for industrial applications. Many investigators reported that the CO_2 emissions from the combustion of biomass are lesser than the combustion of coal. Further utilization of the BECCS concept for biomass combustion for power generation is feasible as CO_2 captured and storage has already been applied effectively in the case of fossil fuels. The membrane separation, adsorption, absorption, chemical looping, and cryogenic distillation methods can be used for post-combustion CO_2 capture both for combustion and gasification processes. The integration of combustion and gasification processes with BECCS within CHP, biomass integrated gasification combined cycle, and chemical looping cycle can be considered as a reliable technology for CO_2 reduction.

4.9 Conclusion

The global energy landscape is undergoing a transformative shift towards renewable sources, driven by the imperative to address climate change and ensure sustainable development. Within this context, biomass energy, owing to its wide availability and versatile applications, holds a significant role in renewable energy power generation. Thermochemical conversion processes, such as torrefaction, carbonization, pyrolysis, liquefaction, and gasification, offer diverse pathways to transform biomass into valuable energy products, spanning solid char, bio-oil, syngas, and more. These processes underscore the potential for optimizing energy extraction while minimizing environmental impact.

The various power generation routes, encompassing steam turbines, gas turbines, IC engines, fuel cells, and cogeneration, showcase the adaptability of biomass in catering to different energy needs. Government initiatives aimed at incentivizing biomass-based power generation play a crucial role in fostering technological innovation and accelerating the adoption of renewable energy solutions. Socio-economic benefits tied to biomass-based power generation, such as rural employment, waste management, and energy access, resonate with broader sustainable development goals. Moreover, commercial biomass-based power generation signifies a growing sector that contributes to the energy mix and promotes rural development and waste reduction in India.

Economic feasibility is a key determinant in the viability of biomass-based power projects, influenced by factors such as resource availability, technology efficiency, and market dynamics. Furthermore, the remarkable characteristic of biomass as a carbon–neutral energy source underscores its potential in reducing CO_2 emissions and contributing to a greener future. The confluence of technology innovation, supportive policies, and a global commitment to sustainability has positioned biomass as a pivotal player in the RE sector. Its multifaceted contributions encompass environmental stewardship, socio-economic progress, and a cleaner energy transition, making biomass-based power generation a vital component of a sustainable energy future.

Acknowledgements This work is a part of the project titled "Studies on efficacy of upgradation and utilization of north eastern coal and biomass for gasification in a plant prototype and its scale up" (Grant number: CPRI/R&D/TC/Thermal/2023) awarded by the Central Power Research Institute (CPRI), Govt. of India to Dr. Pankaj Kalita, Associate Professor, School of Energy Science and Engineering, Indian Institute of Technology, Guwahati, Assam, India. The financial assistance extended by CPRI is acknowledged with gratitude.

References

1. United Nation Sustainable Development Goals (2023) Goal 13-take urgent action to combat climate change and its impacts. https://www.un.org/sustainabledevelopment/climate-change/. Accessed on 5 Aug 2023
2. Prime Minister's Office (2023) National statement by Prime Minister Shri Narendra Modi at COP26 Summit in Glasgow. https://pib.gov.in/PressReleasePage.aspx?PRID=1768712. Accessed on 15 Jul 2023
3. Ministry of New and Renewable Energy (2022) Year-end review 2022. https://pib.gov.in/PressReleasePage.aspx?PRID=1885147#:~:text. Accessed on 31 Oct 2022
4. Shukla KK, Natarajan N, Vasudevan M (2023) Comparison of wind speed probability distribution models for accurate evaluation of wind energy potential: a case study from Kerala, India. J Inst Eng Ser A 104:551–563. https://doi.org/10.1007/s40030-023-00734-9
5. International Energy Agency, Bioenergy. https://www.iea.org/energy-system/renewables/bioenergy. Accessed on 5 May 2023
6. Dahmen N, Lewandowski I, Zibek S, Weidtmann A (2019) Integrated lignocellulosic value chains in a growing bioeconomy: status quo and perspectives. GCB Bioenergy 11:107–117. https://doi.org/10.1111/gcbb.12586
7. Sahoo G, Sharma A, Chandra Dash A (2022) Biomass from trees for bioenergy and biofuels—a briefing paper. Mater Today Proc 65:461–467. https://doi.org/10.1016/j.matpr.2022.02.639
8. Wani AK, Rahayu F, Yustina I, Fatah GSA, Kariada IK, Antarlina SS, Jufri A, Pamungkas D (2023) Contribution of yeast and its biomass for the preparation of industrially essential materials: a boon to circular economy. Bioresour Technol Reports 23:101508. https://doi.org/10.1016/j.biteb.2023.101508
9. International Energy Agency (2022) World energy outlook 2022. https://www.iea.org/reports/world-energy-outlook-2022
10. Soomauroo Z, Blechinger P, Creutzig F (2023) Electrifying public transit benefits public finances in small island developing states. Transp Policy 138:45–59. https://doi.org/10.1016/j.tranpol.2023.04.017
11. Hajam YA, Kumar R, Kumar A (2023) Environmental waste management strategies and vermi transformation for sustainable development. Environ Challenges 13:100747. https://doi.org/10.1016/j.envc.2023.100747
12. Madadian E, Haelssig JB, Pegg M (2020) A comparison of thermal processing strategies for landfill reclamation: methods, products, and a promising path forward. Resour Conserv Recycl 160:104876. https://doi.org/10.1016/j.resconrec.2020.104876
13. Seo MW, Lee SH, Nam H, Lee D, Tokmurzin D, Wang S, Park Y-K (2022) Recent advances of thermochemical conversion processes for biorefinery. Bioresour Technol 343:126109. https://doi.org/10.1016/j.biortech.2021.126109
14. Zhang S, Dong Q, Chen T, Xiong Y (2016) Combination of light bio-oil washing and torrefaction pretreatment of rice husk: its effects on physicochemical characteristics and fast pyrolysis behavior. Energy Fuels 30:3030–3037. https://doi.org/10.1021/acs.energyfuels.5b02968
15. Manatura K (2020) Inert torrefaction of sugarcane bagasse to improve its fuel properties. Case Stud Therm Eng 19:100623. https://doi.org/10.1016/j.csite.2020.100623
16. Bach QV, Skreiberg O (2016) Upgrading biomass fuels via wet torrefaction: a review and comparison with dry torrefaction. Renew Sustain Energy Rev 54:665–677. https://doi.org/10.1016/j.rser.2015.10.014
17. Xiao LP, Shi ZJ, Xu F, Sun RC (2012) Hydrothermal carbonization of lignocellulosic biomass. Bioresour Technol 118:619–623. https://doi.org/10.1016/j.biortech.2012.05.060
18. Hoekman SK, Broch A, Robbins C, Zielinska B, Felix L (2013) Hydrothermal carbonization (HTC) of selected woody and herbaceous biomass feedstocks. Biomass Convers Biorefinery 3:113–126. https://doi.org/10.1007/s13399-012-0066-y
19. Basu P (2013) Biomass gasification, pyrolysis and torrefaction: practical design and theory, 2nd edn. Academic Press

20. Yan W, Hastings JT, Acharjee TC, Coronella CJ, Vásquez VR (2010) Mass and energy balances of wet torrefaction of lignocellulosic biomass. Energy Fuels 24:4738–4742. https://doi.org/10.1021/ef901273n
21. Chang S, Zhao Z, Zheng A, He F, Huang Z, Li H (2012) Characterization of products from torrefaction of sprucewood and bagasse in an auger reactor. Energy Fuels 26:7009–7017. https://doi.org/10.1021/ef301048a
22. Granados DA, Ruiz RA, Vega LY, Chejne F (2017) Study of reactivity reduction in sugarcane bagasse as consequence of a torrefaction process. Energy 139:818–827. https://doi.org/10.1016/j.energy.2017.08.013
23. Kanwal S, Chaudhry N, Munir S, Sana H (2019) Effect of torrefaction conditions on the physicochemical characterization of agricultural waste (sugarcane bagasse). Waste Manag 88:280–290. https://doi.org/10.1016/j.wasman.2019.03.053
24. Pinto F, Gominho J, André RN, Gonçalves D, Miranda M, Varela F, Neves D, Santos J, Lourenço A, Pereira H (2017) Effect of rice husk torrefaction on syngas production and quality. Energy Fuels 31:5183–5192. https://doi.org/10.1021/acs.energyfuels.7b00259
25. Chen WH, Lin BJ, Lin YY, Chu YS, Ubando AT, Show PL, Ong HC, Chang JS, Ho SH, Culaba AB, Pétrissans A, Pétrissans M (2021) Progress in biomass torrefaction: principles, applications and challenges. Prog Energy Combust Sci 82:100887. https://doi.org/10.1016/j.pecs.2020.100887
26. Amer M, Elwardany A (2016) Biomass Carbonization. In: Renewable Energy - Resources, Challenges and Applications. IntechOpen, p 13
27. Elkhalifa S, Al-Ansari T, Mackey HR, McKay G (2019) Food waste to biochars through pyrolysis: a review. Resour Conserv Recycl 144:310–320. https://doi.org/10.1016/j.resconrec.2019.01.024
28. Tripathi M, Sahu JN, Ganesan P (2016) Effect of process parameters on production of biochar from biomass waste through pyrolysis: a review. Renew Sustain Energy Rev 55:467–481. https://doi.org/10.1016/j.rser.2015.10.122
29. Nanda S, Dalai AK, Berruti F, Kozinski JA (2016) Biochar as an exceptional bioresource for energy, agronomy, carbon sequestration, activated carbon and specialty materials. Waste and Biomass Valorization 7:201–235. https://doi.org/10.1007/s12649-015-9459-z
30. Hu X, Gholizadeh M (2019) Biomass pyrolysis: a review of the process development and challenges from initial researches up to the commercialisation stage. J Energy Chem 39:109–143. https://doi.org/10.1016/j.jechem.2019.01.024
31. Yang J, (Sophia)He Q, Yang L (2019) A review on hydrothermal co-liquefaction of biomass. Appl Energy 250:926–945. https://doi.org/10.1016/j.apenergy.2019.05.033
32. Elliott DC, Biller P, Ross AB, Schmidt AJ, Jones SB (2015) Hydrothermal liquefaction of biomass: developments from batch to continuous process. Bioresour Technol 178:147–156. https://doi.org/10.1016/j.biortech.2014.09.132
33. Peterson AA, Vogel F, Lachance RP, Fröling M, Antal MJ, Tester JW (2008) Thermochemical biofuel production in hydrothermal media: a review of sub- and supercritical water technologies. Energy Environ Sci 1:32–65. https://doi.org/10.1039/b810100k
34. Ni J, Qian L, Wang Y, Zhang B, Gu H, Hu Y, Wang Q (2022) A review on fast hydrothermal liquefaction of biomass. Fuel 327:125135. https://doi.org/10.1016/j.fuel.2022.125135
35. Zhang S, Yang X, Zhang H, Chu C, Zheng K, Ju M, Liu L (2019) Liquefaction of biomass and upgrading of bio-oil: a review. Molecules 24. https://doi.org/10.3390/molecules24122250
36. Isa KM, Abdullah TAT, Ali UFM (2018) Hydrogen donor solvents in liquefaction of biomass: a review. Renew Sustain Energy Rev 81:1259–1268. https://doi.org/10.1016/j.rser.2017.04.006
37. Hasan NH, Jazie AA (2021) Hydrothermal liquefaction conversion of lignocelluloses and waste biomass using zeolite catalyst. IOP Conf Ser Earth Environ Sci 790:1–13. https://doi.org/10.1088/1755-1315/790/1/012036
38. Das S, Basumatary S, Kalita P, Kulkarni V, Goswami P, Garg A, Peng X (2020) Bioelectricity production from lignocellulosic biomass. In: Ingle AP, Chandel AK, da Silva SS (eds) Lignocellulosic biorefining technologies, first. Wiley and Sons, West Sussex, pp 87–123

39. Nagappan S, Bhosale RR, Nguyen DD, Chi NTL, Ponnusamy VK, Woong CS, Kumar G (2021) Catalytic hydrothermal liquefaction of biomass into bio-oils and other value-added products—a review. Fuel 285:119053. https://doi.org/10.1016/j.fuel.2020.119053
40. Castello D, Pedersen TH, Rosendahl LA (2018) Continuous hydrothermal liquefaction of biomass: a critical review. Energies 11. https://doi.org/10.3390/en11113165
41. Mathanker A, Das S, Pudasainee D, Khan M, Kumar A, Gupta R (2021) A review of hydrothermal liquefaction of biomass for biofuels production with a special focus on the effect of process parameters, co-solvents and extraction solvents. Energies 14. https://doi.org/10.3390/en14164916
42. Pfeifer C, Koppatz S, Hofbauer H (2011) Steam gasification of various feedstocks at a dual fluidised bed gasifier: impacts of operation conditions and bed materials. Biomass Convers Biorefinery 1:39–53. https://doi.org/10.1007/s13399-011-0007-1
43. Sher F, Pans MA, Sun C, Snape C, Liu H (2018) Oxy-fuel combustion study of biomass fuels in a 20 kWth fluidized bed combustor. Fuel 215:778–786. https://doi.org/10.1016/j.fuel.2017.11.039
44. Zhang W (2010) Automotive fuels from biomass via gasification. Fuel Process Technol 91:866–876. https://doi.org/10.1016/j.fuproc.2009.07.010
45. Sikarwar VS, Zhao M, Clough P, Yao J, Zhong X, Memon MZ, Shah N, Anthony EJ, Fennell PS (2016) An overview of advances in biomass gasification. Energy Environ Sci 9:2939–2977. https://doi.org/10.1039/c6ee00935b
46. Mishra S, Upadhyay RK (2021) Review on biomass gasification: gasifiers, gasifying mediums, and operational parameters. Mater Sci Energy Technol 4:329–340. https://doi.org/10.1016/j.mset.2021.08.009
47. Vejahati F, Mahinpey N, Ellis N, Nikoo MB (2009) CFD simulation of gas-solid bubbling fluidized bed: a new method for adjusting drag law. Can J Chem Eng 87:19–30. https://doi.org/10.1002/cjce.20139
48. Yang S, Wang S, Wang H (2020) Numerical study of biomass gasification in a 0.3 MWth full-loop circulating fluidized bed gasifier. Energy Convers Manag 223:113439. https://doi.org/10.1016/j.enconman.2020.113439
49. Basu P, Cheng L (2000) An analysis of loop seal operations in a circulating fluidized bed. Chem Eng Res Des 78:991–998. https://doi.org/10.1205/026387600528102
50. Göransson K, Söderlind U, He J, Zhang W (2011) Review of syngas production via biomass DFBGs. Renew Sustain Energy Rev 15:482–492. https://doi.org/10.1016/j.rser.2010.09.032
51. Karl J, Pröll T (2018) Steam gasification of biomass in dual fluidized bed gasifiers: a review. Renew Sustain Energy Rev 98:64–78. https://doi.org/10.1016/j.rser.2018.09.010
52. Sesotyo PA, Nur M, Suseno JE (2019) Plasma gasification with municipal solid waste as a method of energy self sustained for better urban built environment: modeling and simulation. IOP Conf Ser Earth Environ Sci 396. https://doi.org/10.1088/1755-1315/396/1/012002
53. Guan G, Fushimi C, Tsutsumi A, Ishizuka M, Matsuda S, Hatano H, Suzuki Y (2010) High-density circulating fluidized bed gasifier for advanced IGCC/IGFC–ADvantages and challenges. Particuology 8:602–606. https://doi.org/10.1016/j.partic.2010.07.013
54. Deshmukh R, Jacobson A, Chamberlin C, Kammen D (2013) Thermal gasification or direct combustion? Comparison of advanced cogeneration systems in the sugarcane industry. Biomass Bioenerg 55:163–174. https://doi.org/10.1016/j.biombioe.2013.01.033
55. Bose A, Saini DK (2022) biomass fired thermal power generation technology- a route to meet growing energy demand and sustainable development. Nat Environ Pollut Technol 21:1307–1315. https://doi.org/10.46488/NEPT.2022.v21i03.037
56. Mertzis D, Mitsakis P, Tsiakmakis S, Manara P, Zabaniotou A, Samaras Z (2014) Performance analysis of a small-scale combined heat and power system using agricultural biomass residues: the SMARt-CHP demonstration project. Energy 64:367–374. https://doi.org/10.1016/j.energy.2013.11.055
57. Ahrenfeldt J, Thomsen TP, Henriksen U, Clausen LR (2013) Biomass gasification cogeneration–a review of state of the art technology and near future perspectives. Appl Therm Eng 50:1407–1417. https://doi.org/10.1016/j.applthermaleng.2011.12.040

58. Aguado R, Escámez A, Jurado F, Vera D (2023) Experimental assessment of a pilot-scale gasification plant fueled with olive pomace pellets for combined power, heat and biochar production. Fuel 344:128127. https://doi.org/10.1016/j.fuel.2023.128127
59. Khattak MA, Mohd Ali NS, Zainal Abidin NH, Azhar NS, Omar MH (2016) Common type of turbines in power plant: a review. J Adv Res Appl Sci Eng Technol 3:77–100
60. Rashidi MM, Aghagoli A, Ali M (2014) Thermodynamic analysis of a steam power plant with double reheat and feed water heaters. Adv Mech Eng 2014. https://doi.org/10.1155/2014/940818
61. Naik RJ (2012) Exergy analysis of 4.5mw biomass based steam power plant. IOSR J Humanit Soc Sci 1:01–04. https://doi.org/10.9790/0837-0110104
62. Breeze P (2005) Power generation technologies. Power Gener Technol 1–276. https://doi.org/10.1016/B978-0-7506-6313-7.X5000-1
63. Mondal P, Ghosh S (2017) Exergo-economic analysis of a 1-MW biomass-based combined cycle plant with externally fired gas turbine cycle and supercritical organic Rankine cycle. Clean Technol Environ Policy 19:1475–1486. https://doi.org/10.1007/s10098-017-1344-y
64. Braimakis K, Magiri-Skouloudi D, Grimekis D, Karellas S (2020) Energy-exergy analysis of ultra-supercritical biomass-fuelled steam power plants for industrial CHP, district heating and cooling. Renew Energy 154:252–269. https://doi.org/10.1016/j.renene.2020.02.091
65. Lee SH, Lee TH, Jeong SM, Lee JM (2019) Economic analysis of a 600 mwe ultra supercritical circulating fluidized bed power plant based on coal tax and biomass co-combustion plans. Renew Energy 138:121–127. https://doi.org/10.1016/j.renene.2019.01.074
66. Srinivas T, Reddy BV (2014) Hybrid solar-biomass power plant without energy storage. Case Stud Therm Eng 2:75–81. https://doi.org/10.1016/j.csite.2013.12.004
67. Martínez JD, Mahkamov K, Andrade RV, Silva Lora EE (2012) Syngas production in downdraft biomass gasifiers and its application using internal combustion engines. Renew Energy 38:1–9. https://doi.org/10.1016/j.renene.2011.07.035
68. Yassin L, Lettieri P, Simons SJR, Germanà A (2009) Techno-economic performance of energy-from-waste fluidized bed combustion and gasification processes in the UK context. Chem Eng J 146:315–327. https://doi.org/10.1016/j.cej.2008.06.014
69. Li CY, Shen Y, Wu JY, Dai YJ, Wang CH (2019) Experimental and modeling investigation of an integrated biomass gasifier–engine–generator system for power generation and waste heat recovery. Energy Convers Manag 199:1–15. https://doi.org/10.1016/j.enconman.2019.112023
70. Soares J, Oliveira AC (2020) Experimental assessment of pine wood chips gasification at steady and part-load performance. Biomass Bioenerg 139. https://doi.org/10.1016/j.biombioe.2020.105625
71. HajiHashemi MS, Mazhkoo S, Dadfar H, Livani E, Naseri Varnosefaderani A, Pourali O, Najafi Nobar S, Dutta A (2023) Combined heat and power production in a pilot-scale biomass gasification system: experimental study and kinetic simulation using ASPEN Plus. Energy 276:127506. https://doi.org/10.1016/j.energy.2023.127506
72. Yaliwal VS, Banapurmath NR, Gireesh NM, Tewari PG (2014) Production and utilization of renewable and sustainable gaseous fuel for power generation applications: a review of literature. Renew Sustain Energy Rev 34:608–627. https://doi.org/10.1016/j.rser.2014.03.043
73. Bridgwater AV, Toft AJ, Brammer JG (2002) A techno-economic comparison of power production by biomass fast pyrolysis with gasification and combustion. Renew Sustain Energy Rev 6:181–248. https://doi.org/10.1016/S1364-0321(01)00010-7
74. Arunachalam A, Olsen DB (2012) Experimental evaluation of knock characteristics of producer gas. Biomass Bioenerg 37:169–176. https://doi.org/10.1016/j.biombioe.2011.12.016
75. Marseglia G, Medaglia CM (2019) Energy efficiency in gasoline direct injection engines. WIT Trans Built Environ 182:85–91. https://doi.org/10.2495/UT180081
76. Sahoo BB, Sahoo N, Saha UK (2012) Effect of H2:CO ratio in syngas on the performance of a dual fuel diesel engine operation. Appl Therm Eng 49:139–146. https://doi.org/10.1016/j.applthermaleng.2011.08.021

77. Dhole AE, Yarasu RB, Lata DB, Priyam A (2014) Effect on performance and emissions of a dual fuel diesel engine using hydrogen and producer gas as secondary fuels. Int J Hydrogen Energy 39:8087–8097. https://doi.org/10.1016/j.ijhydene.2014.03.085
78. Paisley MA, Anson D (1998) Biomass gasification for gas turbine-based power generation. J Eng Gas Turbines Power 120:284–288. https://doi.org/10.1115/1.2818118
79. Emun F, Gadalla M, Majozi T, Boer D (2010) Integrated gasification combined cycle (IGCC) process simulation and optimization. Comput Chem Eng 34:331–338. https://doi.org/10.1016/j.compchemeng.2009.04.007
80. Copa Rey JR, Tamayo Pacheco JJ, da Cruz A, Tarelho L, Silva V, Cardoso JS, Silveira JL, Tuna CE (2021) Evaluation of cogeneration alternative systems integrating biomass gasification applied to a Brazilian sugar industry. Renew Energy 178:318–333. https://doi.org/10.1016/j.renene.2021.06.053
81. Asadullah M (2014) Barriers of commercial power generation using biomass gasification gas: a review. Renew Sustain Energy Rev 29:201–215. https://doi.org/10.1016/j.rser.2013.08.074
82. EG&G Technical Services I (2004) Fuel cell handbook. U.S. Department of Energy, Office of Fossil Energy, National Energy Technology Laboratory, Morgantown, West Virginia, USA
83. Li M, Rao AD, Brouwer J, Samuelsen GS (2010) Design of highly efficient coal-based integrated gasification fuel cell power plants. J Power Sources 195:5707–5718. https://doi.org/10.1016/j.jpowsour.2010.03.045
84. Ghezel-Ayagh H, Way R, Huang P, Walzak J, Jolly S, Patel D, Lukas M, Willman C, Davis K, Stauffer D, Vaysman V, Borglum B, Tang E, Pastula M, Petri R (2010) Integrated coal gasification and solid oxide fuel cell systems for centralized power generation. ECS Trans 26:305–313. https://doi.org/10.1149/1.3429002
85. El-Emam RS, Dincer I (2015) Thermal modeling and efficiency assessment of an integrated biomass gasification and solid oxide fuel cell system. Int J Hydrogen Energy 40:7694–7706. https://doi.org/10.1016/j.ijhydene.2015.02.061
86. Costa P, Pinto F, André RN, Marques P (2021) Integration of gasification and solid oxide fuel cells (SOFCs) for combined heat and power (CHP). Processes 9:1–26. https://doi.org/10.3390/pr9020254
87. Nhuchhen DR, Basu P, Acharya B (2014) A comprehensive review on biomass torrefaction. https://doi.org/10.5171/2014.506376
88. Proskurina S, Heinimö J, Schipfer F, Vakkilainen E (2017) Biomass for industrial applications: the role of torrefaction. Renew Energy 111:265–274. https://doi.org/10.1016/j.renene.2017.04.015
89. Mamvura TA, Pahla G, Muzenda E (2018) Torrefaction of waste biomass for application in energy production in South Africa. South African J Chem Eng 25:1–12. https://doi.org/10.1016/j.sajce.2017.11.003
90. Kuzmina JS, Sytchev GA, Zaychenko VM (2016) Torrefaction prospects and application. Chem Eng Trans 50:265–270. https://doi.org/10.3303/CET1650045
91. Roy P, Dias G (2017) Prospects for pyrolysis technologies in the bioenergy sector: a review. Renew Sustain Energy Rev 77:59–69. https://doi.org/10.1016/j.rser.2017.03.136
92. Huang YF, Syu FS, Te CP, Lo SL (2013) Life cycle assessment of biochar cofiring with coal. Bioresour Technol 131:166–171. https://doi.org/10.1016/j.biortech.2012.12.123
93. Gallucci F, Liberatore R, Sapegno L, Volponi E, Venturini P, Rispoli F, Paris E, Carnevale M, Colantoni A (2020) Influence of oxidant agent on syngas composition: gasification of hazelnut shells through an updraft reactor. Energies 13:102. https://doi.org/10.3390/en13010102
94. Bartoli M, Giorcelli M, Jagdale P, Rovere M, Tagliaferro A (2020) A review of non-soil biochar applications. Materials (Basel) 13:261. https://doi.org/10.3390/ma13020261
95. Ram Mohan MP, Thomas Phillippe GT, Shiju MV (2006) Biofuel laws in Asia: instruments for energy access, security, environmental protection and rural empowerment. Asian Biotechnol Dev Rev 8:51–75
96. Singh R, Setiawan AD (2013) Biomass energy policies and strategies: harvesting potential in India and Indonesia. Renew Sustain Energy Rev 22:332–345. https://doi.org/10.1016/j.rser.2013.01.043

97. Ministry of New and Renewable Energy (2023) What is the ministry all about? https://www.mnre.gov.in/the-ministry/what-does-the-ministry-do/. Accessed on 6 Aug 2023
98. Sinha SK, Subramanian KA, Singh HM, Tyagi VV, Mishra A (2019) Progressive trends in bio-fuel policies in India: targets and implementation strategy. Biofuels 10:155–166. https://doi.org/10.1080/17597269.2018.1522483
99. Saravanan AP, Mathimani T, Deviram G, Rajendran K, Pugazhendhi A (2018) Biofuel policy in India: a review of policy barriers in sustainable marketing of biofuel. J Clean Prod 193:734–747. https://doi.org/10.1016/j.jclepro.2018.05.033
100. Narnaware SL, Panwar NL (2022) Biomass gasification for climate change mitigation and policy framework in India: a review. Bioresour Technol Reports 17:100892. https://doi.org/10.1016/j.biteb.2021.100892
101. Central Electricity Authority (2022) Guidelines for exemption/relaxation from mandatory co-firing of biomass by thermal power plants
102. Ministry of New and Renewable Energy (2021) Promotion of biomass based cogeneration in sugar mills and other industries. https://pib.gov.in/PressReleaseIframePage.aspx?PRID=1742794#:~:text=Underthescheme%2CCentralFinancial,agro-basedindustrial residue%2Ccrop. Accessed on 1 Aug 2023
103. Ministry of New and Renewable Energy, Bio-energy Schemes. https://mnre.gov.in/bio-energy/schemes. Accessed on 10 Jul 2023
104. Ministry of Environment Forest & Climate Change (2022) Union Minister of environment, forest & climate change releases guidelines for grant of financial support for establishment of paddy straw pelletisation and torrefaction plants. https://pib.gov.in/PressReleasePage.aspx?PRID=1867506. Accessed on 5 Aug 2023
105. Ministry of New and Renewable Energy, Bio Energy. https://mnre.gov.in/bio-energy/current-status. Accessed on 2 Aug 2023
106. Somashekhar HI, Dasappa S, Ravindranath NH (2000) Rural bioenergy centres based on biomass gasifiers for decentralized power generation: case study of two villages in southern India. 4:55–63
107. Mukhopadhyay K (2004) An assessment of a biomass Gasiÿcation based power plant in the sunderbans. 27:253–264. https://doi.org/10.1016/j.biombioe.2003.11.008
108. Winrock International India (2010) Access to clean energy: a glimpse of off grid projects in India. New Delhi
109. Vasudha-foundation (2005) Case studies of successful decentralized renewable energy projects that have not only ensure energy access but also improved livelihoods of people
110. Abdul Malek ABM (2016) Economic assessment of biomass based power generation. In: Biomass, biorefineries and bioeconomy. IntechOpen, p 13
111. Purohit P, Chaturvedi V (2018) Biomass pellets for power generation in India: a techno-economic evaluation. Environ Sci Pollut Res 25:29614–29632. https://doi.org/10.1007/s11356-018-2960-8
112. Kaur M, Dhundhara S, Verma YP, Chauhan S (2020) Techno-economic analysis of photovoltaic-biomass-based microgrid system for reliable rural electrification. Int Trans Electr Energy Syst 30:1–20. https://doi.org/10.1002/2050-7038.12347
113. Energy Information Administration (2013) EIA projects world energy consumption will increase 56% by 2040. US Energy Inf. Adm
114. Rahman A, Farrok O, Haque MM (2022) Environmental impact of renewable energy source based electrical power plants: solar, wind, hydroelectric, biomass, geothermal, tidal, ocean, and osmotic. Renew Sustain Energy Rev 161:112279. https://doi.org/10.1016/j.rser.2022.112279
115. Kuparinen K, Lipiäinen S, Vakkilainen E, Laukkanen T (2023) Effect of biomass-based carbon capture on the sustainability and economics of pulp and paper production in the Nordic mills. Environ Dev Sustain 25:648–668. https://doi.org/10.1007/s10668-021-02074-9
116. Shahbaz M, AlNouss A, Ghiat I, Mckay G, Mackey H, Elkhalifa S, Al-Ansari T (2021) A comprehensive review of biomass based thermochemical conversion technologies integrated with CO2 capture and utilisation within BECCS networks. Resour Conserv Recycl 173:105734. https://doi.org/10.1016/j.resconrec.2021.105734

Chapter 5
Integration of Anaerobic Digestion and Algal Cultivation—Prospects and Challenges

Kaustav Nath, Rupak Jana, and Ranjana Chowdhury

Abstract Biogas, generated through anaerobic digestion (AD) of organic feedstocks like bio-wastes, is a sustainable and clean source of energy, capable of generating electricity, and direct thermal use. However, biogas is a mixture of methane and CO_2 and the latter does not contribute towards the overall heating value of the gaseous fuel. The AD process also generates digestate slurry which is generally rich in ammonia-based nitrogen, phosphorous, and potassium (NPK). While the solid part of the effluent is traditionally utilized as fertilizers, the liquid digestate requires proper treatment and management to prevent leach out and contamination of the ground water systems. Algal cultivation (AC), on the other hand, is gaining attention for CO_2 capture and for the generation of algal lipids and biochemicals, like proteins, carbohydrates, and pigments. Algal cultivation systems can be integrated as photo bioreactors for CO_2 scrubbing in a two stage system or as anaerobic membrane reactors combining AD and algal processes along with membrane filtration technology for enhancing nutrient recovery, Therefore, algal cultivation can act as a potential CO_2 trap for upgradation of the quality of biogas by enhancing methane concentration while utilizing the nutrient (NPK) rich effluent from AD process. The algal biomass can be further processed for the production of value-added biochemicals and bioenergy applications, like biodiesel production. The present article focuses on the prospects and challenges of the integration of AD and AC processes which can have potential implications on bioenergy sector.

Keywords Algal cultivation · Anaerobic digestion · Biogas upgradation · Integration · Production of Algal by-products · Utilization of AD effluent

---

K. Nath · R. Jana · R. Chowdhury (✉)
Department of Chemical Engineering, Jadavpur University, 188, Raja S.C Mallick Road, Kolkata, West Bengal 700032, India
e-mail: ranjana.chowdhury@jadavpuruniversity.in

K. Nath
e-mail: kaustavn.chem.rs@jadavpuruniversity.in

R. Jana
e-mail: rupakj.chem.rs@jadavpuruniversity.in

S. De et al. (eds.), *Challenges and Opportunities of Distributed Renewable Power*, Energy, Environment, and Sustainability, https://doi.org/10.1007/978-981-97-1406-3_5

5.1 Introduction

In the pursuit of sustainable energy solutions, biogas production through anaerobic digestion of organic feedstock, such as bio-wastes, has emerged as a promising alternative. This sustainable approach revolves around the efficient utilization of organic waste materials, which can range from agricultural residues to municipal solid waste and wastewater sludge. It not only converts these materials into valuable energy resources but also actively contributes to the reduction of greenhouse gas emissions, aligning with global efforts to transition towards cleaner and more sustainable energy alternatives. Biogas consists primarily of methane (CH_4), a highly combustible and valuable fuel component. Nevertheless, coexisting within this gas mixture is a notable presence of carbon dioxide (CO_2), which is inert and does not contribute to the heat generated during combustion. Addressing this challenge has become a focal point of research and engineering efforts, with various methods and technologies explored to minimize CO_2 content, ultimately enhancing the energy potential of biogas and further promoting its role in sustainable energy production with reduced environmental impact.

In addition to its prowess in biogas production, the anaerobic digestion process yields a valuable by-product: digestate slurry, enriched with essential nutrients such as ammonia-based nitrogen, phosphorus, and potassium (NPK). These nutrients, collectively known as NPK, are vital for plant growth and are fundamental components of commercial fertilizers. The solid fraction of this digestate has long been embraced by agriculture as a potent natural fertilizer, playing a pivotal role in enhancing soil fertility, promoting crop growth, and increasing agricultural yields. This practice aligns harmoniously with the principles of circular agriculture, effectively closing the nutrient loop by returning essential elements to the soil. However, the liquid fraction of the digestate demands vigilant and strategic management. While it contains valuable nutrients, the liquid effluent carries the potential for environmental concerns, particularly the risk of groundwater contamination if not managed with care. To mitigate these risks, precise management practices, including controlled application rates and timing, are employed to ensure responsible utilization of liquid digestate without overwhelming the soil's capacity to absorb and retain nutrients. On a different front, algal cultivation (AC) is attracting considerable attention for its potential to capture CO_2 and produce valuable algal lipids, proteins, carbohydrates, and pigments. The integration of algal cultivation with anaerobic digestion presents an exciting avenue to enhance biogas quality by increasing methane concentration while utilizing the nutrient-rich digestate effluent. Moreover, the algal biomass holds promise for further processing, leading to the production of value-added biochemical and bioenergy applications, such as biodiesel.

The central scientific challenge revolves around effectively merging anaerobic digestion with algal cultivation, with the overarching aim of optimizing resource utilization and bolstering sustainability in wastewater treatment and biomass generation. Within this realm, there exist notable limitations and shortcomings. These encompass the intricate task of maintaining a balanced nutrient profile, notably concerning crucial elements like nitrogen and phosphorus, within the integrated system. Such imbalances can hamper algae growth and compromise the effectiveness of the treatment process. Furthermore, the efficient utilization of carbon dioxide (CO_2) sourced from anaerobic digestion presents a multifaceted issue, influenced by various factors, including the methods of CO_2 capture and the delivery mechanisms to the algae. Moreover, as we consider the scaling-up of these integrated systems from laboratory-scale to industrial or municipal levels, we grapple with intricate engineering and operational complexities that often translate into substantial capital costs. The issue of energy consumption is also salient. Certain integration methods, such as the deployment of photo bioreactors, are recognized for their energy-intensive nature, potentially offsetting the net energy gains achieved through biogas production. Thus, optimizing energy efficiency within these integrated systems is of paramount importance. Additionally, challenges manifest during the harvest and processing phases of algal biomass, where the energy-intensive nature of these processes raises concerns about the overall economic feasibility of integrated systems. It is essential to acknowledge that sustaining the ideal conditions for both anaerobic digestion and algal growth in real-time necessitates intricate and often costly monitoring and control systems, which may not be universally accessible or economically viable. Furthermore, although integration can enhance the removal of CO_2, challenges may persist concerning the quality of the resulting biogas, particularly regarding the presence of hydrogen sulfide (H_2S), requiring supplementary purification measures. Lastly, the variability of wastewater composition across different sources and conditions adds another layer of complexity, demanding adaptable integrated systems capable of consistent performance across this spectrum of inputs. Consequently, it is incumbent upon researchers and practitioners to confront and address these persistent challenges and limitations to advance the integration of anaerobic digestion and algal cultivation into a more effective, efficient, and scalable solution for wastewater treatment, resource recovery, and sustainable biomass production. Thus, in this article, we will explore the possibilities and challenges of combining anaerobic digestion and algal cultivation processes, investigating their potential impact on the rapidly advancing bioenergy sector that will assuage the frequently asked questions regarding such technology by providing a technical overview.

5.2 Biogas Utilization and Production

In this section, we will briefly discuss various uses of biogas and how it is produced via anaerobic digestion process through a series of microbial reactions. Biogas, consisting mainly of methane (CH_4) and carbon dioxide (CO_2) with other trace gases, is a valuable renewable energy resource that can address the issue of increasing energy demand, at the same time reducing waste and greenhouse gas (GHG) emissions.

Biogas offers various applications and uses, as represented in Fig. 5.1. Biogas can be used as a fuel for heat and power generation by burning directly in gas engines or turbines. It can be used in transport sector. Biogas can be further purified to bio

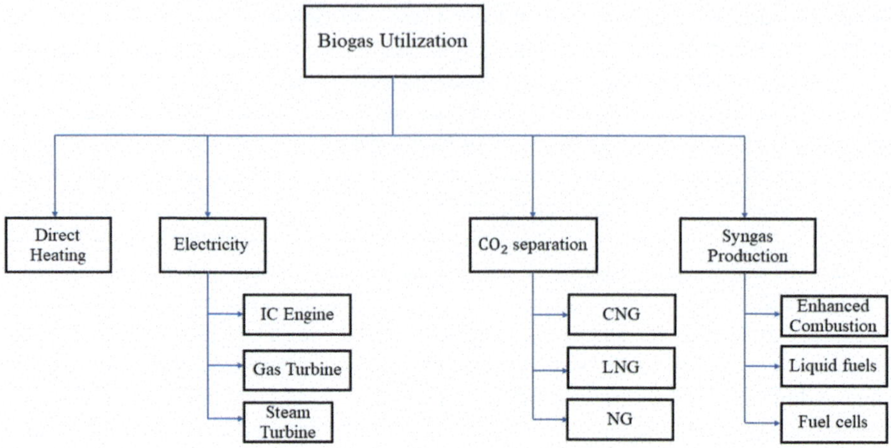

Fig. 5.1 Utilization of biogas

methane and injected into the natural gas grid, blending with conventional natural gas supplies [1]. Another advantage of using biogas is that it can be utilized in decentralized energy systems, providing energy access in remote or off-grid areas. Moreover, biogas can also be used as a material for the production of chemicals (bioethanol), hydrogen synthesis gas, etc., which have their specific applications [2, 3]. It is produced via the anaerobic digestion (AD) process, in which different microorganisms especially bacteria and archaea degrade organic matter through a series of sequential microbial conversions [4]. AD systems have been applied for hundreds of years for waste treatment and renewable energy production in the form of biogas. A wide variety of feedstock ranging from agricultural waste (agro residues, animal and poultry manure, energy crops, algal biomass), food waste, garden waste, municipal waste (solid waste, sludge) to industrial organic waste (waste generated by food processing, dairy, sugar, pharmaceutical, biochemical, pulp and paper industries, etc.) can be utilized in AD to generate biogas [5, 6]. Anaerobic digestion consists of four stages named hydrolysis, acidogenesis, acetogenesis, and methanogenesis, as represented in Fig. 5.2 [7].

In the first stage of AD, namely, hydrolysis, complex organic polymers (i.e., carbohydrates, proteins, and lipids) are decomposed or depolymerized into corresponding soluble monomers (monosaccharides, long chain fatty acids, amino acids). The general formula of one example, that is breakdown of carbohydrate to monosaccharide, is given in the following Eq. (5.1) [8]:

$$(C_6H_{10}O_5)_n + nH_2O \rightarrow nC_6H_{10}O_5 \tag{5.1}$$

In the acidogenesis stage, those organic monomers (sugars and amino acids) obtained from the hydrolysis step are converted into volatile fatty acids (propionic acid, butyric acid, acetic acid) and organic acids (succinic acid, valeric acid, and

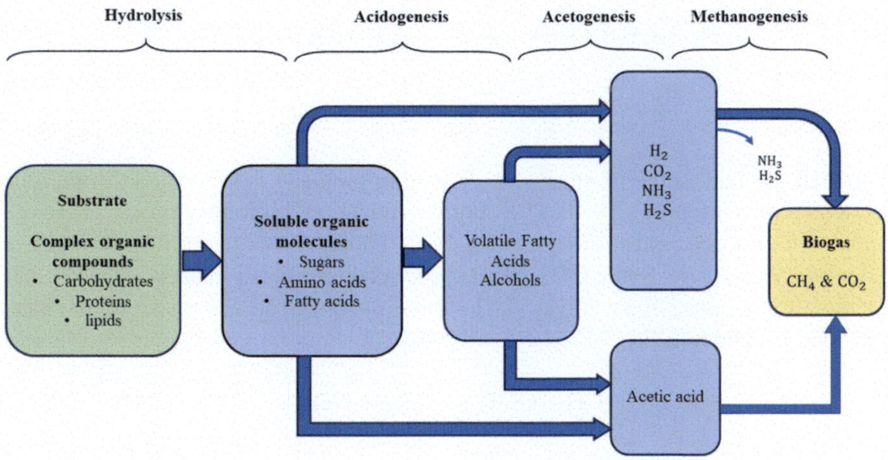

Fig. 5.2 A graphical representation of the various stages in biogas production

lactic acid) [9, 10]. The principal reactions can be interpreted as shown by Eqs. 5.2–5.5. Equations 5.2–5.5 represent the formation of acetic acid, butyric acid, propanoic acid, and ethanol from sugar, respectively.

$$nC_6H_{10}O_5 \rightarrow 3nCH_3COOH \quad (5.2)$$

$$C_6H_{10}O_5 \rightarrow CH_3CH_2CH_2COOH + 2CO_2 + 2H_2 \quad (5.3)$$

$$C_6H_{10}O_5 + 2H_2 \rightarrow 2CH_3CH_2COOH + 2H_2O \quad (5.4)$$

$$C_6H_{10}O_5 \rightarrow 2CH_3CH_2OH + 2CO_2 \quad (5.5)$$

In the next step, namely acetogenesis, microorganisms transform volatile fatty acids to acetic acid, which is called acetogenesis. Equations 5.6–5.9 represent some of the acetogenic reactions [11]. As shown in Eq. 5.2 earlier, some acetic acid is produced directly from sugar, which skips the acetogenesis step. Other higher acids produced in the previous stage are being converted to acetic acid by the microorganisms. Equations 5.6 and 5.8 show the formation of acetic acid from propanoic acid and Eq. 5.7 shows the formation of acetic acid from butyric acid. Most of the CO_2 and H_2 produced in acidogenesis skip this stage, but some organisms can produce acetic acid by consumption of carbon dioxide and hydrogen, which is shown in Eq. 5.9.

$$CH_3CH_2COOH + 2H_2O \rightarrow CH_3COOH + CO_2 + 3H_2 \quad (5.6)$$

$$CH_3CH_2CH_2COOH + 2H_2O \rightarrow 2CH_3COOH + 2H_2 \quad (5.7)$$

$$CH_3CH_2COOH + H_2O \rightarrow CH_3COOH + 2H_2 \tag{5.8}$$

$$2CO_2 + 4H_2 \rightarrow CH_3COOH + 2H_2O \tag{5.9}$$

And in the final stage of anaerobic digestion, methane is generated through the reactions represented by s 10–12 [12]. For the most part, biological process here is a breakdown of acetic acid as shown in Eq. 5.10. Though other forms of reactions can also produce methane. Some CO_2 and H_2 produced previously react with each other to generate methane, given in Eq. 5.11. Some bacteria can also produce methane from CO, as shown in Eq. 5.12.

$$CH_3COOH \rightarrow CH_4 + CO_2 \tag{5.10}$$

$$CO_2 + 4H_2 \rightarrow CH_4 + 2H_2O \tag{5.11}$$

$$4CO + 2H_2O \rightarrow CH_4 + 3CO_2 \tag{5.12}$$

Along with biogas, anaerobic digestion produces a nutrient-rich digestate as a by-product, which can be used as a valuable fertilizer for agricultural purposes, closing the loop in nutrient cycling. This is another reason why AD processes become popular and used worldwide for waste treatment and biogas generation. By returning nutrients from organic waste to the soil, anaerobic digestion supports sustainable agriculture and reduces the need for synthetic fertilizers [13]. However, various physiochemical and biological factors such as pH, temperature, pressure, feedstock characteristics, feedstock particle size, total solid content (TSC), amount of volatile solids (VS), nutrients, carbon/nitrogen (C/N) ratio, presence of inhibitory components (volatile fatty acids and ammonia), and operational factors like agitation (mixing or stirring), organic loading rate (OLR), solid retention time (SRT), and hydraulic retention time (HRT) greatly affecting the performance of AD are needed to be taken into consideration and should be monitored carefully for higher biogas production rate, higher cumulative biogas yield and better process stability [10].

5.3 Biogas Composition

As mentioned earlier, biogas is mainly composed of two major components: CH_4, which ranges from approximately 50 to 70%, and CO_2, which ranges from approximately 30 to 50%. Other minor gases such as 0–3%oxygen (O_2), 0–15%nitrogen (N_2), 0–10,000 ppm_v hydrogen sulfide (H_2S), 0–100 ppm_v ammonia (NH_3), 1–5%water vapor (H_2O), 0–41 mg Si/m^3methyl siloxanes and 0–200 mg/m^3volatile organic compounds (VOCs) can also be present in the biogas depending on the source [14]. This section highlights the need for the removal of impurities like CO_2

Table 5.1 Concentration of biogas constituents [1]

Component	Concentration (%)
Methane	45–70
Carbon dioxide	35–55
Hydrogen sulphide	1–2
Hydrogen	0–3
Ammonia	Trace
Oxygen	0–1
Nitrogen	0–1

and other trace gases like H_2S to obtain enriched biogas. The presence of different constituents and their quantities is largely determined by the oxidation–reduction state of the organic carbon present in the substrate fed to the anaerobic digester and the configuration of the AD process [15]. For example, substrates rich in protein and lipids led to the development of H_2S and NH_3, produced by the reduction of sulfur compounds (amino acids, peptides, sulfates) and ammonium contents present in feedstock, respectively. N_2 can be originated from the air entering the system. O_2 can enter either through leakages or can be present along with the feed as saturated oxygen. Water vapor can be present depending on the digestion temperature (at 35 °C, the water content of biogas is usually 5%) [16]. The presence of siloxanes can result from the introduction of products such as shampoos, cosmetics, deodorants, food additives, soaps, detergents, and cleaning agents along with the waste substrate fed to the AD system [17]. Volatile organic compounds (VOCs) are other constituents of biogas, which include alkanes, alcohols, halogens, and aromatic compounds, which are found at various concentration levels depending on the conditions of anaerobic digestion and type of biomass [18]. The overall composition of the biogas is given in Table 5.1.

5.3.1 *Necessity for CO_2 Removal from Biogas*

Apart from methane, all the other components present in biogas are considered impurities and are not desired. Although being a major constituent of the biogas, CO_2 doesn't contribute to the heating value of the biogas as it acts as a diluent. During combustion, methane reacts with oxygen which results in the formation of CO_2 and water vapor. The net result of breaking C–H bonds and forming C–O bond is the release of heat energy which is harnessed. Carbon dioxide remains inert and non-combustible during this process as you can see it's the product of burning not the reactant. Thus, for the same volume of gas, a mixture with a high proportion of CO_2 will release less amount of heat while combusting. Thus, the presence of CO_2 in higher concentrations reduces the calorific value, significantly lowering the quality of the biogas (The calorific value of natural gas typically lies in the range of

30–35 MJ/m^3, while the calorific value of biogas remains in the range between 20 and 25 MJ/m^3) [19]. For applications prioritizing heating value, like heating systems and power generation, reducing CO_2 through purification processes is vital to enhance biogas efficiency as a fuel source [20]. Another issue with CO_2 is the release of it to air during combustion of biogas, storage, and transportation, which posses a significant problem from the point of view of global warming.

5.3.2 *Why Other Trace Gases Need to Be Removed*

Apart from CO_2 other minor gases present in biogas are also undesired. These gases either decrease the heating value of the gas or can be toxic to human beings and other living organisms. N_2 present in biogas does not contribute to the heating value of the biogas. If present over a certain critical limit, O_2 poses the threat of explosion hazards. H_2S in combination with the inherent moisture present in biogas can cause corrosion in gas storage tanks, compressors, pipelines, and engines by generating H_2SO_4. Likewise, NH_3 can produce corrosive by-products during combustion, which can destroy valves, tubes, monitoring equipment, gas motors, pipelines, and engines [21, 22]. Siloxanes cause problems in gas combustion motors due to the formation of glassy microcrystalline silica and can cause abrasion, and overheating [23]. Although rarely present in biogas, halogens are corrosive and, during the combustion of biogas, can lead to the formation of dioxins and furans, which pose public health risks [24]. The potential impact of the constituent gases on biogas utilization is given in Table 5.2.

Table 5.2 Impact of biogas components [1]

Component	Hazards/effects
Carbon dioxide	Decrease the specific calorific value
Nitrogen	Decrease the specific calorific value
Oxygen	Explosion risk
Hydrogen sulphide	Corrosion in storage tanks, distribution lines, engines Results in SO_2 emissions
Ammonia	Corrosion in storage tanks, distribution lines, engines Results in NO_x emissions
Siloxanes	Silicone oxide deposits Abrasion and malfunctioning of engines

5.3.3 Biogas Treatment

Therefore, instead of directly using it for energy generation, treating raw biogas can be helpful as it will significantly improve the quality of biogas. Upgrading is also required because of some specifications set by manufacturers and certain authorities based on the final use of biogas (given in Table 5.3). Generally, biogas treatment involves two steps: first, biogas cleaning (removal of minor components present in biogas), and second, upgrading (removal of CO_2 present in biogas) [25, 26]. Many physical, chemical, and biological biogas cleaning and upgrading technologies have been developed in recent years both at pilot scale and industrial scale. In practice, some technologies remove several components present at once, achieving cleaning and upgrading at the same time [26]. Alongside their efficiency, these technologies need to be assessed in terms of energy consumption and operating cost. Although physical/chemical methods are the most used technologies for biogas upgrading, they require high energy and are costly. Therefore, biological upgrading methods are gaining more superiority over conventionally used technologies because of their low operational cost and environmental sustainability/friendliness [27]. In this context, this chapter will give a review of the biological upgrading method via algal cultivation. Algae can simultaneously remove CO_2, NH_3, and H_2S in biogas in a single-step process at low energy, cost, low environmental impact, and with minimal O_2 and N_2 contamination. Critical evaluation of the potential and limitations of the process are presented. A brief overview of the main physical/chemical upgrading technologies

Table 5.3 Biogas specifications based on final use of biogas [2]

S. no	Utilization pathway	Specifications
1	Domestic stoves	$H_2S < 10$ ppm
2	Boiler	$H_2S < 1000$ ppm
3	Internal combustion engines for combined heat and power (CHP) generation	$H_2S < 200–1000$ ppm $NH_3 < 32–50$ mg/m^3 Siloxanes < 5–28 mg/m^3
4	Gas turbines/micro-turbine	Siloxanes < 0.03–0.1 ppm
5	Injection into natural gas grid	$CH_4 > 80–90\%$ $CO_2 < 2–3\%$ $O_2 < 0.2–0.5\%$ $H_2S < 5$ mg/m^3 $NH_3 < 3–20$ mg/m^3 Siloxanes <5–10 mg/m^3
6	Vehicle fuel	CO_2 <3% H_2S: 5 mg/m^3
7	Fuel cells	H_2S: 1–5 ppm (MCFC) and 1 ppm (SOFC) Siloxanes: few ppm
8	Stirling engine	$H_2S < 2800$ ppm Siloxanes: about 0.42 ppm Halides: 232 ppm

available is also presented so that a ground can be established for comparison with algal upgrading. Cleaning technologies have also been discussed briefly.

5.4 Biogas Cleaning

Biogas cleaning i.e., removal of trace elements other than CO_2 present in biogas is required prior to biogas upgrading to avoid corrosion and wearing of the equipment present in the upgradation process. If not cleaned, it can also cause problems in storage and distribution. Many physical/chemical technologies are available for biogas cleaning.

Water vapor and hydrogen sulfide are the main impurities that need to be removed. Water vapor removal can be achieved by cooling compression, absorption, or adsorption. Biogas drying simultaneously removes oil and dust particles during absorption of water [19, 22, 25, 28]. The concentration of hydrogen sulfide can be decreased either by precipitation in the liquid or physical/chemical absorption from gas or by membrane separation. For precipitation technologies iron salts are used. Iron effectively reacts with H_2S forming insoluble salts FeS, which can be removed from the system with discharged solid and can act as nutrients o the plants if used as a fertilizer [2, 19]. For adsorption technologies, either Fe_2O_3 or $Fe(OH)_3$ or ZnO or activated carbon is deployed as adsorbents [29]. For absorption either water or organic solvents are used [30]. Membrane separation is based on the selective permeability property of the membranes [31, 32]. Some biological methods of H_2S removal have also been proposed, which are based on biochemical oxidation of sulfide (S^{2-}) to sulfate (SO_4^{2-}), thiosulfate ($S_2O_3^{2-}$), or sulfur (S). However, it requires supplementation of an electron acceptor [33–35].

Likewise, Oxygen and nitrogen, present in the landfill gas, can also be freed by adsorption, molecular sieves, or membranes. Siloxanes can be removed by cooling, adsorption, or by absorption in liquid mixtures of hydrocarbons [26]. However, no single technology can remove all contaminants in biogas. Generally, each technology is capable of removing one or two contaminants. Based on the biomethane requirement, a combination of processes is used to build the biogas upgrading plant.

5.5 Upgradation

Upgrading of biogas refers to either removal of CO_2 or conversion of CO_2 into methane [36]. Anyway around, biogas upgrading increases the specific calorific value of the raw biogas, raising the standard of the fuel. Upgrading can be done in both physical/chemical methods and biological methods. This section elucidates on the various physicochemical methods like physical absorption (water scrubbing, organic solvent scrubbing), chemical absorption, adsorption, membrane separation technology, cryogenic separation, and chemical hydrogenation process, represented in

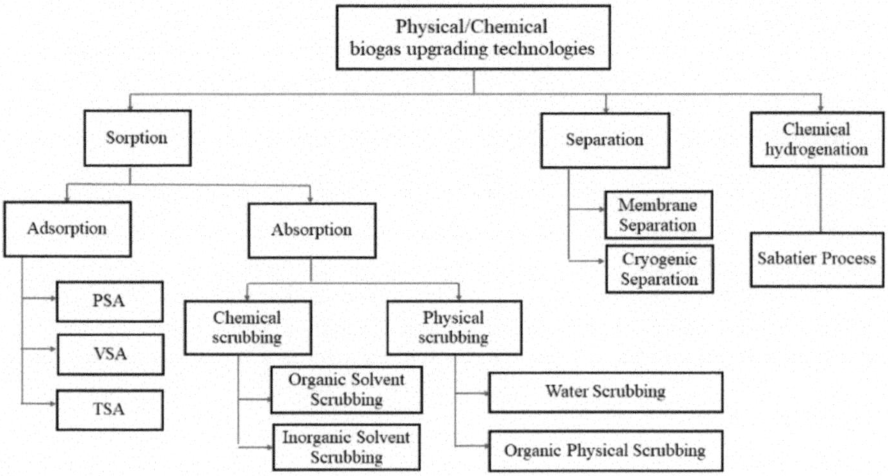

Fig. 5.3 Physical and chemical biogas upgrading technologies

Fig. 5.3. Methane recovery can be reached >96% through physicochemical upgrading processes [25, 26]. Biological methods include chemoautotrophic biogas upgrading, photosynthetic biogas upgrading, and bioelectrochemical biogas upgrading [26, 37].

5.5.1 *Physical and Chemical Methods*

5.5.1.1 Absorption

Solubility of CO_2 and CH_4 is different in different solvents. Based on this principle, absorption technology has been developed for biogas upgrading. If a selective solvent, which has a higher solubility for CO_2 than CH_4, is introduced as a scrubbing liquid a large amount of CO_2 can be dissolved while kept in contact with the raw biogas, thus separating the two gasses. Absorption can be achieved by either physical scrubbing or chemical scrubbing [38]. In case of physical scrubbing no chemical reaction takes place, while in case of chemical scrubbing, the separation is achieved by absorbent reacting with the absorbed substance.

Physical Absorption

Selection of suitable solvent is critical for physical absorption technologies. Available solvents must have suitability for gas solubility, also they are required to be non-hazardous in nature alongside being efficient and economical [2, 25]. Mainly, water or some organic solvent is used for biogas upgrading via physical absorption.

Water Scrubbing

Among all biogas upgrading technologies water scrubbing is the most widely used method applied at industrial scale. In this case, water is used as the selective absorbent. It is driven by the fact that solubility of CO_2 in water is 26 times higher compared to that of CH_4 [26] at 25 °C. CO_2 is absorbed by high pressure water with counter-current flow as untreated gas is injected into the bottom of an absorption tower at a high operation pressure, which is in general packed randomly in order to increase the gas–liquid contact surface [39]. In the desorption column, CO_2 is removed from water by the addition of atmospheric air. The regenerated absorbent is then returned to the absorption unit.H_2S can also be removed together with CO_2 in this method, because both H_2S and CO_2 are more polar with respect to CH_4. However, when the H_2S concentration is higher, pre-removal of H_2S is required to reduce operational hazards [40, 41].

Organic Solvent Scrubbing

This upgrading technology uses the principle similar to that in water scrubbing process, apart from the fact that organic solvents such as methanol and dimethyl ethers of polyethylene glycol (DMPEG) are used instead of water as absorbents [28, 42, 43]. Advantage of these solvents is that they have considerably higher solubility of CO_2 compared to that in water [44]. However, this needs pretreatment of the biogas to separate water and several inter-cooling/-heating stages required to guarantee an efficient operation. In addition, users need to be careful about constructing pipes and equipment due to the corrosive nature of organic compounds [45].

Chemical Absorption

As stated earlier, chemical absorption involves a chemical reaction between the solvent and absorbed substance (CO_2). Following the reaction of CO_2 and the solvent used, the rich liquid formed by the reaction enters the desorption tower to heat and decompose CO_2. Chemical solvents tend to be favored over physical solvents when the concentration of CO_2 is low [26]. Depending on the solvents used chemical absorption method can be divided into Amine scrubbing and Inorganic solvent scrubbing [37].

Amine Scrubbing

Amine scrubbing is the most widely used organic absorption technology. Ethanolamine (C_2H_7NO), diethanolamine ($C_4H_{11}NO_2$), triethanolamine ($C_6H_{15}NO_3$), and methyl-diethanolamine ($C_5H_{13}NO_2$) are used for this purpose [44, 46]. H_2S can also be removed by this process [1, 25]. The reactions involving absorption and subsequent desorption are given in Eqs. 5.13 and 5.14. As shown in these equations, during absorption CO_2 molecules from the gas phase interact with the amine molecules in the liquid phase. Amine molecules form chemical bonds with CO_2 molecules, effectively removing it from the gas stream. And desorption process is achieved by heating CO_2-loaded amine solution. Increased temperature reduces the affinity of the amine for CO_2, breaking the chemical bonds between them. The released CO_2 exits the amine solution as a concentrated CO_2 stream, which can then be compressed and further processed or stored. The amine solution, now depleted of CO_2, can be returned to the adsorption step to capture more CO_2 from the incoming gas stream. This step is crucial for regenerating the amine solution and allowing it to be reused for further CO_2 capture.

$$\textbf{Absorption of } \mathbf{CO_2}\textbf{:}\ RNH_2 + H_2O + CO_2 \rightarrow RNH_3^- + HCO_3^- \tag{5.13}$$

$$\textbf{Desorption of } \mathbf{CO_2}\textbf{:}\ RNH_3^- + HCO_3^- \rightarrow RNH_2 + H_2O + CO_2 \tag{5.14}$$

However, more research is needed in the context of solvent regeneration as it takes a substantial amount of energy [26].

Inorganic Solvent Scrubbing

In this method, alkaline aqueous solution of salts such as NaOH, $Ca(OH)_2$, KOH, Na_2CO_3, and K_2CO_3, is used to convert CO_2 into carbonates, and can be safely returned to the environment by precipitation [47]. The reactions are represented in the Eqs. (5.15)–(5.18). The principle of these reactions are same as amine scrubbing.

$$2NaOH(aq) + CO_2 \rightarrow Na_2CO_3(aq) + H_2O(l) \tag{5.15}$$

$$Na_2CO_3(aq) + Ca(OH)_2(s) \rightarrow 2NaOH(aq) + Ca_2CO_3(s) \tag{5.16}$$

$$2KOH(aq) + CO_2 \rightarrow K_2CO_3(aq) + H_2O(l) \tag{5.17}$$

$$K_2CO_3(aq) + Ca(OH)_2(s) \rightarrow 2KOH(aq) + Ca_2CO_3(s) \tag{5.18}$$

5.5.1.2 Adsorption

This technology is built on the principle of adsorption of gas molecules (adsorbate, in this context CO_2 in general) onto solid surfaces (adsorbent) by means of Van der Waal forces, thus separating CO_2 and CH_4. It is generally operated at high pressure and temperature. The loading of the adsorbent depends on pressure and temperature. Based on the method used for the regeneration of adsorbent adsorption technologies are divided into three categories: reduction of atmospheric pressure level (pressure swing adsorption or PSA), the application of vacuum (vacuum swing adsorption or

VSA), and conversion of a higher temperature at the same pressure level (temperature swing adsorption or TSA) [37]. Among these technologies, PSA is the most used one, where regeneration of adsorbents is achieved by reducing the total pressure in the system.

Porous solids are used as adsorbent materials due to their high specific surface area. Adsorbents should have a higher selectivity for CO_2 to the surface than CH_4, alongside compatible pore size with that of particle size entering in [48]. Moisture needs to be removed prior to adsorption in order to avoid poisoning of the adsorbent material. Typically, activated carbon, zeolites, titanosilicates, silica gels, and carbon molecular sieves are used as adsorbent materials [49]. Depending upon the adsorbent the process can selectively or simultaneously remove moisture, H_2S, and other impurities present in biogas alongside CO_2.

5.5.1.3 Separation

Membrane Separation

In Biogas membrane separation, gasses are being separated by taking advantage of different permeation rates of the constituent gasses in the membrane driven by pressure differences. This technology is one of the newest methods and gaining importance increasingly. CO_2 passes through the membrane to the permeate side, while CH_4 remains in the inlet side [44, 47]. Usually, Polyimide and cellulose acetate-based membranes are used for biogas upgrading [50]. It is an easy process with no chemical requirement. But, one of the main limitations of this process is that some of the CH_4 molecules may also pass through the membrane, so achieving high purity means large CH_4 losses [25]. The biogas needs cleaning for the removal of particles, H_2O, H_2S, VOCs, NH_3, and siloxanes to avoid deterioration and clogging of the membrane. This process may have several membrane stages to achieve greater purity of the biogas.

On the basis of separation media, the process can be grouped into two types: dry (gas/gas separation) and wet (gas/liquid separation) membrane separation techniques. The difference between dry and wet process is that the hydrophobic properties of membranes are utilized in the latter one. Wet process combines membrane separation method with the absorption method. Gas molecules diffuse through the membrane and are absorbed by the solvent flowing counterwise [51].

Cryogenic Separation

The boiling point of CO_2 (−78 °C) is considerably higher than that of CH_4 (−161.5 °C). Using this information, a mixture of gases is separated in the cryogenic separation method. Raw biogas is compressed and cooled so that CO_2 becomes either solid or liquid, while CH_4 remains gaseous, and gets purified. In particular, cryogenic separation represents a promising solution for simultaneous purification and liquefaction [27].

This technology is still under development. It can produce high-purity biomethane, but does not prove as energy efficient as cryogenic separation involves high pressure and low temperature with adequate insulation system [46].

5.5.1.4 Chemical Hydrogenation Process

In this upgrading technology, CO_2 is converted to methane, by reacting with H_2 (shown in Eq. 5.19).

$$4H_2 + CO_2 \rightarrow CH_4 + 2H_2O \quad (5.19)$$

This reduction reaction is carried out either by organisms or by catalysts. The catalyst-based reaction is called Sabatier reaction. Nickel and Ruthenium are generally used as catalysts. However, Nickel catalysts are very sensitive to H_2S, so the raw biogas must be completely free of H_2S to keep the catalyst in order [37]. This has been successfully used to remove CO_2 from contaminated natural gas, however, associated CH_4 removed with CO_2 is still relatively high. Thus, the reduction of energy consumption and methane loss are critical for further development of this technology [52] (Tables 5.4 and 5.5).

5.5.2 Biological Methods

This section briefly discusses the various biological technologies for biogas upgrading which are based on microbial activities that utilize impurities present in biogas as a carbon source. Unlike, physical and chemical biogas upgrading technologies which are somewhat mature and used in industries commercially, biological techniques have mainly been examined at a laboratory or pilot scale [53].

However, in recent years, various types of biological techniques have gained attention for biogas upgrading purposes due to their low cost, low energy requirement, and environment friendliness [25]. Among them, one is called chemoautotrophic biogas upgrading. In this method, hydrogenotrophic methanogens, mainly microorganisms from the archaeal domain such as Methanobacterium sp., Methanococcus sp., Methanothermobacter sp., Methanosarcina sp., Methanosaeta sp., Metanospirilum sp., and Methanoculleus sp., convertCO_2 to CH_4 by using H_2 [27]. Microorganisms use CO_2 as their carbon source and electron acceptor, and H_2 as an electron donor, as shown in the reaction represented by Eq. 5.20 [54]:

$$4H_2 + CO_2 \rightarrow CH_4 + 2H_2O \quad (5.20)$$

Table 5.4 Comparison of physicochemical biogas upgrading technologies [19, 26–28]

	Water scrubbing	Physical scrubbing	Chemical adsorption	PSA	Membrane separation	Cryogenic	Sabatier process
CH_4 losses (%)	<2	2–4	<0.1	<4	<0.6	2	–
CH_4 enrichment (%)	96–98	96–98	96–99	96–98	96–98	97–98	97–99
Pretreatment	Recommended	Recommended	Needed	Needed	Recommended	Needed	Recommended
H_2S Co-removal	Yes	Possible	Contaminant	Possible	Possible	Yes	No
N_2&O_2 Co-removal	No	No	No	Possible	Possible	No	No
Cost	Medium	Medium	High	Medium	High	High	Medium

Table 5.5 Advantages and disadvantages of different physical and chemical biogas upgrading [1, 26, 44]

Technology	Advantages	Disadvantages
Water scrubbing	• High efficiency • Low methane losses • Simple process • No chemical requirement • NH3 and H_2S co-removal possible	• High energy requirement • Low flexibility regrading biogas composition • Difficulty in CO_2 recovery • Clogging due to bacterial growth
Organic solvent scrubbing	• High efficiency • Low methane losses • Greater solubility of CO_2 than water • Regeneration possible	• Expensive • Difficult operation • High energy regeneration
Chemical absorption	• Greater solubility of CO_2 than water • Co-removal of ammonia and H_2S • Easy regeneration	• High equipment costs • High heat required for regeneration • corrosion
PSA	• H_2S co-removal possible • Low energy used • Compact technique	• Methane losses • High capital cost • Pretreatment required
Membrane separation	• Simple construction and operation • Requires little maintenance • High reliability • No heat and chemicals required	• Low membrane selectivity • Multiple stages required to achieve high purity • Methane losses
Cryogenic separation	• Production of liquid methane • No chemical and water • Pure methane is obtained	• High operation and maintenance cost • Pretreatment necessary • High energy required for cooling

The source of H_2 is the hydrolysis of water. Electricity needed in the hydrolysis process comes from renewable sources like solar or wind to ensure the approach is sustainable. Subsequently, H_2 is injected into an anaerobic digester [55]. Researchers have proposed three different types of chemoautotrophic biogas upgrading technology based on the injection of H_2 in the system [36]. One is called "in situ biogas upgrading process", where H_2 is added directly into the anaerobic reactor. This method does not need the construction and installation of new reactor tanks. However, due to the consumption of CO_2, pH can increase remarkably which can affect overall biogas production. The second method is called "ex-situ biological upgrading", where H_2 is injected into an external anaerobic reactor containing hydrogenotrophic culture. This technology does not affect the overall biogas production but needs the construction of external reactor tanks. Also, increased H_2 partial pressure can lead to acetate production. Thirdly, a new "hybrid biological upgrading" has been proposed, which integrates the concepts of both 'in situ" and "ex-situ" upgrading technologies. Initially, H_2 is injected into the biogas reactor capturing a part of CO_2 and then the enriched CH_4 is further upgraded in a separate digester, thus avoiding both the problems of high pH level and potential acetate production [36, 37, 56].

Some researchers have also investigated bio electrochemical systems (BES) for biogas upgrading, where H_2 is produced bioelectrochemically. In a microbial electrolysis cell (MEC), electrons released by bacteria from the oxidation of organic substances in the anode can combine with protons to produce hydrogen in the cathode, which can be used for biogas upgrading [57]. Methanogens can also accept electrons directly from the electrodes to reduce CO_2 into CH_4 (shown in Eq. 5.21).

$$CO_2 + 8H^+ + 8e^- \rightarrow CH_4 + 2H_2O \quad (5.21)$$

Nevertheless, the main challenge for implementing these biotechnologies on an industrial scale is the low aqueous solubility of H_2, which limits the gas–liquid mass transfer. Therefore, lowering the availability of dissolved substrates for methanogens is resulted [53].

Nowadays, another novel biotechnology has been proposed and investigated by researchers that utilizes microalgal photosynthesis to remove CO_2 from biogas and upgrade the CH_4 content. This is the only biotechnology that is capable of removing CO_2, H_2S, and VOCs simultaneously [58]. It also offers the possibility of generating many value-added chemicals as a by-product. The integration of anaerobic digestion (AD) and algal cultivation involves a range of methods and approaches designed to harness the symbiotic potential of these processes. These strategies include sequential processing, where waste undergoes AD first, producing biogas and nutrient-rich digestate, which is then used to cultivate algae. Two-stage systems utilize separate bioreactors for AD and algae, allowing independent process optimization. Photobioreactors and raceway ponds are integrated into wastewater treatment facilities, with digestate directed to algae for efficient nutrient utilization. Anaerobic Membrane Bioreactors (AnMBRs) combine AD with membrane filtration technology, enhancing nutrient recovery. Controlled carbon dioxide injection ensures algae efficiently utilize CO_2 from AD. Monitoring and control systems optimize conditions, while strategies for algal harvesting and biomass utilization complete the integration cycle. These diverse methods demonstrate the adaptability of AD-algae integration for efficient waste treatment, resource recovery, and sustainable biomass production. In the next section, we will focus extensively on this technology of biogas upgrading via microalgal cultivation, also called photoautotrophic technology.

5.6 Evolution of Photosynthesis

This section will provide the reader with an overview of photosynthetic machinery developed over time by algae. Prior to harnessing solar radiation, early prokaryotic organisms on earth during Archean Period obtained energy for their metabolism and movement by oxidizing organic and inorganic materials like H_2S, S, or Fe [59–61]. However, due to the low energy efficiency inherent in chemosynthesis, these primitive cells were required to consume a substantial amount of substrate to meet their metabolic energy demands. Subsequently, these prokaryotic bacteria evolved to utilize multiple pigments termed bacteriochlorophylls for harnessing solar energy [62, 63]. As a consequence, photosynthesis emerged significantly as a more energetically efficient process compared to chemosynthesis. The entirety of such primitive photosynthetic bacteria relied on a lone photosystem, resulting in the absence of oxygen evolution, as depicted in Fig. 5.4 [64].

As the oxygen levels in the early Earth's atmosphere remained exceedingly low, these organisms underwent adaptations to thrive in anoxic conditions [63, 65, 66]. Prokaryotic cyanobacteria demonstrated a remarkable photosynthetic capability with the use of two sequential photosystems. This photosynthesis process resembles that of modern day cyanobacteria and algae. Through these photosystems, fuelled by photosynthetic pigment chlorophyll, the organisms were able to efficiently conduct linear electron transport [67]. They absorbed carbon dioxide (CO_2) and skilfully harnessed ATP energy and NADPH + H^+ to convert CO_2 into sugar, leading to the production of diverse organic materials like starch and cellulose as exhibited in Eq. 5.22 [68]. Notably, the energy generated by the photosystems was sufficiently potent to split water into protons, electrons, and oxygen.

$$6CO_2 + 12H_2O \rightarrow C_6H_{12}O_6 + 6O_2 + 6H_2O \tag{5.22}$$

Similarly, Eukaryotic algae also engage photosynthetic mechanism that employs two photosystems. As shown in Fig. 5.5, algae chloroplasts exhibit various shapes, such as spirals in Spirogyra, ribbons in Ulothrix, or stars in Zygonema [69].

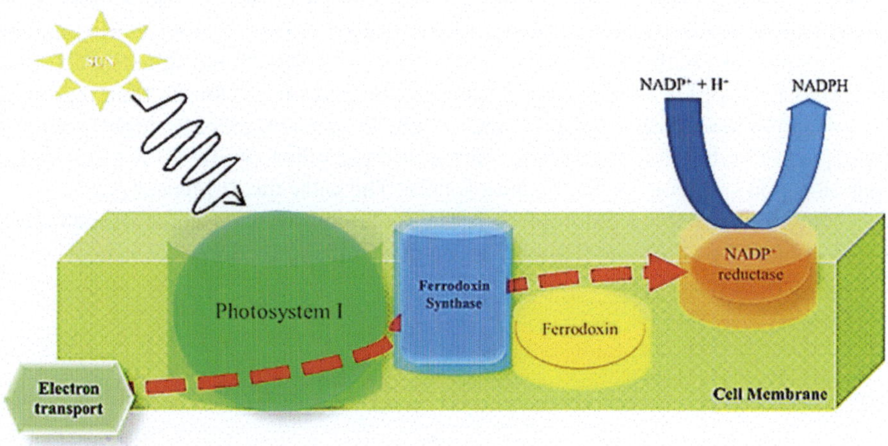

Fig. 5.4 Graphical illustration of primitive photosynthetic mechanism

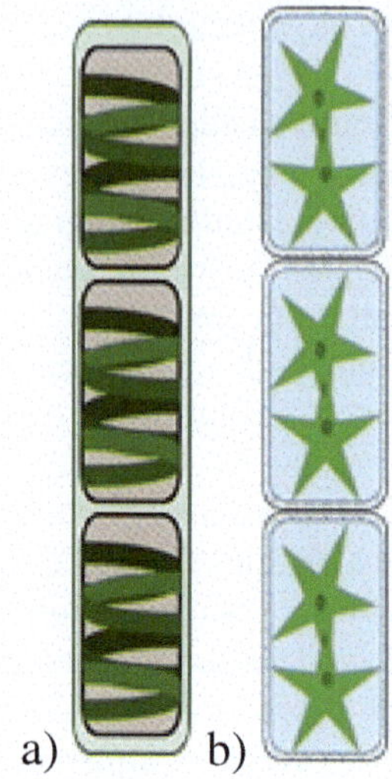

Fig. 5.5 Illustration of different chlorophyll shapes **a** spiral chlorophyll in spirulina **b** star shaped chlorophyll in Zygonema

Within these chloroplasts, photosynthetic pigments are located on thylakoid membranes, forming an interconnected system of vesicles. These vesicles can either stack up as grana thylakoids or traverse through the stroma as stroma thylakoids [70–72]. Additionally, the chloroplasts contain lipid droplets, starch grains, DNA, and ribosomes [73]. These findings suggest the possibility that heterotrophic eukaryotes may have formed symbiotic associations with cyanobacteria-like organisms [74, 75]. The molecule involved in photosynthesis, chlorophyll, illustrated in Fig. 5.6 comprises a tetrapyrrole ring system with a central Mg atom, and it is hydrophilic with a lipophilic chain attached to the lower left ring [76, 77].

Different variations of chlorophylls exist, including Chl A, B, C, and E, as well as bacteriochlorophylls found in photosynthetic bacteria [78]. Cyanobacteria contain Chl A, plants have both Chl A and B, and various algae possess multiple Chl C forms [79, 80]. Chlorophyll A's absorption spectrum displays peaks in the blue region at 435 nm and in the red region at 660 nm, while chlorophyll B exhibits slightly shifted peaks [81]. Notably, the absorption spectra of these chlorophylls do not match the solar emission spectrum under the atmosphere. The solar spectrum peaks at 550 nm, a wavelength that neither chlorophyll A nor B can absorb. Consequently, cells and tissues containing chlorophyll appear green due to the reflection or transmission of this wavelength band.

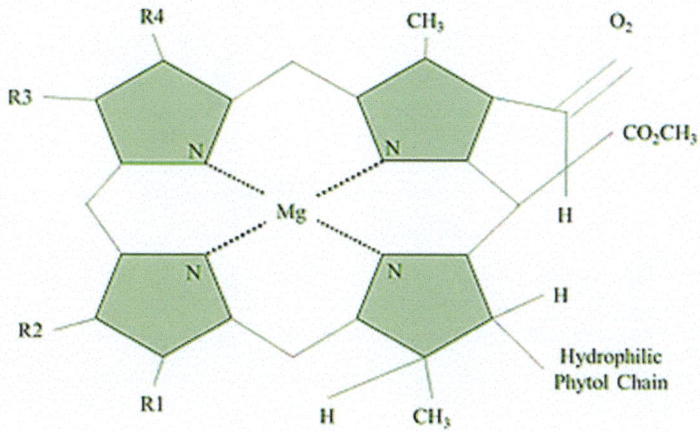

Name	R1	R2	R3	R4
Chlorophyll A	CH3	CH=CH2	CH3	CH2-CH3
Chlorophyll B	CH3	CH=CH2	CHO	CH2-CH3

Fig. 5.6 The basic structure of chlorophyll molecule with alkyl groups for chlorophyll A and B

5.7 CO_2 Sequestration in Algae

As elucidated, in algal cells photosynthesis is the fundamental process that facilitates and sustains their growth ensuring survival. The current section will trace the pathway of CO_2 metabolism in algae. This comprises an intricate series of biochemical reactions occurring within the cells, where light energy absorption is facilitated by pigments like chlorophyll [82]. The absorbed light initiates a cascade of light-dependent reactions, leading to the splitting of water molecules into oxygen and protons as shown in Fig. 5.7.

The electrons generated from the reactions then pass through intricate electron transport chains, establishing a proton gradient across the cell membrane, which in turn drives adenosine triphosphate (ATP) synthesis [83, 84]. The heart of carbon assimilation in algae lies in the Calvin-Benson cycle, the crucial enzymatic process (Fig. 5.8). In his groundbreaking work, Melvin Calvin (1911–1997) elucidated the precise mechanism of photosynthesis in plants and algae for fixingCO_2, leading to the formation of a C6 sugar that serves as the fundamental building block for all other organic materials [85]. The starting point involves ribulose-5-phosphate, which undergoes activation by ATP to yield ribulose-1,5-bisphosphate (RuBP) in the initial step. Subsequently, CO_2 is bound to RuBP by the highly abundant enzyme ribulose-1,5-bisphosphate carboxylase-oxygenase (RubisCo). This step, commonly referred to as carbon fixation, marks the initial assimilation of CO_2 into organic compounds.

In the subsequent steps of Calvin–Benson cycle,3PGA converts to triose-3-phosphate (G3P) through ATP activation and NADPH + H^+ reduction. These

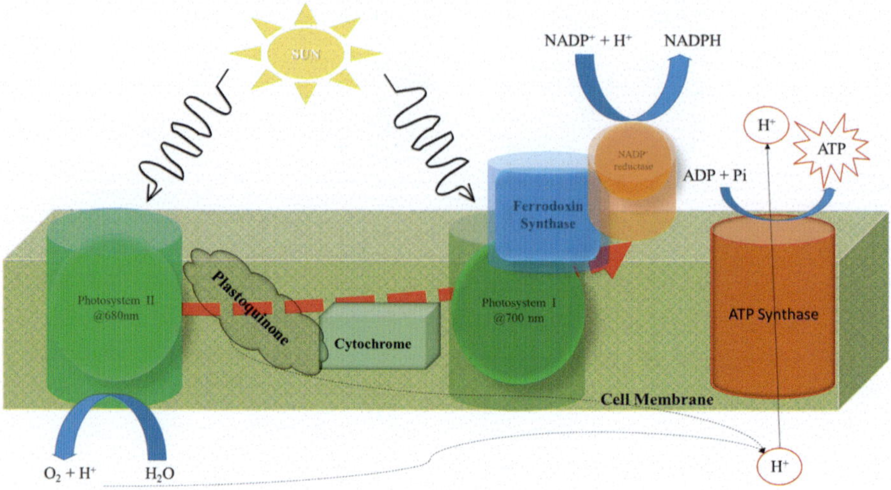

Fig. 5.7 Illustration of photosynthetic mechanism found in present day algae

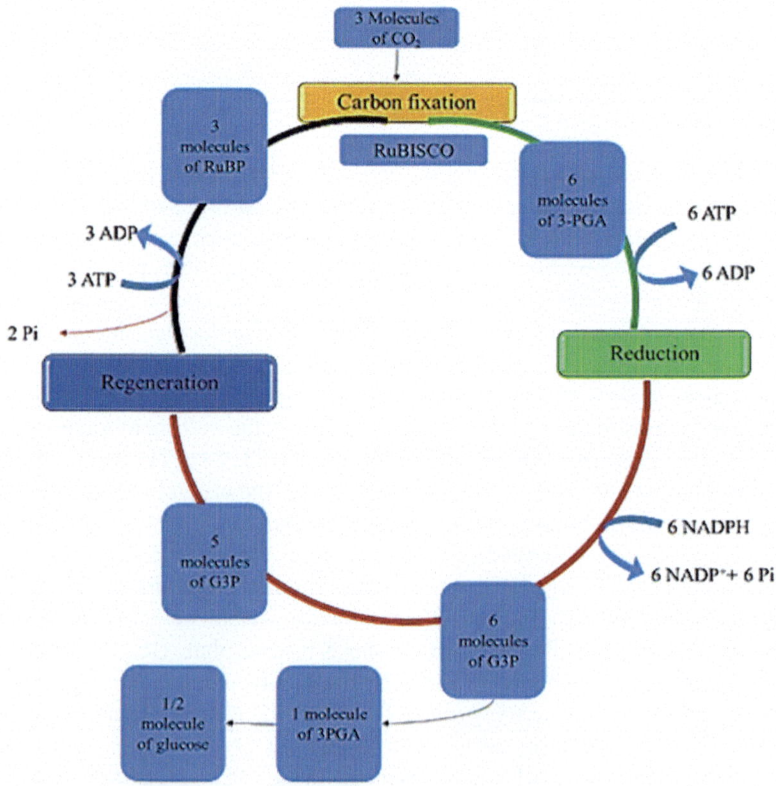

Fig. 5.8 Calvin Benson Bassham cycle

Table 5.6 CO_2 fixation rate of different algal and cyanobacterial strains

Strain	CO_2 uptake efficiency (g L^{-1} day^{-1})	References
Spirulina platensis	0.67	[89]
Anabaena sp. ATCC 33,047	1.45	[90]
Phaeodactylumtricornutum	2.47	[91]
A. platensis NIES-39	0.39	[92]
D. salina NIES-2257	0.22	[92]
Euhalothece sp. ZM-001	0.84	[92]
Trichocoleusdesertorum	1.57	[93]

G3P molecules serve as precursors for the synthesis of glucose and other essential organic compounds required for algal growth and metabolism Starting with six ribulose-5-phosphate, 12 triose-3-phosphate molecules are obtained, leading to fructose-1,6-bisphosphate and subsequent conversions to other organic compounds.

Through light-driven reactions and carbon assimilation, algae effectively capture CO_2 from the environment, helping regulate atmospheric carbon levels and contributing to the maintenance of ecological balance and atmospheric composition [86–88], as showcased in Table 5.6.

5.8 Algae-Based Biogas Upgradation

This section reviews the removal of the biogas impurities using algal cultivation systems.

5.8.1 CO_2 Removal

Due to their carbon capturing mechanism, algae hold a significant promise in the realm of biogas upgradation through the process of photosynthetic biogas upgrading. This bio-based approach involves the cultivation of algal biomass within a photo bioreactor fed with biogas. During this upgrading process, algal cells can utilize CO_2 present in crude biogas for photosynthesis, leading to methane enrichment (Fig. 5.9).

The advantages of utilizing algae for biogas upgrading are multifold. Notably, it is a sustainable process that can be used as a capturing mechanism for CO_2 from biogas, contributing to climate change mitigation. Additionally, the high-value algal biomass produced can serve multiple purposes, including applications in energy, food, and pharmaceutical industries [94].

Several microalgal species have been tested and observed to have promising potential for biogas upgrading, including *Spirulina platensis* [58, 95], *Chlorella vulgaris* [96, 97], *Dunaliella sp.* [98], *Nannochloropsis sp.* [99, 101], and *Scenedesmus sp.*

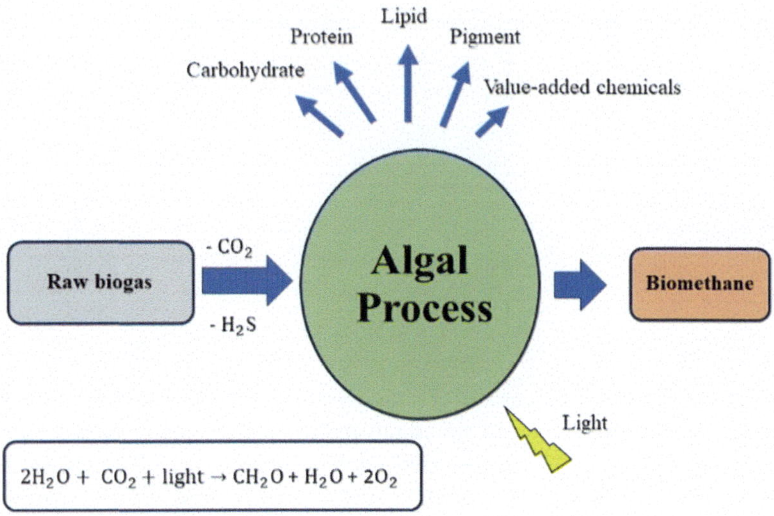

Fig. 5.9 Schematic representation of an integrated biogas upgradation system

[100, 102] as elucidated in Table 5.7. These species thrive in the presence of CO_2 and exhibit the capacity to effectively remove substantial amounts of CO_2 from biogas, while also ensuring relative ease of cultivation and high biomass yield.

Algae-based biogas upgrading may soon emerge as a viable option for biogas producers globally. Its environmentally friendly nature and potential to create new markets for algae biomass further underscore its significance in the biogas industry [103]. Additionally, its implementation could lead to the creation of new job opportunities in the biogas and algae sectors [104]. Furthermore, algae-based biogas upgrading is a versatile technology capable of upgrading biogas from diverse sources, including landfill gas, agricultural waste, and wastewater treatment plants [105]. This adaptability renders it a valuable asset in addressing energy and waste management challenges.

Table 5.7 CO_2 removal efficiency of different algal and cyanobacterial strains

Strain	CO_2 removal efficiency (%)	References
Spirulina platensis SAG 85.79	>98	[95]
Spirulina platensis	90	[58]
Chlorella vulgaris	69	[96]
Chlorella vulgaris	79	[97]
D. tertiolecta	85	[98]
Nannochloropsisgaditana	96	[99]
Scenedesmus sp.	>96	[100]

5.8.2 H_2S Removal

Another advantage of microalgae-based biogas upgrading is the capability of removing H_2S. In general, if the H_2S present in the biogas gets dissolved algae cultivation media (present in equilibrium with H^{S-} and S^{2-} in the liquid phase), it lowers the pH level inhibiting the microalgal culture by reducing photosynthetic oxygen production rate [106]. But, when the sulfide is oxidized to become sulfate, it can act as a sulfur source for microalgae, which is an essential micronutrient for algal growth [107]. Thereby, removing H_2S from the biogas.

Oxidization of sulfide can be done chemically or biologically. As mentioned earlier, sulfur-oxidizing bacteria (SOB) such as the *Thiobacillus* genus oxidize sulfide to sulfate in the presence of higher dissolved oxygen [53]. In conventional biological upgrading techniques, it can be a problem because most of the oxygen present in biogas remains in the gaseous phase rather than dissolved in the liquid phase. But for algae-based systems, SOBs use the required oxygen photo synthetically produced during the bio-fixation of CO_2 via microalgae from the liquid broth. Thus, SOB combined with a microalgae-based system offers the potential for the simultaneous removal of CO_2 and H_2S in a single step [108]. The process is represented as Eq. 5.22:

$$H_2S + CO_2 + \text{nutrients} + O_2 \rightarrow \text{Biomass} + SO_4^{2-}/S + H_2O \quad (5.23)$$

However, under limited oxygen conditions, the biological oxidation of H_2S might produce elemental S, which can cause operational problems [109, 110]. Moreover, the presence of excess HS^- or high concentrations of S^{2-} can inhibit microalgae activity and growth [111, 112]. Thus, despite its potential, it is still in its early developing phase and needs more studies as few reports are available in the literature.

5.8.3 VOCs Removal

Alongside CO_2 and H_2S removal, some researchers have reported the removal of volatile organic compounds (VOCs) present in biogas such as methyl mercaptan, toluene, and hexane using algal or algal–bacterial systems based on *Mychonasteshomosphaera*, *Tetraselmischuii*, and *Nanochloropsisgaditana* [113, 114]. However, interactions between VOCs and microalgal culture are still majorly unknown.

5.9 Value-Added Products Through Algal Upgradation

The biogas upgradation process using algae offers the opportunity to obtain several value-added products, which can contribute to a more sustainable and economically viable system which has been extensively discussed in the current section (Fig. 5.10).

5.9.1 Biomethane

Biomethane stands as the flagship product of photosynthetic biogas upgrading. This transformative process revolves around the meticulous removal of CO_2 from biogas. The resulting enriched methane gas, known as biomethane, emerges as a highly prized renewable energy resource with exceptional versatility. The value of biomethane lies in its remarkable potential to drive positive environmental impact [115]. Its applications span a wide spectrum, from providing efficient heating solutions to serving as a clean and reliable source for electricity generation. Moreover, biomethane can be harnessed as a low-carbon alternative fuel for vehicles, further reducing emissions and promoting sustainable transportation [116]. This lessens society's dependence on traditional fossil fuels. The process contributes significantly to mitigating climate change and advancing sustainable practices. Beyond environmental advantages, the widespread adoption of biomethane supports energy security and independence for communities, industries, and nations [117]. As a locally produced and renewable

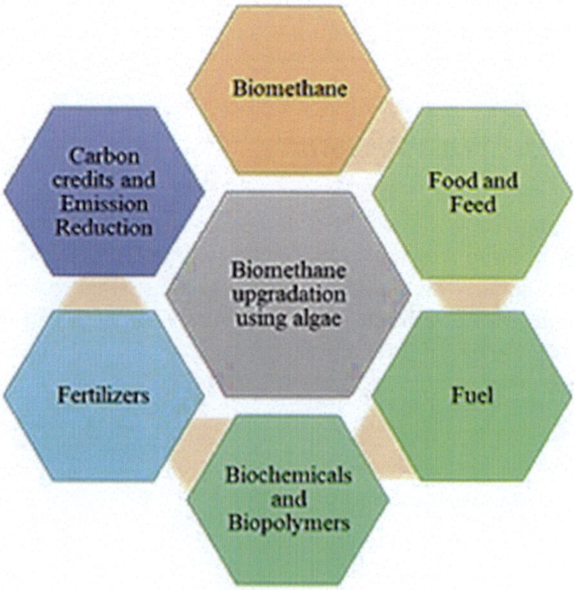

Fig. 5.10 Summary of value-added products emerging from such an integrated system

resource, biomethane empowers regions to reduce their reliance on imported fossil fuels, enhancing energy resilience and self-sufficiency.

5.9.2 Food and Feed

The cultivation of algae during the biogas upgrading process unlocks a treasure trove of potential in the form of high-value algal biomass. This biomass represents a remarkable amalgamation of vital nutrients, including proteins, lipids, carbohydrates, and an array of essential compounds. The richness of algae as a source of nutrition is undeniable [118]. Bursting with protein, these microorganisms offer a sustainable and eco-friendly alternative to conventional protein sources like poultry, meat, and dairy. It can easily cater to the growing demand for protein-rich diets, particularly in regions facing nutritional deficiencies. Essential fatty acids in algae, such as omega-3 and omega-6, are critical for human health, playing vital roles in brain function [119]. Cardiovascular health and immune system regulation are yet another coveted nutritional component found in algae [120]. Algae are also known to be rich in vitamins such as vitamin A, vitamin C, and various B vitamins, along with essential minerals like iron, calcium, and magnesium. These micronutrients can be harnessed to fortify food products, ensuring the provision of a balanced and wholesome diet [121]. In areas of functional foods and nutritional supplements, due to their rich nutrient profile, algae present an opportunity to create functional food products that offer specific health benefits, such as immune support, cardiovascular health, or cognitive enhancement. Moreover, algal biomass can find utility as a feed supplement in livestock and aquaculture industries, bolstering the nutrition of animals and enhancing their overall health and growth [122]. By incorporating algal-derived proteins, vitamins, and minerals into animal feed, it is possible to fortify the end products while enhancing the overall health and growth of livestock and aquaculture species, contributing to sustainable and efficient food production.

5.9.3 Fuel

The lipid content of the algal biomass holds immense significance. Extracting these lipids can yield valuable algal oils, which serve as a renewable source for producing biofuels like biodiesel. Through innovative conversion processes, algae pave the way for the production of two significant biofuels: biodiesel and bioethanol. The extraction of lipids from algal biomass unlocks the pathway to biodiesel production. These lipids, rich in energy content, can be refined and processed into biodiesel which exhibits a lower carbon footprint, contributing to the global efforts to combat climate change [123]. Carbohydrates found in the algal biomass are another invaluable component. These carbohydrates hold promise as potential feedstock for bioethanol production through fermentation [124]. The sustainable generation of bioethanol from algae aligns with the goals of renewable energy advancement and the transition to greener fuel options. The cosmetics and pharmaceutical industries, for instance, can tap into the potent bioactive compounds found in algae, exploring their potential for skincare products and therapeutic formulations. The potential of algal biofuels lies in their versatility and compatibility with existing infrastructure. Biodiesel and bioethanol produced from algae can readily integrate into existing engines and fuel distribution systems, making their adoption seamless and efficient [125]. In conclusion, the integration of algal biomass into the production of biofuels presents a promising and sustainable avenue for renewable energy. With the capability to produce biodiesel and bioethanol, algae offer a renewable alternative to conventional fossil fuels, mitigating climate change, promoting energy security, and reducing the environmental impact. As the world seeks cleaner and greener energy solutions, the transformative potential of algal biofuels emerges as a significant step towards a more sustainable future.

5.9.4 Biochemicals and Bioplastics

Algal biomass serves as a valuable resource for the production of biochemicals. Algae contain a vast array of bioactive compounds, such as antioxidants, pigments, and enzymes, which can be extracted and utilized in various applications [126]. These biochemicals find use in cosmetics, pharmaceuticals, and nutraceuticals, offering safer and greener alternatives to synthetic chemicals. Beyond biochemical, by harnessing the unique properties of algae-derived polymers, a new era of eco-friendly materials emerges, offering a sustainable solution to the global plastic waste challenge [127]. Algae-derived polymers possess inherent biodegradability, a critical characteristic that sets them apart from conventional plastics derived from fossil fuels [128]. These biodegradable materials offer a promising alternative to single-use plastics, leading to a reduction in the accumulation of plastic waste in landfills and marine environments. Moreover, the use of algae-derived polymers fosters innovation in various industries, including packaging, consumer goods, and biomedical applications.

5.9.5 Fertilizer

The leftover biomass from biomethane production emerges as a valuable resource for nutrient-rich organic fertilizers. The residue from anaerobic digestion process is rich in nitrogen, phosphorus, potassium, and other essential elements required for plant growth [129]. Through a process of proper treatment and composting, this nutrient-rich biomass is transformed into organic fertilizer. When applied to agricultural fields, this organic fertilizer serves as a powerful catalyst for enhancing soil health and fertility [130]. Its natural composition enriches the soil with vital nutrients, stimulating microbial activity and fostering a favorable environment for plant growth. The improved soil structure and nutrient content contribute to better water retention, reducing the need for excessive irrigation and conserving water resources. The use of such fertilizers also promotes a closed-loop approach, as the leftover reside from anaerobic digestion is transformed into a valuable resource, rather than being discarded as waste [131]. This circular economy approach aligns with sustainable practices, minimizing resource wastage and maximizing resource efficiency.

5.9.6 Carbon Credits and Emission Reduction

Nations participating in the Paris Agreement submitted their Nationally Determined Contributions (NDCs), outlining their climate action plans and emission reduction targets [132]. The use of sustainable biogas technologies, including algae-based systems, can contribute to countries' efforts to achieve their NDC commitments. Many countries and regions have implemented Renewable Energy Standards or set renewable energy targets to increase the share of renewable sources in their energy mix [133]. Such policies encourage the use of biogas and other renewable technologies for electricity generation and heat production. Various countries and jurisdictions have also adopted carbon pricing mechanisms, such as carbon taxes or cap-and-trade systems [134]. These policies put a price on carbon emissions and create financial incentives for industries to reduce their carbon footprint [135]. By capturing and utilizing CO_2 from biogas, the photosynthetic biogas upgrading system contributes to emission reduction. This can result in the generation of carbon credits or other incentives for carbon offsetting, which can be monetized or used to support sustainable practices [136]. Companies adopting photosynthetic biogas upgrading could benefit from carbon credits or offsets under such policies. Companies, especially in the energy and agriculture sectors, often adopt sustainability commitments, including carbon neutrality goals.

Overall, such an integrated system presents a unique opportunity to obtain a range of value-added products. The ability of algae to absorb CO_2 during their growth further contributes to the sustainability of biofuel production. By capturing carbon dioxide, algae effectively act as carbon sinks, offsetting the emissions generated during the combustion of biofuels, resulting in a near-neutral carbon footprint. These products contribute to sustainability, circular economy principles, and environmental protection while creating economic opportunities in various industries.

5.10 Limitations and Future Perspectives of Algal Process for Biogas Upgrading

Physical/chemical technologies available like PSA, water/chemical/organic solvent scrubbing, and selective membrane separation used individually or in combination have been successfully implemented in industries and demonstrated satisfactory removal of CO_2, H_2S, H_2O, VOCs (including aromatic hydrocarbons, halocarbons, and siloxanes) [27]. But large investment costs, high operating expenses, high energy consumption, environmental unsustainability, and significant CH_4 losses limit these conventional biogas upgrading technologies [39]. In this context, algal-based biogas upgrading technology, which is still at the initial stage of development, possesses a high potential to replace those conventional ones. It can remove CO_2 and H_2S simultaneously, alongside with it can also produce various value-added chemicals. Many factors affecting the system need to be further studied for further development of this technology. First of all, the selection of algal species is very important. Among more than forty thousand various microalgae species identified, only a few such as *C. vulgaris*, *S. obliquus*, *C. sorokiniana*, and *Chlorella kessleri* have been studied and found to have rapid growth rates, strong resistance against pollution risks, tolerance to CO_2, H_2S and metal ions [137–139]. Optimization of algal culture strategies like mono-cultivation, microalgal co-cultivation, and co-cultivation with fungi or bacteria is required to be determined [140]. A suitable microalgal culture medium is needed to be chosen as it might affect the growth rate of algae. Light intensity, photoperiod, and mixing light also play a major role in microalgae growth as light is the source and power of cell synthesis. So, finding optimum light intensity is necessary. The transfer and solubility of CO_2 from gas to liquid, rates of metabolism, and cell chemical compositions of microalgae are affected by temperature. A stable and appropriate temperature thus needs to be set. The L/G ratio also influences CO_2 solubility. Increased L/G ratio leads to stripping of N_2 and O_2, thereby leading to poor biogas content [141]. Generally, CO_2 dissolves in water in three different forms; carbonic acid, bicarbonate, and carbonate. For microalgae to take CO_2 in the form of bicarbonate, the pH level of the algal broth needs to be stayed in the alkaline phase [141]. All these factors require monitoring and optimization for effective and efficient biogas upgrading. Moreover, the performance of this technology in order to remove siloxanes, benzene/toluene/xylene (BTX), or halocarbons is still little known [53]. In addition, large-scale/industrial implementation of this technology requires large lands for algal cultivation, which limits its applicability compared to conventional physical/chemical technologies [53]. If these factors are sorted, due to environmental friendliness, nutrient and carbon recycling ability, along with low cost and energy consumption, will encourage the use of algal biogas upgrading technology. In the foreseeable future, the prospects of algae-based biogas upgrading appear promising, driven by the imperative for sustainable energy solutions and carbon capture. Key developments on the horizon encompass genetic engineering to create advanced algae strains capable of more efficient CO_2 capture and methane production. Additionally, photobioreactor technology will likely see

Table 5.8 Advantages and disadvantages of different algae-based biogas upgrading [1, 26, 44]

Technology	Advantages	Disadvantages
Represents the algal biomass scrubbing	• CO_2 capture • Nutrient utilization • Potential for value-added products	• Energy-intensive • Capital intensive • Complexity • Scale-up challenges
Microalgal photobioreactors	• Precise control • Efficient CO_2 fixation • High biomass production	
Anaerobic digestion with algae (ADA)	• Simultaneous processes • Wastewater treatment • Energy recovery	

enhancements in light and nutrient delivery systems, improving overall efficiency and scalability. Research into algal consortia will continue to optimize nutrient uptake and gas conversion. Innovations in scaling-up technologies will ease the transition from laboratory to industrial scales, facilitating wider adoption. The integration of algae-based upgrading with existing biogas systems and the maximization of resource recovery will also be areas of significant exploration. Supportive policies, financial incentives, and increased public awareness are expected to further drive research and commercial adoption of algae-based biogas upgrading, offering a pathway to reduced emissions and a more sustainable energy landscape. Table 5.8 represents the advantages and disadvantages of algal upgradation of biogas.

5.11 Conclusions

In conclusion the integration of anaerobic digestion and algal cultivation for biogas upgrading stands as a multifaceted and promising approach within the realm of renewable energy and resource recovery. Biogas, a product of anaerobic digestion, is a valuable source of renewable energy primarily composed of methane (CH_4) and carbon dioxide (CO_2), alongside other trace gases and impurities. It is imperative to remove CO_2 is paramount to maximize biogas' energy potential, making it economically viable for various applications. The elimination of other trace gases like hydrogen sulfide (H_2S) and volatile organic compounds (VOCs) is equally essential to ensure biogas quality, safety, and environmental compliance.

Conventional biogas treatment encompasses diverse physical and chemical methods, such as absorption, adsorption, and membrane separation, each carrying its unique advantages and limitations. Cleaning and upgradation processes, including physical and chemical methods, are vital in removing impurities and enhancing biogas quality. One of the biological methods discussed is the leveraging of the remarkable capacity of algae to sequester CO_2, contributing significantly to biogas quality improvement and the generation of value-added products. Algal cultivation presents a promising solution to enhance biogas quality by increasing methane concentration while utilizing the nutrient-rich effluent. Notably, this integration enables the production of a spectrum of valuable products, from biomethane to food, feed, biofuels, biochemicals, bioplastics, fertilizer, and even carbon credits and emission reduction. Integrating anaerobic digestion and algal cultivation processes holds potential implications for the bioenergy sector, although challenges must be addressed to maximize its benefits.

However, challenges persist, including nutrient management, efficient CO_2 utilization, energy-intensive processes, and variable wastewater compositions. Despite these challenges, ongoing research and development efforts are expected to address limitations and establish algae-based biogas upgradation as a sustainable and economically viable solution, aligning with global efforts to reduce greenhouse gas emissions, enhance sustainability, and transition towards a circular and environmentally conscious economy.

References

1. Awe OW, Zhao Y, Nzihou A, Minh DP, Lyczko N (2017) A review of biogas utilisation, purification and upgrading technologies. Waste Biomass Valor. 8:267–283. https://doi.org/10.1007/s12649-016-9826-4
2. Sun Q, Li H, Yan J, Liu L, Yu Z, Yu X (2015) Selection of appropriate biogas upgrading technology-a review of biogas cleaning, upgrading and utilisation. Renew Sustain Energy Rev 51:521–532. https://doi.org/10.1016/j.rser.2015.06.029
3. Damyanova S, Beschkov V (2020) Biogas as a source of energy and chemicals. In: Beschkov V (ed) Biorefinery concepts, energy and products. IntechOpen
4. Kougias PG, Angelidaki I (2018) Biogas and its opportunities—a review. Front Environ Sci Eng 12:14. https://doi.org/10.1007/s11783-018-1037-8
5. Steffen R, Szolar O, Braun R (1998) Feedstocks for anaerobic digestion. University of Agricultural Sciences, Institute of Agrobiotechnology Tulin, Vienna
6. Bong CPC, Lim LY, Lee CT, Ho WS, Klemes JJ (2017) The kinetics for mathematical modelling on the anaerobic digestion of organic waste- a review. Chem Eng Trans 61:1669–1674. https://doi.org/10.3303/CET1761276
7. Aworanti OA, Agbede OO, Agarry SE, Ajani AO, Ogunkunle O, Laseinde OT, Rahman SMA, Fattah IMR (2023) Decoding anaerobic digestion: A holistic analysis of biomass waste technology, process kinetics, and operational variables. Energies 16:3378. https://doi.org/10.3390/en16083378
8. Angelidaki I, Karakashev D, Batstone DJ, Plugge CM, Stams AJM (2011) Biomethanation and its potential. In: Methods in enzymology. Elsevier, pp 327–351
9. Nayono SE (2010) Anaerobic digestion of organic solid waste for energy production. KIT Scientific Publishing, Karlsruhe
10. Sarker S, Lamb JJ, Hjelme DR, Lien KM (2019) A review of the role of critical parameters in the design and operation of biogas production plants. Appl Sci 9:1915. https://doi.org/10.3390/app9091915
11. Felchner-Żwirełło M (2014) Propionic acid degradation by syntrophic bacteria during anaerobic biowaste digestion. KIT Scientific Publishing, Karlsruhe
12. André L, Ndiaye M, Pernier M, Lespinard O, Pauss A, Lamy E, Ribeiro T (2016) Methane production improvement by modulation of solid phase immersion in dry batch anaerobic digestion process: dynamic of methanogen populations. Biores Technol 207:353–360. https://doi.org/10.1016/j.biortech.2016.02.033

13. Koszel M, Lorencowicz E (2015) Agricultural use of biogas digestate as a replacement fertilizers. Agric Agric Sci Proc 7:119–124. https://doi.org/10.1016/j.aaspro.2015.12.004
14. Werkneh AA (2022) Biogas impurities: environmental and health implications, removal technologies and future perspectives. Heliyon 8:e10929. https://doi.org/10.1016/j.heliyon.2022.e10929
15. Jönsson O, Polman E, Jensen JK, Eklund R, Schyl H, Ivarsson S (2003) Sustainable gas enters the European gas distribution system. Danish gas technology center, pp 1–9
16. van Loosdrecht MCM, Nielsen PH, Lopez-Vazquez CM, Brdjanovic D (eds) (2016) Experimental methods in wastewater treatment. IWA Publishing, London
17. Paglini R, Gandiglio M, Lanzini A (2022) Technologies for deep biogas purification and use in zero-emission fuel cells systems. Energies 15:3551. https://doi.org/10.3390/en15103551
18. Papadias DD, Ahmed S, Kumar R (2012) Fuel quality issues with biogas energy—an economic analysis for a stationary fuel cell system. Energy 44:257–277. https://doi.org/10.1016/j.energy.2012.06.031
19. Ryckebosch E, Drouillon M, Vervaeren H (2011) Techniques for transformation of biogas to biomethane. Biomass Bioenerg 35:1633–1645. https://doi.org/10.1016/j.biombioe.2011.02.033
20. Paolini V, Petracchini F, Segreto M, Tomassetti L, Naja N, Cecinato A (2018) Environmental impact of biogas: a short review of current knowledge. J Environ Sci Health Part A 53:899–906. https://doi.org/10.1080/10934529.2018.1459076
21. Petersson A, Wellinger A (2009) Biogas upgrading technologies–developments and innovations. IEA Bioenergy 20:1–19
22. Persson M (2003) Evaluation of upgrading techniques for biogas. Report SGC, 142
23. Abatzoglou N, Boivin S (2009) A review of biogas purification processes. Biofuels Bioprod Bioref 3:42–71. https://doi.org/10.1002/bbb.117
24. Soreanu G, Beland M, Falletta P, Edmonson K, Svoboda L, Al-Jamal M, Seto P (2011) Approaches concerning siloxane removal from biogas–a review. Can Biosyst Eng 53(8):8.1–8.18
25. Angelidaki I, Xie L, Luo G, Zhang Y, Oechsner H, Lemmer A, Munoz R, Kougias PG (2019) Biogas upgrading: current and emerging technologies. In: Biofuels: alternative feedstocks and conversion processes for the production of liquid and gaseous biofuels. Elsevier, pp 817–843
26. Xie L, Xu J, Zhang Y, He Y (2020) Biogas upgrading. In: Advances in bioenergy. Elsevier, pp 309–344
27. Muñoz R, Meier L, Diaz I, Jeison D (2015) A review on the state-of-the-art of physical/chemical and biological technologies for biogas upgrading. Rev Environ SciBiotechnol. 14:727–759. https://doi.org/10.1007/s11157-015-9379-1
28. Bauer F, Persson T, Hulteberg C, Tamm D (2013) Biogas upgrading–technology overview, comparison and perspectives for the future. Biofuels Bioprod Bioref 7:499–511. https://doi.org/10.1002/bbb.1423
29. Hosseini SE, Wahid MA (2014) Development of biogas combustion in combined heat and power generation. Renew Sustain Energy Rev 40:868–875. https://doi.org/10.1016/j.rser.2014.07.204
30. Maia DCS, Cardozo F, Frare L, Gimenes M, Pereira N (2014) Purification of biogas for energy use. Chem Eng Trans 37:643–648. https://doi.org/10.3303/CET1437108
31. Deng L, Hägg M-B (2010) Techno-economic evaluation of biogas upgrading process using CO2 facilitated transport membrane. Int J Greenhouse Gas Control 4:638–646. https://doi.org/10.1016/j.ijggc.2009.12.013
32. Scholz M, Melin T, Wessling M (2013) Transforming biogas into biomethane using membrane technology. Renew Sustain Energy Rev 17:199–212. https://doi.org/10.1016/j.rser.2012.08.009
33. Montebello AM (2013) Aerobic biotrickling filtration for Andrea Monzón Montebello. J Hazard Mater 280:200–208
34. Mora M, Fernández M, Gómez JM, Cantero D, Lafuente J, Gamisans X, Gabriel D (2015) Kinetic and stoichiometric characterization of anoxic sulfide oxidation by SO-NR mixed

cultures from anoxic biotrickling filters. Appl Microbiol Biotechnol 99:77–87. https://doi.org/10.1007/s00253-014-5688-5
35. Iovane P, Nanna F, Ding Y, Bikson B, Molino A (2014) Experimental test with polymeric membrane for the biogas purification from CO2 and H2S. Fuel 135:352–358. https://doi.org/10.1016/j.fuel.2014.06.060
36. Kougias PG, Treu L, Benavente DP, Boe K, Campanaro S, Angelidaki I (2017) Ex-situ biogas upgrading and enhancement in different reactor systems. Biores Technol 225:429–437. https://doi.org/10.1016/j.biortech.2016.11.124
37. Adnan O (2019) Nomanbhay, chew, show: technologies for biogas upgrading to biomethane: a review. Bioengineering 6:92. https://doi.org/10.3390/bioengineering6040092
38. Singhal S, Agarwal S, Arora S, Sharma P, Singhal N (2017) Upgrading techniques for transformation of biogas to bio-CNG: a review: a review on upgradation techniques to transform biogas to bio-CNG. Int J Energy Res 41:1657–1669. https://doi.org/10.1002/er.3719
39. Bauer F, Hulteberg C, Persson T, Tamm D (2013) Biogas upgrading-review of commercial technologies
40. Angelidaki I, Treu L, Tsapekos P, Luo G, Campanaro S, Wenzel H, Kougias PG (2018) Biogas upgrading and utilization: current status and perspectives. Biotechnol Adv 36:452–466. https://doi.org/10.1016/j.biotechadv.2018.01.011
41. Aghel B, Behaein S, Wongwises S, Shadloo MS (2022) A review of recent progress in biogas upgrading: with emphasis on carbon capture. Biomass Bioenerg 160:106422. https://doi.org/10.1016/j.biombioe.2022.106422
42. Aghel B, Maleki M, Sahraie S, Heidaryan E (2021) Desorption of carbon dioxide from a mixture of monoethanolamine with alcoholic solvents in a microreactor. Fuel 306:121636. https://doi.org/10.1016/j.fuel.2021.121636
43. Götz M, Köppel W, Reimert R, Graf F (2011) Optimierungspotenzial von Wäschenzur-Biogasaufbereitung. Teil 1–PhysikalischeWäschen. ChemieIngenieurTechnik 83:858–866. https://doi.org/10.1002/cite.201000211
44. Zhou K, Chaemchuen S, Verpoort F (2017) Alternative materials in technologies for biogas upgrading via CO2 capture. Renew Sustain Energy Rev 79:1414–1441. https://doi.org/10.1016/j.rser.2017.05.198
45. Tock L, Gassner M, Maréchal F (2010) Thermochemical production of liquid fuels from biomass: thermo-economic modeling, process design and process integration analysis. Biomass Bioenerg 34:1838–1854. https://doi.org/10.1016/j.biombioe.2010.07.018
46. Kadam R, Panwar NL (2017) Recent advancement in biogas enrichment and its applications. Renew Sustain Energy Rev 73:892–903. https://doi.org/10.1016/j.rser.2017.01.167
47. Baena-Moreno FM, Rodríguez-Galán M, Vega F, Vilches LF, Navarrete B (2019) Review: recent advances in biogas purifying technologies. Int J Green Energy 16:401–412. https://doi.org/10.1080/15435075.2019.1572610
48. Augelletti R, Conti M, Annesini MC (2017) Pressure swing adsorption for biogas upgrading. A new process configuration for the separation of biomethane and carbon dioxide. J Clean Prod 140:1390–1398. https://doi.org/10.1016/j.jclepro.2016.10.013
49. Sahota S, Shah G, Ghosh P, Kapoor R, Sengupta S, Singh P, Vijay V, Sahay A, Vijay VK, Thakur IS (2018) Review of trends in biogas upgradation technologies and future perspectives. Bioresource Technol Rep 1:79–88. https://doi.org/10.1016/j.biteb.2018.01.002
50. Baker RW (2012) Membrane technology and applications. John Wiley & Sons
51. Harasimowicz M, Orluk P, Zakrzewska-Trznadel G, Chmielewski A (2007) Application of polyimide membranes for biogas purification and enrichment. J Hazard Mater 144(3):698–702
52. Xia A, Cheng J, Murphy JD (2016) Innovation in biological production and upgrading of methane and hydrogen for use as gaseous transport biofuel. Biotechnol Adv 34:451–472. https://doi.org/10.1016/j.biotechadv.2015.12.009
53. Toledo-Cervantes A, Lebrero R, Cavinato C, Muñoz R (2017) Biogas upgrading using algal-bacterial processes. In: Microalgae-based biofuels and bioproducts. Elsevier, pp 283–304
54. Xu H, Wang K, Holmes DE (2014) Bioelectrochemical removal of carbon dioxide (CO 2): an innovative method for biogas upgrading. Biores Technol 173:392–398. https://doi.org/10.1016/j.biortech.2014.09.127

55. Jürgensen L, Ehimen EA, Born J, Holm-Nielsen JB (2014) Utilization of surplus electricity from wind power for dynamic biogas upgrading: Northern Germany case study. Biomass Bioenerg 66:126–132. https://doi.org/10.1016/j.biombioe.2014.02.032
56. Aryal N, Kvist T, Ammam F, Pant D, Ottosen LDM (2018) An overview of microbial biogas enrichment. Biores Technol 264:359–369. https://doi.org/10.1016/j.biortech.2018.06.013
57. Zhang Y, Angelidaki I (2014) Microbial electrolysis cells turning to be versatile technology: recent advances and future challenges. Water Res 56:11–25. https://doi.org/10.1016/j.watres.2014.02.031
58. Bahr M, Díaz I, Dominguez A, González Sánchez A, Muñoz R (2014) Microalgal-biotechnology as a platform for an integral biogas upgrading and nutrient removal from anaerobic effluents. Environ Sci Technol 48:573–581. https://doi.org/10.1021/es403596m
59. Müller B (2023) Iron transport mechanisms and their evolution focusing on chloroplasts. J Plant Physiol 154059. https://doi.org/10.1016/j.jplph.2023.154059
60. Therby-Vale R, Lacombe B, Rhee SY, Nussaume L, Rouached H (2022) Mineral nutrient signaling controls photosynthesis: focus on iron deficiency-induced chlorosis. Trends Plant Sci 27:502–509. https://doi.org/10.1016/j.tplants.2021.11.005
61. Schad M, Konhauser KO, Sánchez-Baracaldo P, Kappler A, Bryce C (2019) How did the evolution of oxygenic photosynthesis influence the temporal and spatial development of the microbial iron cycle on ancient Earth? Free Radical Biol Med 140:154–166. https://doi.org/10.1016/j.freeradbiomed.2019.07.014
62. Nowicka B, Kruk J (2016) Powered by light: phototrophy and photosynthesis in prokaryotes and its evolution. Microbiol Res 186–187:99–118. https://doi.org/10.1016/j.micres.2016.04.001
63. Blankenship RE (2001) Molecular evidence for the evolution of photosynthesis. Trends Plant Sci 6:4–6. https://doi.org/10.1016/S1360-1385(00)01831-8
64. Soo RM, Hemp J, Hugenholtz P (2019) Evolution of photosynthesis and aerobic respiration in the cyanobacteria. Free Radical Biol Med 140:200–205. https://doi.org/10.1016/j.freeradbiomed.2019.03.029
65. Iverson TM (2006) Evolution and unique bioenergetic mechanisms in oxygenic photosynthesis. Curr Opin Chem Biol 10:91–100. https://doi.org/10.1016/j.cbpa.2006.02.013
66. Jones MR, Fyfe PK (2004) Photosynthesis: a new step in oxygen evolution. Curr Biol 14:R320–R322. https://doi.org/10.1016/j.cub.2004.03.055
67. Kirilovsky D, Büchel C (2019) Evolution and function of light-harvesting antenna in oxygenic photosynthesis. In: Advances in botanical research. Elsevier, pp 247–293
68. Zubay G (2000) Evolution of photosynthesis. In: Origins of life on the earth and in the cosmos. Elsevier, pp 451–485
69. Häder D-P (2022) Photosynthesis in plants and algae. Anticancer Res 42:5035–5041. https://doi.org/10.21873/anticanres.16012
70. Dekker JP, Boekema EJ (2005) Supramolecular organization of thylakoid membrane proteins in green plants. Biochim Biophys Acta (BBA) Bioenerget 1706:12–39. https://doi.org/10.1016/j.bbabio.2004.09.009
71. Mechela A, Schwenkert S, Soll J (2019) A brief history of thylakoid biogenesis. Open Biol 9:180237. https://doi.org/10.1098/rsob.180237
72. Chow WS, Kim E-H, Horton P, Anderson JM (2005) Granal stacking of thylakoid membranes in higher plant chloroplasts: the physicochemical forces at work and the functional consequences that ensue. Photochem Photobiol Sci 4:1081–1090. https://doi.org/10.1039/b507310n
73. Jones MR (2007) Lipids in photosynthetic reaction centres: structural roles and functional holes. Prog Lipid Res 46:56–87. https://doi.org/10.1016/j.plipres.2006.06.001
74. Zepernick BN, Wilhelm SW, Bullerjahn GS, Paerl HW (2023) Climate change and the aquatic continuum: a cyanobacterial comeback story. Environ Microbiol Rep. 15:3–12. https://doi.org/10.1111/1758-2229.13122
75. Amaral-Zettler LA, Zettler ER, Mincer TJ (2020) Ecology of the plastisphere. Nat Rev Microbiol 18:139–151. https://doi.org/10.1038/s41579-019-0308-0

76. Kê B (2001) Photosynthesis: photobiochemistry and photobiophysics. Kluwer Academic Publishers, Dordrecht; Boston
77. Fiedor L, Kania A, Myśliwa-Kurdziel B, Orzeł Ł, Stochel G (2008) Understanding chlorophylls: Central magnesium ion and phytyl as structural determinants. Biochim Biophys Acta (BBA) Bioenerget 1777:1491–1500. https://doi.org/10.1016/j.bbabio.2008.09.005
78. Suzuki JY, Bollivar DW, Bauer CE (1997) Genetic analysis of chlorophyll biosynthesis. Annu Rev Genet 31:61–89. https://doi.org/10.1146/annurev.genet.31.1.61
79. Tomitani A, Okada K, Miyashita H, Matthijs HCP, Ohno T, Tanaka A (1999) Chlorophyll b and phycobilins in the common ancestor of cyanobacteria and chloroplasts. Nature 400:159–162. https://doi.org/10.1038/22101
80. Larkum AWD (2020) Light-harvesting in cyanobacteria and eukaryotic algae: an overview. In: Larkum AWD, Grossman AR, Raven JA (eds) Photosynthesis in algae: biochemical and physiological mechanisms. Springer International Publishing, Cham, pp 207–260
81. Breton J, Katoh S (1987) Orientation of the pigments in photosystem II: low-temperature linear-dichroism study of a core particle and of its chlorophyll-protein subunits isolated from Synechococcus sp. Biochim Biophys Acta (BBA) Bioenerget 892:99–107. https://doi.org/10.1016/0005-2728(87)90252-0
82. Nelson N, Ben-Shem A (2004) The complex architecture of oxygenic photosynthesis. Nat Rev Mol Cell Biol 5:971–982. https://doi.org/10.1038/nrm1525
83. Zhao R, Jiang S, Zhang L, Yu Z (2019) Mitochondrial electron transport chain, ROS generation and uncoupling (Review). Int J Mol Med. https://doi.org/10.3892/ijmm.2019.4188
84. Gust D, Moore TA, Moore AL (2001) Mimicking photosynthetic solar energy transduction. Acc Chem Res 34:40–48. https://doi.org/10.1021/ar9801301
85. Bassham JA, Calvin M (1960) The path of carbon in photosynthesis. In: Pirson A (ed) Die CO2-assimilation/the assimilation of carbon dioxide. Springer, Berlin Heidelberg, Berlin, Heidelberg, pp 884–922
86. Wu W, Tan L, Chang H, Zhang C, Tan X, Liao Q, Zhong N, Zhang X, Zhang Y, Ho S-H (2023) Advancements on process regulation for microalgae-based carbon neutrality and biodiesel production. Renew Sustain Energy Rev 171:112969. https://doi.org/10.1016/j.rser.2022.112969
87. Prasad R, Gupta SK, Shabnam N, Oliveira CYB, Nema AK, Ansari FA, Bux F (2021) Role of microalgae in global co2 sequestration: physiological mechanism, recent development, challenges, and future prospective. Sustainability 13:13061. https://doi.org/10.3390/su132313061
88. Yaashikaa PR, Senthil Kumar P, Varjani SJ, Saravanan A (2019) A review on photochemical, biochemical and electrochemical transformation of CO2 into value-added products. J CO2 Utilizat 33:131–147. https://doi.org/10.1016/j.jcou.2019.05.017
89. Singh SK, Rahman A, Dixit K, Nath A, Sundaram S (2016) Evaluation of promising algal strains for sustainable exploitation coupled with CO_2 fixation. Environ Technol 37:613–622. https://doi.org/10.1080/09593330.2015.1075599
90. González López CV, AciénFernández FG, FernándezSevilla JM, Sánchez Fernández JF, CerónGarcía MC, Molina Grima E (2009) Utilization of the cyanobacteria anabaena sp. ATCC 33047 in CO2 removal processes. Bioresource Technol 100:5904–5910. https://doi.org/10.1016/j.biortech.2009.04.070
91. Dias RR, Deprá MC, Severo IA, Zepka LQ, Jacob-Lopes E (2022) Smart override of the energy matrix in commercial microalgae facilities: a transition path to a low-carbon bioeconomy. Sustain Energy Technol Assess 52:102073. https://doi.org/10.1016/j.seta.2022.102073
92. Kishi M, Toda T (2018) Carbon fixation properties of three alkalihalophilicmicroalgal strains under high alkalinity. J ApplPhycol 30:401–410. https://doi.org/10.1007/s10811-017-1226-z
93. Khan F, Malik S, Shahid A, Siddiqui AJ, Musharraf SG, Zhu H, Alkhattabi NA, Gull M, Mehmood MA (2023) Characterization of a newly isolated cyanobacterium Trichocoleusdesertorum BERC08 as a potential feedstock for the algal biorefinery. Biomass Conv Bioref 13:5283–5294. https://doi.org/10.1007/s13399-021-01537-3

94. Ahmad A, Banat F, Alsafar H, Hasan SW (2022) Algae biotechnology for industrial wastewater treatment, bioenergy production, and high-value bioproducts. Sci Total Environ 806:150585. https://doi.org/10.1016/j.scitotenv.2021.150585
95. Bose A, O'Shea R, Lin R, Murphy JD (2021) Design, commissioning, and performance assessment of a lab-scale bubble column reactor for photosynthetic biogas upgrading with *Spirulina platensis*. Ind Eng Chem Res 60:5688–5704. https://doi.org/10.1021/acs.iecr.0c05974
96. Zhang J, Zhao C, Sun S, Zhao Y, Liu J (2021) Performance of different microalgae-based technologies in nutrient removal and biogas upgrading in response to various GR24 concentrations. Int Biodeterior Biodegrad 158:105166. https://doi.org/10.1016/j.ibiod.2020.105166
97. Wang X, Gao S, Zhang Y, Zhao Y, Cao W (2017) Performance of different microalgae-based technologies in biogas slurry nutrient removal and biogas upgrading in response to various initial CO2 concentration and mixed light-emitting diode light wavelength treatments. J Clean Prod 166:408–416. https://doi.org/10.1016/j.jclepro.2017.08.071
98. Roy UK, Radu T, Wagner JL (2021) Carbon-negative biomethane fuel production: integrating anaerobic digestion with algae-assisted biogas purification and hydrothermal carbonisation of digestate. Biomass Bioenerg 148:106029. https://doi.org/10.1016/j.biombioe.2021.106029
99. Meier L, Pérez R, Azócar L, Rivas M, Jeison D (2015) Photosynthetic CO2 uptake by microalgae: an attractive tool for biogas upgrading. Biomass Bioenerg 73:102–109. https://doi.org/10.1016/j.biombioe.2014.10.032
100. Srinuanpan S, Cheirsilp B, Prasertsan P (2018) Effective biogas upgrading and production of biodiesel feedstocks by strategic cultivation of oleaginous microalgae. Energy 148:766–774. https://doi.org/10.1016/j.energy.2018.02.010
101. Srinuanpan S, Cheirsilp B, Kassim MA (2020) Oleaginous microalgae cultivation for biogas upgrading and phytoremediation of wastewater. In: Microalgae cultivation for biofuels production. Elsevier, pp 69–82
102. Granada-Moreno CI, Aburto-Medina A, Los Cobos Vasconcelos D, González-Sánchez A (2017) Microalgae community shifts during the biogas upgrading in an alkaline open photobioreactor. J Appl Microbiol 123:903–915. https://doi.org/10.1111/jam.13552
103. Mehta P, Singh D, Saxena R, Rani R, Gupta RP, Puri SK, Mathur AS (2018) High-value coproducts from algae—an innovational way to deal with advance algal industry. In: Singhania RR, Agarwal RA, Kumar RP, Sukumaran RK (eds) Waste to wealth. Springer Singapore, Singapore, pp 343–363
104. Olabi AG, Shehata N, Sayed ET, Rodriguez C, Anyanwu RC, Russell C, Abdelkareem MA (2023) Role of microalgae in achieving sustainable development goals and circular economy. Sci Total Environ 854:158689. https://doi.org/10.1016/j.scitotenv.2022.158689
105. Leong YK, Chang J-S (2022) Integrated role of algae in the closed-loop circular economy of anaerobic digestion. Biores Technol 360:127618. https://doi.org/10.1016/j.biortech.2022.127618
106. González-Camejo J, Serna-García R, Viruela A, Pachés M, Durán F, Robles A, Ruano MV, Barat R, Seco A (2017) Short and long-term experiments on the effect of sulphide on microalgae cultivation in tertiary sewage treatment. Biores Technol 244:15–22. https://doi.org/10.1016/j.biortech.2017.07.126
107. González-Sánchez, A., Posten, C.: Fate of H 2 S during the cultivation of chlorella sp. deployed for biogas upgrading. Journal of Environmental Management. 191:252–257. https://doi.org/10.1016/j.jenvman.2017.01.023
108. Syed M, Soreanu G, Falletta P, Béland M (2006) Removal of hydrogen sulfide from gas streams using biological processes• a review. Can Biosyst Eng 48:2
109. Rafiee A, Khalilpour KR, Prest J, Skryabin I (2021) Biogas as an energy vector. Biomass Bioenerg 144:105935. https://doi.org/10.1016/j.biombioe.2020.105935
110. Nagarajan D, Lee D-J, Chang J-S (2019) Integration of anaerobic digestion and microalgal cultivation for digestate bioremediation and biogas upgrading. Biores Technol 290:121804. https://doi.org/10.1016/j.biortech.2019.121804

111. Kao C-Y, Chiu S-Y, Huang T-T, Dai L, Hsu L-K, Lin C-S (2012) Ability of a mutant strain of the microalga chlorella sp. to capture carbon dioxide for biogas upgrading. Appl Energy 93:176–183. https://doi.org/10.1016/j.apenergy.2011.12.082
112. González-Sánchez A, Revah S, Deshusses MA (2008) Alkaline Biofiltration of H_2S odors. Environ Sci Technol 42:7398–7404. https://doi.org/10.1021/es800437f
113. Franco-Morgado M, Toledo-Cervantes A, González-Sánchez A, Lebrero R, Muñoz R (2018) Integral (VOCs, CO2, mercaptans and H2S) photosynthetic biogas upgrading using innovative biogas and digestate supply strategies. Chem Eng J 354:363–369. https://doi.org/10.1016/j.cej.2018.08.026
114. Anbalagan A, Toledo-Cervantes A, Posadas E, Rojo EM, Lebrero R, González-Sánchez A, Nehrenheim E, Muñoz R (2017) Continuous photosynthetic abatement of CO2 and volatile organic compounds from exhaust gas coupled to wastewater treatment: evaluation of tubular algal-bacterial photobioreactor. J CO2 Utilizat 21:353–359. https://doi.org/10.1016/j.jcou.2017.07.016
115. Fusi A, Bacenetti J, Fiala M, Azapagic A (2016) Life cycle environmental impacts of electricity from biogas produced by anaerobic digestion. Front Bioeng Biotechnol 4. https://doi.org/10.3389/fbioe.2016.00026
116. Gray N, McDonagh S, O'Shea R, Smyth B, Murphy JD (2021) Decarbonising ships, planes and trucks: an analysis of suitable low-carbon fuels for the maritime, aviation and haulage sectors. Adv Appl Energy 1:100008. https://doi.org/10.1016/j.adapen.2021.100008
117. Martinot E, Chaurey A, Lew D, Moreira JR, Wamukonya N (2002) Renewable energy markets in developing countries. Annu Rev Energy Environ 27:309–348. https://doi.org/10.1146/annurev.energy.27.122001.083444
118. Gonçalves AL (2021) The use of microalgae and cyanobacteria in the improvement of agricultural practices: a review on their biofertilising, Biostimulating and Biopesticide Roles. Appl Sci 11:871. https://doi.org/10.3390/app11020871
119. Saini RK, Keum Y-S (2018) Omega-3 and omega-6 polyunsaturated fatty acids: dietary sources, metabolism, and significance—a review. Life Sci 203:255–267. https://doi.org/10.1016/j.lfs.2018.04.049
120. Bortolini DG, Maciel GM, Fernandes IDAA, Pedro AC, Rubio FTV, Branco IG, Haminiuk CWI (2022) Functional properties of bioactive compounds from Spirulina spp.: Current status and future trends. Food Chem Mol Sci 5:100134. https://doi.org/10.1016/j.fochms.2022.100134
121. Anzani C, Boukid F, Drummond L, Mullen AM, Álvarez C (2020) Optimising the use of proteins from rich meat co-products and non-meat alternatives: nutritional, technological and allergenicity challenges. Food Res Int 137:109575. https://doi.org/10.1016/j.foodres.2020.109575
122. Matassa S, Boon N, Pikaar I, Verstraete W (2016) Microbial protein: future sustainable food supply route with low environmental footprint. Microb Biotechnol 9:568–575. https://doi.org/10.1111/1751-7915.12369
123. Sarwer A, Hamed SM, Osman AI, Jamil F, Al-Muhtaseb AH, Alhajeri NS, Rooney DW (2022) Algal biomass valorization for biofuel production and carbon sequestration: a review. Environ Chem Lett 20:2797–2851. https://doi.org/10.1007/s10311-022-01458-1
124. Mushlihah S, Husain DR, Langford A, Tassakka ACMAR (2020) Fungal pretreatment as a sustainable and low cost option for bioethanol production from marine algae. J Clean Prod 265:121763. https://doi.org/10.1016/j.jclepro.2020.121763
125. Bonenkamp TB, Middelburg LM, Hosli MO, Wolffenbuttel RF (2020) From bioethanol containing fuels towards a fuel economy that includes methanol derived from renewable sources and the impact on European Union decision-making on transition pathways. Renew Sustain Energy Rev 120:109667. https://doi.org/10.1016/j.rser.2019.109667
126. Michalak I, Chojnacka K (2015) Algae as production systems of bioactive compounds. Eng Life Sci 15:160–176. https://doi.org/10.1002/elsc.201400191
127. Cywar RM, Rorrer NA, Hoyt CB, Beckham GT, Chen EY-X (2021) Bio-based polymers with performance-advantaged properties. Nat Rev Mater 7:83–103. https://doi.org/10.1038/s41578-021-00363-3

128. Aswathi Mohan A, Robert Antony A, Greeshma K, Yun J-H, Ramanan R, Kim H-S (2022) Algal biopolymers as sustainable resources for a net-zero carbon bioeconomy. Biores Technol 344:126397. https://doi.org/10.1016/j.biortech.2021.126397
129. Möller K, Müller T (2012) Effects of anaerobic digestion on digestate nutrient availability and crop growth: a review: digestate nutrient availability. Eng Life Sci 12:242–257. https://doi.org/10.1002/elsc.201100085
130. Antonangelo JA, Sun X, Zhang H (2021) The roles of co-composted biochar (COMBI) in improving soil quality, crop productivity, and toxic metal amelioration. J Environ Manage 277:111443. https://doi.org/10.1016/j.jenvman.2020.111443
131. Chodkowska-Miszczuk J, Martinát S, Van Der Horst D (2021) Changes in feedstocks of rural anaerobic digestion plants: external drivers towards a circular bioeconomy. Renew Sustain Energy Rev 148:111344. https://doi.org/10.1016/j.rser.2021.111344
132. Weikmans R, Asselt HV, Roberts JT (2020) Transparency requirements under the Paris Agreement and their (un)likely impact on strengthening the ambition of nationally determined contributions (NDCs). Climate Policy 20:511–526. https://doi.org/10.1080/14693062.2019.1695571
133. Timilsina GR, Shah KU (2016) Filling the gaps: policy supports and interventions for scaling up renewable energy development in small island developing states. Energy Policy 98:653–662. https://doi.org/10.1016/j.enpol.2016.02.028
134. Carl J, Fedor D (2016) Tracking global carbon revenues: a survey of carbon taxes versus cap-and-trade in the real world. Energy Policy 96:50–77. https://doi.org/10.1016/j.enpol.2016.05.023
135. Bocken NMP, Allwood JM (2012) Strategies to reduce the carbon footprint of consumer goods by influencing stakeholders. J Clean Prod 35:118–129. https://doi.org/10.1016/j.jclepro.2012.05.031
136. Buchholz T, Gunn J, Springsteen B, Marland G, Moritz M, Saah D (2022) Probability-based accounting for carbon in forests to consider wildfire and other stochastic events: synchronizing science, policy, and carbon offsets. Mitig Adapt Strateg Glob Change. 27:4. https://doi.org/10.1007/s11027-021-09983-0
137. Ding GT, MohdYasin NH, Takriff MS, Kamarudin KF, Salihon J, Yaakob Z, Mohd-Hakimi NIN (2020) Phycoremediation of palm oil mill effluent (POME) and CO2 fixation by locally isolated microalgae: chlorella sorokiniana UKM2, Coelastrella sp. UKM4 and Chlorella pyrenoidosa UKM7. J Water Proc Eng 35:101202. https://doi.org/10.1016/j.jwpe.2020.101202
138. Chew KR, Leong HY, Khoo KS, Vo D-VN, Anjum H, Chang C-K, Show PL (2021) Effects of anaerobic digestion of food waste on biogas production and environmental impacts: a review. Environ Chem Lett 19:2921–2939. https://doi.org/10.1007/s10311-021-01220-z
139. Hu H, Zhou Q, Li X, Lou W, Du C, Teng Q, Zhang D, Liu H, Zhong Y, Yang C (2019) Phytoremediation of anaerobically digested swine wastewater contaminated by oxytetracycline via Lemnaaequinoctialis: nutrient removal, growth characteristics and degradation pathways. Biores Technol 291:121853. https://doi.org/10.1016/j.biortech.2019.121853
140. Xu M, Xue Z, Liu J, Sun S, Zhao Y, Zhang H (2022) Observation of few GR24 induced fungal-microalgal pellets performance for higher pollutants removal and biogas quality improvement. Energy 244:123171. https://doi.org/10.1016/j.energy.2022.123171
141. Yang W, Li S, Qv M, Dai D, Liu D, Wang W, Tang C, Zhu L (2022) Microalgal cultivation for the upgraded biogas by removing CO2, coupled with the treatment of slurry from anaerobic digestion: a review. Biores Technol 364:128118. https://doi.org/10.1016/j.biortech.2022.128118

Chapter 6
Renewable Thermal Energy Storage in Polymer Encapsulated Phase-Change Materials: A Comprehensive Overview

Sumit Kumar, Chandrani Sarkar, and Sampa Saha

Abstract An urgent need to resolve the unwanted climatic change and transition to renewable energy resources has driven significant development and research in advancing renewable energy storage systems. Energy storing approaches aid in efficiently utilizing renewable resources, maintaining their consistency, and having a reliable energy supply. This book chapter contributes significantly to the topic of renewable energy storage. It provides a detailed overview of thermal energy storage (TES) systems based on phase-change materials (PCMs), emphasizing their critical role in storing and releasing latent heat. Moreover, different types of PCMs and their selection criteria for electricity generation are also described. Meanwhile, the role of polymer in thermal energy storage systems is systematically presented in this chapter. Usages of PCM trapped polymer composites and their different fabrication methods including micro/nanoencapsulation by physical/chemical methods, porous polymeric framework, and shape-stabilization strategy are also discussed in detail. In addition, this chapter covers the wide application of PCMs based systems in solar energy storage including solar thermophotovoltaics, waste heat recovery (stationary waste heat recovery and portable waste heat recovery), solar thermoelectric generators, concentrated solar power plants (CSP) and energy conservation in buildings.

Keywords Sustainable energy · Phase-change material · Renewable energy · Waste energy recovery · Thermal energy storage

Nomenclature

Q	Heat absorbed
m	Mass

S. Kumar · C. Sarkar · S. Saha (✉)
Department of Materials Science and Engineering, Indian Institute of Technology Delhi, Hauz Khaz, New Delhi, India
e-mail: Sampa.Saha@mse.iitd.ac.in

S. De et al. (eds.), *Challenges and Opportunities of Distributed Renewable Power*, Energy, Environment, and Sustainability, https://doi.org/10.1007/978-981-97-1406-3_6

C_{sp} and C_{lp}	Specific heat
a_m	Fraction melted
Δh	Heat of fusion per unit mass
T	Temperature
λ	The latent heat of fusion
ρ	Density
s(t)	Surface position
k	Thermal conductivity
H	Total volumetric enthalpy
h	Enthalpy function
α	Thermal diffusivity

Abbreviation

PCM	Phase-change materials
RES	Renewable energy storage
TES	Thermal energy storage
WEO	World energy outlook
IEA	International Energy Agency
SHS	Sensible Heat Storage
LHS	Latent Heat Storage
TC-TES	Thermochemical Thermal Energy Storage
PPC	PCM trapped polymer composites
PE	Polyethylene
EVA	Ethylene Vinyl Acetate
CNF	Carbon nanofibers
DSC	Differential Scanning Calorimetry
PMF	Poly (melamine formaldehyde)
PUF	Poly (urea formaldehyde)
PMUF	Poly (melamine urea formaldehyde)
BS	Butyl Stearate
POFs	Porous Polymeric Frameworks
PVA	Poly (vinyl alcohol)
PCL	Polycaprolactone
PLA	Poly (lactic acid)
CA	Cellulose Acetate
CET	Cellulose Ether
CMC	Carboxymethyl Cellulose
PEO	Polyethylene oxide
PEG	Polyethylene glycol
PU	Polyurethane
GF	Graphite foam
CNTs	Carbon Nanotubes

SSPCM	Shaped Stabilized Phase-change Materials
xGNP	Exfoliated Graphite Nanoplatelets
CSP	Concentrated Solar Power Plant
HTF	Heat Transfer Fluid
STPVS	Solar Thermophotovoltaics Systems
STEG	Solar Thermoelectric Generator
M-TES	Mobilized Thermal Energy Storage
HVAC	Heating, Ventilation and Air Conditioning

6.1 Introduction

According to WEO (World Energy Outlook) reports issued by IEA (International Energy Agency), the world energy demand will rise by one-third from 2011 to 2035, and simultaneously carbon dioxide (CO_2) emission will also increase by 20 to 37.2% due to energy generation by fossil fuels leading to undesired changes in climate. So, the utilization of fossil fuels must be reduced, or more renewable energy-based technologies must be developed. The major advantage of utilization of renewable energy-based systems is that they are naturally replenished or inexhaustible. So, in recent years, the rapid transition to renewable energy resources has driven significant development in the advancement of technologies based on renewable energy, such as wind and solar energy. However, due to their infrequent nature arising from variability of weather and time of day, they exhibit challenges for energy supply and grid stability [1]. So, scientists are in search of new sources for energy harvesting. One option to fulfill the demand and effective use of existing energy is to develop energy storage devices. These devices store extra amount of energy during heat generation and release when there is a demand. There are different kinds of energy storage devices, for example, mechanical energy storage devices, electrical energy storage devices, and thermal energy storage devices. Due to high energy density, long storage duration, flexibility, and utilization of waste heat, thermal energy storage systems are more popular these days [2]. The various aspects of thermal energy storage devices are discussed below.

6.1.1 Thermal Energy Storage and Its Role in Energy Generation

Thermal energy storage (TES) is an emerging technology in the domain of power generation, and it normally acts as energy buffer. The working principle of TES systems lies in storing thermal energy as a change in the internal energy of a material [3]. Various storage mediums such as water [4], molten salts [5], and phase-change materials (PCM) [6] capture the excess energy during periods of high generation, and

Fig. 6.1 Illustration of power storage using thermal energy. Inspired from [9]

ensure a consistent and reliable supply during the period of peak demand [7]. An illustration is shown in Fig. 6.1, representing the utilization of heat energy in green power generation. Due to their outstanding ability to capture, store, and release thermal energy revolutionized the way of energy generation, management, and utilization [8].

6.1.2 Types of Thermal Energy Storage

The storage materials or systems are classified into three categories based on their heat absorbing and releasing behavior, which are- sensible heat storage (SHS), latent heat storage (LHS), and thermochemical storage (TC-TES) [10].

6.1.2.1 Sensible Heat Storage Systems

In SHS, thermal energy is stored and released by changing the temperature without going into any phase change [11]. Some examples of SHS are water, rock, brick, concrete, and engine oil. These materials have a high specific heat capacity, which allows them to store and release a large amount of energy. The amount of energy stored can be calculated by the following equation (Eq. 6.1).

$$Q = mC_p\Delta T \tag{6.1}$$

where Q (kJ) is the amount of heat gained or lost in the material, m (kg) is the mass of storage material, C_p (kJ/kg. K) is the specific heat capacity of storage material, and ΔT is the temperature change (°C).

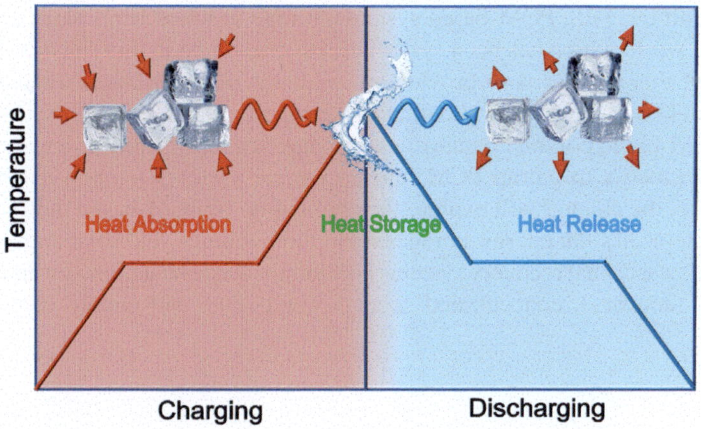

Fig. 6.2 Working principle and temperature profile of PCMs. Inspired from [13]

6.1.2.2 Latent Heat Storage Systems

In some systems, phase transformation of materials occurs during heat absorption and release. Phase transformation can be solid–solid, solid–liquid, solid–gas, and liquid–gas. Those systems are Latent heat storage (LHS) systems. They can absorb and release a large amount of energy due to phase transformation taking place within a specific material known as phase-change material (PCM) [12]. Figure 6.2 represents solid–liquid PCMs system and its phase transformation. In this system, energy storage is associated with the rearrangement of physical bonds in the bulk material. They can store a large amount of energy within a small space. The amount of energy absorbed by the material is calculated by Eq. 6.2.

$$Q = m\left[C_{sp}(T_m - T_i) + a_m \Delta h_m + C_{lp}\left(T_f - T_m\right)\right] \tag{6.2}$$

Q (kJ) is the amount of heat absorbed by the material, m (kg) is the mass of heat storing material, C_{sp} (kJ/kg. K) is the average specific heat capacity between T_i and T_m, C_{lp} is the average specific heat between T_m and T_f, a_m is the fraction melted (J/kg-K), Δh_m is heat of fusion per unit mass (J/kg), T (°C) is the temperature and subscripts 'm', 'i', and 'f' refers to melting, initial and final temperature.

6.1.2.3 Thermochemical Thermal Energy Storage Systems

Another emerging technology in the field of energy storage is thermochemical thermal energy storage (TC-TES). TC-TES utilizes reversible chemical reactions to store energy. It can be defined as the storage of heat through a cycle between two thermodynamic states. Theoretically, TC-TES offers higher energy density with minimum energy losses during long-term storage.

Among all the TES, PCM-based system (storage of latent heat) plays a vital role in practical world which has been systematically presented in this chapter. The main objective of this chapter is to provide an overview of PCM-based thermal energy storing systems, covering their classification, properties, and limitations. It will delve into the PCM-based polymer composites, the significance of using polymer, and their fabrication process to entrap PCM into polymeric systems for power generation. Furthermore, the chapter will explore the application of PCM-based thermal energy storage systems in solar energy storage (solar photovoltaics, and solar thermoelectric generators), waste heat recovery systems (stationary waste heat recovery and portable waste heat recovery), concentrated solar power plants, and energy conversion in buildings.

6.2 Phase-Change Materials

Phase-change materials (PCMs) are materials that have the capability to absorb, store, and release a large amount of energy over a defined range of temperatures during phase transformation [14]. Phase transition refers to the change in the physical state of a substance by absorbing and releasing latent heat. This latent heat is known as either the latent heat of fusion when solid material melts and changes to a liquid state or the latent heat of vaporization when liquid materials change to a gaseous state. It is known that phase transformation occurs in four ways like solid–solid, solid–liquid, solid–gas, and liquid–gas. In solid–solid transition, only the change in the crystal structure of the material occurs while remaining in the solid state [15]. Shape memory alloys are the best examples of solid–solid phase transitions [16]. In solid–liquid transition, the conversion of a substance occurs from a solid state to a liquid state or vice-versa by absorbing a large amount of thermal energy [17]. Paraffin waxes [18], fatty acids [19], and salt hydrates [20] are a few examples. They have very high latent heat capacities, which allow them to be used in TES applications. In solid–gas transition, a material in a solid state changes to a gaseous state or vice-versa [21]. This type of phase transition is less common in TES systems due to leakage problems [22]. These systems are typically used in adsorption-based TES systems. Materials such as zeolites are generally used in solid–gas transition for adsorption and desorption processes [23]. These materials can adsorb gases on their surface, storing thermal energy and release while gas desorption occurs. In liquid–gas phase transition, it involves the conversion of a substance from a liquid to a gaseous state or vice-versa [24]. This type of TES has been generally employed in steam power plants where water undergoes liquid to gas phase transition at its boiling point and absorbs thermal energy during liquid to steam conversion. Generated steam is utilized to drive motors or turbines to generate electricity [25].

6.2.1 Classification of Phase-Change Materials

As discussed above, PCMs are usually categorized into solid–solid, liquid–solid, gas–solid, gas–liquid, and vice-versa. Due to significant volume change and low latent heat during phase change, liquid–gas and solid–gas phase transitions are practically not viable for TES applications. However, solid–liquid PCMs have a broad range of phase change temperatures and high energy storage densities. So, solid–liquid or liquid–solid PCMs are further classified into three categories: organic, inorganic, and eutectic PCMs based on their chemical nature [26]. The classification of PCMs is shown in Fig. 6.3.

6.2.1.1 Organic Phase-Change Materials

Organic PCMs exhibit congruent melting points [27] that means they melt and freeze rapidly without phase separation, and consequent degradation of latent heat of fusion. They are generally non-corrosive in nature [28]. Organic PCMs undergo self-nucleation means they can crystallize with no super cooling [29] (i.e., a process of cooling below the freezing point of liquid). Organic PCMs are classified into paraffin and non-paraffins.

Paraffin

Paraffin wax is a mixture of linear alkanes that release and store a large amount of latent heat during crystallization and melting [30]. Paraffin is the most widely used PCM in TES systems due to its reliability, low cost, non-corrosiveness, chemical inertness, minimum volume change, low vapor pressure, good nucleating properties, and high thermal energy storing capacity [29, 31–33]. A list of paraffin with their

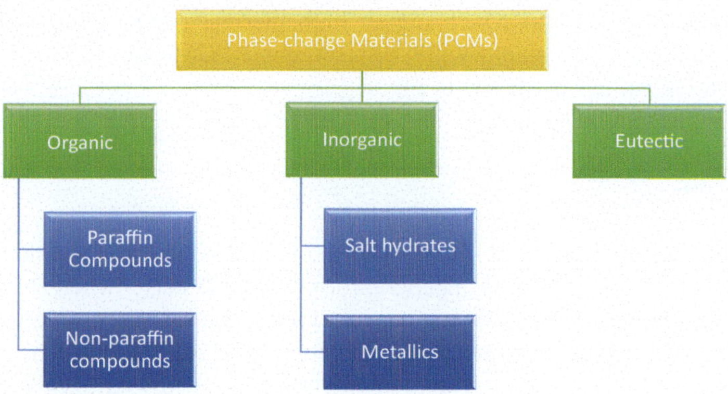

Fig. 6.3 Classification of phase-change materials (PCMs)

Table 6.1 Melting and latent heat of fusion of paraffins [6]

No. of carbon atoms in the alkyl chain	Melting point (°C)	Latent heat of fusion (kJ/kg)
14	5.5	228.00
16	16.7	237.10
18	28.0	246.00
20	36.7	246.00
22	44.0	249.00
24	50.6	255.00

melting and latent heat of fusion is given in Table 6.1. It is observed that the tendency of energy storage capacity and melting point increase with the length of carbon chain or the number of carbon atoms.

Non-paraffins

Non-paraffins are a class of PCMs derived from organic compounds. They also store and release thermal energy during phase transition. Materials such as fatty acids, esters, alcohols, and carbohydrates are examples of non-paraffins PCMs. Non-paraffins PCMs show high heat of fusion, good inflammability, and better thermal conductivity [34]. Among other non-paraffin PCMs, fatty acids have a higher heat of fusion and have reproducible freezing and melting behavior [35]. However, the high cost (approximately 2–2.5 times of paraffin) and mild corrosive nature of non-paraffins make them less usable than paraffin-based PCMs [36].

6.2.1.2 Inorganic Phase-Change Materials

Unlike organic PCMs which are composed of mainly carbon, inorganic PCMs contain elements other than carbon which impart high thermal conductivity and thermal stability with better cycle stability [37, 38]. Inorganic PCMs are classified into salts hydrates, and metallics, as shown in Fig. 6.3.

Salts Hydrates

Salt hydrates mainly consist of salts that form crystalline structures combined with water molecules [39]. They have a general formula of $AB.nH_2O$ where A represents the cation, B represents anion and n represents the number of water molecules. Upon heating, hydrated salt undergoes a phase transition from solid to liquid by absorbing a large amount of thermal energy and, upon cooling again crystallizes and releases stored energy [40]. Salts hydrates based PCMs show three types of melting behavior-congruent, incongruent, and semi-congruent melting [29]. In congruent melting, anhydrous salts are completely solubilized in its water of hydration or crystallized water at the melting point whereas in the case of incongruent melting, complete solubilization of salt is not possible in its water of hydration or crystallized water at melting temperature. Unlikely in the case of semi-incongruent melting, solid and liquid phases are in equilibrium at different melting compositions during phase transition which is due to the conversion of higher hydrates to lower hydrates through loss of water. These salt hydrate based PCMs possess high latent heat of fusion, high dimensional stability, good compatibility with polymers, low toxicity, high thermal conductivity, and mild corrosive in nature [41]. However, the main drawback of salt hydrates based PCMs is their incongruent melting i.e., the released water of crystallization which is not sufficient to solubilize the entire solid phase [29]. As a consequence, density differences occur and anhydrous salt settles down at the bottom. Apart from its incongruent melting, they also have poor nucleating properties [42].

Metallics

Metallics PCMs consist of low melting metals and eutectic metals [43]. Metallics are not yet considered as PCMs due to their high weight, but upon considering their volume they can be used as PCMs in TES systems because of their high heat of fusion per unit volume [44]. Apart from their high heat of fusion, they also possess high thermal conductivity, but the use of metallic PCMs has several unusual engineering problems due to which metallic PCMs are not very popular [45].

6.2.1.3 Eutectics

Eutectic PCMs exhibit eutectic composition that refers to the specific mixture of two or more substances which have minimum melting point among all compositions [46]. Eutectic mixtures nearly freeze and melt without phase separation since they freeze to an intimate mixture of crystals [27]. Eutectic PCMs have sharp and well-defined melting point that allows them for better temperature control during phase transition. Additionally, eutectic PCMs have high thermal conductivity for faster heat transfer [47]. This material finds applications in various fields such as building construction, thermal management systems, and renewable energy systems where they can store and release thermal energy efficiently.

6.2.2 Mathematical Modeling of Phase-Change Materials

Predicting the behavior of PCM is quite challenging because of the complex nonlinear nature of moving interfaces in which latent heat controls the displacement rate which is described by Stephan condition equation:

$$\lambda\rho\left(\frac{ds(t)}{dx}\right) = k_s\left(\frac{\delta T_s}{\delta t}\right) - k_l\left(\frac{\delta T_l}{\delta t}\right) \tag{6.3}$$

where λ is the latent heat of fusion, ρ is the density, s(t) is the surface position, k is thermal conductivity, t is time and T is the temperature. Here, subscript s and l corresponds to solid and liquid phases. Two different phases possess distinct physical properties that generate non-physical discontinuities in the numerical model which should be addressed. Enthalpy method simplifies the phase change problem, as applying the same governing equation to both phases, naturally achieves interface conditions and a mushy zone introduced between them which mitigates the abrupt discontinuities that lead to numerical instability. However, the discretization of mushy zone's quality and thickness is crucial for accurate modeling. The enthalpy method addresses mushy and isothermal phase changes but causes temperature oscillation at grid points over time [48]. This method was successfully applied to various phase change problems [49]. For a phase change process, energy conversion can be expressed in terms of temperature and total volumetric enthalpy for constant thermophysical properties as

$$\frac{\partial H}{\partial t} = \nabla[k_k(\nabla \mathrm{T})] \tag{6.4}$$

where H is total volumetric enthalpy, which is sum sensible and latent heat.

$$(T) = h(T) + \rho_1 f(T) \tag{6.5}$$

where h is the enthalpy function can be expressed as

$$h = \int_{T_m}^{T} \rho_k c_k dT \tag{6.6}$$

In case of isothermal phase change, liquid fraction f is given by

$$= \begin{cases} 0 & T < T_m \ soild, \\ [0, 1] & T = T_m \ mushy, \\ 1 & T > T_m \ liquid. \end{cases} \tag{6.7}$$

An alternate form of two dimensional heat transfers in PCM

$$\frac{\partial h}{\partial t} = \frac{\partial\left(\alpha\frac{\partial h}{\partial x}\right)}{\partial x} + \frac{\partial\left(\alpha\frac{\partial h}{\partial y}\right)}{\partial y} - \rho_1\lambda\frac{\partial f}{\partial t} \tag{6.8}$$

where α is thermal diffusivity.

Over time, numerical techniques for solving phase change problems have evolved. Since its introduction in the 1970s, finite volume and element methods have largely preferred finite-difference methods. In situations where temperature can fluctuate over time, some researchers use the temperature-based capacity method, which transforms the enthalpy-based energy equation into a non-linear equation with a single variable. This method, known as the Temperature Transforming Model (TTM), is both precise and efficient.

6.2.3 Selection Criteria of Phase-Change Materials for Electricity Generation

PCM plays a crucial role in thermal energy storage and management systems that allow efficient and effective utilization of renewable energy sources for electricity generation. During the selection of PCM, several key parameters need to be considered such as

Thermal properties: PCM must possess high latent heat that shows high capacity to absorb and release a large amount of energy during phase transition, which enables efficient energy storage, provides grid stability, and maintains a stable temperature during electricity generation. PCMs having desirable freezing and melting temperatures with wide temperature ranges are always preferred for the storage of renewable energy.

Thermal stability: PCMs must exhibit good thermal stability to withstand multiple heating and cooling cycles without significant loss in performance and degradation. Thermal stability of PCMs also ensures reliability and longevity via maintaining their chemical and physical properties.

Heat transfer characteristics: PCMs should have high thermal conductivity to allow effective heat transfer between heat source and storage medium. More efficient heat transfer allows higher energy conversion and lower thermal losses.

Compatibility: PCMs should have high compatibility with the materials and system components used in the power generation setup.

Cost-effectiveness: For large-scale electricity generation projects, cost is an important factor. PCMs must be economically viable, considering factors such as manufacturing cost, material cost, and scalability.

Safety consideration: PCMs must be non-flammable and non-toxic to ensure the personnel's safety and not harm the surrounding environment. Additionally, PCMs should be stable at operating condition and should not pose any health hazard or safety risk.

Availability: Considering sustainability and environmental source impact, PCMs should be readily available and should not pollute the environment.

As PCMs are promising materials for TES systems they also have some limitations such as poor thermal cycling, limited heat transfer characteristics, short term stability, and performance. Several materials and techniques were used to overcome the above limitations of PCMs in TES systems. Among these, employment of polymers to fabricate PCMs based polymer composites are prominent these days because of their low material and fabrication cost, tunable properties, light weight, non-corrosive insulating nature, good endurance, etc.

6.3 Phase-Change Materials Trapped Polymer Composites

Phase-change materials trapped polymer composites (PPC) are the multiphasic materials in which PCM are incorporated within a polymer matrix that has the capability to store and release large amounts of latent heat at a fixed temperature during phase transition [50]. Hence, the thermal cycling and thermal stability of PCMs have been improved. Thus, PPC have gained tremendous attention in thermal energy storage applications. Some key features of PPC are listed below.

Prevention of the PCM leakage: As PCMs are prone to leakage or loss if not appropriately confined, so the prevention of PCM leakage is most important for the LHS based thermal energy storage systems, particularly solid–liquid PCMs systems [51]. It can be done by incorporating PCMs into a polymer to ensure that the PCMs remain entrapped within the polymeric device, maintaining its performance and stability.

Enhanced durability: Generally, PCMs have poor thermal and chemical stability, so by producing PCM-polymer composites, their thermal as well as chemical stability get improved because polymers act as a barrier against various factors that cause degradation of PCMs.

Structural support: PCMs are soft materials which means they can easily lose their shape, i.e., their dimensional stability is poor [52]. One solution for their poor dimensional stability is to encapsulate them within a polymeric framework to improve their mechanical stability, thereby allowing the composite to mold into various shapes.

Enhanced cycle stability and storage capacity: PCMs have very high latent heat storage capacity and cycle stability. As a consequence, thermal energy storage efficiency will not be reduced even after 1000 cycles. However, due to loss of material during phase transitions performed via thermal cycling, PCMs might be degraded that may affect the efficiency of PCMs. By using suitable polymeric composites, these losses can be minimized, and the cycle stability/integrity can be well maintained.

6.3.1 Role of Polymers in Thermal Energy Storage Systems

Polymers play an important role in thermal energy storage systems. They are utilized to enhance stability, efficiency, and overall performance by acting as encapsulating matrix materials or composite components [53, 54]. Polymeric encapsulation provides an excellent opportunity to stabilize PCMs within a composite that has a unique structure. Additionally, polymers also act as a thermal barrier or insulating material to reduce heat losses which enhances the overall energy efficiency [55]. Moreover, polymers have high tailoring capabilities to have specific thermal properties such as heat capacity by changing their composition. A variety of polymer matrices are available for the fabrication of polymeric phase-change composite which is summarized in Table 6.2. Among them, thermoset polymers are widely used for the encapsulation of PCMs because they can encapsulate PCM at room temperature and prevent its leakage at elevated temperatures. These PPC (PCM-based Polymer composites) are used in compact electronic packaging particularly in military electronics and in structural elements (enclosure skins, acoustical control materials, and frame members) [56]. Furthermore, hybrid TES cooler has stable operating performance which is equivalent to existing systems while having improved reliability at the same time [57].

6.3.2 Fabrication Methods

PPC offers several advantages over bare PCMs for TES applications such as it allows to combine the unique properties of both PCMs and polymer to achieve better performance, efficiency, and structural stability [60]. Moreover, PPC helps in tuning the desired properties such as thermal conductivity, improved heat storage capacity, better heat transfer, and sustainability [61]. PPC can be fabricated by various strategies such as micro or nanoencapsulation, porous polymer framework, shape-stabilized strategy, etc. [60], which are discussed in detail below.

6.3.2.1 Micro or Nanoencapsulation

Micro and nanoencapsulation are the techniques aimed at confining PCMs and preventing them from contamination. Microencapsulation refers to encapsulating phase-change materials within micrometer range particles i.e., typically 1–100 μm [62] whereas in case of nanoencapsulation refers to encapsulating PCM within nanometer range, i.e., 1–100 nm [63]. In the context of PPC, it involves encapsulating PCMs within polymeric shells to yield capsules. Micro and nanoencapsulation strategies are more focused on employing core–shell techniques to achieve the desired energy storage capacity of PCMs and thermophysical properties. In core–shell techniques, the core is PCM, and the shell is polymer which acts as supporting material.

Table 6.2 Variety of polymer matrices and PCM used for the synthesis of PCMs trapped polymer composites [58, 59]

Forming process	Shell materials	Core materials	Particle size (μm)	Encapsulation efficiency (%)	Melting temperature (°C)	Latent heat (J/g)	Application
Interfacial polymerization	Polyurethane	Butyl setarate	10–35.0	95.4	22.3	81.2	–
Interfacial polymerization	Polyurethane	n–octadecane	5–10.00	93.4–94.90	–	110.00	–
Suspension polymerization	n–octadecyl methacrylate–methacrylic acid co–polymer	n–octadecane	1.6–1.70	12–21.0	26.50	93.0	Thermal energy storage
Suspension polymerization	Poly(butyl methacrylate), Poly(butyl acrylate)	n–octadecane	2–75.0	47.7–55.60	29.10	112.0	Thermal energy storage
Suspension polymerization	Poly(styrene)	Paraffin wax	–	75.60	–	153.50	–
Suspension polymerization	Poly(styrene)	Rubitherm®RT27	3.8	43.6	–	79.00	Thermal energy storage
Emulsion polymerization	Poly(styrene)	n-heptadecane	1–20.0	63.30	21.50	136.90	Latent heat thermal energy storage
Emulsion polymerization	Poly(methyl methacrylate)	n-nonadecane	0.1–35.0	60.30	31.20	139.20	Thermal energy storage
Emulsion polymerization	Poly(methyl methacrylate)	Paraffin eutectics	0.01–100.0	50.2–65.40	19.20	171.10	–

Generally, encapsulation can be conducted via physical and chemical methods. Physical methods involve processes like spray drying and solvent evaporation whereas chemical methods involve processes such as *in-situ* polymerization, interfacial polymerization, suspension polymerization, and emulsion polymerization [64]. Details of both physical and chemical methods are discussed below.

Physical Methods

In physical methods, physical processes such as adhesion, dehydration, and drying are responsible for the formation of microcapsule shell. The most common physical methods used for encapsulating PCM are spray drying and solvent evaporation.

Spray drying process includes preparation of oil-in-water emulsion containing shell material, and PCM followed by spraying by using atomizer in a drying chamber, and after drying solid particles are separated by cyclone and filter. Figure 6.4 illustrates the typical setup of spray drying experimental apparatus in which a uniform liquid solution referred to as feed solution is transformed into fine droplets using a stream of carrier gas. Once atomized the solvent present inside the droplets vaporizes and the particles undergo drying due to the influence of another nitrogen gas stream referred to as drying N_2 within the drying chamber. The end product is collected via a collector unit. Hawlader et al. used gelatin and gum arabic to encapsulate PCM via spray drying method. Fabricated microcapsules are spherical and have uniform size. A core–shell microcapsule was prepared by using core–shell ratio of 2:1 and achieved heat release and storage capacities of 221.22 and 216.44 J/g, respectively [65]. Borreguero et al. fabricated polyethylene-ethylene vinyl acetate (PE-EVA) shell and paraffin rubitherm ®RT27 core microcapsules without and with carbon nanofibers (CNF) by using spray drying method. Addition of CNF improved mechanical strength, thermal conductivity, and thermal stability (stable upto 3000 cycles) while maintaining heat storage capacity [66]. Few studies investigated that the mechanical properties, thermal conductivities, and thermal stability have been improved dramatically with the incorporation of nanomaterial in PPCs systems [67, 68].

Chemical Methods

Chemical methods involve the use of a condensation process or polymerization on monomers, oligomers, or pre-polymers to create a shell at the oil–water interface. *In-situ* polymerization, interfacial polymerization, suspension polymerization, and emulsion polymerization are the commonly employed chemical methods to encapsulate PCMs.

In *in-situ* polymerization, a protective shell is formed on droplet surfaces through the polymerization of pre-polymer as shown in Fig. 6.5a. In this process, first, an oil-in-water emulsion of PCMs is prepared into an aqueous surfactant solution. Then,

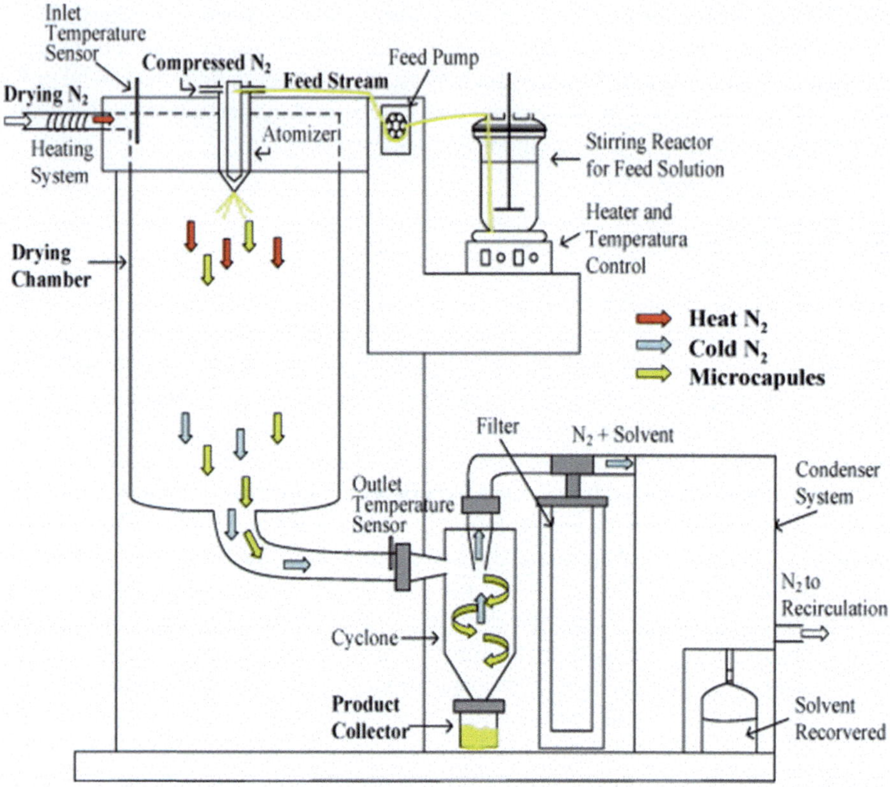

Fig. 6.4 Schematic of encapsulation by spray drying. Reproduced with permission from [59]

a pre-polymer aqueous solution is added into it at an appropriate reaction condition. Subsequently, formation of polymer occurs and wraps the PCMs droplet with polymer layers. Hence, structurally stable PPC is formed by *in-situ* polymerization process.

In interfacial polymerization Fig. 6.5b, two reactive monomers one is water soluble and another is oil soluble are used, and the polymerization takes place at the interface of two immiscible liquids under the influence of an initiator. In suspension polymerization (Fig. 6.5c), oil soluble monomers and initiator are generally used which suspended in continuous medium. Oil-soluble initiators generate radicals to initiate polymerization. Then, monomers convert into polymer and are deposited onto the surface of PCMs droplets. However, in emulsion polymerization (Fig. 6.5d) water-soluble initiators are commonly used and monomers are mostly oil soluble. Polymerization generally occurs when initiators are in contact with monomers and activate them for polymerization.

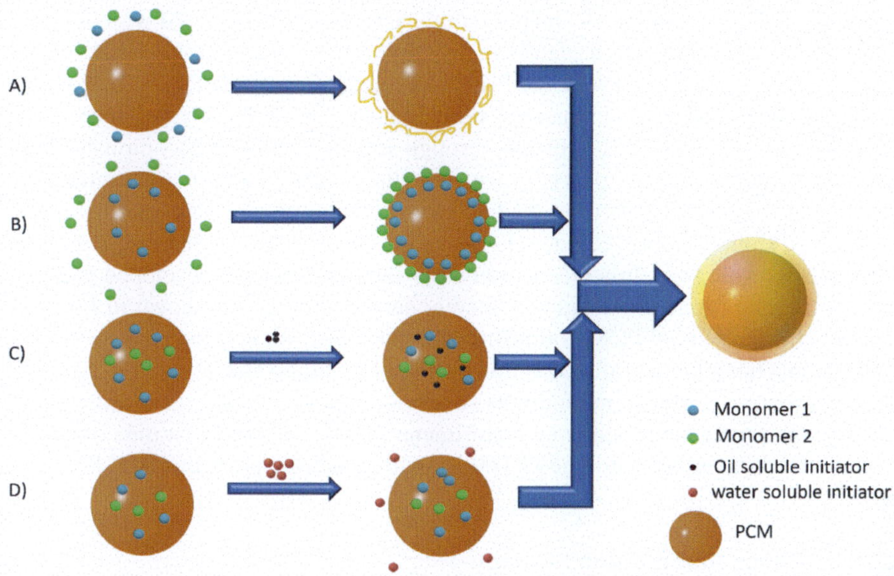

Fig. 6.5 Chemical methods for encapsulation **a** *in-situ* polymerization, **b** interfacial polymerization, **c** suspension polymerization, and **d** emulsion polymerization. Inspired from [71]

In a study conducted by Konuklu et al. dodecanoic acid is encapsulated by using three different polymers: Poly (melamine formaldehyde) PMF, Poly (urea formaldehyde) PUF, and Poly (melamine urea formaldehyde) PMUF by using *in-situ* polymerization [69]. They have concluded that PMUF microcapsules have higher thermal stability with no leakage at 95 °C than PMF and PUF microcapsules [69]. In another study, Yu et al. successfully encapsulated a binary core material consisting of butyl stearate (BS) and paraffin within a polyurea/polyurethane shell via the interfacial polymerization method [70].

6.3.2.2 Porous Polymeric Framework

Porous polymeric frameworks (POFs) are excellent supporting materials to stabilize core PCMs. The results reported in the literatures indicated that the porous polymer foams and aerogels can effectively stabilize the shape of PPC or possess improved dimensional stability [72–75]. Fabrication of POFs offers unique properties such as light weight, high surface area, tunable porosity, outstanding insulating properties, high mechanical strength, and high energy absorbency, making them potential materials for incorporating PCMs [58]. In addition, a series of structural polymers were employed to make foams and aerogels to encapsulate PCMs. Related examples include Poly (vinyl alcohol) (PVA) [76], Cellulose, Polycaprolactone (PCL) [77], Polyacrylates [78, 79] and Poly (lactic acid) (PLA) [80]. Pielichowska et al. use cellulose and its derivatives [cellulose acetate (CA), cellulose ether (CET), carboxymethyl cellulose (CMC) as framework and polyethylene oxide (PEO) as PCM with different compositions. The fabricated blends in 1:1 (w/w) have a heat storage capacity of 136.4 J/g for PEO/CA, 130.8 J/g for PEO/CMC, and 87.8 J/g for PEO/CET [81]. In another study, Entürk et al. conducted a study focusing on TES properties of cellulose and polyethylene glycol (PEG) blends through solvent casting to accomplish a high latent heat storage capacity of 84 J/g with excellent stability and thermal reliability. Furthermore, after 3000 thermal cycles, phase separation of blend was observed, leading to a slightly improved latent heat storage capacity of ~91 J/g during melting and freezing cycle [82]. In a recent study, the same group have developed graphite foam with a 3D (three dimensional) interconnected structure as shown in Fig. 6.6. To assess the stability of PCM composite, a leakage test was conducted by heating it at 65 °C on a hot plate. Paraffin was transformed into liquid state; however, paraffin/graphene foam was able to maintain its structure without any leakage. Further analysis through Differential Scanning Calorimeter (DSC) revealed that the composite exhibited high latent heat storage of 160 J/g with a melting point of 66 °C [52].

6.3.2.3 Shape-Stabilization Strategy

Traditional TES systems have certain limitations due to their low thermal conductivity and PCMs' leakage affecting their performance and energy density. A conventional way to overcome these limitations is blending [52]. Blending involves mixing of polymer and PCM by applying shear through mechanical stirring and creates a uniform mixture of dispersed PCM into the polymer matrix. Moreover, solution mixing, melt blending and *in-situ* polymerization can also be used to prepare the blend [54]. However, it sacrifices the intrinsic energy density of PCMs which is not acceptable [83]. To maintain the balance between thermal performance and intrinsic energy densities, conducting fillers with high surface to volume ratio are usually incorporated in PCMs [84]. Wide range of conductive fillers, for example, carbon nanotubes CNTs [85], graphene [86], and metals [85] are used to achieve shape-stabilization in PCMs. Silva et al. prepared a foam stabilized PCM by hot pressing to form a disc by incorporating a blend of poly (caprolactone) (PCL) and PEO into porous expanded graphite. The resultant disc was subjected to thermal aging for 1700 h at 80 °C. They observed that the aged composite has better heat storage capacity (63 J/g) than that of the pristine composite (69.61 J/g). Figure 6.7 illustrates the manufacturing of SSPCM (shape-stabilized PCM) using three filtered methods [87].

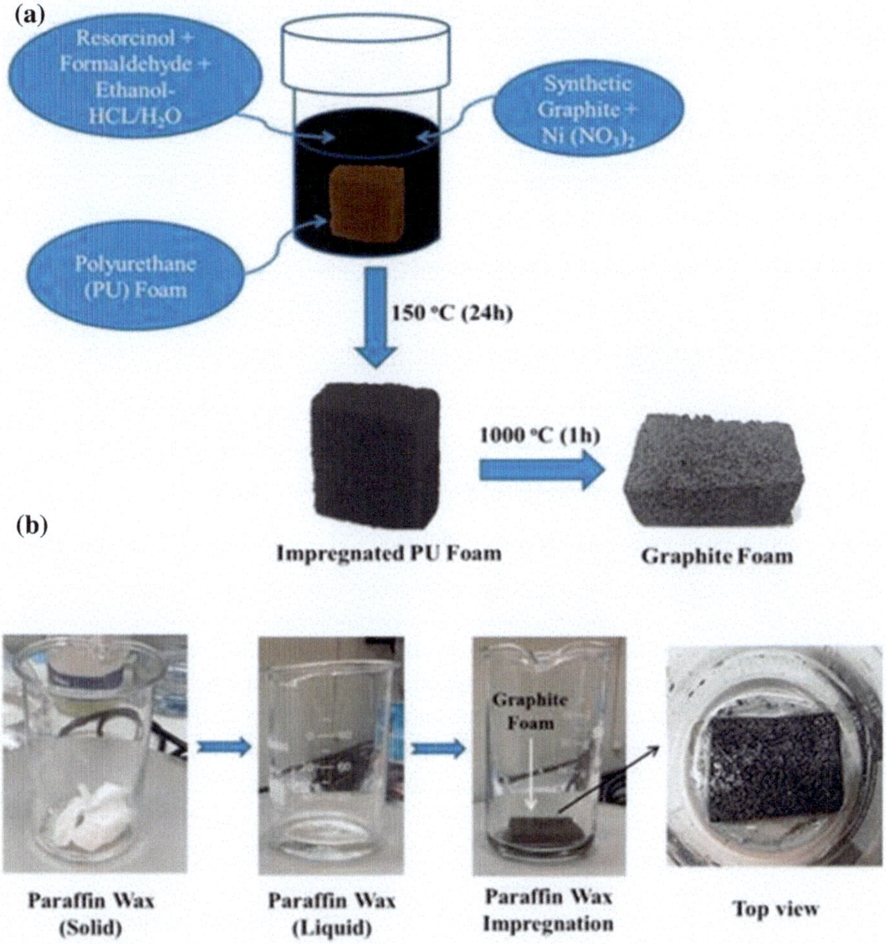

Fig. 6.6 **a** Schematic for synthesis of graphite foam (GF) and **b** GF/paraffin composite fabrication by impregnation method. Reproduced with permission from [52]

6.4 Application of Phase-Change Materials Based Thermal Energy Storage Systems

Microencapsulated PCMs have gained a lot of attention due to their excellent storage capacity and isothermal properties [59]. MPCM-based systems are being used in different fields, including buildings, textiles, and food storage systems. The applications of MPCM-based systems are tabulated in Table 6.3. This MPCM-based group in thermal energy storage application is still in the growing stage. Researchers and

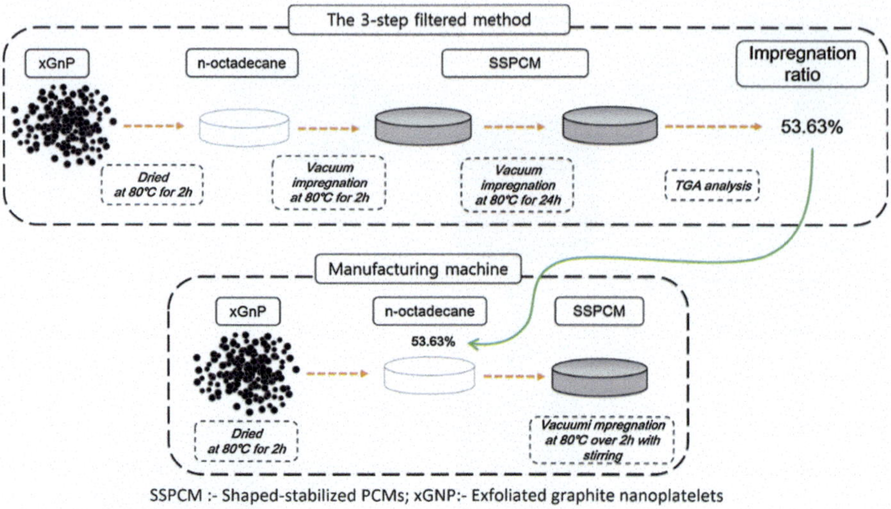

Fig. 6.7 Fabrication of gypsum-based shape-stabilized PCM using vacuum impregnator. Reproduced with permission from [87]

industrialists are continuously working on these for providing better systems for thermal energy storage devices. Few applications of PCM-based thermal energy storage devices in solar energy storage systems, waste energy recovery systems, and energy conservation in buildings are discussed below (Table 6.3).

Table 6.3 Applications of the MPCM [58, 59]

Application fields of the MPCMs	Product
Buildings	Concrete mixes
	Cement mortar
	Wallboards
	Sandwich panels
	Gypsum plaster
	Slabs
Textiles	Shoe insoles
	Cold water diving garments
	Leather products
	Military uniforms
Food storage systems	Smart food packaging
	Cold storage

Table 6.4 Dollar saving in electricity use and demand charges due to TES system as compared to conventional system under different load scenarios. (Estimates of savings are based on PG&E's E-20 rate schedule) [117]

	Weekly saving due to energy savings				Monthly savings due to peak reduction	
	On-peak 12:00–6:00, Mon-Fri	Part-peak 8:30–12:00, 18:00–21:30, Mon-Fri	Off-peak all other times	Total	On-peak	24-h
	$	$	$	$	$	$
Warm week in winter (first week of December 1989)	230	−163	−1544	−1477	846	297
Average week in August (third week of August 1990)	1326	−326	−2850	−1850	6204	2178
Warm week in September (fourth week of September 1990)	2006	42	−2042	6	8460	2970

6.4.1 Solar Energy Storage

6.4.1.1 Concentrated Solar Power Plant

Concentrated solar power plant (CSP) is a power generation plant to generate electrical energy using solar energy. Generally, solar energy is accumulated by reflecting sunlight using mirrors which are concentrated on a receiver where solar energy is stored in the form of heat [88]. The stored heat is then conveyed by heat transfer fluid (HTF) to generate steam, which is further used by the turbine to generate electrical energy [89]. In conventional CSP, HTF is directly used to generate steam due to which conventional CSP has certain limitations such as in colder regions, water as HTF should be avoided due to freezing at 0 °C. So, sometimes CSPs are coupled with PPC systems [90, 91]. PPC coupled CSP systems provide round trip efficiency (~95%) by recovering the initially stored thermal energy [13] and generating additional value to conventional CSP. Moreover, utilization of PPC increases the power generation capacity via improving grid stability. PPC systems store excess energy during day hours and extend the power generation in cloudy weather or during night. The second benefit of PPC integration is that it reduces the cost of power generation via managing the demand and response. The third benefit of PPC coupling is the improvement in the overall reliability of the system by lowering the load, and storing excess energy which in turn reduces the risk of damage during peak hours, hence enhancing the lifetime of the system. PPC integrated CSP systems mainly consist of two parts: heat transfer components and thermal energy storage materials. Thermal energy storage materials store thermal energy whereas heat transfer unit supplies and extracts stored thermal energy. Figure 6.8 illustrates the parabolic trough system which consists of an integrated steam turbine, interconnected linear parabolic troughs, and an electrical generator for power generation. These systems employed cylindrical parabolic reflectors which focus on the vacuum-sealed glass tubes, and the working fluids are positioned within these tubes to maximize solar irradiation and minimize heat losses [92]. Absorption tubes are selectively colored to match the spectral profile of solar radiation. The absorption tube's high absorption coefficient, combined with its location at the focal point of the parabolic trough, results in efficient heating of the working fluid. The low emission features of the absorption tube further reduce heat losses, making the parabolic troughs highly efficient. For CSP, inorganic PCMs such as molten salts, metal, and metal alloys are the most suitable PCMs for the working temperature range (200–600 °C). Salt based PCMs are preferentially considered due to their cost-effectiveness and also, they absorb a significant amount of sensible heat before phase transition. The salt based PCMs are proved to be the most cost-effective materials including metal chlorides, fluorides, nitrites, nitrates, and their eutectic mixture, but their performance suffers due to super cooling and corrosion which are the two fundamental drawbacks of salt based PCMs. Supercooling does not lead to a drastic reduction in overall energy storage performance because it allows utilization of sensible heat of supercooled liquid PCMs, but corrosion remains a serious issue, which can only be overcome by encapsulating salt based PCMs.

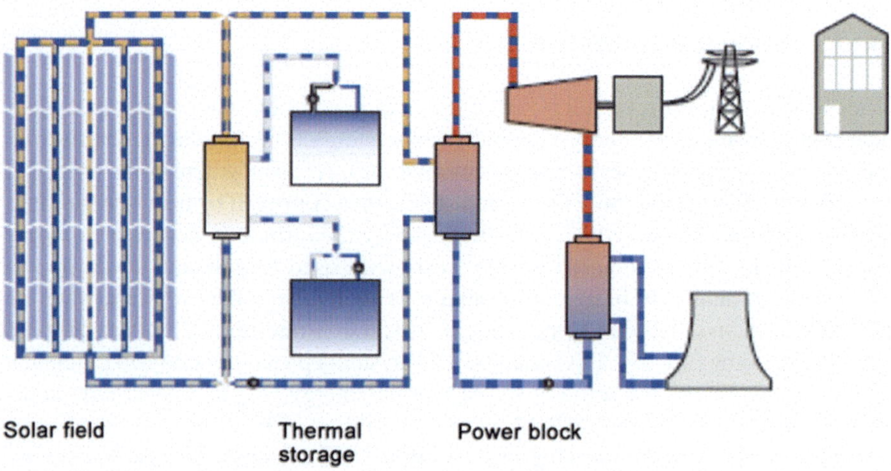

Fig. 6.8 Concentrated solar power plant. Reproduced with permission from [92]

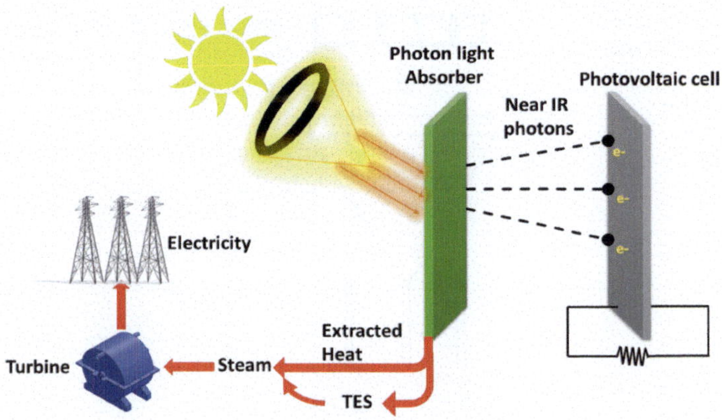

Fig. 6.9 Working illustration of thermophotovoltaic system. Inspired from [13]

6.4.1.2 Solar Thermophotovoltaics

Another electricity generation system that uses solar energy via photovoltaics converter is solar thermophotovoltaics systems (STPVS) (Fig. 6.9). Just like CSP systems, integration of PPC improves the dispatchable electricity generation characteristics of STPVS [93, 94]. Incorporating PPC within STPVS has been achieved by two types of designs. In the first design, PPC materials are placed between the thermal emitter and solar absorber. During daytime, the receiver harnesses solar energy and converts it into thermal energy. This thermal energy is subsequently transferred to PPC material to activate emitter [95]. When sufficient solar energy flux is available, the generated thermal energy heats the emitter and charges the PPC materials simultaneously. Due to storage of excess thermal energy by PPC during daytime, it releases the stored energy at night and maintains a continuous supply of energy for the persistent operation of STPVS [96]. For switchable power generation, photovoltaic modules can be modified within stationary absorber/emitter subsystem that allow collection of thermal radiation. In the second design, solar energy is collected and stored as thermal energy within the separator [13]. HTF extracts this thermal energy and supplies it for electricity generation through photovoltaic converters [97]. Seyf et al. conducted a system-level design study where they have employed graphite-based photo absorber to efficiently convert solar energy to thermal energy. In their study, Tin (Sn) was used as HTF which was responsible for extracting heat from the absorber and transferring it to TES material. The stored heat in TES material was further conveyed to photovoltaic subsystem [98].

6.4.1.3 Solar Thermoelectric Generators

Solar thermoelectric generator (STEG) is a two stage energy conversion system. First step involves heat energy production by converting concentrated solar power

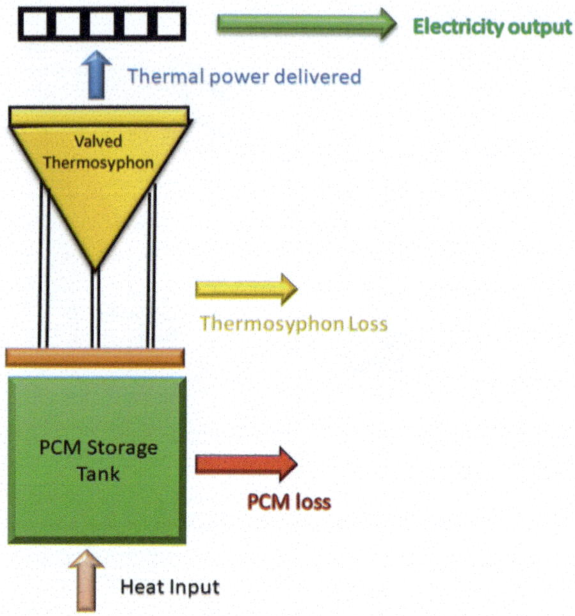

Fig. 6.10 Performance of heat flow of a prototype solar thermoelectric system. Inspired from [13]

via photo absorbers. In the second step, the harvested heat energy creates a temperature gradient profile across the thermoelectric generator to generate electricity from heat energy [99] shown in Fig. 6.10. The STEG technology draws inspiration from the impressive performance of thermoelectric materials that lead to a solar-electric conversion efficiency of 7.4% [13]. However, based on theoretical prediction, the maximum conversion efficiency of STEG systems is limited to 15% [100]. To harness power by using STEG, Rea, and their colleagues [101] developed a prototype in which sunlight is collected at a power block which is placed on the top of a tower. This power block directly stores sunlight as latent heat which might be utilized to generate electricity through either a thermoelectric generator or stirling heat engine.

6.4.2 Waste Heat Recovery Systems

The application of waste heat recovery technology in power generation involves storing and utilizing heat produced as a byproduct of industrial processes [102]. In many industries, a significant amount of heat is released into the environment as waste. Nowadays, many industries work on the concept of zero wastage [103] has led to the integration of PCMs to store latent heat with waste heat recovery technology. This allows the efficient utilization of recovered heat in various residential and industrial sectors [104] resulting in reduced energy consumption and improved process efficiency. Waste heat recovery systems can be categorized as portable and stationary waste heat recovery systems based on the location of heat sinks and sources [105].

6.4.2.1 Portable Waste Heat Recovery

Portable waste heat recovery systems involve the extraction of waste heat from different industrial units and transport it to another location where it is needed. Different approaches were developed for the transportation of heat depending upon heat source and demand. Among these, mobilized thermal energy storage (M-TES) technologies gained substantial attention owing to an improved flexibility and lower investment costs. M-TES materials are designed to compactly store waste heat within heat storage materials, allowing for its efficient transport and distribution to end-users as depicted in Fig. 6.11. Advantages of using M-TES systems include the efficient use of waste heat over a long distance, and cost-effectiveness in terms of setup and maintenance [106]. Wang et al. conducted an experimental study on the thermal charging and discharging performance of a M-TES system based on erythritol. They used storage containers with 74 and 60 kg of erythritol for direct and indirect contact, respectively; and the systems were able to store 9.15 kW h and 6.8 kW h of heat, respectively [107]. However, the performance of thermal charging of direct contact M-TES was limited due to heat blocking effect.

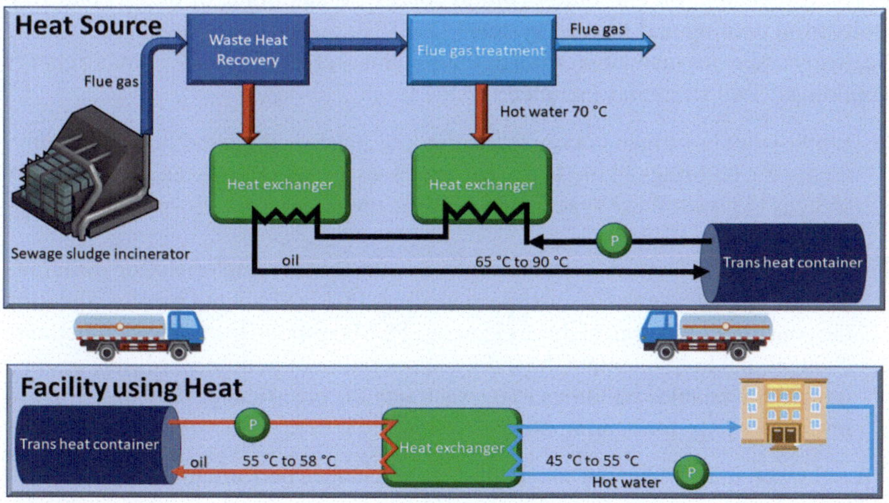

Fig. 6.11 A flow diagram of typical Mobilized thermal energy storage for waste heat recovery. Inspired from [108]

6.4.2.2 Stationary Waste Heat Recovery

Internal combustion engines and cooling systems, such as refrigerators and air conditioners, release a significant amount of exhaust and condensed heat to the atmosphere that leads to energy losses and environmental pollution. To address this issue, PCMs can offer an efficient solution by absorbing waste heat from cold storage systems and the recovered heat can be utilized for other purposes like domestic water heating, space heating, and industrial hot water supply [109, 110]. Li et al. studied a multi-PCM system for conventional AC in order to assess the performance of TES systems. They used PCMs having phase transition temperatures of 10, 6.5, and 5.3 °C along with water as heat transfer fluid. Simulation results revealed that the multi-PCM systems have improved heat storage capacity by approximately 32.22% than the single PCM system [111]. In another study conducted by Diaconu et al. involved development of a PCM microencapsulated slurry, using 45 w/w % paraffin wax RT6 (commercial grade) for air conditioning application [112]. They experimentally evaluated the storage properties and heat transfer characteristics of slurry, with the aim of integrating it into solar air conditioning system operating at low temperature.

6.4.3 Energy Conservation in Buildings

Energy conservation in buildings is an utmost importance, as the construction sector alone is responsible for one-third of world energy consumption, of which 60 to 70% results from cooling and heating systems [113]. To promote sustainable infrastructure, many countries mandate certain developments and regulations for saving energy in buildings. Two strategies included:

(1) Employment of green energy technology to meet cooling and heating requirements for buildings [114, 115]. This involves incorporating renewable energy sources to power these systems, and lowers the reliance on conventional energy sources.
(2) Integrating TES systems in buildings as construction materials for designing windows, walls, and roofs for space heating. PCMs are being employed during the construction design of these elements. PCMs respond to external temperature changes and help maintain the indoor temperature of buildings. Moreover, their integration enables buildings to interact with the electrical grid by utilizing low level electricity for heating purposes.

A light weight PCM having excellent solar reflectance performance can be considered for roof structure cooling applications. Due to their thermal buffering capabilities and solar reflectance of PPC installed at rooftops, it lowers heat gain and reduces building's indoor temperature during summers. In window applications, PCMs are sandwiched between glass plates for insulation against variations in diurnal temperature. During the daytime, PCM stores heat, melts, and becomes optically transparent, thus meeting the fundamental requirement to be used in windows whereas during night, stored heat is released [13] An illustration is shown in Fig. 6.12 which describes the difference between buildings having conventional HVAC (heating, ventilation, and air conditioning) and PCM integrated buildings. It was observed that PCM integrated buildings are able to reduce indoor temperature fluctuations, have better control on temperature, and maintain thermal comfort all day.

In 1989, a study was conducted by Gillespie et al. from PG&E (pacific gas and electric company) for the evaluation of TES system's performance in a building (2,000,000 ft^2 four story building) located at 15 miles east of San Francisco. According to the reported owner's criterion, consumption of energy was 30,000

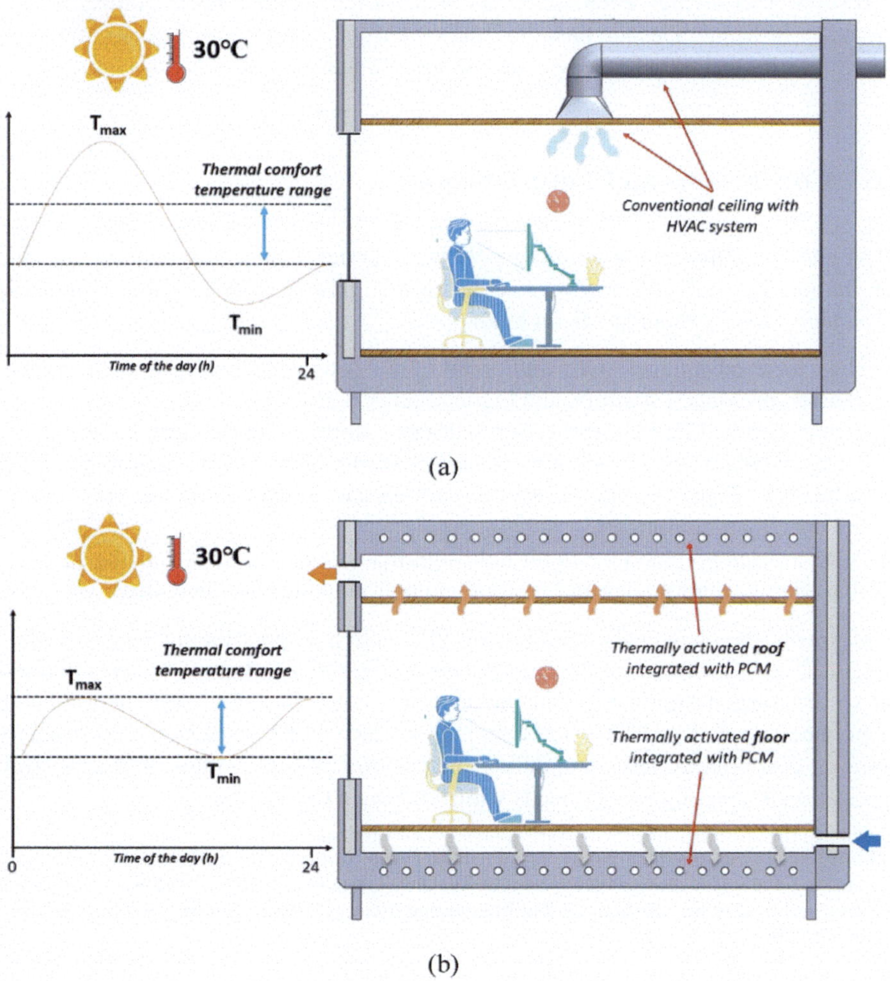

Fig. 6.12 Illustration of **a** conventional HVAC cooled office and **b** TES integrated office. Reproduced with permission from [116]

BTU/ft^2 per year (including all appliances, lighting, heating, ventilation, and air conditioning). To optimize energy cost, an ice storage system was incorporated in building design to shift the electric consumption (kWh) and demand (kW) from on-peak (high cost) to partial peak (lower cost) or off-peak time [117]. On summarizing the analysis done by PG&E, it was found that the TES systems, excluding a few problems related to maintenance and expansion of the system, were successfully installed into the building. It achieved superior peak demand savings compared to non-TES systems on peak days. Dollar Savings in electricity use and demand charges due to the TES system are presented in Table 6.4. This estimation of savings is on the basis of PG&E's E-20 rate schedule.

It shows that savings were significant on a normal summer day as compared to a warm day in winter. Despite this, the losses incurred during several weeks in winter were offset by the saving achieved during the summer season.

6.5 Conclusions and Future Scope

PCM-based thermal energy storing polymeric systems open a promising avenue for harnessing and utilizing renewable energy. These systems offer numerous advantages; few of them are given below-

- They have advantages over traditional TES systems including improved heat transfer, prolonged stability, and higher energy density.
- Incorporation of PCM into polymer matrices enables efficient heat storage and on-demand release across multiple cycles, enhancing intermittent renewable energy utilization. Tunability of these systems allows customization to meet preferences and specific requirements.
- PCM-infused polymeric systems hold significant potential for advancing sustainable energy technologies and promoting environmental sustainability.

However, the main challenge of using PPC system is poor thermal conductivity, and it can be improved with the incorporation of highly conductive fillers and innovative structural designs. A promising approach is to use nano-fillers incorporated biopolymers to enclose PCMs for better performance. Further research is essential to optimize cost-effectiveness, scalability, and performance, expanding their impactful application in meeting energy needs while curbing CO_2 emissions.

Acknowledgements Authors would like to acknowledge Department of Materials Science and Engineering, Indian Institute of Technology, Delhi. One of the authors (Chandrani Sarkar) expresses her deep gratitude to SERB (Science and Engineering Research Board), India for NPDF Scheme (PDF/2022/000679).

Conflict of Interest Authors declare that they have no conflict of interest.

References

1. Kennedy KM, Ruggles TH, Rinaldi K et al (2022) The role of concentrated solar power with thermal energy storage in least-cost highly reliable electricity systems fully powered by variable renewable energy. Adv Appl Energy 6:100091
2. Guelpa E, Verda V (2019) Thermal energy storage in district heating and cooling systems: a review. Appl Energy 252:113474
3. Zhang Y, Zhou G, Lin K et al (2007) Application of latent heat thermal energy storage in buildings: state-of-the-art and outlook. Build Environ 42:2197–2209
4. Shukla A, Buddhi D, Sawhney RL (2009) Solar water heaters with phase change material thermal energy storage medium: a review. Renew Sustain Energy Rev 13:2119–2125
5. Chieruzzi M, Cerritelli GF, Miliozzi A, Kenny JM (2013) Effect of nanoparticles on heat capacity of nanofluids based on molten salts as PCM for thermal energy storage. Nanoscale Res Lett 8:448
6. Pielichowska K, Pielichowski K (2014) Phase change materials for thermal energy storage. Prog Mater Sci 65:67–123
7. Pasupathy A, Velraj R, Seeniraj RV (2008) Phase change material-based building architecture for thermal management in residential and commercial establishments. Renew Sustain Energy Rev 12:39–64
8. Yuan K, Shi J, Aftab W et al (2020) Engineering the thermal conductivity of functional phase-change materials for heat energy conversion, storage, and utilization. Adv Funct Mater 30:1904228
9. 'Beceiro SJ, Saeterli RK (2022) Thermochemical energy storage: the next generation thermal batteries? https://blog.sintef.com/sintefenergy/thermochemical-energy-storage-the-next-generation-thermal-batteries/#:~:text=Thermochemical%20energy%20storage%20(TCES)%20materials,and%20latent%20heat%20TES%20compounds
10. Kanojia N, Kaushik S, Ali S, et al (2022) Review on sustainable thermal energy storage technology with packed bed regenerator
11. Dincer I, Dost S, Li X (1997) Performance analyses of sensible heat storage systems for thermal applications. Int J Energy Res 21:1157–1171
12. Sarbu I, Dorca A (2019) Review on heat transfer analysis in thermal energy storage using latent heat storage systems and phase change materials. Int J Energy Res 43:29–64
13. Aftab W, Usman A, Shi J et al (2021) Phase change material-integrated latent heat storage systems for sustainable energy solutions. Energy Environ Sci 14:4268–4291
14. Mondal S (2008) Phase change materials for smart textiles–an overview. Appl Therm Eng 28:1536–1550
15. Fallahi A, Guldentops G, Tao M et al (2017) Review on solid-solid phase change materials for thermal energy storage: molecular structure and thermal properties. Appl Therm Eng 127:1427–1441
16. Usman A, Xiong F, Aftab W et al (2022) Emerging solid-to-solid phase-change materials for thermal-energy harvesting, storage, and utilization. Adv Mater 34:2202457
17. Fatih Demirbas M (2006) Thermal energy storage and phase change materials: an overview. Energy Sources Part B 1:85–95
18. Zhang Z, Fang X (2006) Study on paraffin/expanded graphite composite phase change thermal energy storage material. Energy Convers Manag 47:303–310
19. Rozanna D, Chuah TG, Salmiah A et al (2005) Fatty acids as phase change materials (PCMs) for thermal energy storage: a review. Int J Green Energy 1:495–513
20. Zhang N, Yuan Y, Cao X et al (2018) Latent heat thermal energy storage systems with solid-liquid phase change materials: a review. Adv Eng Mater 20:1700753
21. Tyagi VV, Kaushik SC, Tyagi SK, Akiyama T (2011) Development of phase change materials based microencapsulated technology for buildings: a review. Renew Sustain Energy Rev 15:1373–1391

22. Magendran SS, Khan FSA, Mubarak NM et al (2019) Synthesis of organic phase change materials (PCM) for energy storage applications: a review. Nano-Struct Nano-Objects 20:100399
23. Morris RE, Wheatley PS (2008) Gas storage in nanoporous materials. Angewandte Chemie Int Ed 47:4966–4981
24. Reji Kumar R, Samykano M, Pandey AK et al (2020) Phase change materials and nano-enhanced phase change materials for thermal energy storage in photovoltaic thermal systems: a futuristic approach and its technical challenges. Renew Sustain Energy Rev 133:110341
25. Zhao P, Dai Y, Wang J (2014) Design and thermodynamic analysis of a hybrid energy storage system based on A-CAES (adiabatic compressed air energy storage) and FESS (flywheel energy storage system) for wind power application. Energy 70:674–684
26. Ben HG, Younsi Z, Naji H, Salaün F (2021) A comprehensive review of microencapsulated phase change materials synthesis for low-temperature energy storage applications. Appl Sci (Switzerland) 11:11900
27. Rathod MK, Banerjee J (2013) Thermal stability of phase change materials used in latent heat energy storage systems: a review. Renew Sustain Energy Rev 18:246–258
28. Okogeri O, Stathopoulos VN (2021) What about greener phase change materials? A review on biobased phase change materials for thermal energy storage applications. Int J Thermofluids 10:100081
29. Prabhu PA, Shinde NN, Patil PS (2012) Review of phase change materials for thermal energy storage applications. 2:871–875
30. Agarwal A, Sarviya RM (2017) Characterization of commercial grade paraffin wax as latent heat storage material for solar dryers. 4:779–789
31. Zhang L, Dong J (2017) Experimental study on the thermal stability of a paraffin mixture with up to 10,000 thermal cycles. Thermal Sci Eng Prog 1:78–87
32. Zhong Y, Guo Q, Li S et al (2010) Heat transfer enhancement of paraffin wax using graphite foam for thermal energy storage. Sol Energy Mater Sol Cells 94:1011–1014
33. Hasnain SM (1998) Review on sustainable thermal energy storage technologies, Part I: heat storage materials and techniques. 39:1127–1138
34. Kumar A, Shukla SK (2015) A review on thermal energy storage unit for solar thermal power plant application. Energy Procedia 74:462–469
35. Kant K, Shukla A, Sharma A (2016) Performance evaluation of fatty acids as phase change material for thermal energy storage. J Energy Storage 6:153–162
36. Sharma A, Tyagi VV, Chen CR, Buddhi D (2009) Review on thermal energy storage with phase change materials and applications. Renew Sustain Energy Rev 13:318–345
37. Li C, Xie B, Chen J et al (2019) Emerging mineral-coupled composite phase change materials for thermal energy storage. Energy Convers Manag 183:633–644
38. Peng S, Huang J, Wang T, Zhu P (2019) Effect of fumed silica additive on supercooling, thermal reliability and thermal stability of $Na_2HPO_4 \cdot 12H_2O$ as inorganic PCM. Thermochim Acta 675:1–8
39. Li Y, Li C, Lin N et al (2021) Review on tailored phase change behavior of hydrated salt as phase change materials for energy storage. Mater Today Energy 22:100866
40. Socaciu LG (2015) Leonardo electronic journal of practices and technologies thermal energy storage with phase change material. Leonardo Electron J Pract Technol 26:75–98
41. Soares N, Costa JJ, Gaspar AR, Santos P (2013) Review of passive PCM latent heat thermal energy storage systems towards buildings' energy efficiency. Energy Build 59:82–103
42. Safari A, Saidur R, Sulaiman FA et al (2017) A review on supercooling of phase change materials in thermal energy storage systems. Renew Sustain Energy Rev 70:905–919
43. Costa SC, Kenisarin M (2022) A review of metallic materials for latent heat thermal energy storage: thermophysical properties, applications, and challenges. Renew Sustain Energy Rev 154:111812
44. Farid MM, Khudhair AM, Razack SAK, Al-Hallaj S (2004) A review on phase change energy storage: materials and applications. Energy Convers Manag 45:1597–1615

45. Kousksou T, Bruel P, Jamil A et al (2014) Energy storage: applications and challenges. Sol Energy Mater Sol Cells 120:59–80
46. Liu M, Omaraa ES, Qi J et al (2021) Review and characterisation of high-temperature phase change material candidates between 500 C and 700°C. Renew Sustain Energy Rev 150:111528
47. Li G, Hwang Y, Radermacher R, Chun HH (2013) Review of cold storage materials for subzero applications. Energy 51:1–17
48. Ma Z, Zhang Y (2006) Solid velocity correction schemes for a temperature transforming model for convection phase change. Int J Numer Methods Heat Fluid Flow 16:204–225
49. Dutil Y, Rousse DR, Ben SN et al (2011) A review on phase-change materials: Mathematical modeling and simulations. Renew Sustain Energy Rev 15:112–130
50. Shen Z, Qin M, Xiong F et al (2023) Nanocellulose-based composite phase change materials for thermal energy storage: status and challenges. Energy Environ Sci 16:830–861
51. Sharma RK, Ganesan P, Tyagi VV et al (2015) Developments in organic solid-liquid phase change materials and their applications in thermal energy storage. Energy Convers Manag 95:193–228
52. Umair MM, Zhang Y, Iqbal K et al (2019) Novel strategies and supporting materials applied to shape-stabilize organic phase change materials for thermal energy storage–a review. Appl Energy 235:846–873
53. Mohamed SA, Al-Sulaiman FA, Ibrahim NI et al (2017) A review on current status and challenges of inorganic phase change materials for thermal energy storage systems. Renew Sustain Energy Rev 70:1072–1089
54. Wu MQ, Wu S, Cai YF et al (2021) Form-stable phase change composites: preparation, performance, and applications for thermal energy conversion, storage and management. Energy Storage Mater 42:380–417
55. Li T, Zhu M, Yang Z et al (2016) Wood composite as an energy efficient building material: guided sunlight transmittance and effective thermal insulation. Adv Energy Mater 6:1601122
56. Peng S, Fuchs A, Wirtz RA (2004) Polymeric phase change composites for thermal energy storage. J Appl Polym Sci 93:1240–1251
57. Lyu Y, Siddique ARM, Majid SH et al (2019) Electric vehicle battery thermal management system with thermoelectric cooling. Energy Rep 5:822–827
58. Huang X, Chen X, Li A et al (2019a) Shape-stabilized phase change materials based on porous supports for thermal energy storage applications. Chem Eng J 356:641–661
59. Huang X, Zhu C, Lin Y, Fang G (2019b) Thermal properties and applications of microencapsulated PCM for thermal energy storage: a review. Appl Therm Eng 147:841–855
60. Hu H (2020) Recent advances of polymeric phase change composites for flexible electronics and thermal energy storage system. Compos B Eng 195:108094
61. Sari A, Karaipekli A (2007) Thermal conductivity and latent heat thermal energy storage characteristics of paraffin/expanded graphite composite as phase change material. Appl Therm Eng 27:1271–1277
62. Sris J, Prabha K (2011) Microencapsulation: a review. Int J Pharm Bio Sci 3:509–531
63. Zhu S, Nguyen MT, Yonezawa T (2021) Micro- And nano-encapsulated metal and alloy-based phase-change materials for thermal energy storage. Nanoscale Adv 3:4626–4645
64. Jamekhorshid A, Sadrameli SM, Farid M (2014) A review of microencapsulation methods of phase change materials (PCMs) as a thermal energy storage (TES) medium. Renew Sustain Energy Rev 31:531–542
65. Hawlader MNA, Uddin MS, Khin M (2003) Microencapsulated PCM thermal-energy storage system. 74:195–202
66. Borreguero AM, Valverde JL, Rodríguez JF et al (2011) Synthesis and characterization of microcapsules containing Rubitherm®RT27 obtained by spray drying. Chem Eng J 166:384–390
67. Baig N, Kammakakam I, Falath W, Kammakakam I (2021) Nanomaterials: a review of synthesis methods, properties, recent progress, and challenges. Mater Adv 2:1821–1871
68. Vashchuk A, Motrunich S, Lishchuk P et al (2023) Thermal conductivity and mechanical properties of epoxy vitrimer nanocomposites reinforced with graphene oxide. Appl Nanosci (Switzerland) 13:4675–4683

69. Konuklu Y, Paksoy HO, Unal M, Konuklu S (2014) Microencapsulation of a fatty acid with Poly(melamine-urea-formaldehyde). Energy Convers Manag 80:382–390
70. Yu A, Song X, Lu W et al (2022) Facile and safety synthesis of highly loaded phase change microcapsules with paraffin/butyl stearate core and their feasible application in polymer composite. Sol Energy Mater Sol Cells 247:111955
71. Peng G, Dou G, Hu Y et al (2020) Phase change material (PCM) microcapsules for thermal energy storage. Adv Polym Technol 2020:9490873
72. Hu W, Shi X, Gao M, et al (2021) Light-actuated shape memory and self-healing phase change composites supported by MXene/waterborne polyurethane aerogel for superior solar-thermal energy storage. Compos Commun 28:100980
73. Li M, Zhang X (2023) Heat storage and conductive properties of the polypyrrole aerogel-based new shape-stabilized phase change composite. Sol Energy 252:380–390
74. Nazari N, Bahramian AR, Allahbakhsh A (2022) Thermal storage achievement of paraffin wax phase change material systems with regard to novolac aerogel/carbon monofilament/zinc borate form stabilization. J Energy Storage 50:104741
75. Song M, Jiang J, Zhu J et al (2021) Lightweight, strong, and form-stable cellulose nanofibrils phase change aerogel with high latent heat. Carbohydr Polym 272:118460
76. Luo L, Luo W, Chen W et al (2023) Form-stable phase change materials based on graphene-doped PVA aerogel achieving effective solar energy photothermal conversion and storage. Sol Energy 255:146–156
77. Prajapati DG, Kandasubramanian B (2019) Biodegradable polymeric solid framework-based organic phase-change materials for thermal energy storage. Ind Eng Chem Res 58:10652–10677
78. Khan A, Saikia P, Saxena R et al (2020) Microencapsulation of phase change material in water dispersible polymeric particles for thermoregulating rubber composites—a holistic approach. Int J Energy Res 44:1567–1579
79. Pandey K, Ali SF, Gupta SK et al (2021) Facile technique to encapsulate phase change material in an amphiphilic polymeric matrix for thermal energy storage. Appl Energy 292:116917
80. Darzi ME, Golestaneh SI, Kamali M, Karimi G (2019) Thermal and electrical performance analysis of co-electrospun-electrosprayed PCM nanofiber composites in the presence of graphene and carbon fiber powder. Renew Energy 135:719–728
81. Pielichowska K, Pielichowski K (2011) Biodegradable PEO/cellulose-based solid-solid phase change materials. Polym Adv Technol 22:1633–1641
82. Entürk SB, Kahraman D, Alkan C, Göke I (2011) Biodegradable PEG/cellulose, PEG/agarose and PEG/chitosan blends as shape stabilized phase change materials for latent heat energy storage. Carbohydr Polym 84:141–144
83. Abdullahi Hassan Y, Hu H (2020) Current status of polymer nanocomposite dielectrics for high-temperature applications. Compos Part A Appl Sci Manuf 138:106064
84. Wu S, Li T, Tong Z et al (2019) High-performance thermally conductive phase change composites by large-size oriented graphite sheets for scalable thermal energy harvesting. Adv Mater 31:1905099
85. Al-Ahmed A, Mazumder MAJ, Salhi B et al (2021) Effects of carbon-based fillers on thermal properties of fatty acids and their eutectics as phase change materials used for thermal energy storage: a review. J Energy Storage 35:102329
86. Qi GQ, Yang J, Bao RY et al (2015) Enhanced comprehensive performance of polyethylene glycol based phase change material with hybrid graphene nanomaterials for thermal energy storage. Carbon N Y 88:196–205
87. Lee J, Wi S, Yun BY et al (2019) Development and evaluation of gypsum/shape-stabilization phase change materials using large-capacity vacuum impregnator for thermal energy storage. Appl Energy 241:278–290
88. Tariq Chaichan M, Ibrahim Abass K, Kazem HA, et al (2018) Energy yield loss caused by dust and pollutants deposition on concentrated solar power plantsin iraq weathers alcohol-gasoline blends in spark ignition engines performance view project modeling the dust view project energy yield loss caused by dust and pollutants deposition on concentrated solar power plants in Iraq weathers 3:160–169

89. Vignarooban K, Xu X, Arvay A et al (2015) Heat transfer fluids for concentrating solar power systems–a review. Appl Energy 146:383–396
90. Bilal Awan A, Khan MN, Zubair M, Bellos E (2020) Commercial parabolic trough CSP plants: research trends and technological advancements. Sol Energy 211:1422–1458
91. Mehos M, Turchi C, Jorgensen J et al (2016) On the path to sunshot: advancing concentrating solar power technology, performance, and dispatchability. United States. https://doi.org/10.2172/1344199
92. Pavlović TM, Radonjić IS, Milosavljević DD, Pantić LS (2012) A review of concentrating solar power plants in the world and their potential use in Serbia. Renew Sustain Energy Rev 16:3891–3902
93. Chan WR, Bermel P, Pilawa-Podgurski RCN et al (2013) Toward high-energy-density, high-efficiency, and moderate-temperature chip-scale thermophotovoltaics. Proc Natl Acad Sci U S A 110:5309–5314
94. Lenert A, Bierman DM, Nam Y et al (2014) A nanophotonic solar thermophotovoltaic device. Nat Nanotechnol 9:126–130
95. Hu M, Zhao B, Suhendri, et al (2022) Applications of radiative sky cooling in solar energy systems: progress, challenges, and prospects. Renew Sustain Energy Rev 160:112304
96. Zhang J, Wang C, Shi J et al (2022) Solar selective absorber for emerging sustainable applications. Adv Energy Sustain Res 3:2100195
97. Pacio J, Singer C, Wetzel T, Uhlig R (2013) Thermodynamic evaluation of liquid metals as heat transfer fluids in concentrated solar power plants. Appl Therm Eng 60:295–302
98. Seyf HR, Henry A (2016) Thermophotovoltaics: a potential pathway to high efficiency concentrated solar power. Energy Environ Sci 9:2654–2665
99. Settaluri KT, Lo H, Ram RJ (2012) Thin thermoelectric generator system for body energy harvesting. J Electron Mater 41:984–988
100. Baranowski LL, Snyder GJ, Toberer ES (2012) Concentrated solar thermoelectric generators. Energy Environ Sci 5:9055–9067
101. Rea JE, Oshman CJ, Singh A et al (2018) Experimental demonstration of a dispatchable latent heat storage system with aluminum-silicon as a phase change material. Appl Energy 230:1218–1229
102. Ortega-Fernández I, Calvet N, Gil A et al (2015) Thermophysical characterization of a by-product from the steel industry to be used as a sustainable and low-cost thermal energy storage material. Energy 89:601–609
103. Gulghane AA, Khandve PV (2015) Management for construction materials and control of construction waste in construction industry: a review. 5:59–64
104. Chua KJ, Chou SK, Yang WM (2010) Advances in heat pump systems: a review. Appl Energy 87:3611–3624
105. El Hage H, Ramadan M, Jaber H et al (2020) A short review on the techniques of waste heat recovery from domestic applications. Energy Sources Part A Recov Utilizat Environ Eff 42:3019–3034
106. Du K, Calautit J, Eames P, Wu Y (2021) A state-of-the-art review of the application of phase change materials (PCM) in mobilized-thermal energy storage (M-TES) for recovering low-temperature industrial waste heat (IWH) for distributed heat supply. Renew Energy 168:1040–1057
107. Wang W, Guo S, Li H et al (2014) Experimental study on the direct/indirect contact energy storage container in mobilized thermal energy system (M-TES). Appl Energy 119:181–189
108. Miró L, Gasia J, Cabeza LF (2016) Thermal energy storage (TES) for industrial waste heat (IWH) recovery: a review. Appl Energy 179:284–301
109. Pabon JJG, Khosravi A, Belman-Flores JM et al (2020) Applications of refrigerant R1234yf in heating, air conditioning and refrigeration systems: a decade of researches. Int J Refrig 118:104–113
110. Riffat SB, Afonso CF, Oliveira AC, Reay DA (1997) Natural refrigerants for refrigeration and air-conditioning systems. 1:33–42

111. Li XY, Yang L, Wang XL et al (2018) Investigation on the charging process of a multi-PCM latent heat thermal energy storage unit for use in conventional air-conditioning systems. Energy 150:591–600
112. Diaconu BM, Varga S, Oliveira AC (2010) Experimental assessment of heat storage properties and heat transfer characteristics of a phase change material slurry for air conditioning applications. Appl Energy 87:620–628
113. Adolph M, Kopmann N, Lupulescu B, Müller D (2014) Adaptive control strategies for single room heating. Energy Build 68:771–778
114. Chel A, Kaushik G (2018) Renewable energy technologies for sustainable development of energy efficient building. Alex Eng J 57:655–669
115. Hara Chakravarty K, Sadi M, Chakravarty H et al (2022) A review on integration of renewable energy processes in vapor absorption chiller for sustainable cooling. Sustain Energy Technol Assess 50:101822
116. Faraj K, Khaled M, Faraj J et al (2020) Phase change material thermal energy storage systems for cooling applications in buildings: a review. Renew Sustain Energy Rev 119:109579
117. Akbari H (1992) Case studies of thermal energy storage (TES) systems: evaluation and verification of system performance. https://www.osti.gov/servlets/purl/10179001

Chapter 7
Challenges and Opportunities of Hybrid Photovoltaic Thermal Collectors and Their Integration into Small-Scale Energy Systems for Prosumers

Madalina Barbu, Monica Siroux, and George Darie

Abstract Hybrid Photovoltaic-Thermal panels combine the two traditional solar energy production technologies (photovoltaic and solar thermal) in a single compact piece of micro-cogeneration equipment. This technology is in line with the growing trend of decentralization and energy self-reliance, by producing locally and simultaneously both electricity and thermal energy. Photovoltaic-thermal collectors are in particular advantageous in settings where roof space is restricted, such as urban locations, due to the fact that the rate of solar energy conversion per unit of area is increased. In addition, the extraction of thermal energy results in the cooling of the PV cells, which improves their efficiency. This study looks at the main challenges and opportunities of photovoltaic thermal collectors. In the introduction the functioning principles of this technology are presented, including motivation for development, types, and methods of integration in the energy mix. Next, a literature review of the current research developments is carried out, followed by a presentation of the global market situation. The result of this analyses indicate that the main challenges of this technology are the cost of production, which can be higher than traditional PV panels, lack of standardization and policies. Some recommendations are that the development of additional demonstrative projects, improving the user awareness and additional proof of concept would be required in order to launch this technology further on the mass market.

Keywords Photovoltaic thermal panels · Solar energy · PVT · Micro-cogeneration · Prosumers

M. Barbu (✉) · G. Darie
University Politehnica Bucharest, 060042 Bucharest, Romania
e-mail: madalinaioanabarbu@gmail.com

M. Barbu
Faculty of Engineering, University Vasile Alecsandri, Bacau 600115, Romania

M. Siroux
INSA Strasbourg ICUBE, University of Strasbourg, 67000 Strasbourg, France

S. De et al. (eds.), *Challenges and Opportunities of Distributed Renewable Power*, Energy, Environment, and Sustainability, https://doi.org/10.1007/978-981-97-1406-3_7

Abbreviations

PV	Photovoltaic
PVT	Photovoltaic thermal
ST	Solar Thermal
DHW	Domestic Hot Water
AM	Air mass
EU	European Union
IEC	International Electrotechnical Commission
EVA	Ethylene vinyl acetate
CPV	Concentrated photovoltaic
m-Si	Mono crystalline silicon
p-Si	Poly crystalline silicon
a-Si	Amorphous silicon
GaAs	Gallium arsenide
CIGS	Copper indium gallium selenide
CdTe	Cadmium telluride
PCM	Phase-change materials
LCA	Life cycle analysis
CHP	Combined heat and power
BAPVT	Building applied PVT
BIPVT	Building integrated PVT
FiT	Feed in Tariff
HP	Heat pump
ORC	Organic Rankine Cycle
ESS	European Social Survey
IRENA	International Renewable Energy Agency
IEA	International Energy Agency
HVAC	Heating, Ventilation, and Air Conditioning
NPV	Net present value
SHC	Solar heating and cooling
SWOT	Strengths, Weaknesses, Opportunities and Threats

7.1 Introduction

Since the 1990s, electricity consumption in the European Union has increased by about 1% every year. It only started to show a decrease since 2017 (0.3%), when significant measures have been applied in order to increase energy efficiency in all sectors of activity. It is estimated that without the introduction of new energy efficiency policies, the annual increase in consumption will continue to tend to 1% by 2050 [1].

Even in the situation of stagnant growth of electricity demand, there are two essential factors that stir great interest in renewable energy production: the technological factor, related to the fact that most of the energy equipment in large conventional power plants are reaching the end of their life cycle, and closely related to this, the climate factor and implicitly the direction of the strategies of the EU member states to increase the share of renewables in the energy mix by 2030 to 32%. Almost 80% of total greenhouse gas emissions are resulted from the use of fossil fuels in the energy sector [2]. Increasing the share of renewable energy is expected to help the EU achieve its goal of reducing greenhouse gas emissions by 40% by 2030 [3] and by 80–95% by 2050 [4]. In addition, greater use of renewable energy sources to meet the energy demand will reduce the EU's dependence on fossil fuels [1].

As a result of the global energy policies that are being put in place for the future, the development of renewable energy technologies is essential for achieving these goals. In the long run, renewable energy will start to dominate the world energy mix in the detriment of fossil fuels. Due to their intermittence, renewable sources will have to be combined together to provide the full spectrum of energy demand. Solar energy is an abundant source, easily available mostly everywhere in the world, and thus it has a great potential of taking over a large share of the renewable energy mix in the future, combined with other renewables such as wind and biomass. In theory, there is enough solar radiation around the world to supply 10,000 times the energy demand of the world [5].

The importance of implementation and mass adoption of efficient renewable technologies is crucial. PVT collectors are a relatively new technology, with little research and comprehensive reviews available in the literature compared to the other well-established solar technologies (solar thermal and photovoltaics). While other reviews have been carried out on the technological aspects, there is little work available that correlates the technological development with the market position and looks at the future opportunities of PVT. This work aims to review the potential impact the PVT technology can have in the energy mix of the future. The objectives of this chapter are to carry out an in-depth study on the PVT technologies and to highlight the current global development state, both from research and from market perspective. From here, the main challenges and opportunities of this technology are established and discussed, in particular in connection with a future decentralized energy system based on renewable sources.

7.2 Background on PV Technology and the Motivation for Developing PVT Collectors

7.2.1 *Background on Solar Energy*

The sun is a hot gas sphere with an internal temperature of 15×10^6 °C. Its internal energy is transferred by convection to the external layer of the Sun (photosphere) and is subsequently emitted as radiation to the Earth, which, at the entrance into the atmosphere, receives about 1366 W/m^2 of solar radiation, or a total power of 1.7 × 1014 kW, with an annual yield of 1.5 × 1018 kWh.

However, not all solar radiation has the same quality, varying according to the wavelength of the photons emitted. The sun emits radiation in a wide range of wavelengths, from cosmic rays to long-range radio waves. The visible spectrum is in the wavelength range of 400–700 nm. The distribution of solar radiation reaching the Earth is illustrated qualitatively, as a function of the photon wavelength, in Fig. 7.1. The radiation is described by the parameter called spectral emissivity P(λ), which is measured in power over the surface and over the wavelength unit. The air mass (AM) defines the direct optical path length through the Earth's atmosphere, expressed as a ratio relative to the path length vertically upwards at the zenith.

In Fig. 7.1, the red line represents the solar radiation emitted by the Sun's photosphere that can be observed upon entering the Earth's atmosphere (AM = 0), and the blue line is the reference solar radiation, through an air mass of 1.5 (AM = 1.5). An idealized black body with a temperature of 5726.85 °C emits the radiation described

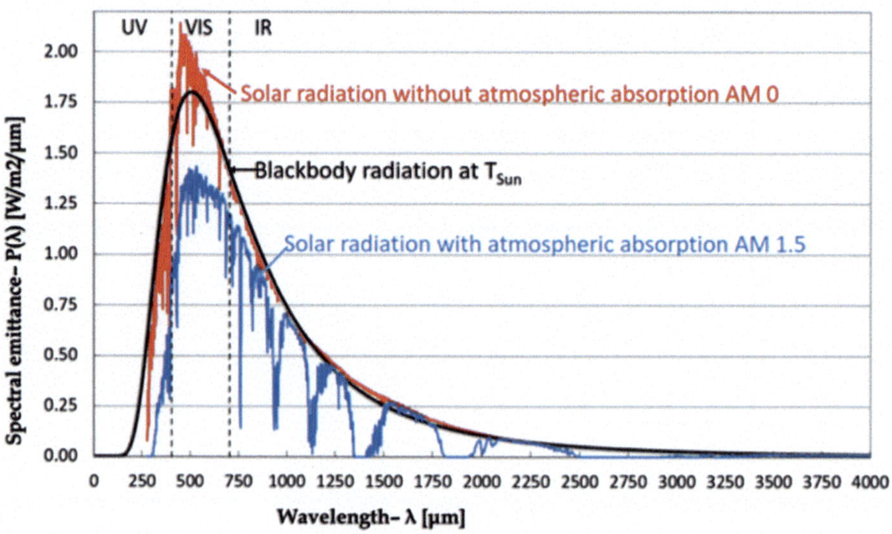

Fig. 7.1 Solar radiation distribution as a function of wavelength [6]

by the black curve; it can be seen that it approximates quite well the curve of solar radiation at AM = 0.

7.2.2 *Energy Conversion in a PV Cell*

From the total solar radiation incident on a PV cell, only a fraction can be captured and converted into electricity. This is primarily due to the specific properties of the semiconductor material, especially the distance between the valence band and the conduction band, called the band-gap. Any photon of light with an energy less than this distance fails to excite the electron in a higher state (to move it from the valence band to the conduction band), so the photon passes without being captured, and the phenomenon is called non-absorption. In addition, photons with a higher energy than the band-gap can excite the electrons with all their energy, but later the electrons return to their default level of the conduction band, and the excess energy is transformed into heat and energy loss. This phenomenon is known as thermalization or thermal relaxation, and results in the heating of the PV cell, which has negative effects on its efficiency.

Figure 7.2 illustrates a Sankey diagram a PV cell. Generally speaking, a Sankey diagram is a visualization used to depict a flow of energy from one point in a system to another. Figure 7.2 shows the share of incident solar radiation that can be extracted in the form of electricity from a silicon cell. The green area represents the excess energy of the photons, which leads to thermalization, and the gray area illustrates the non-absorption, i.e., photons with a wavelength too long, so with insufficient energy to excite the electrons. The red line indicates the specific band-gap, which in this case silicon is 1.12 eV, which corresponds to a wavelength of $\lambda_{Si} = 1107\ \mu m$.

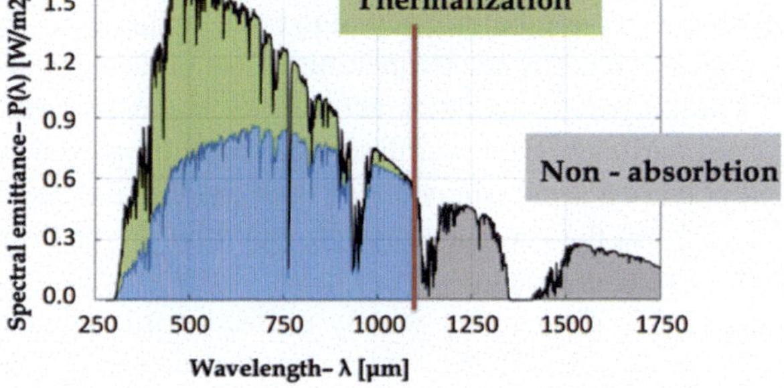

Fig. 7.2 Sankey diagram for a PV collector

Beside the fundamental losses of the semiconductor, there are several other aspects that lead to a decrease in efficiency of the PV cell [7]:

- Light reflection at the surface of the module;
- Recombination of electron-gap pairs before they are separated;
- Serial resistances of the circuit, Joule effect;
- Shading caused by electrodes and conductive bars;
- Degradation caused by environmental conditions—high temperatures, wind, humidity.

7.2.3 *The Effect of High Operating Temperatures on a PV Cell*

The operating temperature is one of the most important parameters and closely related to the electrical efficiency of the PV panel. The most important feature of hybrid PVT systems is the cooling of the photovoltaic cells. To better understand the benefits of PV panel cooling, it is essential to establish the inter-connection between operating temperature and system efficiency.

In a PV cell, thermal energy is produced by several internal phenomena [6, 7]:

- Non-absorption of photons with less energy than the band-gap of the material;
- Radiative or Auger recombination of electron-pair pairs;
- Joule effect on the current passing through the resistance at the junction p-n.

The electrical efficiency of PV cells drops as the operating temperature increases. This means that, as the material is heating up, the electricity output is reduced. Numerous analytical models have been proposed to express the dependence of electrical efficiency as a function of operating temperature. A comprehensive review was carried out by Dubey and Sarvaiya [8]. One of the first correlations between efficiency and temperature, still used today, was proposed by Florschuetz and Evans in 1977 [9] and is shown in Eq. 7.1.

$$\eta_{EL(T)} = \eta_{STC}\left[1 - \beta_{PV}\left(T_c - T_{ref}\right)\right] \tag{7.1}$$

where $\eta_{EL(T)}$ is the electrical efficiency of the cell at temperature T_c, η_{STC} is the reference electrical efficiency of the cell evaluated at the standard testing conditions (STC) and at reference temperature T_{ref}, and β_{PV} is the temperature coefficient.

The temperature coefficient of a PV cell, denoted as β_{PV}, possesses a distinct value corresponding to the specific semiconductor material employed. In the case of mono and polycrystalline silicon cells, both empirical data found in the existing literature [10–13] and the specifications provided by PV cell manufacturers affirm a temperature coefficient ranging from 0.3% to 0.9% above the standard operating temperature of 25 °C. Another widely utilized semiconductor material is copper-indium gallium selenide (CIGS), which demonstrates an efficiency decrease of 0.29–0.53%

Table 7.1 Temperature coefficients for various types of cells [10–15]

Semiconductor	Temperature coefficient (−%/°C)
Mono-c-Si	0.30–0.90
Multi-c-Si	0.30–0.90
a-Si	0.20
a-Si/μc-Si	0.26
CIGS	0.29–0.53
CdTe	0.17–0.25

per Celsius degree [10, 14]. Furthermore, an additional study [15] conducted on cadmium tellurium semiconductor (CdTe) reports an efficiency reduction of 0.17% per Celsius degree. The temperature coefficients for various semiconductors are summarized in Table 7.1.

To explain why the increase in operating temperature causes a decrease in power in a PV cell, the physical phenomena that take place within the cell can be broken down into the following components:

- Increased thermal vibrations in the semiconductor structure in the form of phonons, which leads to lateral dispersion of particles. This creates additional collisions between the load-bearing particles.
- The internal voltage at the p–n junction decreases, therefore the ability to separate the electrons from the gaps decreases, which leads to a smaller potential difference, and a weaker current generated.
- Decreased carrier mobility. Mobility (μ) is the ability of a load-bearing particle to move in an environment as an effect of an electric field and is defined as the ratio between magnitude of the electron drift velocity (v_d) and the magnitude of the electric field (ξ):

$$\mu = v_d/\xi. \tag{7.2}$$

Within the same material, the positive and negative particles have different values of mobility. At standard test temperature (25 °C), the negative particles (electrons) have mobility $\mu_n \approx 1360\ cm^2V^{-1}\ s^{-1}$, and the positive particles (gaps) $\mu_p \approx 450\ cm^2V^{-1}\ s^{-1}$.

This effect is illustrated in Fig. 7.3a, b, where it can be observed that the influence of temperature varies depending on the level of particle doping. In other words, for a highly doped cell with a concentration of 10^{19} particles per cm^3 the effect of temperature on particle mobility is not significant.

In addition to the negative effect on power in the electrical circuit, discussed above, the high operating temperatures in a PV panel also has physical effects of wear on the component materials. An extensive experimental study on the damage mechanisms of PV panels, performed on 1740 panels [17] indicates that 39% of the damage phenomena are associated with high operating temperatures. Mainly, two recurring phenomena occur:

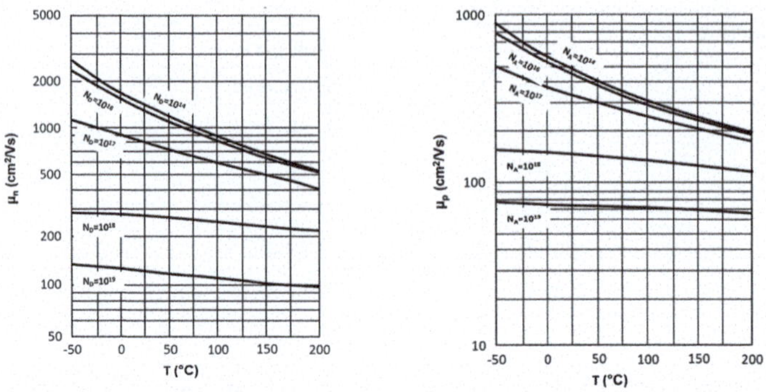

a) The effect of temperature on the mobility of electrons

b) The effect of temperature on the mobility of holes

Fig. 7.3 The effect of temperature on the mobility for various doping concentrations [16]

- Local overheating of the components induces thermal stress and can lead to burning of metal contacts (Fig. 7.4a) or delamination of the semiconductor material (Fig. 7.4b);
- The thermal heating–cooling cycles also induce thermo-mechanical stress in the components. Cyclic temperature tests are part of the testing standards of PV panels and are defined by IEC 61215: 2005. One study [17] shows that for 6 out of 7 panels, at 200, 400, 600, and 800 thermal cycles, the maximum power decreases between 3 and 9%.

Due to the negative effects of high operating temperatures on the PV cell and to the increased rate of solar energy conversion, PVT technology has been developed, and its main characteristics will be discussed in the next section.

Fig. 7.4 Effects of thermal stress in a PV cell **a** Burning of metal contacts **b** Delamination of the semiconductor material [17]

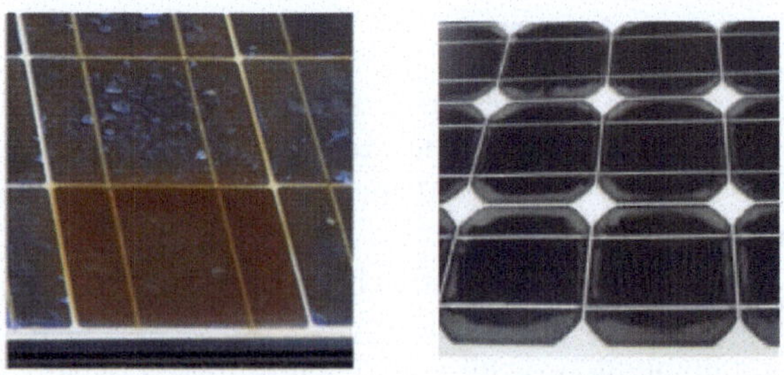

a) b)

7.3 PVT Technology

Currently, traditional solar energy technologies can be broadly classified into solar thermal (ST) collectors and photovoltaic (PV) collectors. Solar thermal collectors transform solar radiation into thermal energy, a concept that has been used for centuries for heating water. Photovoltaic systems use the photovoltaic effect, which means generation of an electric potential in a semiconductor exposed to light radiation. Both of these technologies have already found popularity for a variety of industrial and domestic applications. Solar thermal collectors can be utilized for space heating and domestic hot water directly for the end user or integrated in district heating systems, solar assisted air conditioning (especially due to the incidence of high ambient temperatures with the need for cooling down inside homes), swimming pools heating or water desalination. PV systems can be utilized for producing electrical energy off grid, connected to the local electric grid, or for storage (in fuel cells, batteries).

In order to take full advantage of both technologies, hybrid photovoltaic-thermal (PVT) collectors have been developed. They combine and maximize the benefits of the two traditional types of energy extraction technologies into one single equipment, producing simultaneously electrical and thermal energy (Fig. 7.5). PVT panels can be considered to be a type of micro-cogeneration technology.

7.3.1 Main Characteristics of PVT Panels

Combining solar thermal and photovoltaic technologies into a hybrid PVT panel has a number of advantages.

First of all, significant heat losses occur during the photovoltaic conversion. Most commercial PV cells use silicon as the semi-conductor material, which is very convenient due to it being the second most abundant material in the Earth's mass. However, crystalline silicone has a band-gap of 1.1 eV. Any photon of light with less amount of energy will not be able to move the electrons, therefore no energy is produced. Furthermore, photons of light with higher energy compared to the bandgap will move the electrons but the excess energy is lost, typically in the form of thermal energy.

Fig. 7.5 PV + ST = PVT

Thus, harvesting the thermal energy from the collectors leads to an increase in the operation capacity of the solar collector and an improvement in the rate of solar energy conversion.

The decrease in efficiency of PV cells as the operating temperature rises has been addressed in a previous section. As the temperature of the cell increases, there is a decline in the electrical power generated within the circuit. Numerous analytical models have been put forward to assess how the power efficiency is affected by the operating temperature of different types of cells. A comprehensive review of these models has been conducted by Dubey et al. [8]. One simple example of correlation for the relationship between temperature and efficiency was proposed by Florschuetz [9] is shown in the Eq. (7.1). Studies show that PV cells from PVT collectors have an electrical efficiency of about $4\% - 12\%$ higher than an equivalent cell in a standalone PV collector [18].

The thermal energy extracted from the collector can be used for various applications (residential or industrial) and as a result a higher global energy efficiency can be obtained for the entire energy system. The thermal efficiency of solar collectors η_t is defined by the equation below [19]:

$$\eta_t = \frac{Q_u}{G_{STC} A_c} \tag{7.3}$$

where Q_u is the useful thermal energy obtained, G_{STC} is the solar radiation on the collector at STC, and A_C is the area of the collector. η_{el} is the PV electrical efficiency at STC, shown in Eq. 7.4. The standard test conditions STC are described as irradiance G = 1000 W/m^2, cell temperature t = 25 °C, and air mass AM = 1.5,

$$\eta_{el} = \frac{P_{STC}}{G_{STC} A_C} \tag{7.4}$$

where P_{STC} is the PV nominal power at STC, and it can be described as the voltage V_{STC} times the current I_{STC} at the point of maximum power during the analysis. For the hybrid collector the global efficiency is η_{gl} and can be described as the sum between the thermal and electrical efficiency:

$$\eta_{gl} = \eta_t + \eta_{el} \tag{7.5}$$

As a result, in the energy flow diagram of the hybrid system, shown in Fig. 7.6 there is an additional thermal energy component, which did not appear in the diagram in Fig. 7.2 of a simple PV system. The thermal energy that is otherwise lost in thermalization and non-absorption can now be recovered.

This solution is suitable for urban consumers where roof and facade space are limited, and where electricity and heat are required simultaneously. The installation time as well as the costs related to installation and maintenance are considerably decreased by installing one piece of equipment compared to two separate ones.

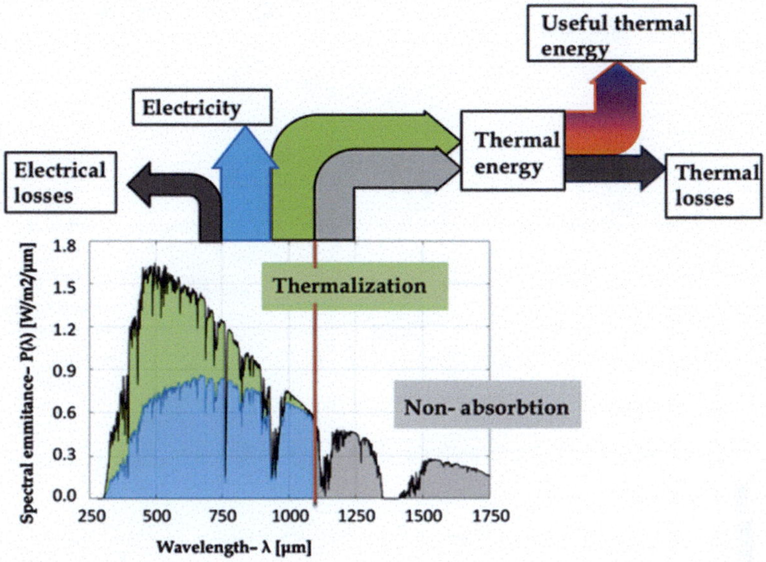

Fig. 7.6 Sankey diagram for a PVT collector

The two components of a hybrid system (thermal collector and photovoltaic cell) generally have the same operating principles as the stand-alone equipment. The PV module has the role of converting solar energy into electricity through PV cells that are inter-connected and encapsulated in glass or polymers for protection. The role of the thermal collectors is to capture solar energy in the form of thermal energy and to transfer this heat to a thermal agent flowing through a system of pipes or channels. From a technical point of view, the combination can have multiple configurations. However, an essential element present in any installation is the PVT layer (Fig. 7.7), which makes possible the heat transfer between the thermal agent and the PV cell.

Solar thermal component

The solar thermal component consists of a thermal absorber (a material with the capacity to collect solar energy), a system of pipes for the circulation of the thermal agent and an insulator material for thermal insulation. The thermal absorber usually consists of a slender material, which can be aluminum, copper, steel, or a thermally stable polymer like silicone. These materials are typically coated with a matte black finish to enhance the absorption of solar radiation. The network of pipes can be circulated with air, water, a combination of air and water, or alternative thermal agents like refrigerants or saline (brine) [21].

Liquid-based collectors are the preferred choice for PVT panels. Air-based collectors are primarily used for space heating, which is not a common requirement in hot climates. However, there are exceptions such as hotels, hospitals, or industrial applications [21]. The thermal efficiency of solar thermal (ST) systems can vary between 45 and 80% depending on the operating conditions and configuration,

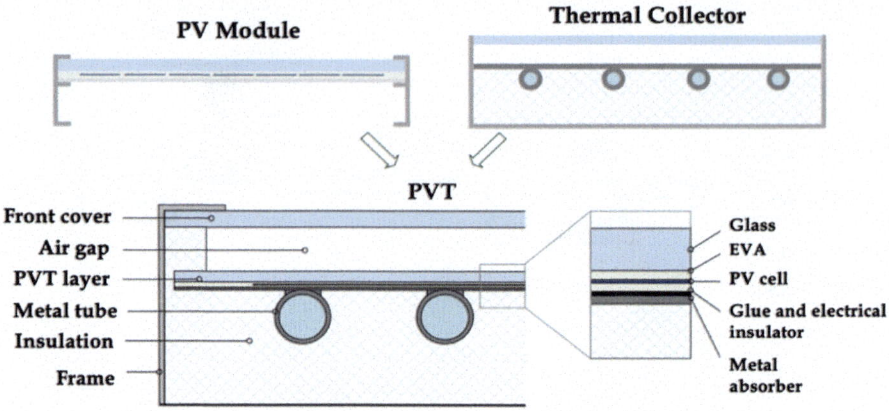

Fig. 7.7 Schematic diagram of the components of a PVT collector [20]

with the highest recorded efficiency reaching approximately 97% in large collectors equipped with concentrating mirrors [22]. In comparison, a typical commercial flat plate solar collector typically achieves a thermal efficiency in the range of 60–70% [23].

Solar thermal collectors can be categorized based on various criteria.

1. According to the type of thermal agent:
   - Air-based
   - Liquid-based
   - Air–liquid mix
   - Phase-changing materials (PCM).

Air-based systems can be natural or by forced convection; the latter has a better heat transfer coefficient than natural convection but has the disadvantage of requiring external energy for ventilation. In the literature, it has been found that the thermal efficiency of these systems varies between 38 and 75% [24]. The main application of air circulation systems is ventilation, room heating, or hot water preheating systems.

Liquid-based systems are generally more efficient than air-based systems, due to a considerably higher heat transfer coefficient and higher heat capacity. The liquid agent can be circulated by gravity or by pumping, and the power required for this is less than that for a fan in an air-based system [24]. The thermal agent may be water, salt water, or a mixture of glycol and water. There are also bi-fluid configurations, consisting of a water–air mixture, but they are not yet developed due to the high production costs. The thermal efficiency of liquid-based systems varies in the literature between 60 and 80% [25].

A novelty in thermal collectors is the use of phase change materials as a thermal agent, especially those with solid–liquid transformation, such as calcium chloride hexahydrate, paraffin wax, sodium sulfate decahydrate (Glauber's salt), or various nanofluids [26]. Although these materials are relatively affordable, technical difficulties in processing, encapsulating and maintaining the systems to ensure proper operation lead to high costs [27].

2. According to the configuration of the cooling system:

The solar thermal panel can be configured as a flat plate (Fig. 7.8a) or with evacuated tubes (Fig. 7.8b) depending on the purpose of the application and local conditions. The use of concentrating mirrors is also possible to amplify the solar ration in parabolic systems (Fig. 7.8c).

3. According to the quality of thermal energy:

- Low temperature—under 50 °C
- Medium temperature—between 50 and 80 °C
- High temperature—above 80 °C.

For residential consumers, solar collectors reach medium temperatures, which can cover the need for domestic hot water for consumption, heating, or even cold generation for high-capacity systems (usually collectors with concentrating mirrors). Depending on the thermal level of the energy produced, industrial applications can be technological processes in the agricultural field, product drying, water distillation,

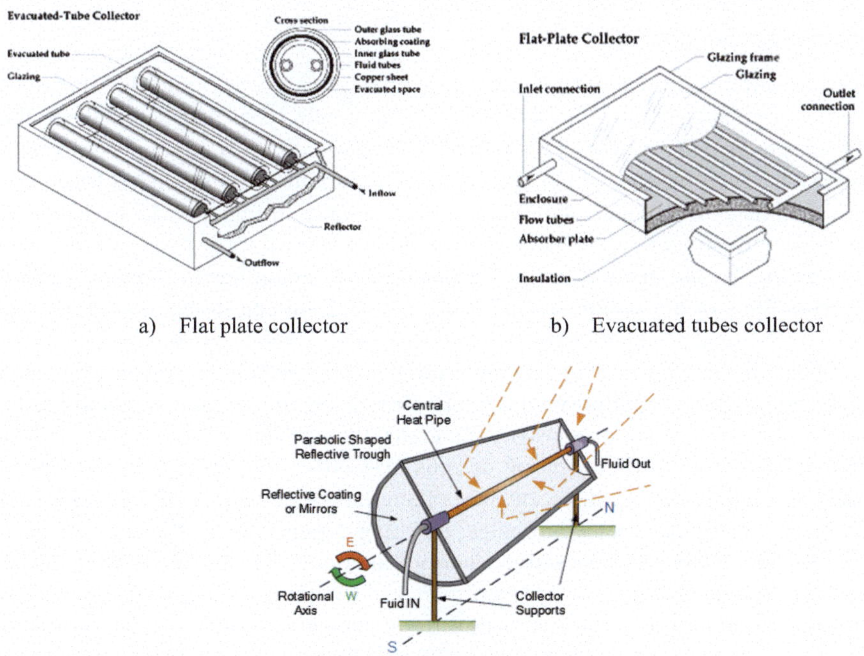

a) Flat plate collector

b) Evacuated tubes collector

c) Parabolic collector

Fig. 7.8 Types of solar thermal collectors

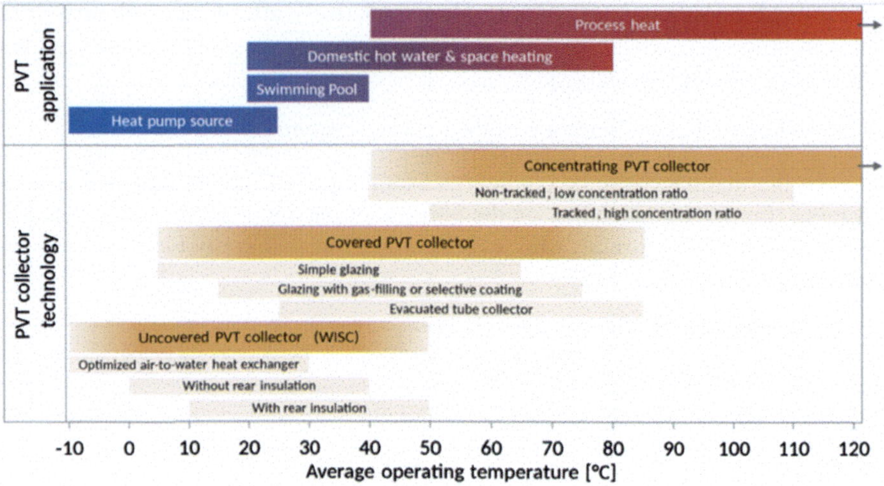

Fig. 7.9 Temperature dependent applications [28]

processing in the food industry, heating of industrial spaces. Figure 7.9 shows the possible applications of the heat produced by the solar thermal component of a PVT system depending on the temperature level.

Photovoltaic component

The photovoltaic component comes in a wide variety of typologies and classifications. In terms of geometric configuration, the most common types of PV panels are flat, but there are also concentrated PV panels (CPV), which use a system of mirrors to concentrate sunlight, and which are the most commonly recommended for arid or tropical climates where the solar intensity is constantly strong, in order to justify the high initial investment.

Traditionally, photovoltaic cells are made of silicon, either in monocrystalline, polycrystalline, or amorphous form. These are predominant in the market, making up 95% of the total production of PV panels in 2020. The mono-crystalline cells cover about 66% of the total market (compared to 45% in 2018) [29]. A new generation of cells are made of thin film with a thickness of a few micrometers, which has the advantage of being flexible and easily integrated into various building materials (e.g. façade, glass). The third generation of cells uses either ink, special types of paint, plastic, or organic matter. In recent years there has been a successful development of Silicon heterojunction technologies, reaching a conversion efficiency of up to 26% [30]. The National Renewable Energy Laboratory has an extensive database, constantly updated to monitor the performance of the latest technologies. Figure 7.10 illustrates a graph of the efficiencies of different cell types over time; it can be seen that a record experimental efficiency of 48% was obtained by a tandem concentrator with 4 junctions [31].

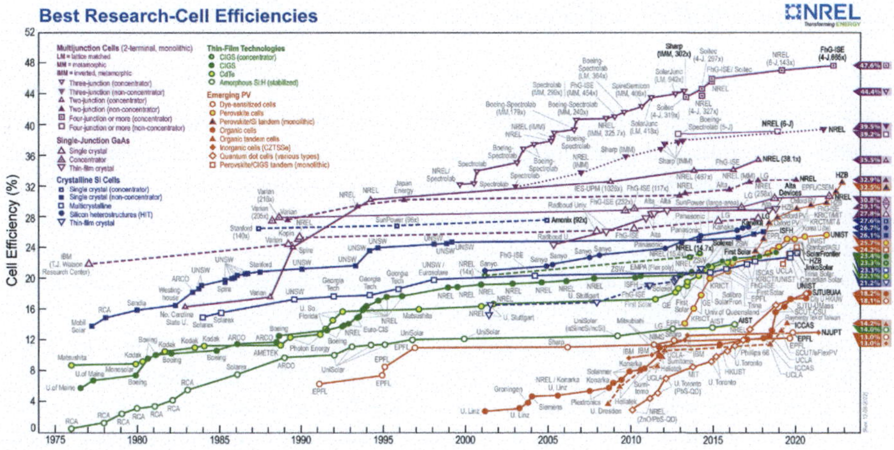

Fig. 7.10 Best research PV cells efficiency [31]

Table 7.2 shows the most common and researched cell types, along with their theoretical performance in the laboratory and on the market, as well as the main fields of applicability.

Table 7.2 Efficiency and applications of PV cells

Cell	Theoretical efficiency (%)	Lab efficiency (%)	Real efficiency (%)	Applications
m-SI	27	24.7	14–16	Large sized roof and facade panels, low power appliances, aerospatial applications
p-SI	27	19.8	12–14	Large sized roof and facade panels, appliances
a-Si	25	13	6–8	Small sized roof and facade panels, appliances
GaAs	29	27.5	18–20	Concentrated PV systems, aerospatial applications
CIGS	27.5	18.2	10–12	Small sized roof and façade panels, low power appliances
CdTe	28.5	16	8	Building integrated modules

7.3.2 Types of PVT Collectors and Their Applications

The type of a PVT panel is dependent on several factors, with the most crucial ones being the choice of thermal agent, the characteristics of the cooling system (including materials and pipe diameter), the type of photovoltaic cells, and whether there is an anti-reflective layer (glazed or unglazed). Various literature reviews have explored the potential types and arrangements of PVT panels [25, 26, 32]. Typically, the ultimate configuration is determined by the selection of the solar collector and the type of PV cell, leading to a multitude of potential combinations. A non-exhaustive classification of the types of PVT systems is illustrated in Fig. 7.11.

Similar to standalone PV panels, c-Si cells are the predominant choice for PVT systems [20]. Mono-crystalline cells have a greater solar absorption coefficient compared to polycrystalline cells. In contrast, thin-film technologies exhibit a lower temperature coefficient, making them well-suited for high-temperature applications [20]. In situations where solar concentration systems function at extremely high temperatures, the cooling of cells becomes necessary, making PVT applications suitable for such conditions.

Depending on the chosen application and the consumers' energy priorities, three PVT configuration can be identified. In the first instance, thermal energy can be considered as the main useful product for various applications (space heating, water, ventilation, etc.). Traditionally, solar thermal systems rely on external electricity to power the pumping and circulation of fluids through pipes. With the integration of PV cells, the requirement for an external energy source is eliminated, rendering the system self-sufficient. In other applications, where electricity is the main useful product, the heating system can be installed for the sole purpose of cooling the mode

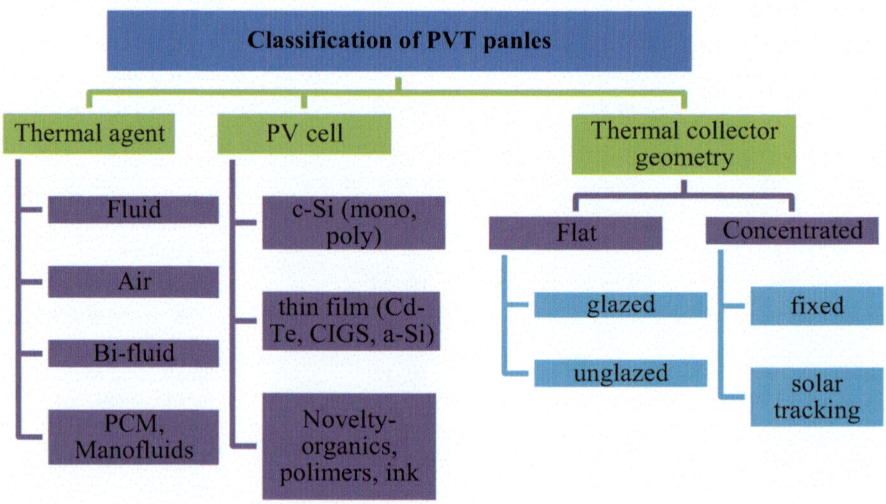

Fig. 7.11 Classification of PVT collectors

and thus improving the efficiency of the cell. To maximize the benefits of the two technologies, it is possible to extract both thermal and electrical energy, and use both by consumers.

In terms of applications, PVT panels are ideal in systems where electricity and heat can be used simultaneously. Unlike individual PV modules, which can be installed in large PV parks, PVT systems are more suitable close to the consumer, to avoid considerable losses with thermal energy transportation. Thus, integration in the urban or industrial environment, close to the end-user is the most common.

7.4 Research and Integration of PVT Collectors

The initial exploration of PVT collectors dates back to the 1970s [33]. Substantial advancements have occurred since then, with considerable efforts directed towards enhancing their efficiency through the analysis and optimization of their structural and operational aspects. This progress has been particularly notable in the past decade, coinciding with the surge in renewable energy initiatives.

7.4.1 Current Developments and Research on PVT Collectors

Numerous investigations have been conducted to assess PVT collectors in comparison to individual PV and solar collectors. Euh et al. [34] reported that, in an experimental PVT system with no transparent coating, the thermal efficiency was lower by 13%, and the electrical efficiency was 0.8% higher when compared to separate installations. The overall conversion efficiency of the system could increase by as much as 40% [35]. Experimental studies on various cooling configurations for PV cell cooling have demonstrated an enhancement in electrical efficiency from 13.9 to 15.9%. [36].

A comprehensive investigation [37] delves into the environmental considerations surrounding PVT systems, particularly focusing on the time it takes to recover the energy invested in their production and the corresponding CO2 emissions associated with the manufacturing process. The findings reveal that the energy recovery periods for their production vary widely, spanning from 1 to 14 years, contingent on the specific configuration (BIPVT, BAPVT, CPVT). Life cycle analysis (LCA) is also an important indicator and is performed for various types of modules, illustrating a wide variety of levels of environmental impact depending on materials and production methods.

The weather condition (solar radiation, wind speed, ambient temperature) and the characteristics of the PVT collectors (channel width and type, insulation, packing factor, glazing) are some of the main factors that impact the performance.

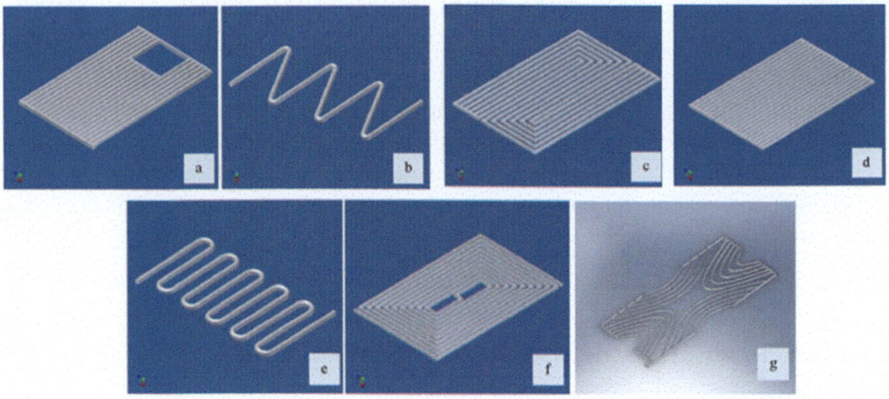

Fig. 7.12 Types of pipe configurations **a** direct flow **b** serpentine design **c** parallel serpentine design **d** modified parallel serpentine design **e** oscillatory flow **f** spiral flow **g** web design [38]

Major research has been dedicated to the evaluation of the impact of various geometrical configurations on the thermal, electrical output, and global energy performance [38, 39]. Various configurations of the piping system are possible (Fig. 7.12) and the most commonly used are 'direct flow' (or 'harp design') and 'serpentine flow'. A mathematical study [38] on a p-Si cell PVT panel indicates that the most thermally efficient design is the spiral configuration, with a thermal efficiency of 68%, and the lowest efficiency was observed to be 45% for the serpentine type design.

The pressure loss that occurs with the thermal agent has a significant influence on the overall system efficiency. A higher pressure drop means a higher pumping energy requirement for a given flow. Experimental tests were performed on aluminum collectors, comparing different classical harp and coil geometries with a new model, called Bionic configuration (Fig. 7.13, right), and for the latter a pressure loss of 400% was obtained than for serpentine, and 50% smaller than the harp [40]. The bionic model was created based on the concepts of fractal geometry, inspired by models in nature (the system of blood vessels or leaf veins), in order to obtain a uniform flow and minimize pressure losses.

Fractal geometry has been studied and successfully applied in various heat exchangers [41, 42]. Fractal-type channels (Fig. 7.14) can outperform the hydraulic performance of traditional coil or harp models due to uniform flow and lack of steep bifurcations. A review study [42] gathered multiple promising results, with up to 80% improvements in fractal heat exchangers compared to classical geometries. Most are based on theoretical and numerical studies, as the main disadvantage of fractal models is the difficulty of manufacturing the complex geometry. However, there is a growing interest in this area, especially in the future perspective of the commercial use of 3D printing methods.

The collector material is an important component, as it transfers solar thermal energy to the heating medium. The thermal conductivity of the material must be as high as possible to permit the transfer of thermal energy, and the specific heat

Fig. 7.13 Experimentally tested configurations for assessing the pressure loss [40]

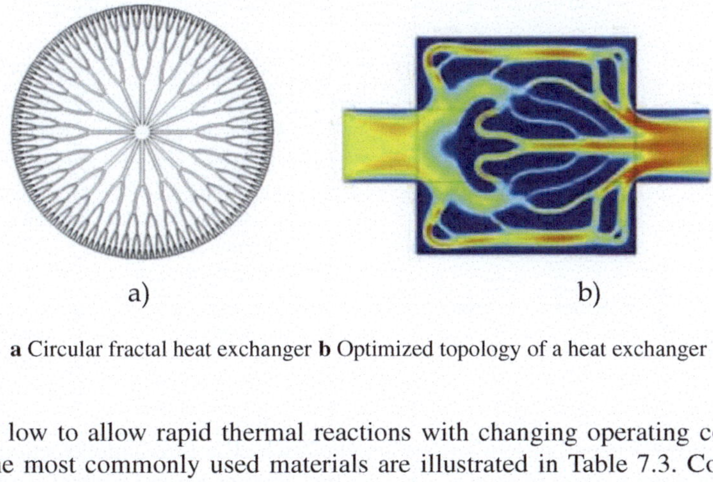

a) b)

Fig. 7.14 **a** Circular fractal heat exchanger **b** Optimized topology of a heat exchanger [43, 44]

must be low to allow rapid thermal reactions with changing operating conditions [25]. The most commonly used materials are illustrated in Table 7.3. Copper has the most suitable properties for this application, and the older generations of solar collectors are made mainly of copper. However, due to the high price in recent years (an increase of about 400% between 2004 and 2012), aluminum is also used, having a cost 4 times lower [45].

The section of the collector channels has also been studied and is closely related to the manufacturing process and the choice of material. Three types of channels were studied and compared in a study by Aste et al. [25] illustrated in Fig. 7.15, highlighting the advantages and disadvantages of each (Table 7.4).

The entire panel can be in direct contact with the environment, without a protective layer (unglazed), or protected by one or more transparent layers (glazed). In the case of an uncovered collector, the incident radiation is higher, but simultaneously the thermal losses through convection and radiation are higher. That is why a transparent

Table 7.3 Thermal collector material properties

Material	Thickness (mm)	Density (kg/m^3)	Thermal conductivity (W/mK)	Specific heat (J/kgK)
Copper	0.3	8920	380	350
Aluminum	1	2700	160	900
Steel	2	7860	50	450
Polymer	2–3	900–1500	0.2–0.8	1200–1800

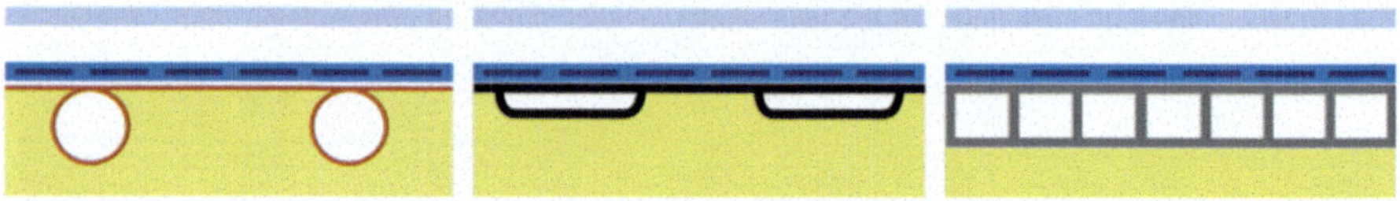

Fig. 7.15 Tubular, trapezoidal, and rectangular channel collector [25]

Table 7.4 Advantages and disadvantages of tubular, trapezoidal and rectangular channel collectors [25]

Configuration	Material	Geometry	Advantages	Disadvantages
Sheet and tube	Copper, aluminum	Serpentine, harp	Simple production, thin materials	Small contact surface Variable costs
Roll bond)	Aluminum	Any	Flexibility regarding geometry, large surface area	Minimum thickness 1 mm
Rectangular channel	Polymer	Harp	Rectangular channel	Special component required

protective layer is usually used, most often glass, which acts as a unidirectional filter, having high transparency in the visible spectrum, and opacity for longer wavelengths, in the infrared area. Alternatives to glass are polymers such as polycarbonate or polyvinyl fluoride, but which show considerable thermal and mechanical degradation over the life of the module [25].

An active study topic is the introduction of a low emissivity layer, which aims to reduce radiation losses, such as zinc, indium oxide, and tin oxide. A detailed study on the effect of various types of low-e coatings [20] indicates, in most cases, a thermal gain at the expense of an electrical loss. Another study [46] on a low emissivity layer of SiO_2 indicates a decrease in emissivity from 80 to 20%. This, in combination with an antireflective layer, results in an increase in both electrical and thermal production, although only by 1% in the latter.

In general, when introducing protective or low-emissivity layers, there is a trade-off between electrical and thermal efficiency. Figure 7.16 illustrates a comparison of

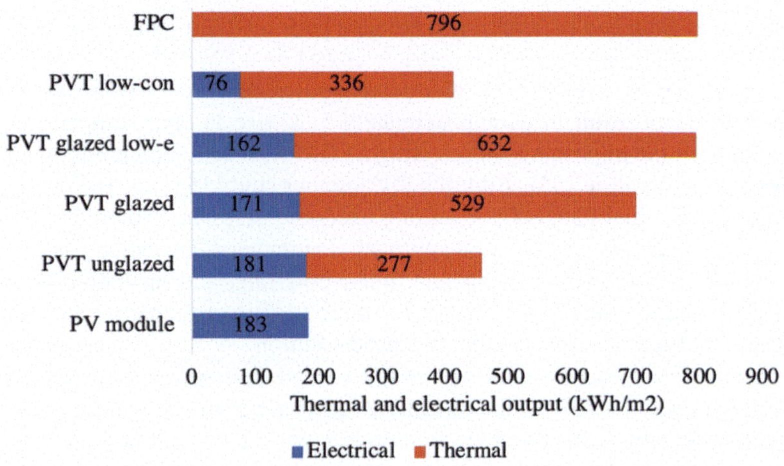

Fig. 7.16 Thermal and electrical output for various collectors at a reference temperature of 25 °C [20]

gross electricity and heat production for climatic conditions in Würzburg, Germany [20], for five types of systems: simple PV systems, thermal collectors, and three types of PVT solutions, without external protection, with glass state and with a layer of glass and low-emissivity coating. It can be observed that, with the introduction of each layer, there is gain on the thermal side, but decreases the electricity production.

The manner in which the photovoltaic cell layer is connected to the thermal collector is a significant factor to consider. One method of connection involves using a thermally conductive adhesive that can withstand high temperatures. However, when using this adhesive, the potential drawback is the formation of air bubbles, leading to an uneven transfer surface and a lower heat transfer coefficient [47]. An alternative approach is the lamination of all elements into a single component, incorporating glass, the PV layer, electrical insulation, and the thermal collector. This lamination method can result in a thermal enhancement of 10% when compared to the adhesive connection [48]. The disadvantages of lamination are the occurrence of possible deformations due to the different coefficients of expansion of the component elements, and the technological difficulty to achieve the lamination for sheet and tube type systems.

Other aspects that can be considered for optimizing the design of a hybrid solar collector are the flow of thermal agent, the type of electrical and thermal insulation, the diameter and distance between the collector tubes, the series or parallel connection of the cells, and the level of cell doping. All these represent the optimization parameters with impact on the design and final performance of a PVT system.

Numerous parametric studies have explored the topic of PVT collectors by optimizing key construction parameters. These include factors such as the packing factor of the PV cells [49], flow rate [50], spacing of tubes [51], tilt angle [49], glazing [52], and insulation thickness [51]. Most of these studies suggest that the optimal performance is achieved by reaching a balance between thermal and electrical efficiency. One study conducted a parametric analysis on a PV collector incorporating a phase-changing material (PCM) and demonstrated that a 100 W/m^2 increase in solar radiation can result in an approximately 5 °C rise in peak temperature. The results highlight the importance of selecting a PCM with a melting point higher than the ambient temperature [53]. Additionally, another study [54] carried out an optimization of individual panels by assessing exergetic assessing exergetic efficiency using a multi-objective optimization approach. Parameters considered for optimization included inlet temperature, air gap, mass flow rate, and insulation thickness. This study identified the optimal design for three PVT operational modes: Thermal exergy-driven, Electric-driven, and a balanced solution.

Other significant factors that impact the overall performance are the thermal parameters of the full PVT system, especially the temperature of the liquid at the inlet of the collector, which has the role of cooling down the PV cell and collecting the excess heat, and which depends on the temperature of the thermal agent at the outlet of the thermal storage tank. The temperature of the thermal storage tank is influenced several factors: type of tank and heat exchanger, volume of tank, water draw (which is determined by the rate of domestic hot water consumption), and the cold water main temperature that replaces the water drawn from the tank [52]. An interesting parameter is the rate of water consumption at the end user not only in terms of total daily quantity but also the distribution over the duration of the day (also referred to as the demand curve). The daily demand curve can vary significantly varying according to the type of consumer (offices, schools, household, industrial).

An interesting paper [55] examines the effects of the behavior of the consumer on the efficiency of a system of solar thermal panels located in South Africa. The results revealed that even small behavioral changes can cause a significant efficiency improvement. Another parametric study of a full system [56] investigates in terms of cost savings a solar-based heating and cooling system designed for an independent, single-family residence located in a hot climate. In another study [57], a parametric analysis was conducted on a solar-assisted home heating system that is integrated with a seasonal underground thermal energy storage tank. This analysis focused on assessing the annual energy production while varying key system parameters such as the insulation thickness and size of the tank. Furthermore, a parametric study was carried out for a solar cooling system in Athens, Greece, aiming to optimize it, and it achieved a solar coverage ratio of up to 89.85% [58]. Additionally, a parametric analysis of a PV system was undertaken from a technical and economic perspective [59], and it examined how factors like weather conditions, cost of the system, electricity prices, location, electrical specifications, and feed-in tariffs influence the economic viability of grid-connected system of PV and batteries.

As explained previously, an increase in the temperature of the PV cell T_c leads to a drop in the electrical efficiency. On the other hand, a decrease in T_c results in an improved electrical efficiency. Therefore, the primary determinants affecting this temperature are a relevant subject for research. The correlation between thermal and electrical efficiency holds a critical role in any co-generation system. For instance, in a combined heat and power (CHP) facility, the electricity generation increases when there is a substantial thermal energy dissipation source available [60]. Likewise, in the case of a PVT collector, if the captured thermal energy is not effectively dissipated from the storage tank, the PV cell remains inadequately cooled, resulting in a notable decline in electrical efficiency, potentially up to 15% [61]. While there is extensive research on the interplay between electrical and thermal efficiency in traditional CHP systems [62], the investigation in the context of PVT collectors remains relatively limited.

7.4.2 Integration of PVT Collectors

Both PV and thermal panels have a large global installed capacity. However, despite the popularity of these two technologies used separately, there are a limited number of installations of PVT reported worldwide. Due to their novelty, little centralized data is available except for some locations. For example, in the UK there are only a few hundred installed collectors with very limited performance monitoring data, most of them installed by the company 'Newform Energy' [63]. A housing complex in Switzerland has 1000 m^2 of installed PVT panels, with a capacity of 190 MWh per year. In Europe there are around 56 manufacturers that produce and install PVT collectors [28].

Due to the global trend of reducing dependence on fossil fuels, many construction projects in recent years integrate solar energy systems directly from the design stage. Thus, instead of building applied PVT (BAPVT) as exemplified in Fig. 7.17a, there are more and more examples of building integrated panels (BIPVT), which are used directly as building material in facades or roofs. As opposed to having the equipment installed on a standard roof, BIPVT utilizes the technology straight from the design stages of a building, integrating it as a building technology through roofing or façade materials (Fig. 7.17b). A review by Debbarma et al. [64] identified a number of case studies that successfully use BIPVT, and an experimental study by Kim et al. [65] indicates promising results.

A significant emerging market segment within renewable energy is comprised of self-sufficient consumers, which can be either businesses, educational institutions, or households. Self-consumption represents the practice of generating and using energy at the same location, making it the most efficient utilization of renewable resources. As a result, transportation costs, and electrical/thermal losses associated with it are reduces, and the intermittence of injecting renewable sources in the electrical grid is avoided. A newly emerging name for these self-generating entities is 'prosumers,' which merges the words 'producer' and 'consumer.' It's worth noting that the notion of energy prosumption is not a recent idea and can find historical examples in practices like biomass or wood combustion. Nevertheless, in the contemporary era, the prevailing business model between energy producers and consumers has primarily functioned as a unidirectional process, with energy companies supplying power to paying customers. Consequently, the 'prosumer' concept is undergoing a reevaluation, increasingly bolstered by the heightened social responsibility seen in the

a) b)

Fig. 7.17 **a** BAPVT **b** BIPV

twenty-first century. More and more consumers prefer to produce their own clean renewable energy, driven by several motivations:

- Autonomy from the main grid and protection against the fluctuation of energy prices;
- Reduction of losses and costs associated with energy transport;
- Financial aids from various governmental subsidies (i.e. Feed in Tariffs, green certificates).

The term 'energy prosumer' started gaining an increased scientific interest. In 2010, there were merely 12 published papers on Science Direct that included the phrase 'energy prosumer.' However, by 2020, this count had surged to 1251 papers on "Web of Science Core Collection" [66]. This indicated the fact that introducing PVT technologies on the market can be regarded as a great potential niche. Some examples of particular scenarios where integrating PVT collectors as part of the energy mix can be highly recommended and yields most benefits are enumerated below:

- New buildings (energy positive or energy neutral buildings)—For home owners, the outcome is a reduction of utility bills and enhancing the off-grid security, while for the residential investors and developers, renewable energy is an important selling point, despite the higher initial cost.
- Retrofitting—the substitution of outdated or malfunctioning solar thermal or PV collectors—makes use of existing space and offers a substantial boost in energy production by employing hybrid PVT collectors in lieu of individual PV and ST collectors.
- Social housing—Implementing PVT collectors can drastically reduce the energy bills of the tenants.

- Schools and business buildings—Renewable energy promote and raises awareness of social responsibility but also reduces the utility bills.
- Remote areas—Implementing renewable is highly advantageous in locations where electrical grid and district heating connection is problematic and expensive.

Furthermore, today's 'prosumer' isn't solely focused on on-site energy generation but also pays attention to their carbon footprint and energy consumption, with a goal to strike a balance between personal energy needs and environmental impact. Consequently, there is a growing demand for demand-side management, with an increasing number of individuals actively advocating for carbon footprint reduction. Monitoring peak energy production periods and aligning them with peak demand through improved planning strategies would further enhance the adoption of PVT collectors among consumers. Many novels, state of the art ideas and technologies are constantly emerging in the scientific community; however, regardless of their ingenious solutions and high efficiencies, more often than none the economic feasibility test is decisive when it comes to market scalability and penetration. This also applies to the PVT technology, which needs to bring economic benefits for consumers to adopt it.

PVT panels can be integrated into energy systems in different configurations, depending on consumer needs, climatic conditions, and the local energetic system. The simplest configuration is illustrated in Fig. 7.18 and consists of a heat exchanger connected directly to the domestic hot water consumption or heating system, or both.

Another frequently employed approach involves combining a heat pump with the solar collector, where the solar collector serves as an evaporator (as illustrated in Fig. 7.19). This system comprises a PVT collector, a condenser, a compressor, an expansion valve, and a storage tank. A coolant functions as the thermal agent, and it evaporates due to exposure to solar radiation [67]. An experimental study [68] evaluated a PVT-HP (Heat Pump) system (Fig. 7.20) with 2.25 m^2 manifolds, a 150 l

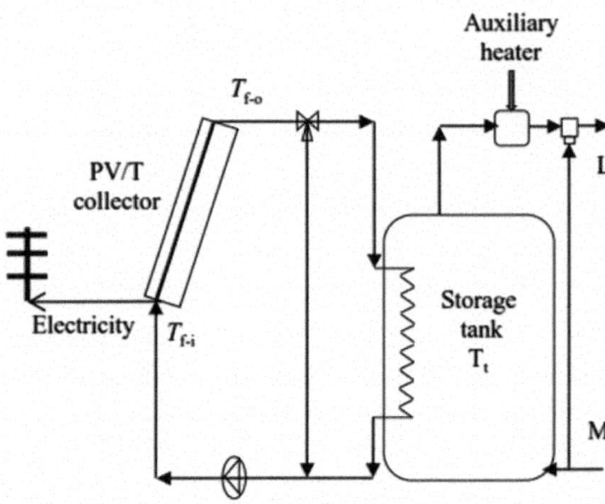

Fig. 7.18 Basic PVT system connected through a coiled heat exchanger to the DHW/ heating storage tank of a dwelling [24]

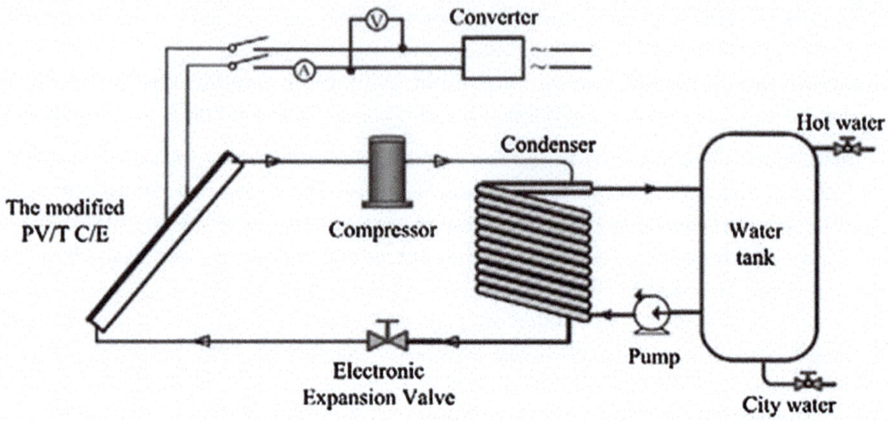

Fig. 7.19 PVT system coupled with a heat pump [68]

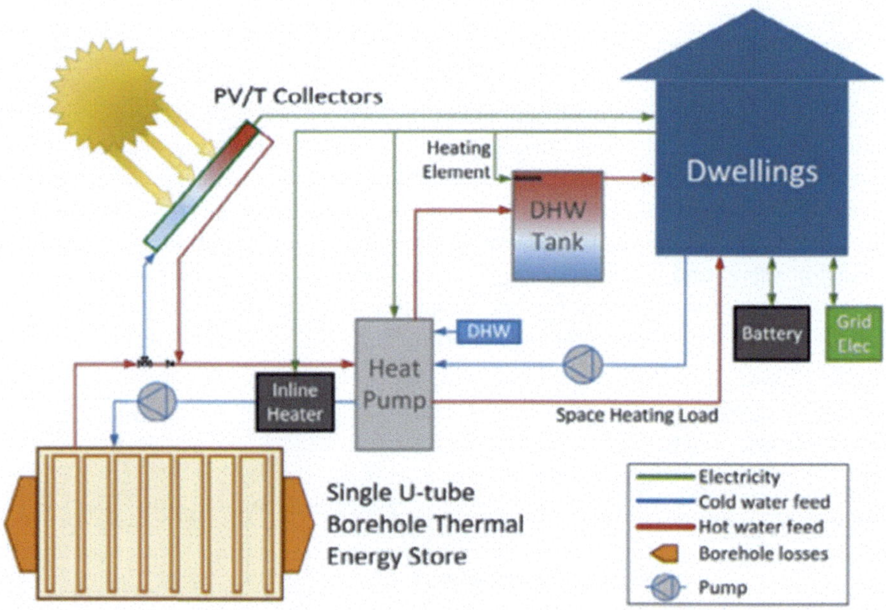

Fig. 7.20 PVT system coupled with a geothermal source [70]

storage tank, and R22 as coolant, and obtained a COP of 3.42 in January and 4.70 in July, weather conditions in Nanjing, China.

In more intricate systems, it is also possible to incorporate a geothermal source. This approach can attain elevated temperatures and harness a substantial portion of renewable energy to fulfill the building energy requirements. An example is illustrated in Fig. 7.20. These systems are complex and expensive but have the advantage of seasonal heat storage, ideal for locations with climatic variations throughout the year [69]. In this case, there is also the demand-control problem, i.e. the attempt to change the demand curve to maximize consumption during periods of production or low electricity prices. Salpakari and Lund [70] developed a non-linear energy cost optimization algorithm for a NYEB home in Helsinki considering a geothermal source, the change in consumption and an integrated battery storage system, achieving an improvement in the consumption rate by 30% and reduced annual energy costs by 25%.

Another innovative application of PVT collectors is the coupling with an organic Rankine cycle (ORC) engine (Fig. 7.21). A thermal energy storage tank (TES) acts as an interface between the ORC motor and the solar collectors and is also an evaporator in the ORC cycle. The study was conducted on an engine with a rated power of 1 kW_{el} combined with 15 m^2 of PVT collectors and a storage requirement of 100 L in Cyprus and 400 L in the UK. Various energy storage materials were examined, including phase-change hydrated salt, quartz, oil, granite, and water. In this exploration, the utilization of hydrated salt yielded solar-electric conversion efficiencies ranging from 4.4 to 6.4% in the UK climate and 6.3–7.3% in Cyprus [71].

Connecting renewable sources directly to end-users presents an appealing option but can pose challenges for the electricity and district heating grid. This integration needs technical solutions like centralized control, adaptable functioning of traditional exchanges, and real-time monitoring and responsiveness. These barriers are less significant at a very small level, in the home or small community, where heat and electricity can be stored. Examples of market penetration of distributed electricity and heat production can be seen in countries such as Germany, Denmark, and Sweden. An example of an interconnected system at the community level is shown in Fig. 7.22.

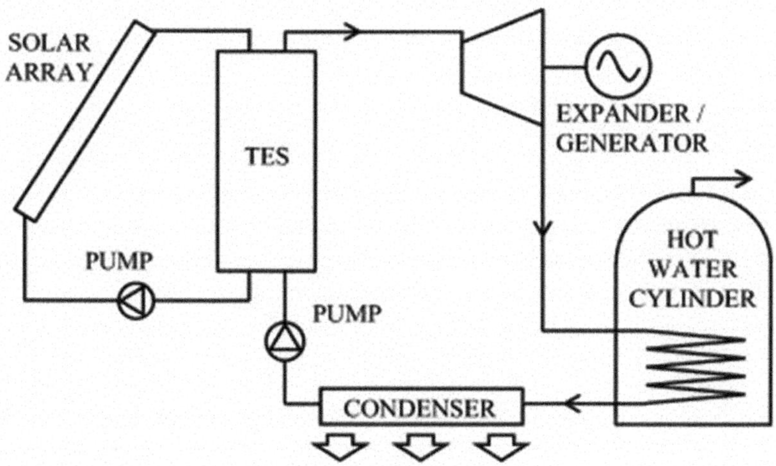

Fig. 7.21 PVT system coupled with an ORC engine [71]

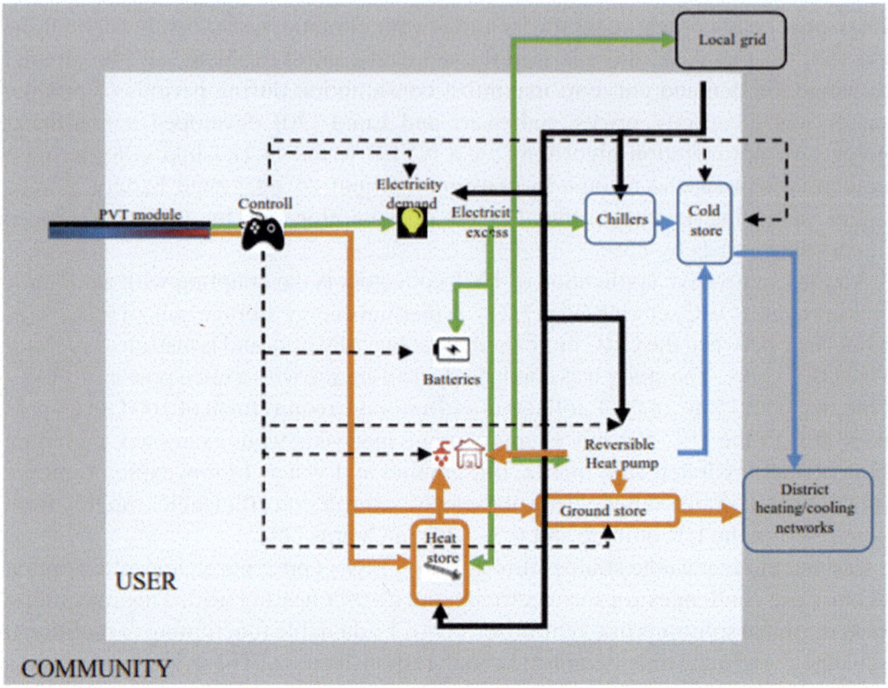

Fig. 7.22 PVT system integrated in a community grid [24]

Case study—A PVT system in University Politehnica Bucharest

An experimental system was installed on the preschool located in the campus of University Politehnica Bucharest for producing domestic hot water and electricity for the building. The system is shown in Fig. 7.23 and is formed of two solar thermal panels (yellow), two PV panels (red), and two PVT panels (green).

The system's performance was assessed over a span of four months, and the cumulative energy production is depicted in Fig. 7.24. In general, it's apparent that the electrical energy generated by the PVT system surpasses that of the PV collectors. This can be attributed to the removal of heat from the PV cells, which enhances their efficiency. This effect is particularly visible in July, the warmest month, where cell overheating is more significant. May witnesses the highest energy output from the PV collectors, while July is the peak for PVT collectors. This suggests that PVT collectors are better suited for high-temperature conditions compared to traditional PV collectors. In summary, this small experimental setup fulfills approximately 20–40% of the building's hot water and electricity demands during the examined months.

Fig. 7.23 Photography of the solar energy system on the preschool of Politehnica Bucharest [30]

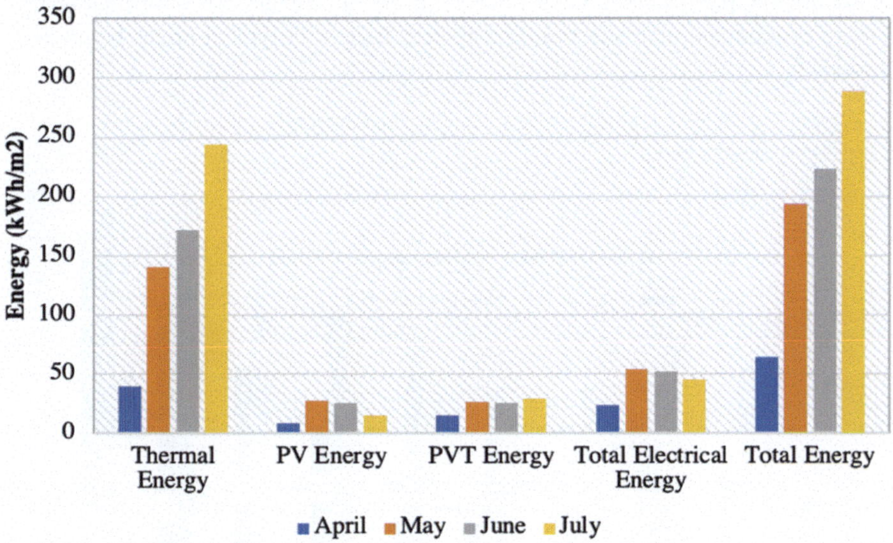

Fig. 7.24 Thermal and electrical energy collected over four months [30]

7.5 Barriers and Opportunities of PVT Technology

In the last ten years, solar energy has become one of the most popular forms of renewable energy, both globally and in Europe. It is playing an increasingly important role in terms of environmental protection and preservation, and in terms of the economic efficiency given by the latest solar technologies. In addition to ensuring self-production and self-consumption of energy among individual consumers, solar and photovoltaic panels also have significant industrial and agricultural potential. Figure 7.25 shows the development of global installed capacity of solar thermal heat, photovoltaics, and wind energy between 2010 and 2019.

In the face of global warming, the public has started to gain more and more confidence in solar energy. A study conducted by the European Social Survey (ESS) on the attitude of European citizens towards climate change and various forms of energy showed that solar energy is the most popular among renewable energies, with nine out of ten people considering it is the most appropriate technology to achieve the EU's decarbonization targets by 2050 [73].

The report on "Costs of electricity production", published by IRENA in 2019, shows a considerable decrease in costs related to the manufacture of photovoltaic panels, up to 96% in the period 2010–2019 [74] (Fig. 7.26). The report also predicts that by 2024 prices will continue to drop by 15% to 35%. Also, a report on the solar energy market, published by Solar Power Europe, shows that currently solar energy is more economical than any fossil or nuclear fuel, with typical prices below 0.04 Euro/kWh.

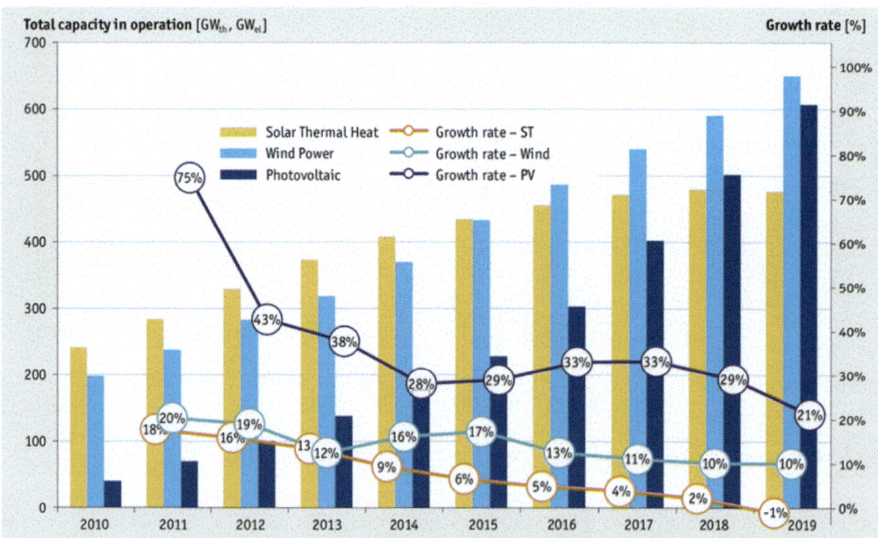

Fig. 7.25 Global capacity in operation and market growth rates between 2010 and 2019 [72]

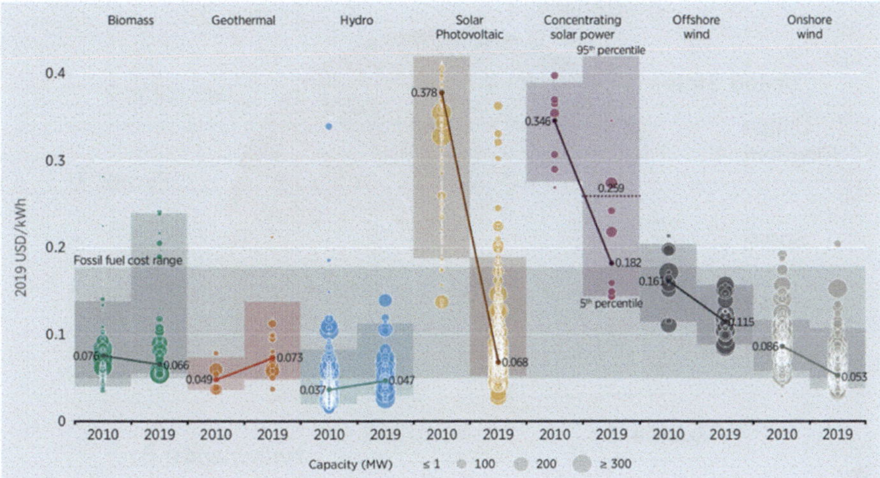

Fig. 7.26 Global levelized cost of electricity from newly built utility-scale renewable energy technologies for 2010–2019 [74]

Currently, the PVT panels market is limited and they are scarcely adopted worldwide. This is due to limited performance data, no standardization, and few producers with limited marketing possibilities, which leads to limited end-user awareness regarding the product.

7.5.1 Market Development

Presently, there is a growing global emphasis on the research and advancement of PVT collectors. The International Energy Agency (IEA) has introduced the research initiative Task 35 PV-Thermal Solar Systems [75], with the primary objective of expediting the development and widespread adoption of competitive, high-quality commercial collectors. Additionally, it seeks to contribute to the establishment of international standards for assessing the performance, monitoring, and testing of PVT collectors. Furthermore, since May 2018, another research endeavor known as Task 60-PVT Systems, which focused on exploring novel PVT technologies, with a particular emphasis on HVAC solutions [76]. Overall, this project has identified and surveyed a total of 24 manufacturers worldwide. Their distribution is shown in Fig. 7.27.

China, the current leader in solar thermal energy, dominates with 86% of the global energy production market and boasts an installed capacity of 64% of the total. China is also displaying a keen interest in PVT technology, particularly in systems linked with geothermal pumps [63], and it has the potential to become the primary solar technology integrated into new projects [78].

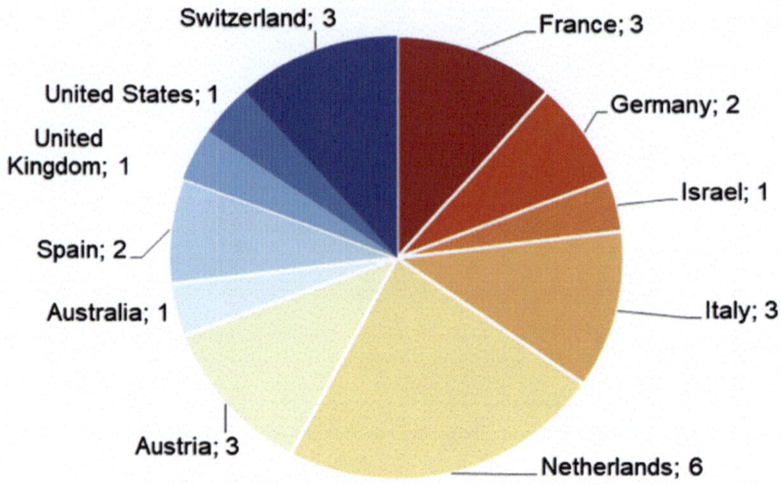

Fig. 7.27 PVT manufacturers by country [77]

As part of Task 35, a pricing analysis was conducted, revealing that unglazed modules with a thermal water agent had an average cost of 300 euro/m^2, which is notably higher in comparison to the 120 euro/m^2 for PV panels and 220 euro/m^2 for traditional thermal collectors. These mean price figures show that, even at the current commercial stage, installing a PVT system proves to be a more cost-effective option than utilizing two separate PV and solar thermal panels. This study is carried out on commercially available components of similar specifications. In theory, the exact same specifications of a solar and thermal component in both cases of the comparison would lead to a better PV efficiency due to the cooling of the cells, and a lower thermal efficiency due to less solar energy available for conversion.

In addition to module costs, another crucial economic factor is the impact of subsidy programs. The ‘Feed-in Tariffs’ (FiT) scheme, commonly employed in Europe, operates as a fixed-rate subsidy system for feeding surplus electricity into the grid. Figure 7.28 below illustrates the major impact that the tariff rate has on the net present value (NPV) after 20 years, and the investment payback period of a PVT project.

Because they are relatively new, PVT collectors have been incorporated into the Solar Heat Worldwide annual energy survey only since 2019. A survey carried out in 2023 identified a total installed PVT collector area of 1,524,945 m^2 (789 MW$_{th}$, 276 MW$_{el}$), with 567 MW$_{th}$ and 194 MW$_{el}$ of the total installed capacity. A significant share is in Europe (950,155 m^2) followed by Asia excluding China (316,653 m^2) and China (146,926 m^2).

France leads the EU market with an installed collector area of 608,172 m^2, followed by Germany with 146,729 m^2, and the Netherlands with 111,342 m^2. Switzerland, Italy, and Spain have collector areas ranging from 18,000m^2 to

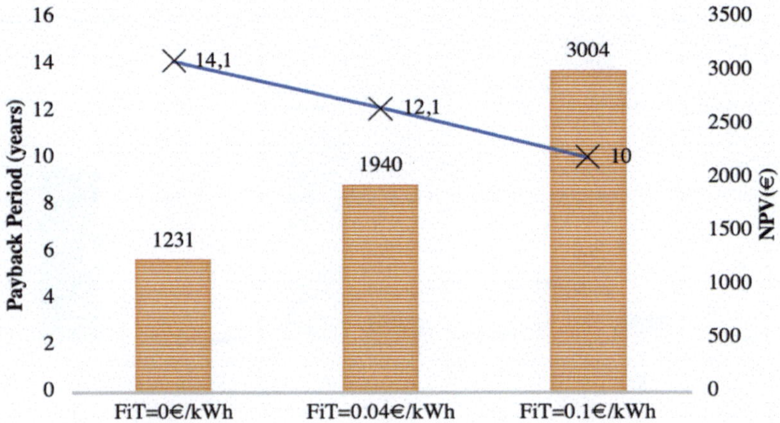

Fig. 7.28 Case study of the impact of feed in tariffs on the payback period and net present value [79]

26,000m^2, and less than 19,000 m^2 in the rest of the European countries [80]. The global market development and distribution by location is shown in Fig. 7.29.

A survey involving 43 PVT manufacturers reveals that the market for PVT collectors exhibited a consistent average growth of 9% between 2017 and 2020, reaching its peak in 2021 at +13%. However, in 2022, the market experienced a substantial decline of −52% compared to the previous year. The newly added capacity in 2022 was 42.4 MW_{th} and 21.7 MW_{el}.

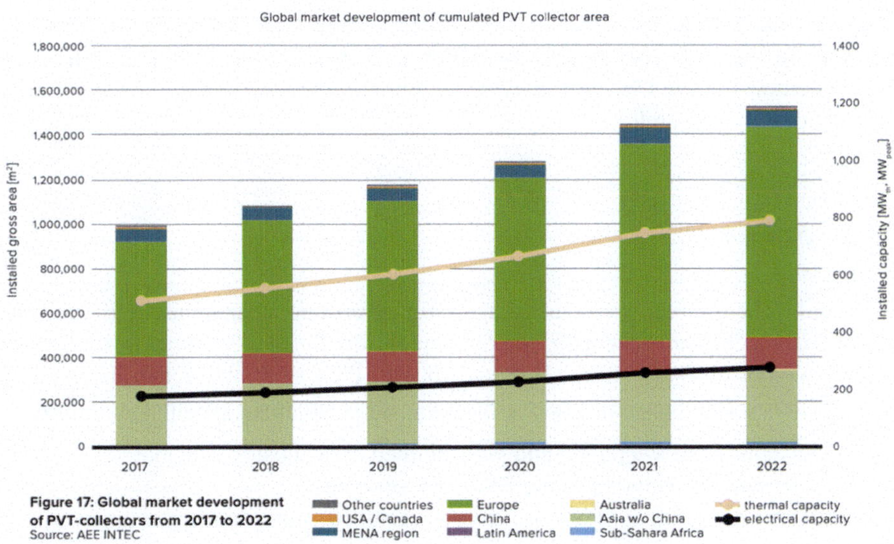

Fig. 7.29 Distribution of the total installed PVT area by location in 2022 [80]

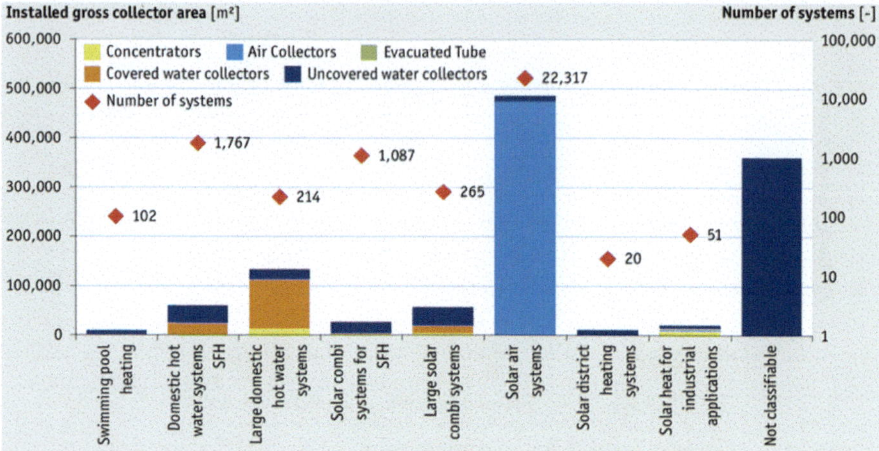

Fig. 7.30 PVT systems applications and installed areas in 2019 [72]

In terms of typology, the most common types are uncovered liquid based PVT collectors, with 55% of the total thermal installed capacity, followed by air-based PVT collectors with 43%, covered liquid based collectors with only 2%, and a very minor installed capacity of concentrated PVT and evacuated tube PVT [72].

The main applications identified for PVT collectors are domestic hot water production (small or large scale), swimming pool heating, solar air heating, solar district heating, and combined systems. Their distribution as a function of the type of collector is shown in Fig. 7.30.

The installed capacity of PVT panels is considerably small compared to the solar thermal and PV collectors separately, which is a direct result of their novelty and limited demonstration. However, they are gaining momentum in the EU and provide an area of active research.

7.5.2 Subsidies Schemes

Renewable energy has been encouraged by government subsidies and funding schemes in order to enhance their market penetration. Subsidy schemes are common for photovoltaic and solar thermal systems, but for PVT's the situation is unclear and yet to be defined in many European countries. There are three possible situations:

- PVT systems have a separate and well-defined own subsidy scheme;
- PVT systems are included in the PV or ST schemes;
- No scheme is available / PVT are clearly excluded from funding schemes.

A survey was carried out by SHC Task 60 in 7 European countries with the leading PVT markets in order to investigate the type of subsidies available [81]. The

Table 7.5 Available subsidy schemes in EU countries for PV, ST, and PVT (adapted from [81])

Country	PV schemes			ST schemes			Number of specific PVT schemes
	Total	PVT included	Rate (%)	Total	PVT included	Rate (%)	
Austria	1	0	0	7	2	29	0
France	3	3	100	3	0	0	0
Germany	1	1	100	1	0	0	1
Italy	2	2	100	4	2	50	0
Netherlands	3	3	100	2	0	0	0
Switzerland	2	2	100	1	0	0	0
UK	–	–	–	2	1	50	0

results are summarized in Table 7.5. In Austria, there is one PV subsidy scheme that explicitly excludes PVT collectors, and 7 solar thermal schemes, 2 of which allow for inclusion of PVT panels. In France, all 3 PV subsidy schemes allow for PVT, and none of the 3 solar thermal include the PVT collectors. In Germany, the one subsidy scheme available also includes PVT panels, while the one solar thermal does not. In Italy there are two PV schemes, both of which allow PVT panels, and 4 solar thermal schemes, out of which only half include PVT panels. In Netherlands all 3 PV schemes include PVT, while none of the 2 ST do. In Switzerland both PV schemes include PVT collectors, and the one solar thermal scheme excludes them. Finally, in UK there is no information on the PV schemes, and out of the two available ST schemes only one allows for PVT collectors to be included. In terms of standalone PVT schemes, the only country that has one in place is Germany, since 2019.

The main conclusions of this investigation are:

- Many of the schemes do not indicate specifically if PVT are accepted and require further clarifications;
- In general, the PV schemes have higher acceptance of PVT in their funding than solar thermal schemes;
- The type of PVT collector (air-based, liquid-based, covered, uncovered) is also a criterion of exclusion in some cases;
- There is only one standalone PVT subsidy scheme;
- Further clarification and standardization are necessary in all the investigated countries and also in the rest of Europe.

In terms of monetarily quantifying the subsidy contributions, the survey concluded that they are difficult to compare in different countries due to the currency and manner of being offered—feed in tariff, percentage of tax relief, or ceiling amount.

Therefore, the two aspects of major financial importance for the commercial development of PVT systems are:

- Decreasing the price of modules and energy storage systems (including transport, installation, and maintenance).
- Implementation of subsidy schemes such as Feed in Tariffs, with an optimal tariff for the purchase of electricity.

An economic analysis [79] revealed that investing in PVT collectors is not financially advantageous on a small scale due to the relatively high cost of PVT in comparison to standard PV and ST collectors. However, when dealing with a larger installed capacity and with an appropriate subsidy program in place, the co-generation solution becomes profitable, the investment is recovered in a short period, and the net present value (NPV) over the equipment's lifetime is substantial. Consequently, it can be deduced that, given the current market conditions, PVT collectors are profitable only at larger scales and specifically when supported by subsidy programs. Moreover, in cases where available space is limited, the investment in PVT collectors can still be justified against standard PV and ST for maximization of the solar conversion.

7.5.3 Barriers and Opportunities

Barriers:

- Cost: PVT panels tend to be more expensive than traditional photovoltaic or solar thermal systems due to the combined technology. The initial investment can be a significant barrier for potential users, especially when compared to the alternatives.
- Efficiency trade-offs: Combining photovoltaic and thermal functionalities in a single panel can lead to compromises in efficiency for both. While it enhances overall energy utilization, it might not match the peak performance of standalone PV or solar thermal systems.
- Design complexity: The integration of two technologies in a single panel makes the design and engineering of PVT systems more complex, which could increase maintenance and installation costs.
- Lack of awareness: Many consumers and businesses might not be aware of the existence and benefits of PVT panels, leading to limited demand and market penetration.

Opportunities:

- Simultaneous energy generation: PVT panels offer the advantage of generating both electricity and heat energy from the same footprint, maximizing the utilization of solar resources and providing a more consistent energy supply throughout the day.

- Space efficiency: By combining two functions in one, PVT panels can save space, making them particularly suitable for applications with limited installation areas.
- Increased overall efficiency: The thermal component of PVT panels allows for better heat management, reducing the operating temperature of photovoltaic cells and enhancing their electrical efficiency.
- Diverse applications: PVT panels are versatile and can be used in various settings, such as residential, commercial, and industrial, providing electricity and hot water or space heating simultaneously.
- Environmental benefits: PVT panels contribute to reducing greenhouse gas emissions by generating clean electricity and displacing the need for fossil fuels used for heating purposes.
- Technological advancements: Ongoing research and development in the field of PVT technology may lead to improved efficiencies and cost reductions in the future, making them more competitive with other solar technologies.
- Incentives and policies: Government incentives, subsidies, and supportive policies focused on promoting renewable energy and energy-efficient technologies can encourage the adoption of PVT panels.

Therefore, photovoltaic-thermal panels offer a promising and efficient approach to harness solar energy for both electricity and thermal needs. While challenges related to cost, efficiency, and awareness exist, the numerous opportunities, such as simultaneous energy generation and space efficiency, present a compelling case for further research, development, and investment in PVT technology. With the right support from policymakers, increased public awareness, and advancements in technology, PVT panels could play a vital role in the global transition toward sustainable and renewable energy systems.

7.5.4 SWOT Analysis

An alternative approach to assess the viability of this technology involves conducting a SWOT analysis. SWOT stands for Strengths, Weaknesses, Opportunities, and Threats. This analysis is commonly employed during the strategic planning of a project or as a method for evaluating a product. A SWOT analysis is employed to identify and define the internal and external factors that can either facilitate or pose risks to achieving a particular objective. Decision-makers can use this analysis to assess the extent of alignment between the internal and external attributes pertaining to the objective [82].

The analysis focuses on attributes that provide a competitive edge over others (Strengths) or conversely, aspects that may put one at a disadvantage (Weaknesses). It also considers aspects of the surrounding environment that could be leveraged as advantages (Opportunities) or may pose potential challenges (Threats) in the pursuit of the objective. These categories are interconnected, and it is valuable to examine them both independently and in relation to each other. For instance, when exploring opportunities, it is beneficial to assess strengths, as new opportunities might emerge, and also weaknesses, to ascertain if their mitigation could create new opportunities [82]. SWOT analysis offers several advantages over other strategy planning tools, making it a popular choice for organizations, decision-makers and government bodies. It provides a holistic view by considering both internal and external factors, such as market trends, competition, and potential risks, unlike some other tools that focus solely on internal aspects. SWOT analysis is a simple and straightforward tool that can be easily understood and applied by individuals at various levels within an organization. Its simplicity makes it accessible to a wide range of users, enabling effective communication and collaboration between all levels of stakeholders of a project.

Strengths

- Efficient use of space—two separate collectors can be replaced with a single one;
- Better solar conversion—maximizing the solar energy harvested;
- Aesthetic advantage—have a more appealing look from an architectural point of view;
- On-site energy production for both types of energy demands—electrical and thermal.

Weaknesses

- High cost of equipment and installation;
- Lack of standardization;
- Lack of long-term performance data;
- Efficiency needs to improve;
- Lack of demonstration (pilot projects).

Opportunities

- Design optimization;
- Better emerging photovoltaic materials;
- Can benefit from combined subsidy;
- Niche market of prosumers;
- Can be integrated in building materials from design stage.

Threats

- Reduced scientific-industrial collaboration;
- Lack of awareness about the product among the consumers/prosumers;
- Currently there are two separate industries;
- Markets are divided;
- Financial schemes and incentives need to be clarified;
- Subsidies schemes might be removed in the future.

7.6 Conclusions

PVT collectors are still a novel technology, not yet fully penetrated on the market, with little reliable performance data available for designers and planners. In terms of technology and economy, the two separate industries (PV and solar thermal) need to converge more, and an interdisciplinary effort from industry and academia is required in order to accelerate the development of PVT collectors and reach a competitive market price that it is accessible to the large mass of prosumers.

An examination of the economic setting in the PVT market, along with the existing subsidy programs, has shown a lack of uniform standards at the European level. Furthermore, there is a degree of ambiguity regarding how PVT collectors are classified within the realm of solar energy technologies. In terms of economic assessment, subsidies represent a crucial factor for evaluating the advantages of any investment.

The identified barriers, including higher initial costs, efficiency trade-offs, design complexity, and limited awareness, pose significant hurdles for widespread adoption. Overcoming these barriers will require targeted efforts from researchers, manufacturers, policymakers, and the renewable energy industry as a whole.

However, the opportunities presented by PVT panels are equally compelling. The ability to generate both electricity and heat from the same installation footprint offers space efficiency and enhanced overall energy utilization. Additionally, the potential for diverse applications in various settings, coupled with the environmental benefits of reducing greenhouse gas emissions, makes PVT panels a viable solution for sustainable energy generation.

To capitalize on these opportunities, continued research and development in PVT technology are essential to improve efficiency and drive down costs. Government incentives and supportive policies can incentivize investments in PVT installations, while public awareness campaigns can help educate consumers and businesses about the benefits of PVT panels.

In general, further PVT research should be focused on improving the materials, the efficiency and driving down the costs of this technology. For this chapter in particular, given the fluctuating energy prices in recent years, a comprehensive cost benefit analysis should be carried out to establish be able to compare the PVT technology not only with the ST and PV but also with other renewables such as wind, geothermal, biomass, but also with conventional fuels.

To sum up, small-scale PVT technology presents a promising approach for optimizing solar energy conversion, offering significant advantages over conventional PV and thermal collectors, particularly in residential and urban settings. To advance this technology in the mainstream market, additional showcase projects, increased consumer awareness, standardization efforts, and concrete proof of concept are essential. PVT panels represent a promising avenue for harnessing solar energy in a more efficient and multifunctional manner. By addressing the barriers and seizing the opportunities, we can foster the widespread adoption of PVT technology and contribute significantly to a cleaner, more sustainable energy future. Adopting PVT panels will not only enhance energy security and reduce environmental impact but also accelerate the global transition to a greener and more sustainable world for generations to come.

References

1. Curtea de Conturi Europeana (2020) Raportul special nr. 8/2019: Energia eoliana si solara pentru producerea de energie electrica. https://doi.org/10.2865/343610. [Online]. eca.europa.eu
2. Agentia Europeana de Mediu (2017) Eea greenhouse gas—data viewer, [Online]. https://www.eea.europa.eu/data-and-maps/data/dataviewers/ greenhouse-gases-viewer/. visited on 9/07/2023
3. European Comission (2019) Communication from the commission to the european parliament, the council, the European economic and social committee and the committee of the regions -a policy framework for climate and energy in the period from 2020 to 2030, [Online]. Available: http://publications.europa.eu/resource/cellar/88484dd0-8947-11e3-9b7d-01aa75ed71a1.0001.04/DOC_1. visited on 12/07/2023
4. European Comission (2019) Communication from the commission to the european par- liament, the council, the european economic and social committee and the committee of the regions a roadmap for moving to a competitive low carbon economy in 2050, [Online]. http://publications.europa.eu/resource/cellar/5db26ecc-ba4e-4de2-ae08-dba649109d18.0002.02/DOC_1 (visited on 12/07/2023)
5. The European Renewable Energy Council (2015) Renewable Energy in Europe, 2nd edn. EarthScan
6. Isabella O, Smets A, Van Swaaij R, Zeman M, Klaus J (2014) Solar energy-fun-damentals, technology, and systems. Delft University of Technology
7. Radziemska E (2009) Performance analysis of a photovoltaic-thermal integrated system. Int J Photoenergy 2009, Bhattacharya RN (ed) pp 732 093. ISSN: 1110-662X. https://doi.org/10.1155/2009/732093. [Online]. https://doi.org/10.1155/2009/732093
8. Dubey S, Sarvaiya JN, Seshadri B (2013) Temperature dependent photovoltaic (pv) efficiency and its effect on pv production in the world—a review. Energy Procedia 33:311–321. PV Asia Pacific conference 2012, ISSN: 1876-6102. https://doi.org/10.1016/j.egypro.2013.05.072. [Online]. http://www.sciencedirect.com/science/article/pii/S1876610213000829
9. Florschuetz L (1979) Extension of the hottel-whillier model to the analysis of combined photovoltaic/thermal flat plate collectors. Solar Energy 22(4):361–366. ISSN: 0038-092X. https://doi.org/10.1016/0038-092X(79)90190-7. [Online]. http://www.sciencedirect.com/science/article/pii/0038092X79901907
10. Kaldellis JK, Kapsali M, Kavadias KA (2014) Temperature and wind speed impact on the efficiency of pv installations. Experience obtained from outdoor measurements in Greece. Renew Energy 66:612–624
11. Singh P, Ravindra N (2012) Temperature dependence of solar cell performance—an analysis. Sol Energy Mater Sol Cells 101:36–45. ISSN: 0927-0248. https://doi.org/10.1016/j.solmat.2012.02.019. [Online]. http://www.sciencedirect.com/science/article/pii/S0927024812000931
12. Chander S, Purohit A, Sharma A, Nehra S, Dhaka M (2015) Impact of temperature on performance of series and parallel connected mono-crystalline silicon solar cells. Energy Reports 1:175–180. ISSN: 2352-4847. https://doi.org/10.1016/j.egyr.2015.09.001. [Online]. http://www.sciencedirect.com/science/article/pii/S2352484715000293
13. Radziemska E (2003) The effect of temperature on the power drop in crystalline silicon solar cells. Renew Energy 28(1):1–12. ISSN: 0960-1481. https://doi.org/10.1016/S0960-1481(02)00015-0. [Online]. http://www.sciencedirect.com/science/article/pii/S0960148102000150.

14. Meneses-Rodriguez D, Horley PP, Gonzalez-Hernández J, Vorobiev YV, Gorley PN (2005) Photovoltaic solar cells performance at elevated temperatures. Solar Energy 78(2):243–250. ISES Solar World Congress 2003
15. Dash P, Gupta N (2015) Effect of temperature on power output from different commercially available photovoltaic modules. Int J Eng Res Appl
16. Pierret R, Neudeck G (1982) Advanced semiconductor fundamentals, Volume VI of modular series on solid state devices. Pearson Education
17. Kurtz S, Packard U, Berger K, Kato K, Friesen T, Liu H, Van Iseghem M (2014) Performance and reliability of photovoltcaic systems–subtask 3.2: review of failures of photovoltaic modules. International Energy Agency (IEA), Photovoltaic Power Systems Programme, Technical report
18. BRE National Solar Centre (2016) Evidence gathering—low carbon heating technologies—hybrid solar photovoltaic thermal panels. Deltaee
19. Duffie J, Beckman W (2006) Solar engineering of thermal processes. Wiley
20. Lämmle M (2018) Thermal management of pvt collectors—development and modelling of highly efficient glazed, flat plate pvt collectors with low-emissivity coatings and overheating protection. PhD thesis, Fraunhofer Institute for Solar Energy Systems ISE. [Online]. https://www.researchgate.net/publication/327882787
21. Daghigh R, Ruslan M, Sopian K (2011) Advances in liquid based photovoltaic/thermal (pv/t) collectors. Renew Sustain Energy Rev 15(8):4156–4170. ISSN: 1364–0321. https://doi.org/10.1016/j.rser.2011.07. 028. [Online]. http://www.sciencedirect.com/science/article/pii/S1364032111002681
22. Vorrath S (2017) Reneweconomyanu team cracks solar thermal efficiency of 97%—a world record., [Online]. http://reneweconomy.com.au/anu-team-cracks-solar-thermal-efficiency-of-97-a-world-record-34199/. visited on 10/07/2023
23. Weiss W, Spörk-Dür M, Mauthner F (2017) Solar heat worldwide. iea solar heating and cooling programme. IEA, AEE-Institute for Sustainable Technologies. [Online]. http://www.iea-shc.org/data/sites/1/publications/Solar-Heat-Worldwide-2017.pdf
24. Guarracino I (2017) Hybrid photovoltaic-thermal solar systems for combined heating. PhD thesis, Imperial College London. https://doi.org/10.1016/j.enconman. 2017.03.024
25. Aste N, del Pero C, Leonforte F (2014) Water flat plate PV-thermal collectors: a review. Solar Energy 102:98–115. ISSN: 0038092X. https://doi.org/10.1016/j.solener.2014.01.025
26. Sathe T, Dhoble A (2017) A review on recent advancements in photovoltaic thermal techniques. Renew Sustain Energy Rev 76:645–672. https://doi.org/10.1016/j.rser.2017.03.075
27. Alghoul M, Azmi B, Wahab M (2005) Review of materials for solar thermal collectors. Anti-Corros Methods Mater 52:199–206. https://doi.org/10.1108/00035590510603210
28. IEA Solar Heating and Cooling Technology Collaboration Programme (2020) Existing pvt systems and solutions. International Energy Agency, SHC Task 60/Report A1
29. Fraunhofer Institute for Solar Energy Systems (2020) Photovoltaics report. Fraunhofer Institute. [Online]. https://www.ise.fraunhofer.de/content/dam/ise/de/documents/publications/studies/Photovoltaics-Report.pdf
30. Barbu M (2021) Study and optimisation of hybrid photovoltaic thermal collectors and integration in small-scale prosumer energy systems. Ph.D. thesis, University Politehnica of Bucharest and INSA Strasbourg, Bucharest, Romania
31. NREL (2020) Best research-cell efficiency chart, [Online]. https://www.nrel.gov/pv/cell-efficiency.html. visited on 10/29/2020
32. Zondag H (2008) Flat-plate pv-thermal collectors and systems: a review. Renew Sustain Energy Rev 12(4):891–959. ISSN: 1364-0321. https://doi.org/10.1016/j.rser.2005.12.012. [Online]. http://www.sciencedirect.com/science/article/pii/S1364032107000020
33. Kern Jr EC, Russell MC (1978) Combined photovoltaic and thermal hybrid collector systems. Photovoltaic specialists conference, 13th, Washington, D.C. June 5–8
34. Euh SH, Lee J-B, Choi Y, Kim D-H (2012) The performance and efficiency analysis of a pvt system compared with a pv module and a solar collector. J Korean Sol Energy Soc 32. https://doi.org/10.7836/kses.2012.32.2.001

35. Ramos A, Guarracino I, Mellor A (2017) Solar-thermal and hybrid photovoltaic-thermal systems for renewable heating. Imperial College, London, UK, Grantham Institute Briefing paper No 22
36. Nižetic S, Coko D, Yadav A, Grubišic´ F (2016) Waterspraycoolingtechnique applied on a photovoltaic panel: The performance response. Energy Convers Manag 108:287–296. ISSN: 0196-8904. https://doi.org/10.1016/j.enconman.2015.10.079. [Online]. http://www.sciencedirect.com/science/article/pii/S0196890415010018
37. Lamnatou C, Chemisana D (2017) Photovoltaic/thermal (pvt) systems: a review with emphasis on environmental issues. Renew Energy 105:270–287. ISSN: 0960-1481. https://doi.org/10.1016/j.renene.2016.12.009. [Online]. http://www.sciencedirect.com/science/article/pii/S0960148116310588
38. Ibrahim A, Othman M, Ruslan M, Alghoul M, Yahya M, Zaharim (2009) Performance of photovoltaic thermal collector (pvt) with different absorbers design. WSEAS Trans Environ Dev 5
39. Aste N, Leonforte F, Del Pero C (2015) Design, modeling and performance monitoring of a photovoltaic-thermal (PVT) water collector. Sol Energy 112:85–99. ISSN: 0038092X. https://doi.org/10.1016/j.solener.2014.11.025
40. Hermann M, Lunz K, Keyl H, Koch L, Stryi-Hipp G (2010) Fluid flow investigations of bionic absorbers made from aluminum and steel. EuroSun conference 28 September–1 October 2010
41. Zhiwei H, Hwang Y, Aute V, Radermacher R (2016) Review of fractal heat exchangers review of fractal heat exchangers. International refrigeration and air conditioning conference
42. Huang Z, Hwang Y, Radermacher R (2017) Review of nature-inspired heat exchanger technology. Int J Refrig 78:1–17. ISSN: 0140-7007. https://doi.org/10.1016/j.ijrefrig.2017.03.006. [Online]. http://www.sciencedirect.com/science/article/pii/S0140700717301019
43. Yaji K, Yamada T, Kubo S, Izui K, Nishiwaki S (2015) A topology optimization method for a coupled thermal–fluid problem using level set boundary expressions. Int J Heat Mass Transf 81:878–888. ISSN: 0017-9310. https://doi.org/10.1016/j.ijheatmasstransfer. 2014.11.005. [Online]. http://www.sciencedirect.com/science/article/pii/S0017931014009776
44. Pence D (2010) The simplicity of fractal-like flow networks for effective heat and mass transport. Exp Therm Fluid Sci 34(4):474–486. ECI International Conference on Heat Transfer and Fluid Flow in Microscale. ISSN: 0894-1777. https://doi.org/10.1016/j.expthermflusci.2009.02.004. [Online]. http://www.sciencedirect.com/science/article/pii/S0894177709000338
45. European Comission (2020) Internal market, industry, entrepreneurship and smes, [Online]. https://ec.europa.eu/growth/sectors/raw-materials/specific-interest/critical_en/. visited on 10/29/2020
46. Santbergen R, Rindt C, Zondag H, van Zolingen R (2010) Detailed analysis of the energy yield of systems with covered sheet-and-tube pvt collectors. Solar Energy 84(5):867–878. ISSN: 0038-092X. https://doi.org/10.1016/j.solener.2010.02.014. [Online]. http://www.sciencedirect.com/science/article/pii/S0038092X10001015
47. Charalambous P, Kalogirou S, Maidment G, Yiakoumetti K (2011) Optimization of the photovoltaic thermal (pv/t) collector absorber. Solar Energy 85(5):871 –880. ISSN: 0038-092X. https://doi.org/10.1016/j.solener. 2011.02.003. [Online]. http://www.sciencedirect.com/science/article/pii/S0038092X11000314
48. Dupeyrat P, Ménézo C, Rommel M, Henning H-M (2011) Efficient single glazed flat plate photovoltaic–thermal hybrid collector for domestic hot water system. Solar Energy 85(7):1457–1468. ISSN: 0038-092X. https: //doi.org/https://doi.org/10.1016/j.solener.2011.04.002. [Online]. http://www.sciencedirect.com/science/article/pii/S0038092X11001125
49. Singh I, Singh D, Singh M (2019) Thermal modeling and performance evaluation of photovoltaic thermal (pv/t) systems: a parametric study. Int J Green Energy 16:1–7. https://doi.org/10.1080/15435075.2019.1584103
50. Chow TT (2003) Performance analysis of photovoltaic-thermal collector by explicit dynamic model. Solar Energy 75:143–152. ISSN: 0038092X. https://doi.org/10.1016/j.solener.2003.07.001

51. Rejeb O, Dhaou H, Jemni A (205) Parameters effect analysis of a photovoltaic thermal collector: case study for climatic conditions of Monastir, Tunisia. Energy Convers Manag 89:409–419, 2015. ISSN: 01968904. https://doi.org/10.1016/j.enconman.2014.10.018. [Online]
52. Guarracino I, Mellor A, Ekins-Daukes NJ, Markides CN (2016) Dynamic coupled thermal-and-electrical modelling of sheet-and-tube hybrid photovoltaic/thermal (PVT) collectors. Appl Therm Eng 101:778–795. 13594311. https://doi.org/10.1016/j.applthermaleng.2016.02.056
53. Ma T, Zhao J, Li Z (2018) Mathematical modelling and sensitivity analysis of solar photovoltaic panel integrated with phase change material. Appl Energy 228:1147–1158. ISSN: 0306–2619. https://doi.org/10.1016/j.apenergy.2018.06.145
54. Kallio S, Siroux M (2020) Energy analysis and exergy optimization of photovoltaic-thermal collector. Energies 13. https://doi.org/10.3390/en13195106
55. Ijumba P, Sebitosi A (2010) Evaluating the impact of consumer behavior on the performance of domestic solar water heating systems in south Africa. J Energy South Afr 21:25–34. https://doi.org/10.17159/2413-3051/2010/v21i1a3247
56. Arsalis A, Alexandrou AN (2015) Parametric study and cost analysis of a solar- heating-and-cooling system for detached single-family households in hot climates. Solar Energy 117:59–73. ISSN: 0038–092X. https://doi.org/10.1016/j.solener.2015.04.024. [Online]. http://www.sciencedirect.com/science/article/pii/S0038092X15002121
57. Nhut L, Raza W, Park Y (2020) A parametric study of a solar-assisted house heating system with a seasonal underground thermal energy storage tank. Sustainability 12:8686. https://doi.org/10.3390/su12208686
58. Tzinnis S, Bellos E, Tzivanidis C (2016) Parametric analysis of a solar cooling system designed for athens climate. IConference: EinB2016–5th international conference ENERGY in BUILDINGS
59. Khalilpour KR, Vassallo A (2016) Technoeconomic parametric analysis of pv-battery systems. Renew Energy 97:757–768. ISSN: 0960-1481. https://doi.org/10.1016/j.renene.2016.06.010. [Online]. http://www.sciencedirect.com/science/article/pii/S0960148116305225
60. Raj NT, Iniyan S, Goic R (2011) A review of renewable energy based cogeneration technologies. Renew Sustain Energy Rev 15(8):3640–3648. ISSN: 1364–0321. https://doi.org/10.1016/j.rser.2011.06.003. [Online]. http://www.sciencedirect.com/science/article/pii/S1364032111002346
61. Slimani MEA, Amirat M, Kurucz I, Bahria S, Hamidat A, Chaouch WB (2017) A detailed thermal-electrical model of three photovoltaic/thermal (pv/t) hybrid air collectors and photovoltaic (pv) module: Comparative study under algiers climatic conditions. Energy Convers Manag 133:458 –476. ISSN: 0196-8904. https://doi.org/10.1016/j.enconman. 2016.10.066. [Online]. http://www.sciencedirect.com/science/article/pii/S0196890416309487
62. Onovwiona H, Ugursal V (2006) Residential cogeneration systems: review of the current technology. Renew Sustain Energy Rev 10(5):389–431. ISSN: 1364-0321. https://doi.org/10.1016/j.rser. 2004.07.005. [Online]. http://www.sciencedirect.com/science/article/pii/S1364032104001340
63. Good C, Chen J, Dai Y, Hestnes AG Hybrid photovoltaic-thermal systems in buildings—a review. Energy Procedia 70:683–690 (2015). International Conference on Solar Heating and Cooling for Buildings and Industry, SHC 2014, ISSN: 1876-6102. https://doi.org/10.1016/j.egypro.2015.02.176. [Online]. http://www.sciencedirect.com/science/article/pii/S1876610215002982
64. Debbarma M, Sudhakar K, Baredar P (2017) Comparison of bipv and bipvt: a review. Resour Effic Technol 3(3):263–271. ISSN: 2405-6537. https://doi.org/10.1016/j.reffit.2016.11.013. [Online]. http://www.sciencedirect.com/science/article/pii/S240565371630207X
65. Kim J-H, Park S-H, Kang J-G, Kim J-T (2014) Experimental performance of heating system with building-integrated pvt (bipvt) collector. Energy Procedia 48:1374–1384. Proceedings of the 2nd International Conference on Solar Heating and Cooling for Buildings and Industry (SHC 2013), ISSN: 1876-6102. https://doi.org/10.1016/j.egypro.2014.02.155. [Online]. http://www.sciencedirect.com/science/article/pii/S1876610214004172
66. Zhang H, Dai C, Cheng K, Lei Y, Yang Y (2020) Research hotspots and evolution of energy prosumer: a literature review and bibliometric analysis. Mathematical problems in engineering.

International conference on solar heating and cooling for buildings and industry, SHC 2014. https://doi.org/10.1155/2020/5703101
67. Kamel R, Fung A, Dash P (2015) Solar systems and their integration with heat pumps: a review. Energy Build 87:395–412. https://doi.org/10.1016/j.enbuild.2014.11.030
68. Xu G, Deng S, Zhang X, Yang L, Zhang Y (2009) Simulation of a photovoltaic/thermal heat pump system having a modified collector/evaporator. Solar Energy 83(11):1967–1976. ISSN: 0038-092X. https://doi.org/10.1016/j.solener.2009.07.008. [Online]. http://www.sciencedirect.com/science/article/pii/S0038092X09001765
69. Sommerfeldt N, Madani H (2016) Review of solar pv/thermal plus ground source heat pump systems for European multi-family houses. 1–12. https://doi.org/10.18086/eurosun.2016.08.15
70. Salpakari J, Lund P (2016) Optimal and rule-based control strategies for energy flexibility in buildings with pv. Appl Energy 161:425–436. ISSN: 0306-2619. https://doi.org/10.1016/j.apenergy.2015.10.036. [Online]. http://www.sciencedirect.com/science/article/pii/S0306261915012635
71. Freeman J, Guarracino I, Kalogirou S, Markides C (2017) A small-scale solar organic rankine cycle combined heat and power system with integrated thermal energy storage. Appl Therm Eng 127:1543–1554. ISSN: 1359–4311. https://doi.org/10.1016/j.applthermaleng.2017.07.163. [Online]. http://www.sciencedirect.com/science/article/pii/S1359431117348287
72. Weiss W, Spork-Dur M (2020) Solar heat worldwide-global market development and trends in 2019 | detailed market figures 2018. [Online]. https://www.iea-shc.org/solar-heat-worldwide-2020
73. Poortinga W, Fisher S, Bohm G, Steg L, Whitmarsh L, Ogunbode C (2018) European attitudes to climate change and energy: topline results from Round 8 of the European Social Survey, English, ser. University of London, ESS Topline Results Series
74. International Renewable Energy Agency (2020) Renewable power generation costs in 2019. [Online]. https://www.irena.org/publications/2020/Jun/Renewable-Power-Costs-in-2019
75. International Energy Agency (2018) IEA SHC task 35, [Online]. http://task35.iea-shc.org. visited on 10/07/2023
76. Hadorn J (2017) French pvt market is picking up, [Online]. http: //www.iea-shc.org/article?NewsID=205. visited on 10/07/2023
77. SHC IEA (2019) Task 60-pvt systems: application of pvt collectors and new solutions in hvac systems. http://task60.iea-shc.org. Accessed 01 July 2023
78. Fang X, Li D (2013) Solar photovoltaic and thermal technology and applications in china. Renew Sustain Energy Rev 23:330–340. ISSN: 1364-0321. https://doi.org/10.1016/j.rser.2013.03.010. [Online]. http://www.sciencedirect.com/science/article/pii/S136403211300169X.
79. Barbu M, Patrascu R, Darie G, Tutica˘ D (2019) A technical-economical analysis of the implementation of hybrid solar energy systems in small energy prosumer applications. Qual Access Success 20:134–138
80. Weiss W, Spork-Dur M (2023) Solar heat worldwide - global market development and trends in 2023 | detailed market figures 2022 [Online]. https://www.iea-shc.org/solar-heat-worldwide-2023
81. Resch A (2020) 2020 subsidies for pvt collectors in selected countries -shc task 60/re- port d6. Sol Heat Cool Program Int Energy Agency. https://doi.org/10.18777/ieashc-task60-2020-0005.[Online].Available:https://task60.iea-shc.org/Data/Sites/1/publications/IEA-SHC-Task60-D6-Subsidies-PVT.pdf
82. Lupu AG, Dumencu A, Atanasiu MV, Panaite CE, Dumitrașcu G, Popescu A (2016) SWOT analysis of the renewable energy sources in Romania—case study: solar energy. IOP Conf Ser Mater Sci Eng 147:012138. https://doi.org/10.1088/1757-899x/147/1/012138

Part III
Potential Technologies

Chapter 8
Older Oilfields as Distributed Geothermal Energy Resources

Subha Mondal and Chitta Sahana

Abstract The global share of renewable energy uses is to be increased to reduce the carbon footprint. Geothermal-based renewable resources are preferred for satisfying localized energy needs due to their steady and uninterrupted supply. However, geothermal-based energy systems are cost-intensive unless the geothermal heat is readily available. Many of the oilfields produce hot geothermal water at the stage of high water cut. Removing a larger amount of water produced at the high water cut stage of an oilfield is a compulsion. Thus, the hot water produced from the oilfield at the high water cut stage can be used to satisfy the localized energy needs fully or partially. On the other hand, many of the researchers proposed to recover geothermal heat from depleted oilfields. Recovering geothermal heat from depleted oil reservoirs would be an attractive option as there would be no additional cost of drilling the geothermal well. In this chapter, various techniques to recover and utilize geothermal heat are presented, along with a few relevant case studies. From the summarized case studies, it appears that recovering geothermal heat from older oilfields may emerge as one of the future carbon–neutral distributed energy options for delivering electricity and other energy utilities. It is necessary to mention that additional study and appropriate policy formulation are required to address potential obstacles and challenges in the practical application of geothermal heat recovery from oilfields.

Keywords High water cut stage · Depleted oil field · Geothermal heat · Energy utilities

S. Mondal (✉)
Department of Mechanical Engineering, Aliah University, Kolkata, India
e-mail: subhamondal53@gmail.com

C. Sahana
Department of Mechanical Engineering, Dr. B.C. Roy Engineering College, Durgapur, India

S. De et al. (eds.), *Challenges and Opportunities of Distributed Renewable Power*,
Energy, Environment, and Sustainability, https://doi.org/10.1007/978-981-97-1406-3_8

8.1 Introduction

A substantial part of global secondary energy needs is still satisfied by fossil fuel-based power plants. It is necessary to mention that fossil fuel-based power plants are greatly responsible for the emission of greenhouse gases. Thus, fossil fuel-based power plants are to be gradually replaced by renewable ones to address the global warming issue. It is worthwhile mentioning that due to higher reliability and consistent supply, geothermal-based renewable sources are widely applied to satisfy localized energy needs. However, the average drilling cost of a 500 m depth geothermal well is about 5 million dollars [1]. Due to the higher well drilling cost, a geothermal-based energy system with a smaller capacity is not economically feasible unless geothermal heat is readily available as the by-product of some processes or from a hot spring.

A major part of the global energy needs is catered by oil and gas. However, continuing oil production from an older oilfield is usually a costly affair. When recovering oil from an oilfield is not profitable, the well is usually plugged and abandoned. The abandonment process of an oilfield is also complex and cost-intensive.

Due to the presence of requisite infrastructure, researchers are recommending geothermal heat recovery from many of the abandoned gas/oilfields. It was also reported that requisite capital investment for retrofitting an abandoned gas/oil well to a geothermal well is about 50% lower than that of commissioning a new geothermal well [2]. Barbacki [3] recommended geothermal heat recovery from four oil/gas wells in the eastern zone of the Grobla field. Oil/gas production from these wells were expected to cease with in a very short period. Sanyal and Bulter [4] reported that geothermal heat recovery from an abandoned gas well in Texas would generate 340 kW output power.

It is necessary to note that at the mature stage, most of the gas/oilfields produce hydrocarbon accompanied by water. According to the [5], in the US, there are about 823,000 wells coproducing hot water with oil and gases. The volume of water production increases with time and removing this water from the gas/oilfield becomes a compulsion. The temperature of water produced at this high water cut stage of the oilfield usually varies between 60 and 140 °C. Thus, some of the energy needs of the local community may be satisfied by the low-grade heat of this water [6]. It is necessary to note that the types of deliverable energy output are usually decided by the temperature and quantity of the recoverable geothermal heat.

In this chapter, different methods applied to harvest geothermal heat from gas/oil fields are presented, along with a few relevant case studies.

8.2 Methods to Recover Geothermal Heat from Oilfield

To recover geothermal heat from an oilfield, either an open loop system or a closed loop system can be employed [7] as summarized below:

8.2.1 Open Loop Geothermal Heat Recovery System

The open loop geothermal heat recovery system directly circulates the ground water to transmit the geothermal heat. An open loop heat recovery system needs to have two or more oil wells—at least one well for geothermal heat production and one well for re-injection. Reducing reservoir pressure over time is a drawback of the naturally occurring geothermal heat recovery system. In many cases, the naturally occurred open loop heat recovery system coproduces hydrocarbon liquid with the hot water. Thus, extra care is needed to separate the hydrocarbon liquid from the hot water. An enhanced geothermal system (EGS) can be adopted to overcome above mentioned drawbacks of the naturally occurring open loop heat recovery system. In an EGS, a subsurface fracture is created artificially, on the hot dry rock. The subsurface fracture helps in geothermal heat recovery by continuous water circulation between the production well and re-injection well as shown in Fig. 8.1. The hot water exiting the production well after delivering utility heat to end-users is re-injected into the injection well. EGS technology is capable of supplying geothermal heat at a higher temperature. However, due to involved uncertainties and higher costing, its applications are limited. Numerous pilot projects with EGS technologies were initiated in Japan, Australia, and Switzerland [8]. However, they either failed to reach the projected capacity or were terminated due to lost circulation and seismic activity. Only one large scale EGS is currently in operation in Soultz of France.

8.2.2 Closed Loop Geothermal Heat Recovery System

It is worthwhile mentioning that a closed loop system is suitable for an individual oil well. In the closed loop system, a heat recovery fluid is continuously circulated through a borehole heat exchanger (or well bore heat exchanger) in a closed circuit. U-tube heat exchangers and coaxial heat exchangers are the two most common types of borehole heat exchangers. Different arrangements of wellbore heat exchangers are presented in Fig. 8.2. The heat transfer area of a U-tube heat exchanger is usually lower than that of a coaxial heat exchanger. Thus, a U-tube heat exchanger permits a lesser flow rate of the heat recovery fluid. It is necessary to note that instead of a single U-tube, a multiple U-tube heat exchanger can be used to permit a higher flow rate of the heat transport fluid.

The coaxial arrangement of the borehole heat exchanger is a preferred technology due to its greater heat transfer efficiency [9]. The suitable thermal design of the inner column of a coaxial borehole heat exchanger is very important for successful geothermal heat recovery from an oil well. It was reported that the use of an insulated inner column with very small thermal conductivity would effectively enhance the geothermal heat recovery efficiency of a down-borehole heat exchanger [10]. Thus, the central tube of the coaxial borehole heat exchanger is thermally insulated to restrict heat loss to the cold re-injected heat transport fluid. The efficiency of the

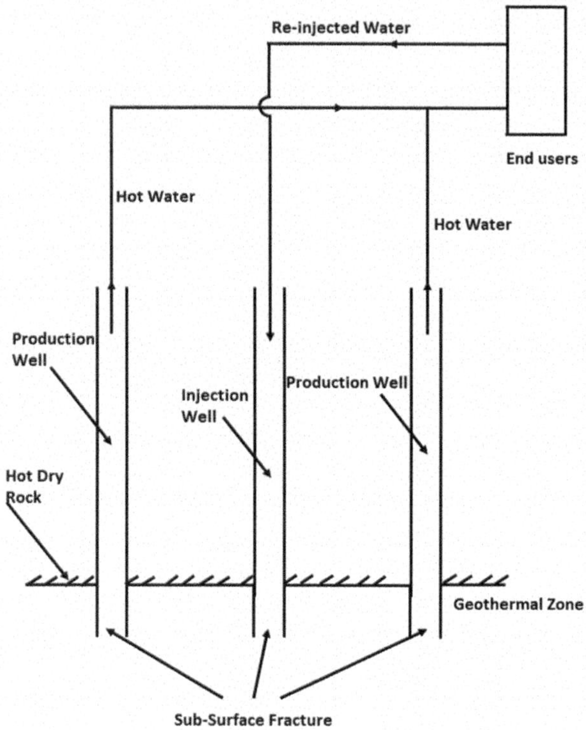

Fig. 8.1 Enhanced open loop geothermal heat recovery system

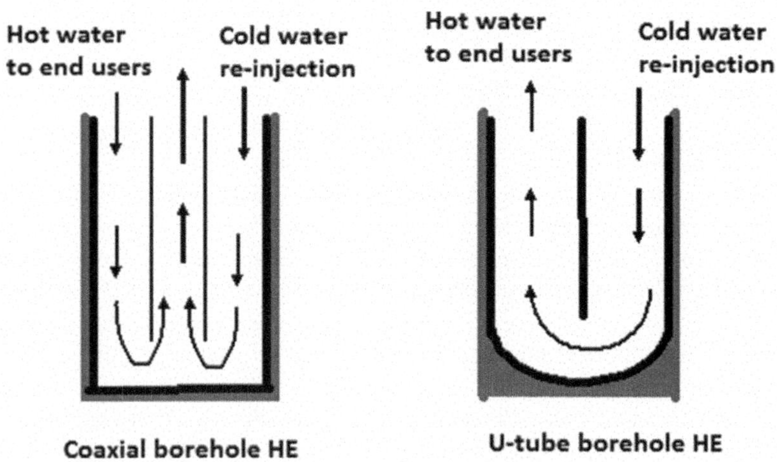

Fig. 8.2 Different arrangement of the closed loop borehole heat exchanger

borehole heat exchanger can be further enhanced by using a vacuum insulated tube (VIT) as the central column [11]. The VIT consists of two coaxial steel pipes—the air between these pipes is evacuated by using a vacuum pump during the manufacturing process of the VIT. The void space between the pipes acts as thermal insulation that minimizes heat flow between the fluid in the central tube and the fluid in the annulus. Usually, the L80 type of steel is used to construct the outer pipe of a VIT. Due to the corrosive environment, high chrome content steels (such as 13CR85, 13CR95, 13CR110) are employed to construct the inner pipe of a VIT.

The closed loop arrangement of the borehole heat exchangers also permits the use of heat transport fluids other than water. Recently it was observed that while recovering geothermal heat from a wellbore, CO_2 exhibited superior performance compared to water [12]. Sun et al. [13] also recommended circulation of super critical CO_2 through a closed horizontal wellbore to recover geothermal heat. Due to larger compressibility, the density difference between the CO_2 mass in the central tube and the CO_2 mass in the annulus generates a large buoyant drive that eases the movement of CO_2 in the closed loop [14–16].

8.3 Possible Energy Outputs from Recoverable Geothermal Heat

It is necessary to note that deliverable energy utilities are greatly dependent on the quality and quantity of the recoverable geothermal heat. Usually the temperature of achievable heat from an oilfield remains below 150 °C. Power generation from this low-grade heat through a steam based Rankine cycle is not techno-economically feasible. However, if geothermal heat is obtained at or above 100 °C an organic Rankine cycle (ORC) can be employed to generate electricity. Transcritical CO_2 power cycle [17], Organic flash cycle [18], Organic flash cycle with ejector [19] are some recently reported technologies capable of producing secondary energy (i.e., electricity) from the available geothermal heat at or above 100 °C. It is necessary to note that the energy conversion efficiency of any power cycle reduces with increasing ambient temperature. Thus, producing power generation from recovered geothermal heat will not be feasible for a higher value of local ambient temperature. However, for a higher local ambient temperature (i.e., above 30 °C), available geothermal heat can be used to produce cooling effects through either an absorption refrigeration chiller or an ejector refrigeration cycle. For geothermal heat available between 70–100 °C, combining mechanical compressor and ejector would be effective in reducing electrical power consumption of a vapour compression refrigeration cycle [20]. For low-grade geothermal heat available between 60–80 °C, humidification-dehumidification (HDH) desalination is an effective technology to deliver desalinated water for localized uses [21]. Low grade geothermal heat available below 60 °C may be directly used for district heating. A heat pump can also be employed to enhance the temperature of the available geothermal heat to usable level. It is necessary to note that producing two or more energy utilities from a single unit often ensures greater flexibility with higher utilization efficiency [22] Thus, employing a cogeneration/ trigeneration/poly generation unit would also be a preferable option for geothermal heat recovery.

8.4 A Few Case Studies

A few recently reported studies on geothermal heat recovery from oil fields are summarized in this section. These studies would be helpful in assessing the applicability of the older oil-well as the source of geothermal heat. As discussed earlier, either an abandoned oilfield can be converted into a geothermal field, or hot water produced at the high water cut stage of an oilfield can be considered as the source of geothermal heat. Thus, all reported case studies are presented in two different subsections.

8.4.1 Geothermal Heat Recovery from Abandoned Oilfields

It is already discussed that employing a binary power cycle is a convenient way to generate electricity from a low-grade heat source. Usually, a low GWP organic fluid is used as the secondary working fluid (SWF) of the binary power cycle. Wight and Bennett [23] evaluated operations of binary power cycles using water as the primary fluid to recover the geothermal heat from abandoned oilfields of Texas. The layout of the proposed binary geothermal cycle is presented in Fig. 8.3. They employed double pipe arrangement of the wellbore heat exchanger due its greater efficiency [24]. In a binary power cycle, the heated water coming out of the central tube is circulated through the heat recovery unit (HRU) to heat a low GWP secondary working fluid. The secondary working fluid exits the HRU at saturated/superheated vapour states and delivers a certain power output while undergoing an expansion process in the turbine. Wight and Bennett [23] reported that a wellbore of 4200 mm depth was capable of supplying hot water at 130 °C. For 2.5 kg/s of mass flow rate of this hot water, deliverable output power of aR134a based binary power cycle was about 109 kW. The achievable output power would be close to 630 kW for a 15 kg/s mass flow rate of the primary working fluid.

Cheng et al. [25] simulated a 6000 m dry hole abandoned oil well. It was observed that initially the temperature of deliverable geo-fluid was decreasing steadily with the increasing time. However, after certain time of operation the temperature of the geo-fluid became more or less constant. It was also observed that the temperature of the deliverable geo-fluid could be enhanced appreciably by increasing the well depth. It is necessary to note that Cheng et al. [25] employed isobutane as the heat transport fluid, instead of water. They installed a double pipe heat exchanger in the oil well to recovery the geothermal heat. Isobutane was pressurized by a pump and circulated

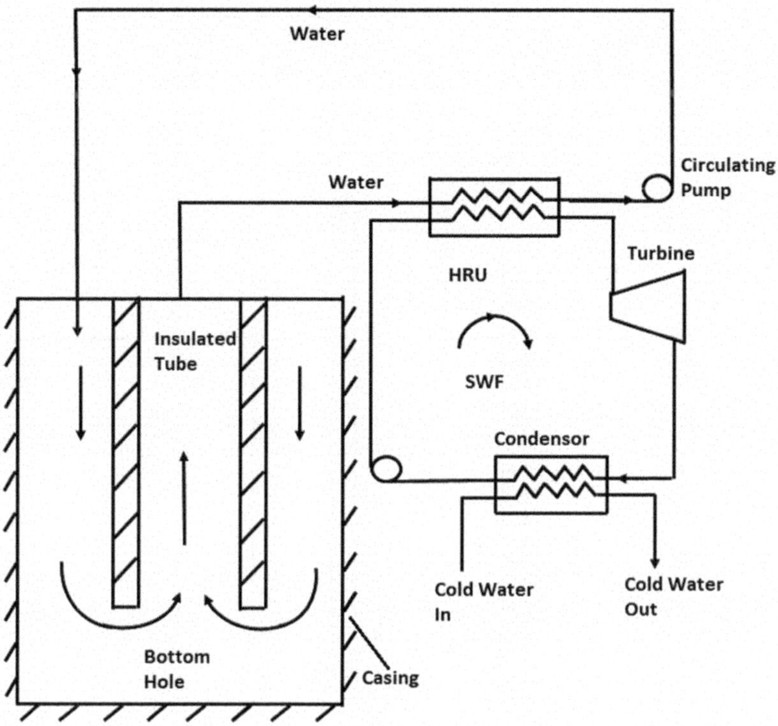

Fig. 8.3 Binary power cycle for geothermal heat recovery from oil well [23]

through the double pipe heat exchanger. In the double pipe heat exchanger sub-cooled isobutane was heated to supercritical vapour stage by the geothermal heat of the oil well. Heated isobutane was expanded in turbine to generate the output power. Exhaust isobutane exiting the turbine was condensed to sub-cooled liquid state in the condenser. After being condensed the isobutane mass was fed to the circulating pump as shown in Fig. 8.4. It was observed that electricity production from a single abandoned oil well was not very appreciable. Maximum achievable output power per abandoned well was about 154 kW. The achievable capacity of electricity productions could be appreciably enhanced by recovering geothermal heat from abandoned oil wells cluster. Thus, if group oil wells are employed for geothermal heat recovery the economic feasibility of the heat recovery scheme would be justified.

Yang et al. [26] designed a geothermal-ORC system to generate secondary energy from the abandoned wells in the Huabei oilfield of China. The achievable mass flow rate and temperature of geo-fluid from the abandoned oil wells in the Huabei oilfield were 20 t/h and 110 °C, respectively. The geo-fluid was a mixture of oil, water, and gas with 98% water content. The layout of the geothermal-ORC system proposed by Yang et al. [26] is presented in Fig. 8.5 in a simplified manner. It appears that the geo-fluid coming out of the oil well was directed through a gas separator to remove the dissolved gases from the geo-fluid. Liquid geo-fluid exiting the gas separator

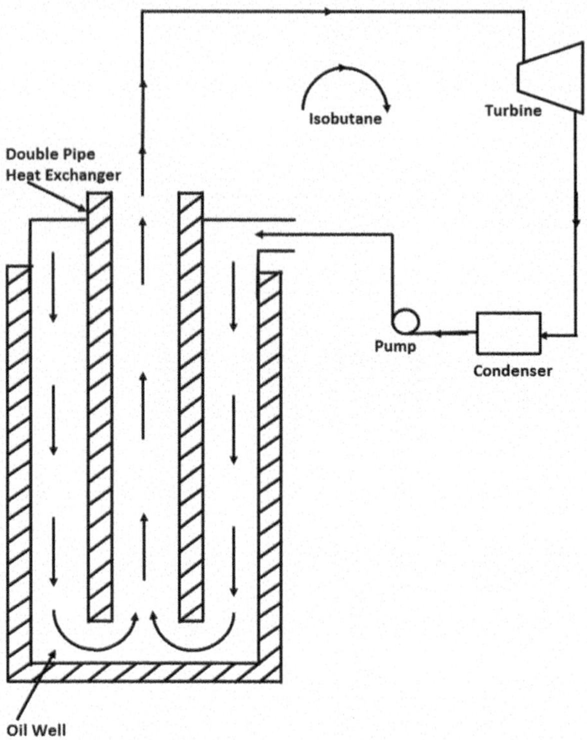

Fig. 8.4 Geothermal heat recovery from oil well using isobutane [25]

was split into two streams. One of the streams was used in some industrial processes. Another stream was used to heat water in a secondary water circuit. The Secondary water circuit consisted of three heat exchangers recovering the geothermal heat. The heated secondary water was used to preheat and vaporize R245fa in the pre-heater and the evaporator, respectively. R24fa vapour exiting the evaporator was expanded in the turbine to produce secondary energy (i.e., electricity). Liquid geo-fluid exiting the secondary water circuit was allowed to enter a settling tank. In the settling tank, oil was separated from the geo-fluid and finally was stored in the oil tank. The remaining geo-fluid was re-injected into the oil well. Yang [27] also conducted an onsite experimental study to generate the preliminary data. It was observed that 84 °C turbine inlet temperature could be achieved. R245fa mass flow rate was about 33.04 kg/s. The achievable thermal efficiency of the ORC was close to 5.33%. It was reported that the efficiency of the Geothermal-ORC system was substantially reduced due to the presence of the secondary water circuit. However, the secondary water circuit helped to avoid blockage in different important equipment and pipe lines.

Moustafa et al. [28] conducted feasibility studies on geothermal heat recovery from existing abandoned oil wells of Egypt. The ASPEN HYSYS software was employed to estimate the deliverable mass flow rate and temperature of the heat

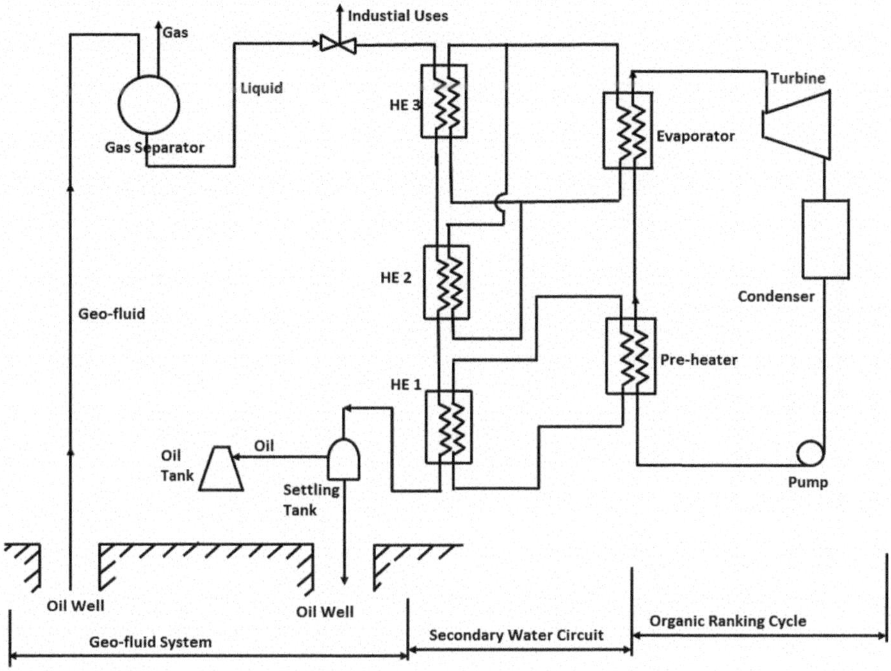

Fig. 8.5 Geothermal heat recovery from the Huabei oilfield of China [26]

transport fluid. It was observed that the achievable outlet temperature and mass flow rate of the heat transport fluid varied appreciably over different locations of Egypt. It was observed that most of the abandoned oil wells in the eastern desert region of Egypt were capable of delivering low-grade geothermal heat. Thus, for this region binary power cycles were employed to generate electricity from the recoverable low-grade geothermal heat. It is necessary to note that water was used as the heat transport fluid and R134a was used as the binary working fluid of the secondary working fluid. However, they proposed to use a steam flash cycle in regions like Gulf of Suez. The schematic representation of steam flash cycle employed in this region is shown in Fig. 8.6. It was reported that in the Eastern region, the outlet temperature of heat transport fluid would be close to 125 °C, and by running binary power cycles, about 3.063 MW power could be generated. On the other hand, in the regions like Gulf of Suez, the total yielded output power of steam flash cycles would be close to 44.6 MW. In this region, the achievable steam/water temperature could reach closer to 300 °C.

Nian and Cheng [29] proposed to use recovered geothermal heat from an oil well for space heating in a virtual building. They considered a 3000 m deep abandoned oil well of Shengli oilfield as the source of geothermal heat. They employed a coaxial double pipe wellbore heat exchanger to recover the geothermal heat. The virtual building was assumed to have 1000 m^2 of heating area. The considered heating period was from15th November to 15th March. The initial room temperature and

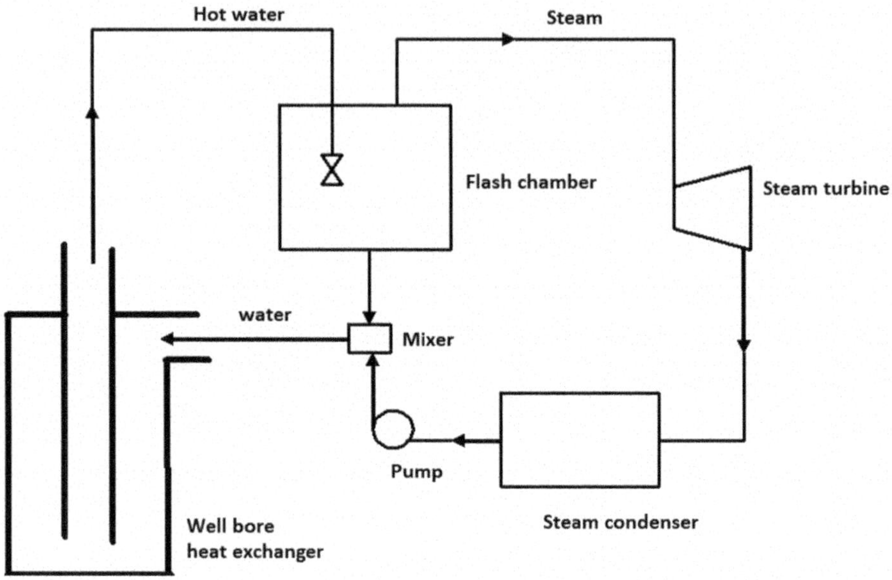

Fig. 8.6 Steam flash cycle for geothermal heat recovery in regions like Gulf of Suez [28]

initial water injection temperature into the oil well were assumed to be 10 °C each. The desirable room temperature was set close to 26 °C. It was observed that for maintaining an indoor room temperature of 26 °C, the desirable water injection rate to the oil well was 24 m^3/h. The associated injection pressure was 2 MPa. It was also observed that though at the start of the operation, the geothermal system was capable of producing hot water at 90 °C; the water outlet temperature was decreasing with increasing operating hours. After about 240 h of operation, the water outlet temperature became more or less fixed at 55 °C. The geothermal heat recovery system would save about 185 tons of coal, each year. Thus, by running the geothermal space heating system, about 678 tons of CO_2 emissions could be avoided in a year. It was also reported that the annualized cost of space heating by this geothermal heat recovery was half of that of the conventional space heating system.

An efficient design of the wellbore heat exchanger is desirable to maintain the desirable temperature and flow rate of the deliverable geo-fluid. Zhu et al. [12] employed COMSOL Multiphysics to simulate the operation of a coaxial wellbore heat exchanger recovering geothermal heat from an abandoned oil well. Based on the main oil field data of China, the geothermal gradient of the oil well was assumed to be 0.3 K/m in the simulation. Considered inner diameter and outer diameter of the casing were 15 cm and 19.6 cm, respectively. A central pipe with 4 cm inner diameter and 2 cm insulation thickness was employed as the extraction well. The considered depth of the extraction well was about 400 m. Water is considered as the heat transport fluid in the simulation. According to the reported output of the simulation model, there was an optimum water injection velocity for which outlet temperature of the geo-fluid (i.e., water) became a maximum. The maximum achievable geo-fluid temperature was close to 170 °C at 0.6 m/s of water injection velocity. It was also reported that using deliverable geo-fluid, about 164.6 kW power could be produced from a single oil well. Zhu et al. [12] also recommended combining situ combustion with the geothermal energy production unit to make the project economically feasible. It was observed that combining situ combustion with the geothermal energy production unit would yield a payback period of 4.5 years with a cumulative cash flow of $4.94 million.

8.4.2 Geothermal Heat Recovery at the High Water Cut Stage of the Oilfield

An oil field operating at a high water cut stage coproduces a large amount of water with the oil. Disposal of this coproduced water puts an extra financial burden on the oil producing organization. It was reported that the temperature of deliverable hot water at the high water cut stage of an oilfield would be between 65–150 °C [30]. The coproduced hot water at the high water cut stage of an oilfield can be utilized to satisfy some of the localized energy needs as summarized below:

Recently Sahana et al. [31] proposed an integrated energy system recovering geothermal heat of hot water produced at the high water cut stage of an oilfield. The assumed temperature of the hot water was 140 °C. The proposed integrated energy system consisted of a supercritical CO_2 power cycle, an ejector expansion CO_2 refrigeration cycle, and a humidification-dehumidification (HDH) desalination unit. The integrated energy system was capable of producing cooling effects and desalinated water, simultaneously. The output power of the CO_2 turbine was expensed to run the power cycle compressor and refrigeration cycle compressor. All the compressors and the turbine were mounted on the same shaft as shown in Fig. 8.7. An ejector was used as the main expansion device in the refrigeration cycle to minimize the throttling loss. The state of CO_2 at the ejector exit was two-phase vapour. It appears from Fig. 8.7 that a vapour separator was employed to separate the liquid and vapour states of the two-phase CO_2 vapour. Liquid CO_2 exiting the vapour separator after undergoing a throttling process entered into the evaporator to produce the desired cooling effect. The CO_2 vapour coming out of the vapour separator was pressurized by a compressor and allowed to mix with the exhaust CO_2 mass exiting the CO_2 turbine. After the mixing, the total CO_2 mass was directed through the saline water heater of the HDH desalination unit. The CO_2 mass exiting the saline water heater was further cooled and condensed in the condenser. The saturated liquid CO_2 exiting the condenser was split into two streams. One of the streams was used to drive the ejector and another stream was pressurized by the power cycle compressor to the CO_2 heater inlet pressure. During the parametric study, the evaporator temperature was varied from −15 °C to 0 °C. It was reported that for the specified range of evaporator

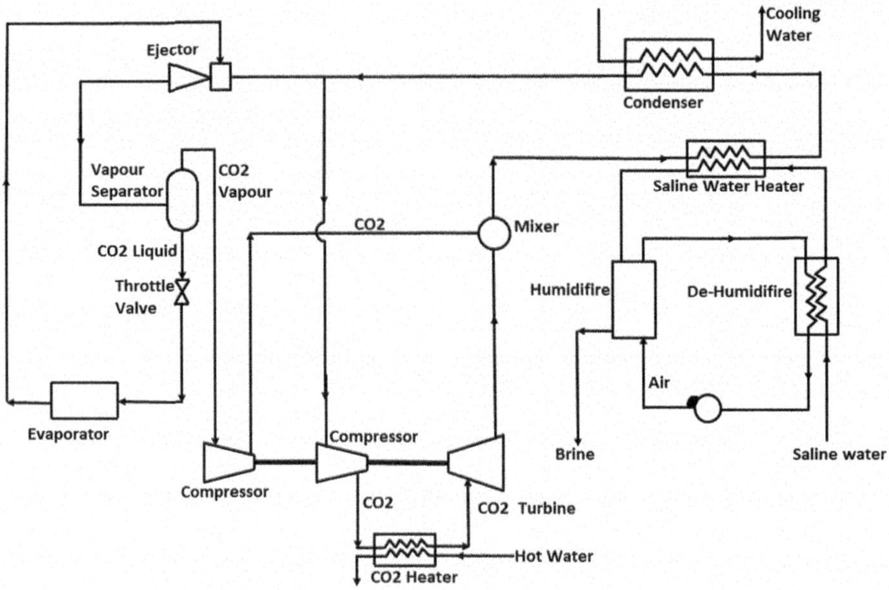

Fig. 8.7 CO_2-based combined energy system recovering geothermal heat at the high water cut stage of an oilfield [31]

temperature, the payback period of the optimized combined energy system would be between 6.65 and 8.25 years.

Li et al. [32] analyzed the operation of a polygeneration unit recovering geothermal heat from the hot water delivered by an oilfield at the high water cut stage. The temperature of the hot water was assumed to be varied between 100 and 150 °C. The polygeneration unit was capable of producing power, cooling and heating, simultaneously. The geo-fluid obtained from the oil well was directed through a three-phase separator unit to separate the oil and gas from the geothermal water. The hot geothermal water exiting the three-phase separator was split into two streams. The heat of one stream was utilized to run a two-stage series organic Rankine cycle (ORC) and the heat of the other stream was expensed to run a water-lithium bromide type absorption chiller. R245fa was used as the working fluid of the ORC. Geothermal water streams exiting the ORC and the absorption chiller were mixed further. After the mixing process temperature of geothermal water became about 50 °C. This temperature of geothermal water was sufficient for heating the crude well to reduce its viscosity and ease the transport process. Geothermal water exiting the crude oil heater was re-injected after recovering some radiant heat. The simplified layout of the proposed polygeneration by Li et al. [32] is presented in Fig. 8.8. According to the study conducted by Li et al. [32], this polygeneration unit was capable of recovering 75% of the available geothermal heat. Li et al. [33] also conducted a techno-economic assessment of this polygeneration unit. During the techno-economic assessment, it was observed that the payback period of the proposed polygeneration unit was greatly

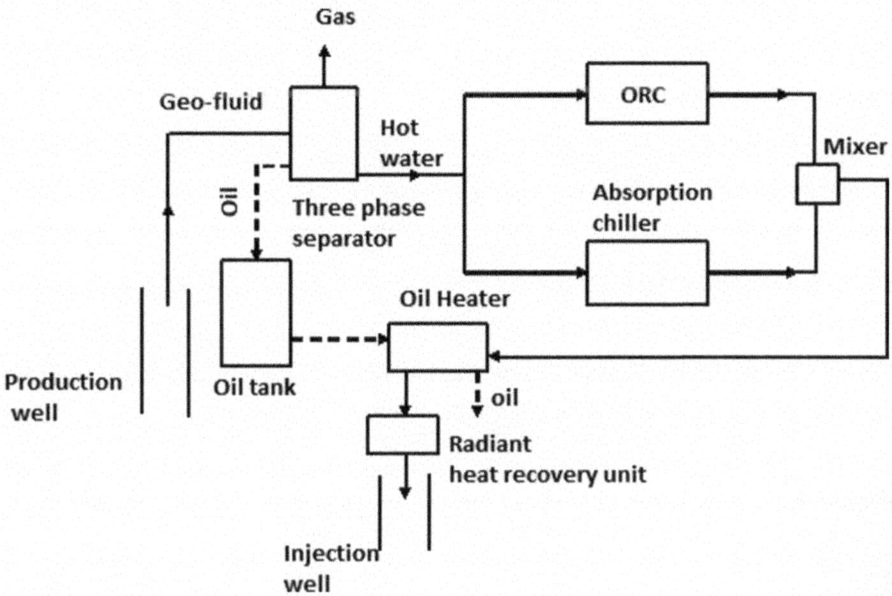

Fig. 8.8 Polygeneration system proposed by Li et al. [32] for geothermal heat recovery at the high water cut stage of an oilfield

influenced by the temperature of the available geothermal water. The minimum payback period was about 3.07 years, at 90 °C driving heat source temperature.

8.5 Challenges and Issues in Recovering Geothermal Heat from Older Oilfields

Although older and abandoned oil fields are capable of generating geothermal energy, the main obstacle to converting them into geothermal fields is low energy conversion efficiency. It is necessary to note that many power and refrigeration cycles capable of utilizing low-grade heat have been proposed in recent times. However, most of them are in the developing stages. Equipment (such as turbo-machinery, efficient ejectors, etc.) needed for the practical operation of the presented thermodynamic cycles is either in the developing stage or yet to be developed. As extensive research is ongoing to develop suitable turbomachinery and ejectors, it is expected that technological barriers will be overcome in the near future.

Most of the oil and gas fields are usually located in distant places with low population densities [15]. Thus, the harvested geothermal energy (transported in the form of electricity or heat) may suffer from appreciable transmission losses during the distribution among the end-users. Development of co-generation or polygeneration

Table 8.1 Pilot projects on electricity production from oilfields at high water cut stage

Project location	Prime mover	Power output capacity (kW)
L RMOTC Wyoming of USA [34]	ORC	110
Laurel oilfield of USA [35]	ORC with twin screw expander	50
Huabei oilfield of China [27]	ORC using R123 as the working fluid and screw expander as the turbo machine	310
Williston basin of USA [36]	ORC	121

units may be useful to address the issue of transmission losses by allowing the entire harvested energy to be consumed by the local community [22].

It is necessary to formulate suitable policies and guide lines for assessing the geothermal heat recovery potentials of different older and abundant oilfields and retrofitting them as sources of recoverable geothermal heat. No definite guidelines and policies have been formulated in this regard till date [2].

8.6 Current Status of Geothermal Heat Recovery from Oilfield

Geothermal heat recovery from older gas and oilfields may emerge as one of the future sustainable energy options. Thus, in recent times, researchers have been recommending geothermal heat recovery from older and abandoned gas/oilfields. Although no such commercial heat recovery plants currently exist, there are several pilot projects to recover geothermal heat from fluids produced along with oil and gas. Data related to a few such pilot projects generating electricity using recovered geothermal heat are summarized in Table 8.1. It appears from Table 8.1 that in all the pilot projects organic Rankine cycle (ORC) was selected as the prime mover due to its technological maturity.

8.7 Conclusions

Usually, an oil well should be permanently plugged and abandoned after the completion of its life. These plugging and abandoned processes are complex and cost intensive. An older oilfield often co-produces a large volume of water with the oil, which affects the profitability of the oil producing organization. As the temperature of the co-produced water usually varies between 65 and 140 °C, an abandoned oilfield may be retrofitted to recover the geothermal heat. Geothermal heat recovered from an older oilfield can also be utilized to satisfy some of the localized energy needs. For a lower ambient temperature, geothermal heat available at or above 100 °C can be used to run a suitable power cycle satisfying the local electricity needs. ORC, OFC, TLC, and T-CO_2 power cycle are a few of such power cycles. For a moderate or higher ambient temperature, the available heat should be expensed to achieve cooling or heating effects instead of electricity. In this chapter, possible technologies employed to recover the geothermal heat from older oil wells are summarized. A few relevant case studies are also presented to assess the different techno-economic aspects of geothermal heat recovery from older oil well. From the summarized case studies, it appears that recovering geothermal heat from an older oilfield is one of the techno-economically feasible carbon neutral energy options. As there are large numbers of matured and abandoned oilfields all over the world and appreciable numbers of oilfields are going to be abandoned in the near future, recovering geothermal energy from these oilfields would emerge as one of the attractive distributed energy options.

References

1. Tester CAJW, Anderson B (2006) A comparison of geothermal with oil and gas well drilling costs. Thirty-first workshop on geothermal reservoir engineering Stanford University, Stanford, California
2. Kurnia JC, Shatri MS, Putra ZA, Zaini J, Caesarendra W, Sasmito AP (2021) Geothermal energy extraction using abandoned oil and gas wells: techno-economic and policy review. Int J Energy Res 46:28–60. https://doi.org/10.1002/er.6386
3. Barbacki AP (2000) The use of abandoned oil and gas wells in Poland for recovering geothermal heat. Proceedings, World geothermal congress. Kyushu-Tohoku Japan 1–5
4. Sanyal S, Bulter S (2010) Geothermal power capacity for petroleum wells—some case histories of assessment. Proceedings, world geothermal congress. Bali, Indonesia
5. US Department of Energy (2012) Geothermal power/oil and gas coproduction opportunity
6. Alimonti C, Gnoni AA (2015) Harnessing the fluids heat to improve mature oil field: the Villafortuna-Trecate case study. J Petrol Sci Eng 125:256–262
7. Santos L, Taleghani AD, Elsworth D (2022) Repurposing abandoned wells for geothermal energy: current status and future prospects. Renew Energy 194:1288–1302. https://doi.org/10.1016/j.renene.2022.05.138
8. Ziagos J, Phillips BR, Boyd L, Jelacic A, Stillman G, Hass E (2013). A technology roadmap for strategic development of enhanced geothermal systems. https://doi.org/10.2172/1219933
9. Wight NM, Bennett NS (2015) Geothermal energy from abandoned oil and gas wells using water in combination with a closed wellbore. Appl Therm Eng 89:908–915. https://doi.org/10.1016/j.applthermaleng.2015.06.030
10. Śliwa T, Rosen MA (2015) Natural and artificial methods for regeneration of heat resources for borehole heat exchangers to enhance. Sustainability 7:13104–13125. https://doi.org/10.3390/su71013104
11. Śliwa T, Kruszewski M, Zare A, Assadi M, Sapińska-Śliwa A (2018) Potential application of vacuum insulated tubing for deep borehole heat exchangers. Geothermics 75:58–67. https://doi.org/10.1016/j.geothermics.2018.04.001
12. Zhu Y, Li K, Liu C, Mgijimi MB (2019) Geothermal power production from abandoned oil reservoirs using in situ combustion technology. Energies 12:4476. https://doi.org/10.3390/en12234476
13. Sun X, Liao Y, Wang Z, Sun B (2019) Geothermal exploitation by circulating supercritical CO2 in a closed horizontal wellbore. Fuel 254:115566. https://doi.org/10.1016/j.fuel.2019.05.149

14. Wang CL, Cheng WL, Nian YL, Yang L, Han B, Liu M (2018) Simulation of heat extraction from CO2-based enhanced geothermal systems considering CO2 sequestration. Energy 142:157–167. https://doi.org/10.1016/j.energy.2017.09.139
15. Wang K, Yuan B, Ji G, Wu X (2018) A comprehensive review of geothermal energy extraction and utilization in oilfields. J Petrol Sci Eng 168:465–477. https://doi.org/10.1016/j.petrol.2018.05.012
16. Wang X, Liu D, Zhu T, Zhu T, Li F, Wang D, Yang Z, Lin R (2023) An optimization of acquiring geothermal energy from a deep geothermal well by downhole coaxial open loop (DCOL). Pet Sci Technol. https://doi.org/10.1080/10916466.2023.2183964
17. Mondal S, De S (2015) CO2 based power cycle with multi-stage compression and intercooling for low temperature waste heat recovery. Energy 90:1132–1143. https://doi.org/10.1016/j.energy.2015.06.060
18. Ho T, Mao SS, Greif R (2012) Comparison of the organic flash cycle (OFC) to other advanced vapor cycles for intermediate and high temperature waste heat reclamation and solar thermal energy. Energy 42:213–223. https://doi.org/10.1016/j.energy.2012.03.067
19. Mondal S, Alam S, De S (2018) Performance assessment of a low grade waste heat driven organic flash cycle (OFC) with ejector. Energy 163:849–862. https://doi.org/10.1016/j.energy.2018.08.160
20. Mondal S, De S (2020) Performance assessment of a low-grade heat driven dual ejector vapour compression refrigeration cycle. Appl Therm Eng 179:115782. https://doi.org/10.1016/j.applthermaleng.2020.115782
21. Narayan GP, Sharqawy MH, Lienhard VJH, Zubair SM (2010) Thermodynamic analysis of humidification dehumidification desalination cycles. Desalin Water Treat 16:339–353. https://doi.org/10.5004/dwt.2010.1078
22. Mondal S, De S (2022) Polygeneration systems in industry. Polygeneration systems design, process and technologies. pp 411–430. https://doi.org/10.1016/B978-0-12-820625-6.00002-5
23. Xin S, Liang H, Hu B, Li K (2012) Electrical power generation from low temperature co-produced geothermal resources at Huabei oilfield. In Thirty seventh workshop on geothermal reservoir engineering. Stanford University, Stanford, California, p 6
24. Singh HK (2020) Geothermal energy potential of Indian oilfields. Geomech Geophys Geoenerg Georesour 6:19. https://doi.org/10.1007/s40948-020-00148-y
25. Cheng W-L, Li T-T, Nian Y-L, Wang C-L (2013) Studies on geothermal power generation using abandoned oil wells. Energy 59:248–254. https://doi.org/10.1016/j.energy.2013.07.008
26. Zhong C, Xu T, Gherardi F, Yuan Y (2022) Comparison of CO2 and water as working fluids for an enhanced geothermal system in the Gonghe Basin, Northwest China. Gondwana Res https://doi.org/10.1016/j.gr.2022.05.014
27. Yang Y, Huo Y, Xia W, Wang X, Zhao P, Dai Y (2017) Construction and preliminary test of a geothermal ORC system using geothermal resource from abandoned oil wells in the Huabei oilfield of China. Energy 140:633–645. https://doi.org/10.1016/j.energy.2017.09.013
28. Moustafa AM, Shehata AS, Shehata AI, Hanafy AA (2022) Reuse of abandoned oil and gas wells for power generation in western dessert and gulf of Suez fields of Egypt. Energy Rep 8:1349–1360. https://doi.org/10.1016/j.egyr.2022.07.067
29. Nian Y-L, Cheng YL (2018) Evaluation of geothermal heating from abandoned oil wells. Energy 142:592–607. https://doi.org/10.1016/j.energy.2017.10.062
30. Liu X, Falcone G, Alimonti C (2018) A systematic study of harnessing low-temperature geothermal energy from oil and gas reservoirs. Energy 142:346–355. https://doi.org/10.1016/j.energy.2017.10.058
31. Sahana C, De S, Mondal S (2021) Integration of CO2 power and refrigeration cycles with a desalination unit to recover geothermal heat in an oilfield. Appl Therm Eng 189:116744. https://doi.org/10.1016/j.applthermaleng.2021.116744
32. Li T, Xu Y, Wang J, Kong X, Zhu Z (2018) Poly-generation energy system driven by associated geothermal water for oilfield in high water cut stage: a theoretical study. Geothermics 76:242–252. https://doi.org/10.1016/j.geothermics.2018.07.013

33. Li T, Liu Q, Xu Y, Dong Z, Meng N, Jia Y, Qin H (2021) Techno-economic performance of multi-generation energy system driven by associated mixture of oil and geothermal water for oilfield in high water cut. Geothermics 89:101991. https://doi.org/10.1016/j.geothermics.2020.101991
34. Reinhardt T, Johnson LA, Popovich N (2011) Systems for electrical power from coproduced and low temperature geothermal resources. In Thirty-sixth workshop on geothermal reservoir engineering. Stanford University, Stanford, California, p 5
35. ElectraTherm (2011) Mississippi Oilfield Generates Low-Temperature Emission Free Geothermal Energy at the Wellhead
36. Gosnold W (2015) Electric power generation from low to intermediate temperature resources. https://doi.org/10.2172/1186077

Chapter 9
Distributed Renewables, Smart Solutions: A Blueprint for Sustainable Buildings

Kundayi Muchafangeyi and Harpreet Kaur Channi

Abstract Renewable energy and smart technology are crucial to sustainable and eco-friendly structures, as the chapter explains. This comprehensive plan empowers building owners, architects, energy managers, and legislators to use distributed renewables and smart solutions for environmental, energy, and financial benefits. Distributed renewables, their benefits and drawbacks, and their relevance and principles are covered in this chapter. Next, it discusses distributed renewables' crucial role in sustainable buildings, including environmental, economic, and green building regulations. As smart technologies drive contemporary sustainability, the chapter examines those in detail. Smart buildings need IoT, building automation, data analytics, and energy management systems, which readers will learn about. Solar photovoltaics, wind power, battery storage, and microgrid solutions are extensively covered in the distribution of renewables in sustainable buildings. The guidance also emphasizes energy efficiency in building design and retrofitting. Examining building energy management, lighting control, HVAC automation, and security systems highlights smart building automation. With real-world case studies, data and analytics optimize energy use and achieve sustainability targets. This chapter addresses funding distributed renewable projects and using government incentives and rebates. To aid decision-making, ROI and payback times are explained. Addressing grid integration, cybersecurity, and implementation challenges, solutions are suggested. Future sustainable building themes include renewable technology, blockchain's involvement in energy trading, smart cities, and sustainable building certification standards. This chapter helps develop sustainable buildings using distributed renewables and smart technology. Knowledge and a commitment to a sustainable and resilient future are shared.

Keywords Smart technologies · Energy efficiency · Smart building · Energy technology advancements · Sustainable technologies

K. Muchafangeyi · H. K. Channi (✉)
Department of Electrical Engineering, Chandigarh University, Mohali, India
e-mail: harpreetchanni@yahoo.in

K. Muchafangeyi
e-mail: muchafangeyik@gmail.com

S. De et al. (eds.), *Challenges and Opportunities of Distributed Renewable Power*, Energy, Environment, and Sustainability, https://doi.org/10.1007/978-981-97-1406-3_9

9.1 Introduction

In an era marked by climate change concerns and a growing commitment to reducing carbon footprints, the transformation of our built environment into sustainable structures has emerged as an imperative goal. Central to this transformation is the concept of "Distributed Renewables, Smart Solutions: A Blueprint for Sustainable Buildings". As we grapple with transitioning away from fossil fuels, embracing clean energy sources, and optimising energy consumption, this blueprint provides a strategic framework for achieving these critical objectives [1].

Sustainable buildings are the cornerstone of a greener future, and they hold the potential to significantly mitigate the environmental impact of urbanization and industrialization. The integration of distributed renewable energy sources, such as solar panels, wind turbines, and small-scale hydroelectric systems, into building designs has become a linchpin in this endeavor. Simultaneously, the implementation of smart energy management solutions, encompassing advanced technologies like Internet of Things (IoT) sensors, energy analytics, and automation, empowers buildings to intelligently harness, store, and allocate energy resources [2]. This introduction sets the stage for a comprehensive exploration of how distributed renewables and smart solutions converge in the blueprint for sustainable buildings. By examining the synergy between renewable energy generation and intelligent energy management, we illuminate a path towards buildings that reduce their carbon footprint and enhance efficiency, resilience, and economic viability. In the following chapters, we will delve deeper into the strategies, benefits, and challenges associated with this transformative approach, ultimately demonstrating that sustainable buildings are not just a vision for the future but a blueprint we can realize today [3].

A smart building is an intelligently designed and connected structure that employs sensors, automation, and data analytics to optimize its operations, enhance occupant comfort, and reduce energy consumption as shown in Fig. 9.1. These buildings utilize sensor networks to monitor environmental conditions like temperature, occupancy, and lighting, automatically adjusting systems such as HVAC and lighting for efficiency. They prioritize energy sustainability through the integration of renewable sources, data-driven energy management, and real-time analytics. Smart buildings also enhance security with integrated access control and surveillance systems, offering a safe and efficient environment. Overall, smart buildings aim to improve functionality, reduce operating costs, and provide a better experience for occupants in a connected, data-driven ecosystem [2, 3].

Sustainability is now a worldwide necessity, not a slogan. Climate change, resource constraint, and urbanization need structures to adapt. Traditional construction and energy management concepts fail. Sustainable structures are the foundation of a sustainable future. The present work is intended to offer a complete overview of the fundamental components that make sustainable buildings a reality. It introduces distributed renewables and their potential to change our energy landscape. It then examines the environmental, economical, and resilience impacts of distributed renewable sources in sustainable buildings. Smart solutions are then discussed,

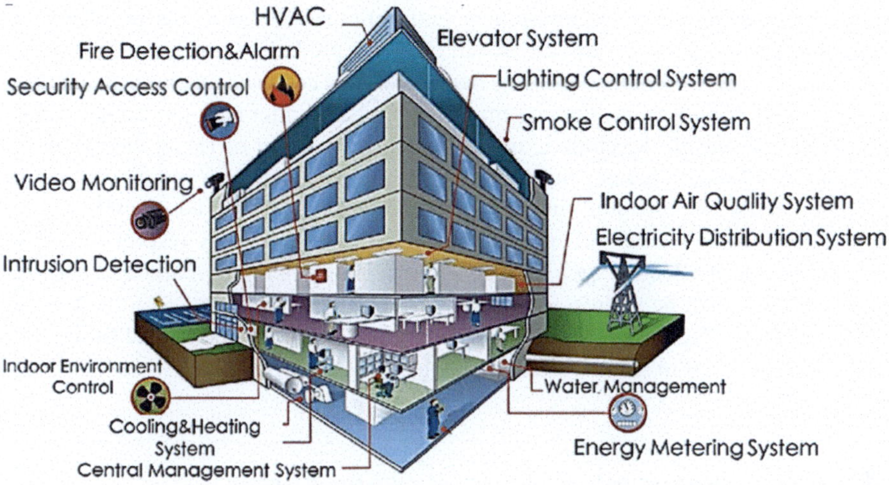

Fig. 9.1 Smart building (CC BY 4.0) [2]

showing how cutting-edge technology may make buildings more efficient, intelligent, and adaptable. It includes IoT, data analytics, energy management, and more. The handbook emphasizes real-world application of these ideas through case studies and successful implementations. Sustainable practices are economically viable by considering funding sources and return on investment.

With increasing urbanization and the need to curb emissions of greenhouse gases, smart building energy management has become more important. Since conventional structures are responsible for significant worldwide energy consumption, improving building energy efficiency is crucial to achieving global sustainability goals. A smart building uses a network of sensors, computers, and other components to collect and analyze data in real-time to make informed decisions. Smart buildings optimize energy use by coordinating several technologies to respond to variable factors like occupancy rates, weather, and energy costs [4].

The potential for significant reductions in energy usage and carbon footprints is one of the main advantages of energy management in smart buildings. Smart buildings are equipped with technology that monitors and controls lighting, heating, cooling, and ventilation to ensure that just the amount of energy required is used. This helps with the worldwide efforts to save energy, protect the environment, and save building owners money [5].

Energy management in smart buildings benefits occupants' comfort, well-being, and productivity beyond only its effect on the environment. With the help of IoT devices and smart sensors, interior settings may be managed individually, considering factors such as comfort, energy efficiency, and health. Smart buildings enhance the experience of their residents by providing environments that are both healthier and more pleasant than conventional structures. In addition, the knowledge acquired through analytics on continuously collected data allows facility managers to do preventative maintenance on equipment and anticipate breakdowns [6]. Reduced energy use and upkeep costs directly result from this predictive maintenance strategy, shortening the period when building systems are out of commission. Energy management in smart buildings has the potential to completely alter how our built environment is conceived of, developed, and managed in the face of rising energy needs and fears of climate change. Smart buildings are essential to a more sustainable, environmentally friendly, and future-proof society because they adopt cutting-edge technological innovations and are committed to environmentally responsible practices. However, challenges such as data security, interoperability, and early investment expenses must be addressed if smart building energy management is to reach its full potential. Together, we can create a greener and more sustainable future, and energy management in smart buildings is where it all starts [7].

9.1.1 Key Components of Smart Building Energy Management Systems

There are several parts to an intelligent building's energy management system, and they all collaborate to improve the building's efficiency and functionality [8]. Some of these most important parts are:

- **Building Automation Systems (BAS)**: BAS is essential for efficient energy management in buildings. Heating, Ventilation, Air Conditioning (HVAC), lighting, security, and access control are just some of the building systems that this system may integrate and control. Through BAS's centralized monitoring and remote management, building managers may optimize energy use in response to changes in occupancy, weather, and other factors.
- **Internet of Things (IoT) Devices and Sensors**: The building is outfitted with Internet of Things (IoT) sensors and gadgets to track things like occupancy, temperature, humidity, and lighting levels in real-time. These tools allow for informed, data-driven decisions on optimizing a building's energy use.
- **Data Analytics and Artificial Intelligence (AI)**: Data analytics and AI technologies process massive amounts of data generated by IoT devices and sensors. Patterns in energy use may be identified, trends in energy usage predicted, and effective energy conservation measures suggested using AI algorithms. Algorithms trained with machine data continually optimize a building's operation in response to shifting environmental and occupant demands.
- **Energy Monitoring and Metering**: Modern energy monitoring and metering technologies are at the heart of smart building energy management systems. Managers of commercial buildings may keep tabs on energy use in real-time with the help of real-time energy monitoring systems. Sub-metering collects data on electricity use in discrete zones or by individual machinery, yielding detailed information that may inform efficient energy conservation measures.

- **Energy Efficiency Measures**: Energy management systems for smart buildings include several strategies for conserving power. Examples are smart windows that change their transparency in response to environmental factors, occupancy-based HVAC management, and efficient operation of power-hungry machinery.
- **Energy Storage and Battery Systems**: Using batteries and other energy storage technologies in smart buildings is rising. They make it possible to save energy from renewable resources, such as solar panels, that are produced in excess and utilized later when demand is high or renewable output is low. This allows for more efficient control of energy use and distribution.
- **Demand Response and Load Management**: Structures may participate in demand-side management initiatives through demand response and load management methods. Smart buildings may aid grid stability and get financial incentives for energy savings by altering energy use during peak demand times or in response to system signals.
- **Building Energy Management Platforms**: The building energy management platform is Central to the administration of all the smart building features. Accessing real-time energy data, customizing settings, and gaining insights into energy performance are all made possible by these systems' intuitive user interfaces.
- **Renewable Energy Integration**: Renewable energy sources, such as solar panels, wind turbines, and geothermal systems, are often included in intelligent building energy management systems. Using renewable energy decreases the structure's reliance on fossil fuels and carbon impact.

Smart building energy management systems provide a unique method to optimize energy use, lower expenses, and make a more sustainable future by including these important components [9–11].

9.2 Literature Review

Liu et al. (2021) offered an additional technique for boosting the capacity of photovoltaic cells to generate power when incorporated into the structure's windows. In this study, a thermotropic hydrogel membrane based on hydroxypropyl cellulose (HPC) was made and specified for a BICPV smart window system. The system was built with a verified wavelength-dependent optical model based on the Monte-Carlo ray-tracing approach. The visual and thermal characteristics of the thermotropic layer were collected via experimental measurement and applied to the optical model prediction [12]. Kaixin Lin et al. (2021) suggested that a smart building must reduce the energy used for its operations. A daytime radiative chiller and a thermochromic smart window are key passive and energy-free technologies for building envelopes and facades in this innovative design. As a result, it is possible to drastically reduce the energy used for interior heating, cooling, and lighting. Model homes are developed to assess energy-saving performance [13]. One innovative notion in the paper is the concept of "Autonomous Cycles of Data Analysis Jobs," which asserts that an autonomous management system needs specialized tasks, such as tracking, assessment, and judgment, to attain goals. The job categorization based on the "Autonomous Cycles of Data Analysis Tasks" concept is an innovative aspect of the work. How the material is arranged allows them to position the research and see each subject of inquiry's advantages and disadvantages [14]. A group of academics has suggested a method for managing the operation of air conditioners to reduce energy usage. They employed a deep learning-based people recognition system that counts the number of individuals in a given region using the YOLOv3 algorithm [15]. It is estimated that electrical lighting accounts for a large portion of total energy utilization. In this study, Bright Infrastructure Photo Vision (BIPV) sunlight collecting in an office space might cut energy consumption while improving tenants' visual preferences. The energy savings potential of an artificial lighting system using BIPV as the building's envelope is investigated. Compared to the current utility grid, the levelized cost of energy (LCOE) is shown to be competent [16]. As a research sample, a 7-story structure in Jakarta, Indonesia, was used for this investigation. This study employed both descriptive and analytical techniques. The advanced control system reduces home energy usage by up to 37%, with the lighting sector experiencing the greatest reduction in energy use. An energy consumption decrease of 20% to 27% will be possible using a sophisticated management system [17]. Buildings utilize more than 40% of all power required globally, with warming, airflow, and air conditioning always being the primary energy users. Conventional heating controllers are wasteful because they cannot adapt to changing user preferences and weather trends. A DRL-based smart controller outperforms a traditional thermostat regulator by enhancing thermal comfort by 15 to 30%, reducing energy use [18]. Smart building is one of the most important IoT technology sectors. Support from connected, IoT-level smart buildings is a solution that is both efficient and affordable. Major considerations for commercial space include comfort, accessibility, security, and energy management. IoT-based technologies can naturally meet these needs. Every device has a unique address, and the Constrained Application Mechanism is a crucial web transmission protocol (CoAP). It is an application layer protocol that does not send data through secured channels. Using the CoAP in a smart building saves around 30.86% of energy, less than the Message Queuing Telemetry Transport scenario (MQTT). To boost energy efficiency, a smart building architecture that employs IoT to regulate the performance of all technical components was proposed [19]. Systems are created in the best way to reduce total costs associated with the original capital, operational and maintenance expenses, and other charges. Last, the one-slot-look-ahead (OSLA) optimization strategy works well for handling DERs [20].

9.3 Understanding Distributed Renewables

The goal of greener buildings necessitates an education in distributed renewables. An innovative strategy for producing electricity, distributed renewables have the potential to alter the current energy landscape. This portion of the book provides a firm footing for what comes later by exploring the fundamental ideas and principles that drive the field of distributed renewables. Decentralized renewable energy production, often on a small to medium size and frequently positioned near to the point of consumption, is what is meant by the term "distributed renewables". Distributed renewables, such as solar, wind, and hydroelectric power, generate electricity without using fossil fuels, in contrast to traditional centralized power plants. These distributed infrastructures help us become more energy independent, cut down on carbon dioxide emissions, and the term "distributed renewables" refers to a broad category that includes many different types of renewable energy sources. The most well-known method of converting sunlight into power is via solar photovoltaic (PV) systems. The kinetic energy of wind is converted into electricity by wind turbines. Hydroelectric power plants use the energy of moving water. The diversity of distributed renewables is further shown by the inclusion of biomass and geothermal systemsstrengthen energy security. Distributed renewable energy systems have several benefits. Diversifying energy sources help minimize carbon emissions, lessen the negative effects on the environment, and increase energy security. It also has the potential to provide considerable savings in the long run. To fully take use of distributed renewables, however, we must overcome obstacles including intermittency, grid integration, and early costs. The energy used in buildings accounts for around one third of the global final energy demand, yet renewables are progressing only slowly in the sector (only 14.7% of energy use in buildings), representing only a 4 percentage-point increase from a decade before.

9.3.1 Harnessing Distributed Renewable Power for Sustainable Buildings

Harnessing distributed renewable power for sustainable buildings involves a multifaceted approach that integrates various technologies and strategies to optimize energy generation, distribution, and consumption within a building. Here are key steps and methods for achieving this goal:

- **Assess Energy Needs**: Conduct an energy audit to understand the building's energy consumption patterns and requirements.
- **Evaluate Renewable Resources**: Identify the most suitable renewable energy sources for your location, such as solar, wind, hydro, or geothermal energy, based on resource availability and building design.
- **Install Renewable Energy Systems**: Invest in renewable energy systems like solar panels, wind turbines, or micro-hydropower systems to generate electricity

on-site. These systems should be appropriately sized to meet the building's energy demand.

- **Energy Storage**: Implement energy storage solutions, such as batteries, to store excess energy generated during periods of high production for use during periods of low production or high demand.
- **Microgrid Integration**: Consider integrating the building into a local microgrid to enhance energy resilience and enable energy sharing with neighboring buildings.
- **Energy Efficiency Measures**: Implement energy-efficient building design and technologies, including LED lighting, efficient HVAC systems, and smart thermostats, to reduce overall energy consumption.
- **Smart Building Management**: Deploy smart building automation systems that monitor and control energy use in real-time. These systems can optimize lighting, heating, cooling, and other systems based on occupancy and weather conditions.
- **Demand Response**: Participate in demand response programs to adjust energy consumption during peak demand periods and take advantage of incentives offered by utility companies.
- **Grid Interaction**: Establish a connection with the electrical grid to sell excess energy back to the grid (net metering) or draw power from the grid when renewable generation is insufficient.
- **Data Analytics and AI**: Utilize data analytics and artificial intelligence to analyze energy usage data and make predictive decisions for optimizing energy consumption and renewable energy generation.
- **Maintenance and Monitoring**: Regularly monitor and maintain renewable energy systems to ensure optimal performance and longevity.
- **Compliance and Regulations**: Comply with local regulations, building codes, and safety standards for renewable energy installations and energy efficiency measures.
- **Education and Engagement**: Educate building occupants and stakeholders about the benefits of distributed renewable power and engage them in energy-saving practices.
- **Continuous Improvement**: Continuously seek ways to improve energy efficiency and renewable energy generation through technology upgrades and process improvements.
- **Financial Analysis**: Conduct a financial analysis to assess the return on investment (ROI) of your distributed renewable power and sustainability initiatives.
- **Environmental Impact**: Consider the environmental benefits of your efforts, including reduced greenhouse gas emissions and resource conservation.

Harnessing distributed renewable power for sustainable buildings requires a holistic and integrated approach that combines renewable energy generation, energy efficiency, smart technologies, and a commitment to long-term sustainability goals. This approach benefits the environment and can lead to cost savings and increased energy resilience for the building and its occupants.

9.4 Energy Monitoring and Metering in Smart Buildings

Effective energy management in smart buildings relies heavily on accurate monitoring and metering. Data on energy usage in buildings may be gleaned via these methods, which include constant monitoring and analysis. Smart buildings can optimize energy efficiency and save waste thanks to real-time monitoring and granular metering made possible by cutting-edge technology [21]. The following are some of the most important components of smart building energy monitoring and metering:

- **Real-Time Energy Monitoring**: The lighting, heating, ventilation, appliance, and outlet loads in a smart building are only a few examples of systems and equipment whose energy usage may be monitored in real time by sensors and meters installed throughout the structure. Building managers may then use this information to track energy use in real-time and spot patterns or outliers by sending the data to centralized energy management systems.
- **Smart Meters and Sub-metering**: Whole-building energy usage may be monitored using smart meters. The energy consumption of a building may be broken down into parts, such as rooms, floors, or even pieces of equipment, using sub-metering. Sub-metering provides more detailed information on energy use, which is useful for developing efficient energy-saving initiatives and pinpointing high-energy-use zones.
- **Energy Auditing and Benchmarking**: Data from energy monitoring systems is required to undertake energy audits and measure performance against energy efficiency criteria. Smart building owners may compare their energy performance to comparable buildings or industry norms by conducting an audit and then benchmarking.
- **Identifying Energy Usage Patterns**: Smart buildings can monitor energy use to determine peak demand times and usage trends. If building managers know when and how much energy is being used, they may take steps to minimize consumption and expenses at peak times via demand response methods, load shifting, and energy conservation.
- **Fault Detection and Diagnostics**: Rapid troubleshooting of building systems is made possible by real-time continuous energy monitoring. Issues like equipment breakdowns or leaks may be quickly remedied and energy waste avoided with the help of automated fault detection and diagnosis systems.
- **Occupant Engagement and Awareness**: To raise awareness and inspire energy-efficient behaviour, data from energy monitoring systems may be presented to a building's inhabitants in an approachable style. Building occupants may actively engage in energy saving by making educated decisions thanks to visualizations, real-time feedback, and reports on energy consumption.
- **Integration with Energy Management Systems**: The smart building's energy management system incorporates monitoring and metering data. Combined with data analytics and AI algorithms, this knowledge allows the system to optimize energy use by making real-time adjustments to the building's systems.

- **Cost Allocation and Billing**: Energy metering aids in the equitable and accurate distribution of utility expenses in buildings with many tenants or shared energy use. This may provide an incentive for residents to reduce their energy use.

Measurement and tracking of used energy are essential features of intelligent building energy management systems. Modern buildings may be operated more sustainably and ecologically responsible, thanks to the information provided by these systems [22–24].

9.4.1 Energy Auditing and Benchmarking

Auditing and comparing buildings' energy performance to others is crucial to smart building energy management. They are helpful because they reveal hidden inefficiencies in a building's energy use and allow for comparisons to norms or other structures of a similar kind [25].

9.4.1.1 Energy Auditing

An energy audit is an in-depth evaluation of a facility's energy consumption habits, energy-using systems, and overall energy performance. The following are the usual stages involved:

- **Data Collection**: Collecting information on energy use, building features, occupant habits, and machine specs.
- **On-Site Inspection**: Inspection of the building's insulation, windows, lights, heating, ventilation, air conditioning (HVAC), and other energy users.
- **Analysis and Evaluation**: The gathered information will be used to examine typical behaviours, spot instances of energy waste, and evaluate the effectiveness of various building technologies.
- **Recommendations**: Energy auditors evaluate a building's energy use and efficiency and suggest upgrades and changes based on their findings.
- **Cost–Benefit Analysis**: Estimating possible energy and cost savings and the economic viability of adopting the suggested changes.

There are a variety of levels at which an energy audit may be performed, from a basic walk-through to a more in-depth level II or level III examination. To optimize energy use, building owners and managers may learn a great deal by doing one of these audits [4].

9.4.1.2 Benchmarking

A building's energy performance is measured and compared to that of other comparable structures or predetermined industry norms through benchmarking. To benchmark energy use against similar buildings and establish goals for improvement [26]. Standard procedures for conducting benchmarks consist of the following:

- **Data Collection**: Information on the building's use of electricity, gas, and water sources is gathered.
- **Normalization**: Energy use data may be comparable if adjusted for building size, occupancy, and climate.
- **Comparison**: Assessing the building's energy efficiency concerning other structures of a similar kind or to industry standards.
- **Analysis and Improvement**: Comparing the building's energy performance against benchmarks and developing plans to fix the gaps found [27].

9.5 Implementing Energy Efficiency Measures in Smart Buildings

Smart building energy efficiency is essential for sustainable and cost-effective operations. Many tactics and technologies are used to reduce energy usage while preserving or improving occupant comfort and productivity. Advanced sensors, automation, and data analytics improve energy use in real-time in smart buildings. Smart lighting systems reduce power waste by adjusting brightness depending on natural light and occupancy [28]. Smart building HVAC systems adjust to changing circumstances to manage temperature and ventilation while saving energy. Another important factor is energy-efficient building design. Smart buildings use high-performance insulation, efficient windows, and natural lighting and ventilation to save energy. Smart meters and sub-metering offer exact energy use tracking down to specific devices or systems, identifying opportunities for improvement. Continuous monitoring and data analysis are essential for energy efficiency. AI algorithms can spot trends and abnormalities, enabling building managers optimize systems and fix energy hogs [29]. Machine learning-enabled predictive maintenance optimizes equipment performance, saving energy and prolonging asset life. In addition to technological factors, tenant behavior affects energy efficiency. Smart buildings utilize interactive displays and real-time feedback to educate residents about energy consumption. This lets users make energy-saving decisions [30–32]. Smart buildings' energy saving strategies are ecologically and economically advantageous. Lower running costs, carbon emissions, and a more sustainable and pleasant building environment result. Smart technology and techniques are changing how we construct and maintain contemporary buildings as energy-efficient building demand rises. Optimizing energy usage and improving smart buildings' overall performance requires energy efficiency strategies. Smart building owners and managers use cutting-edge technology and tactics to reduce energy waste and operating expenses while providing tenants with more sustainable and pleasant spaces [33–36]. Some of the most important energy-saving features found in modern smart buildings as shown in Fig. 9.2.

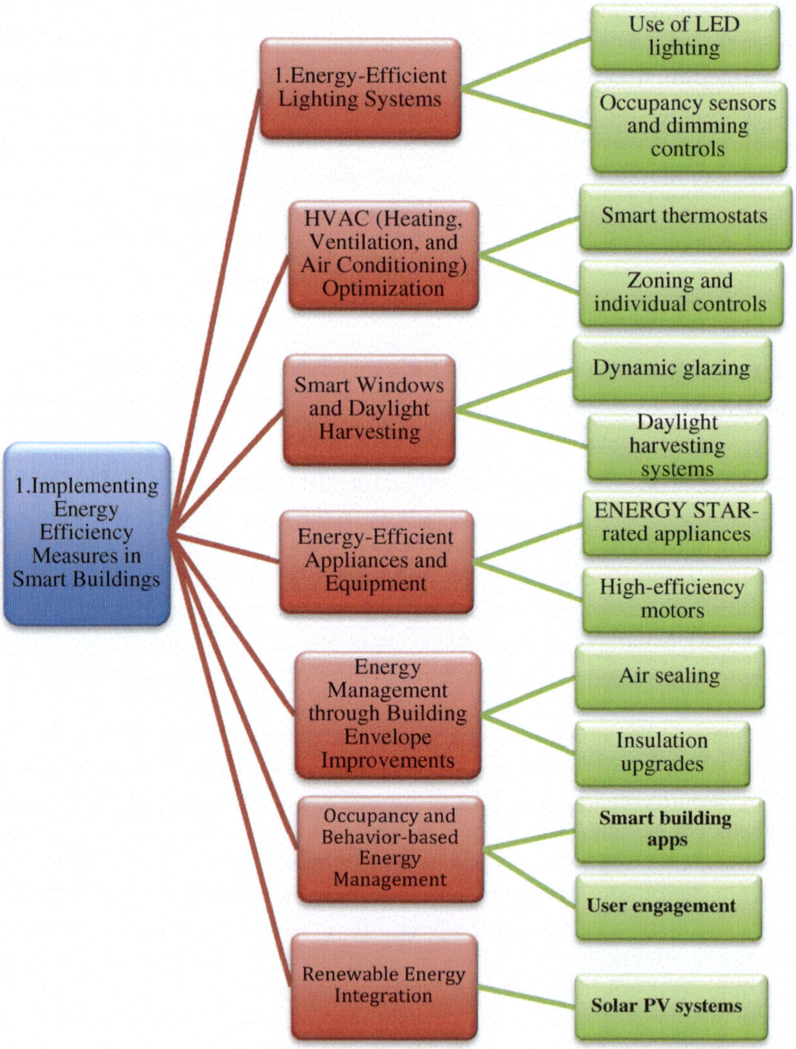

Fig. 9.2 Implementing energy efficiency measures in smart buildings

9.6 Advanced Energy Management Strategies

To maximize energy efficiency and sustainability, smart buildings use cutting-edge energy management systems that employ advanced technology and data analytics in novel ways. These approaches go above and beyond conventional methods by allowing structures to react intelligently to changing environmental conditions, occupant patterns, and external influences. Figure 9.3 depicts some advanced energy management solutions that will be addressed shortly [37].

- **Predictive Analytics and AI**: To anticipate energy consumption trends, identify abnormalities, and optimize building operations in real-time, cutting-edge energy management systems use predictive analytics and artificial intelligence. Machine learning algorithms constantly examine data to spot patterns and make preventative modifications, leading to more effective utilization of resources and less waste.
- **Demand Response and Load Shifting**: By modifying energy use during peak demand hours or in response to grid signals, smart buildings may participate in demand response programs. Load shifting entails moving power-hungry activities to off-peak hours to alleviate grid pressure and reduce utility bills.
- **Energy Storage and Battery Management**: Smart buildings may store surplus energy from renewable sources like solar panels by integrating energy storage technologies like batteries. Increased energy independence and grid stability may result from using the stored energy during peak demand or when renewable supply is constrained.

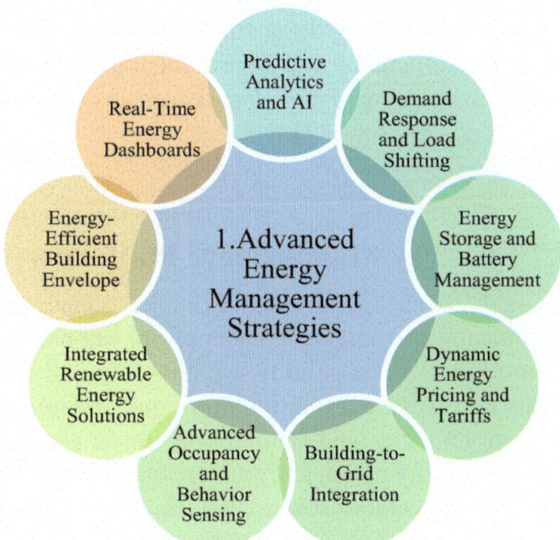

Fig. 9.3 Advanced energy management strategies

- **Dynamic Energy Pricing and Tariffs**: Smart buildings can respond to changes in energy costs by using advanced energy management systems that can track real-time energy prices and tariffs. Building owners may then make cost-effective choices concerning energy use throughout the day.
- **Building-to-Grid Integration**: Besides being just electricity consumers, "smart" structures may also contribute by producing more power and sending it back into the grid. The benefits of this two-way communication improved grid stability and more renewable energy source integration [38].
- **Advanced Occupancy and Behavior Sensing**: Smart buildings may adapt their energy use to match the tastes of their occupants in real-time by using sophisticated occupancy sensors and behaviour monitoring systems. One way to drastically reduce energy use is to turn on lights, air conditioners, and other appliances only in the rooms where people are present.
- **Integrated Renewable Energy Solutions**: The coordinated use of many renewable energy sources enhances energy efficiency in smart buildings. These sources may include solar, wind, and geothermal. By maximizing the use of renewable power, this holistic strategy helps cut down on the need for fossil fuels.
- **Energy-Efficient Building Envelope**: By lowering heat input or loss and the consequent demand for additional heating or cooling, energy efficiency is increased when modern building materials and designs are used.
- **Real-Time Energy Dashboards**: Energy dashboards that are both user- and administrator-friendly are a common feature of today's smart buildings. These dashboards give real-time data on energy use, allowing for more calculated decision-making and encouraging residents to live more sustainably [39–41].

Smart buildings may reduce their carbon footprint, save money, and help the environment by adopting these cutting-edge energy management practices. In the quest for energy efficiency and environmental stewardship, these solutions are invaluable thanks to the synergy of data-driven decision-making, advanced technology, and tenant participation [42].

9.7 Energy Management Platforms and Software Solutions

Optimization of energy consumption, improvement of operational efficiency, and promotion of sustainability are now impossible without energy management platforms and software solutions for smart buildings. Combining data analytics, real-time monitoring, and automation, these cutting-edge solutions provide building owners, facility managers, and tenants the information and power they need to make their buildings more sustainable. Energy management systems provide a center for collecting and analyzing data from a wide variety of sensors and devices installed throughout a structure [43–45]. Among the many variables covered by this data is power use, heating and cooling levels, lighting levels, and occupancy rates. These systems can detect trends, outliers, and potential energy savings thanks to the use of clever algorithms and machine learning. These systems' primary value is that they reveal energy use in real time, allowing for instantaneous reactions to changes in demand or equipment efficiency. When motion detectors detect that a room is empty, the system may turn off the lights and the air conditioner. Equally useful for lowering energy prices and avoiding system overloads, demand response solutions may be used during peak demand times to minimize power usage. Predictive maintenance relies heavily on energy management software solutions. By closely monitoring everything, they may catch inefficiencies or impending failures before they cause major problems. This preventative maintenance method not only reduces costs associated with repairs and replacements but also increases the longevity of essential building systems. Building managers may monitor energy performance over time, establish objectives for energy reduction, and evaluate their progress toward sustainability with the help of the in-depth information and analytics provided by these systems. This data-driven strategy not only helps with meeting code requirements but also encourages lower energy use overall. In conclusion, energy management systems and software solutions are crucial to converting older structures into modern, eco-friendly ones. In addition to lowering operating costs and carbon footprints, they allow for more precise energy use monitoring and analysis. The importance of these solutions in the administration of smart buildings is expected to grow as the need for environmental friendliness and economic prudence increases [46, 47].

9.7.1 Advancements in AI and Machine Learning

Smart buildings have entered a new age of intelligence and efficiency because of the developments in AI (Artificial Intelligence) and machine learning. Comfort, sustainability, and efficiency in building operations are all improving as a consequence of these innovations in technology. With the use of artificial intelligence (AI), smart buildings may have their HVAC, lighting, and security systems monitored and controlled in real time by analyzing massive volumes of data from sensors and devices. By learning from their surroundings and the tastes of their users, machine learning algorithms may reduce energy use without compromising comfort [48–50]. By monitoring and adjusting temperature, ventilation, and lighting based on predicted occupancy levels, they can cut down on energy waste in vacant spaces. Predictive maintenance powered by AI is also changing the way infrastructure in buildings is maintained. These systems are able to identify abnormalities and early symptoms of equipment deterioration by continually evaluating data from equipment and sensors. Because of this, building systems experience less downtime and need fewer repairs over time. Improvements in AI have also improved the quality of life for building occupants. Lighting, temperature, and other aspects of the atmosphere in smart buildings may be customized using AI to meet the needs of the occupants. Occupants' ability to interact and get things done is boosted by voice-activated assistants and chatbots powered by artificial intelligence. When applied to the larger topic of sustainability, AI and machine learning may evaluate data from the past to spot patterns and maximize efficiency. By enabling predictive energy purchase, integrating renewable energy sources, and supporting demand response tactics, these systems may aid buildings in lowering their carbon footprint and energy expenditures. In conclusion, smart buildings that use AI and ML are revolutionizing our understanding of what it means for a structure to be intelligent. In addition to helping achieve environmental objectives, these technologies are also making buildings more flexible, energy-efficient, and user-focused [50]. Advances in artificial intelligence and machine learning promise greener, more efficient structures in the near future. Important developments in artificial intelligence and machine learning for building energy management are shown in Fig. 9.4.

These AI and ML improvements in energy management allow for smarter, more adaptable, and more energy-efficient structures. Smart buildings, which optimize energy use and contribute to global environmental objectives, will play an increasingly important part in developing a sustainable and resilient future as these technologies progress [51].

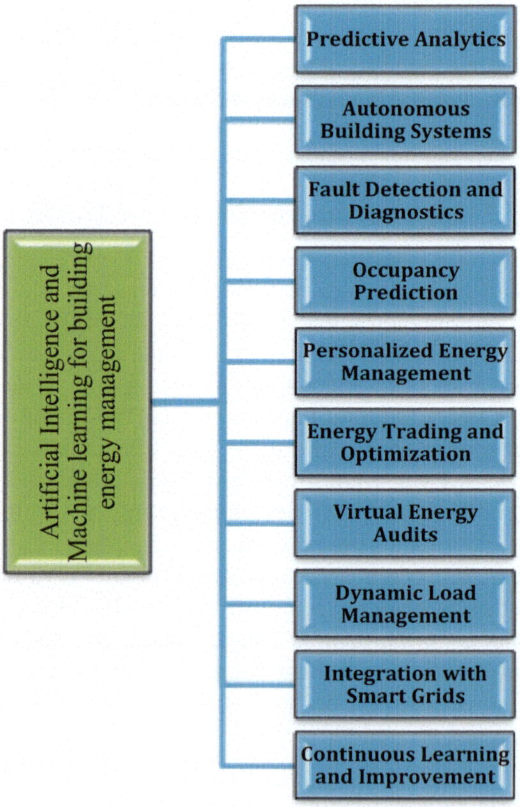

Fig. 9.4 Artificial intelligence and machine learning for building energy management

9.8 Smart Grid Integration and Demand-Side Management

Together, smart grid integration and demand-side management form the bedrock of today's energy infrastructure, allowing for the development of power grids that are more efficient, dependable, and sustainable. Two-way communication between power companies and their customers is made possible by a smart grid, an enhanced electrical distribution system that uses digital connectivity, intelligent devices, and data analytics. Since power flow can be monitored and controlled in real-time, renewable energy sources, demand response programs, and demand-side management methods may be easily integrated [52]. Optimizing the use of renewable energy sources is greatly aided by the use of smart grid technologies. Variability is an intrinsic property of renewable energy sources like solar and wind power. The smart grid can accommodate these variations in power production because of its adaptability. The smart grid can balance supply and demand by gathering real-time data on electricity usage and the availability of renewable energy. It prevents the waste of renewable energy by channelling it to regions with a greater need or storing it in energy storage devices for later use. Therefore, smart grid integration helps raise the proportion of renewable energy in the grid, lessening the need for fossil fuels and reducing greenhouse gas emissions [53]. By concentrating on ways to reduce energy waste on the consumer end, demand-side management is a useful complement to smart grid integration. It entails a variety of tactics for changing energy use in response to changes in grid circumstances and consumer demand. Demand-side management isn't complete without demand-response programs. With the help of demand response programs, power consumption may be reduced or moved to less busy times of the day. In exchange for limiting their energy use during peak hours, consumers may actively join these programs and get incentives or reduced power costs. Demand-side management helps ease grid stress, avert power outages, and postpone the installation of expensive peak-load infrastructure by actively controlling energy use [54].

Demand-side management includes energy efficiency. LED lighting, smart appliances, and building insulation cut power use without affecting comfort or productivity. Another demand-side management method, time-of-use pricing, encourages users to alter their power consumption based on energy costs. Off-peak power use lowers rates and balances the system. Smart grid integration and demand-side management synergistically improve energy system reliability and resilience. The smart grid's real-time monitoring and data analytics let utilities make educated choices and react quickly to system disruptions and crises. Demand-side management also encourages energy saving and environmental responsibility by empowering customers. These efforts create a more sustainable, efficient electrical ecology for future generations [55].

9.8.1 Net-Zero Energy Buildings

Sustainable and energy-efficient net-zero energy buildings (NZEBs) lead the way. These ingenious constructions create as much energy as they use annually, creating a net-zero energy balance. Net-zero power requires energy conservation, efficiency, and on-site renewable energy production [56]. NZEBs help reduce climate change, the built environment's environmental effects and energy use.

- **Energy Efficiency**: Energy efficiency drives NZEB design. These buildings reduce energy losses through sophisticated insulation, glass, and HVAC systems. Smart building controls, LED lighting, and energy-efficient appliances optimize energy use.
- **Renewable Energy Integration**: Renewable energy systems power NZEBs. PV panels turn sunshine into power in NZEBs. Depending on the building's location and resources, wind turbines, geothermal systems, and small-scale hydropower may also be included.
- **Energy Monitoring and Management**: NZEBs need energy monitoring and management systems. To maximize building efficiency, real-time energy data is gathered and analyzed. Real-time energy use is optimized via smart building controls and automated systems.
- **Passive Design Strategies**: NZEBs use passive design. Building orientation, shade, natural ventilation, and thermal mass passively manage inside temperatures. Passive solar architecture heats in winter and cools in summer.
- **Energy Storage Solutions**: Energy storage helps NZEBs reconcile intermittent renewable energy supply with variable energy demand. Battery storage devices store extra solar and wind energy during low renewable energy output.
- **Grid Interaction and Net Metering**: NZEBs are usually grid-connected. The building may feed extra power to the grid to balance electricity demand during reduced renewable energy supply periods. Net metering provides a net-zero energy balance.
- **Life Cycle Assessment**: NZEBs assess the life cycle environmental effect of building materials and construction: sustainable materials, low-carbon concrete, and recycled content lower the building's embodied carbon footprint.
- **Occupant Engagement**: NZEB energy performance depends on occupant behaviour. Energy conservation, sustainability, and energy-saving practices are promoted to building inhabitants.
- **Economic and Environmental Benefits**: Benefits of NZEBs are several. These buildings cut greenhouse gas emissions, fossil fuel usage, and operational expenses by lowering energy demand and grid interactions. They boost energy security by producing power on-site and minimizing grid vulnerability. NZEBs also serve as examples of sustainable construction, encouraging green building practices, and speeding the transition to a more sustainable and resilient built environment. Net-zero energy buildings will be crucial to establishing a more sustainable and ecologically friendly future as technology advances and renewable energy prices drop [57–59].

The integration of distributed renewables in Indian smart buildings is promising yet developing. Distributed renewables are becoming essential to smart building projects as the country faces rising energy consumption, environmental concerns, and a sustainability push. India's sunlight and favorable government regulations have made solar power popular. Rooftop solar installations have increased in residential, commercial, and industrial buildings due to subsidies and net metering incentives. Grid integration and intermittency management remain difficult. Renewable energy sources strain India's power infrastructure, making smart building technology critical for energy supply and demand. Microgrids and energy storage are being used more to enhance grid dependability and resilience. Distributed renewables in smart buildings are driven by government programs like the Smart Cities Mission and renewable energy objectives. Tax breaks and low-interest loans encourage building owners to invest in these technology. Energy management and monitoring systems improve energy use, reduce waste, and boost efficiency. Distributed renewables support India's climate change efforts. Smart buildings with distributed renewables support national sustainability objectives by lowering fossil fuel usage and greenhouse gas emissions. In conclusion, India's distributed renewables in smart building journey has obstacles and gains. Solar and other renewables are changing the energy landscape, but grid integration remains a challenge. Distributed renewables in smart buildings may improve energy sustainability, carbon emissions, and resilience and efficiency in India as technology and legislation grow.

9.9 Analysis of Distributed Renewables in Smart Building in India

India has made considerable and ambitious strides in adopting distributed renewables in smart buildings. The country has made fast strides in increasing its solar power capacity, with plans to have more than 36 GW of installed solar capacity by 2022 and an even more impressive 175 GW of renewable energy capacity by 2022 and 450 GW by 2030. There has been a dramatic increase in the amount of rooftop solar capacity added, now exceeding 5 GW. Decentralized solar power production has expanded thanks to government incentives like net metering rules. The government's $31 billion Smart Cities Mission is also accelerating the use of distributed renewables and other energy-saving technology in construction. Buildings are responsible for a significant amount of India's total energy usage, hence these programs are crucial. Energy management systems and Internet of Things (IoT) sensors are two examples of smart building technologies that are helping to solve this problem. India plans to cut greenhouse gas emissions and increase its non-fossil fuel capacity to 40% by 2030, and its development of distributed renewables is consistent with these goals. To get there, however, the United States must overcome obstacles like grid integration and intermittency difficulties, which threaten the consistent functioning of distributed renewable energy systems that power smart buildings.

9.9.1 Case Studies

9.9.1.1 The Edge, Amsterdam, Netherlands

It is also known as the "greenest building in the world," The Edge incorporates advanced sustainable features. It uses a smartphone app to allow employees to customize their workspace environment, including lighting and temperature. The Edge has achieved remarkable energy efficiency, with over 28,000 sensors optimizing building operations. It generates energy through rooftop solar panels, stores it in batteries and even shares excess energy with neighboring buildings.

9.9.2 One Bryant Park (Bank of America Tower), New York City, USA

This LEED Platinum-certified skyscraper employs an innovative ice storage system to optimize cooling and reduce energy consumption. It features advanced lighting controls, daylight harvesting, and rainwater harvesting for irrigation. The building has reduced its energy consumption by around 50% compared to similar buildings, saving millions of dollars in operational costs. It serves as a prime example of sustainable high-rise construction.

9.9.3 The Crystal, London, United Kingdom

The Crystal is an iconic sustainable building that showcases Siemens' smart technology solutions. It employs a Building Management System (BMS) to monitor and control energy usage, lighting, and HVAC systems. The building has achieved BREEAM Outstanding certification and is a living example of sustainable urban planning. It hosts exhibitions on urban sustainability and serves as a learning center.

9.9.4 Shanghai Tower, Shanghai, China

Shanghai Tower is one of the tallest skyscrapers globally and features a range of sustainability and smart building technologies, including advanced double-skin facades for natural ventilation, energy-efficient lighting systems, and regenerative elevators. The building is LEED Platinum certified and has reduced energy consumption significantly compared to traditional skyscrapers of its size. It serves as a model for skyscraper sustainability in densely populated urban areas.

9.9.5 Pixel, Melbourne, Australia

Pixel is a sustainable residential development known for its integrated technology solutions. It employs IoT devices to enable residents to control lighting, blinds, heating, and cooling through a smartphone app. Residents have experienced increased comfort and energy savings, with the ability to remotely monitor and control their home systems. Pixel demonstrates the potential for smart home solutions in multi-unit residential buildings.

These case studies highlight the diverse range of technologies and strategies employed in smart buildings, from energy efficiency and sustainable design to occupant comfort and advanced control systems. Each project showcases the benefits of integrating smart technologies into building design and management to create more efficient, sustainable, and user-friendly environments [60–62].

9.10 Conclusion

Ultimately, "Smart Energy Management: Harnessing Distributed Renewable Power for Sustainable Buildings" provides a dynamic and inventive method to revolutionize the built environment and advance sustainability. Smart energy management systems and distributed renewable power solutions may alter creating energy efficiency, climate change, and environmental stewardship. Intelligent technologies help buildings optimize energy use, avoid waste, and adapt to changing energy needs. Real-time energy monitoring, control, and data-driven decision-making reduce operating costs and carbon emissions. Buildings may become energy self-sufficient and resilient by seamlessly integrating dispersed renewable power sources. On-site renewable energy production boosts energy independence, grid stability, and power infrastructure. Demand-response methods and energy storage improve renewable energy use. Smart buildings may dynamically regulate energy usage depending on renewable power availability and store extra energy for low-generation times, guaranteeing a continuous and sustainable energy supply. Smart energy management using distributed renewable power has economic benefits as well as environmental ones. Sustainable construction practices provide lower operating costs, incentives, and protection against growing energy prices. Smart energy management using distributed renewable electricity requires cooperation between governments, corporations, industry professionals, and consumers. Policies, financial incentives, and public awareness are needed to accelerate the transition to sustainable and smart buildings. Smart energy management systems and distributed renewable power will increasingly integrate as technology advances, and renewable energy becomes more available. Sustainable buildings will accelerate, affecting global energy efficiency, carbon reduction, and environmental preservation. Smart Energy Management creates a cleaner, more resilient, and more sustainable built environment. Smart buildings will shape a greener and wealthier future for future generations as they become the norm. This paradigm shift embraces a vision of sustainable living, where buildings actively contribute to a more peaceful cohabitation with the earth, providing a brighter and more sustainable future for everybody.

References

1. Aliero MS, Asif M, Ghani I, Pasha MF, Jeong SR (2002) Systematic review analysis on smart building: challenges and opportunities. Sustainability 14(5):3009
2. https://www.researchgate.net/figure/An-illustration-of-related-systems-in-Buildings-Adapted-from-a-source-picture-on_fig1_326633723. Accessed on 07 Nov 2023
3. Hakimi SM, Hasankhani A (2020) Intelligent energy management in off-grid smart buildings with energy interaction. J Clean Prod 244:118906
4. Mir U, Abbasi U, Mir T, Kanwal S, Alamri S (2021) Energy management in smart buildings and homes: current approaches, a hypothetical solution, and open issues and challenges. IEEE Access 9:94132–94148
5. Ejaz W, Naeem M, Shahid A, Anpalagan A, Jo M (2017) Efficient energy management for the internet of things in smart cities. IEEE Commun Mag 55(1):84–91
6. Zipperer A, Aloise-Young PA, Suryanarayanan S, Roche R, Earle L, Christensen D, Bauleo P, Zimmerle D (2013) Electric energy management in the smart home: Perspectives on enabling technologies and consumer behavior. Proc IEEE 101(11):2397–2408
7. Lee D, Cheng C-C (2016) Energy savings by energy management systems: a review. Renew Sustain Energy Rev 56:760–777
8. Sartori I, Hestnes AG (2007) Energy use in the life cycle of conventional and low-energy buildings: a review article. Energy Build 39(3):249–257
9. Iqbal N, Kim DH (2002) Iot task management mechanism based on predictive optimization for efficient energy consumption in smart residential buildings. Energy Build 257:111762
10. Yan D, Hong T, Dong D, Mahdavi A (2014) Definition and simulation of occupant behavior in buildings. Int Energy Agency EBC Annex 66: 1–14
11. McDowall R (2007) Fundamentals of HVAC systems: SI edition. Academic Press
12. Liu X, Wu Y (2021) Design, development and characterization of a Building Integrated Concentrating Photovoltaic (BICPV) smart window system. Sol Energy 220:722–734
13. Lin K, Chao L, Lee HH, Xin R, Liu S, Ho TC, Huang B, Yu KM, Tso CY (2021) Potential building energy savings by passive strategies combining daytime radiative coolers and thermochromic smart windows. Case Stud Therm Eng 28:101517
14. Aguilar J, Garces-Jimenez A, R-Moreno MD, García R (2021) A systematic literature review on the use of artificial intelligence in energy self-management in smart buildings. Renew Sustain Energy Rev 151:111530
15. Elsisi M, Tran MQ, Mahmoud K, Lehtonen M, Darwish MMF (2021) Deep learning-based industry 4.0 and internet of things towards effective energy management for smart buildings. Sensors 21(4):1038
16. Shankar A, Vijayakumar K, Babu BC (2021) Energy saving potential through artificial lighting system in PV integrated smart buildings. J Build Eng 43:103080
17. Majdi A, Dwijendra NKA, Muda I, Chetthamrongchai P, Sivaraman R, Hammid AT (2022) A smart building with integrated energy management: Steps toward the creation of a smart city. Sustain Energy Technol Assess 53:102663
18. Gupta A, Badr Y, Negahban A, Qiu RG (2021) Energy-efficient heating control for smart buildings with deep reinforcement learning. J Build Eng 34:101739
19. Kumar A, Sharma S, Goyal N, Singh A, Cheng X, Singh P (2021) Secure and energy-efficient smart building architecture with emerging technology IoT. Comput Commun 176:207–217
20. Ahmad A, Khan JY (2022) Optimal sizing and management of distributed energy resources in smart buildings. Energy 244:123110

21. Manic M, Wijayasekara D, Amarasinghe K, Rodriguez-Andina JJ (2016) Building energy management systems: the age of intelligent and adaptive buildings. IEEE Ind Electron Mag 10(1):25–39
22. Hannan MA, Faisal M, Ker PJ, Mun LH, Parvin K, Mahlia TMI, Blaabjerg F (2018) A review of internet of energy based building energy management systems: issues and recommendations. IEEE Access 6:38997–39014
23. Mariano-Hernández D, Hernández-Callejo L, Zorita-Lamadrid A, Duque-Pérez O, Santos García F (2021) A review of strategies for building energy management system: model predictive control, demand side management, optimization, and fault detect & diagnosis. J Build Eng 33:101692
24. Silva BN, Khan M, Han K (2020) Futuristic sustainable energy management in smart environments: a review of peak load shaving and demand response strategies, challenges, and opportunities. Sustainability 12(14):5561
25. Huseien GF, Shah KW (2022) A review on 5G technology for smart energy management and smart buildings in Singapore. Energy AI 7:100116
26. Aslam S, Iqbal Z, Javaid N, Khan ZA, Aurangzeb K, Haider SI (2017) Towards efficient energy management of smart buildings exploiting heuristic optimization with real time and critical peak pricing schemes. Energies 10(12):2065
27. Capozzoli A, Piscitelli MS, Brandi S, Grassi D, Chicco G (2018) Automated load pattern learning and anomaly detection for enhancing energy management in smart buildings. Energy 157:336–352
28. Agarwal Y, Balaji B, Gupta R, Lyles J, Wei M, Weng T (2010) Occupancy-driven energy management for smart building automation. In: Proceedings of the 2nd ACM workshop on embedded sensing systems for energy-efficiency in building, pp 1–6
29. Hakimi SM, Hasankhani A (2020) Intelligent energy management in off-grid smart buildings with energy interaction. J Clean Prod 244:118906
30. Afuosi MB, Zoghi MR (2020) Indoor positioning based on improved weighted KNN for energy management in smart buildings. Energy Build 212:109754
31. Bellagente P, Ferrari P, Flammini A, Rinaldi S (2015) Adopting IoT framework for energy management of smart building: a real test-case. In: 2015 IEEE 1st international forum on research and technologies for society and industry leveraging a better tomorrow (RTSI). IEEE, pp 138–143
32. Foroozandeh Z, Ramos S, Soares J, Lezama F, Vale Z, Gomes A, Joench LR (2020) A mixed binary linear programming model for optimal energy management of smart buildings. Energies 13(7):1719
33. Fotopoulou E, Zafeiropoulos A, Terroso-Sáenz F, Şimşek U, González-Vidal A, Tsiolis G, Skarmeta A (2017) Providing personalized energy management and awareness services for energy efficiency in smart buildings. Sensors 17(9):2054
34. Nazari S, Borrelli F, Stefanopoulou A (2020) Electric vehicles for smart buildings: a survey on applications, energy management methods, and battery degradation. Proc IEEE 109(6):1128–1144
35. Allerding F, Mauser I, Schmeck H (2014) Customizable energy management in smart buildings using evolutionary algorithms. In: Applications of evolutionary computation: 17th European conference, evoapplications 2014, Granada, Spain, 23–25 Apr 2014, Revised Selected Papers 17. Springer, Heidelberg, pp 153–164
36. Ma L, Liu N, Wang L, Zhang J, Lei J, Zeng Z, Wang C, Cheng M (2016) Multi-party energy management for smart building cluster with PV systems using automatic demand response. Energy Build 121:11–21
37. Chen L, Xu Q, Yang Y, Song J (2021) Optimal energy management of smart building for peak shaving considering multi-energy flexibility measures. Energy Build 241:110932
38. Wang Y, Li Y, Cao Y, Tan Y, He L, Han J (2018) Hybrid AC/DC microgrid architecture with comprehensive control strategy for energy management of smart building. Int J Electr Power Energy Syst 101:151–161

39. Yu L, Qin S, Zhang M, Shen C, Jiang T, Guan X (2021) A review of deep reinforcement learning for smart building energy management. IEEE Internet Things J 8(15):12046–12063
40. Elsisi M, Tran MQ, Mahmoud K, Lehtonen M, Darwish MM (2021) Deep learning-based industry 4.0 and internet of things towards effective energy management for smart buildings. Sensors 21(4):1038
41. Arun SL, Selvan MP (2017) Intelligent residential energy management system for dynamic demand response in smart buildings. IEEE Syst J 12(2):1329–1340
42. Aguilar J, Garces-Jimenez A, R-moreno MD, García R (2021) A systematic literature review on the use of artificial intelligence in energy self-management in smart buildings. Renew Sustain Energy Rev 151:111530
43. Rocha P, Siddiqui A, Stadler M (2015) Improving energy efficiency via smart building energy management systems: a comparison with policy measures. Energy Build 88:203–213
44. Chaouch H, Bayraktar AS, Çeken C (2019) Energy management in smart buildings by using M2M communication. In: 2019 7th International Istanbul smart grids and cities congress and fair (ICSG). IEEE, pp 31–35
45. Majdi A, Dwijendra NKA, Muda I, Chetthamrongchai P, Sivaraman R, Hammid AT (2022) A smart building with integrated energy management: steps toward the creation of a smart city. Sustain Energy Technol Assess 53:102663
46. Foroozandeh Z, Ramos S, Soares J, Vale Z, Dias M (2022) Single contract power optimization: a novel business model for smart buildings using intelligent energy management. Int J Electr Power Energy Syst 135:107534
47. Di Piazza MC, La Tona G, Luna M, Di Piazza A (2017) A two-stage energy management system for smart buildings reducing the impact of demand uncertainty. Energy Build 139:1–9
48. Kermani M, Adelmanesh B, Shirdare E, Sima CA, Carnì DL, Martirano L (2021) Intelligent energy management based on SCADA system in a real Microgrid for smart building applications. Renew Energy 171:1115–1127
49. Nezhad AE, Rahimnejad A, Gadsden SA (2021) Home energy management system for smart buildings with inverter-based air conditioning system. Int J Electr Power Energy Syst 133:107230
50. Ock J, Issa RR, Flood I (2016) Smart building energy management systems (BEMS) simulation conceptual framework. In: 2016 Winter simulation conference (WSC). IEEE, pp 3237–3245
51. Ain QU, Iqbal S, Khan SA, Malik AW, Ahmad I, Javaid N (2018) IoT operating system based fuzzy inference system for home energy management system in smart buildings. Sensors 18(9):2802
52. Fahim M, Sillitti A (2018) An anomaly detection model for enhancing energy management in smart buildings. In: 2018 IEEE international conference on communications, control, and computing technologies for smart grids (SmartGridComm). IEEE, pp 1–6
53. Morvaj B, Lugaric L, Krajcar S (2011) Demonstrating smart buildings and smart grid features in a smart energy city. In: Proceedings of the 2011 3rd international youth conference on energetics (IYCE). IEEE, pp 1–8
54. Liu Y, Yang C, Jiang L, Xie S, Zhang Y (2019) Intelligent edge computing for IoT-based energy management in smart cities. IEEE Netw 33(2):111–117
55. Haidar N, Attia M, Senouci SM, Aglzim EH, Kribeche A, Asus ZB (2018) New consumer-dependent energy management system to reduce cost and carbon impact in smart buildings. Sustain Cities Soc 39:740–750
56. Chaouch H, Çeken C, Arı S (2021) Energy management of HVAC systems in smart buildings by using fuzzy logic and M2M communication. J Build Eng 44:102606
57. Chinchero HF, Alonso JM (2020) A review on energy management methodologies for LED lighting systems in smart buildings. In: 2020 IEEE international conference on environment and electrical engineering and 2020 IEEE industrial and commercial power systems Europe (EEEIC/I&CPS Europe). IEEE, pp 1–6
58. Revati G, Hozefa J, Shadab S, Sheikh A, Wagh SR, Singh NM (2021) Smart building energy management: load profile prediction using machine learning. In: 2021 29th Mediterranean conference on control and automation (MED). IEEE, pp 380–385

59. Hernández Uribe O, San Martin JP, Garcia-Alegre MC, Santos M, Guinea D (2015) Smart building: decision making architecture for thermal energy management. Sensors 15(11):27543–27568
60. Dong B, Prakash V, Feng F, O'Neill Z (2019) A review of smart building sensing system for better indoor environment control. Energy Build 199:29–46
61. King J, Perry C (2017) Smart buildings: using smart technology to save energy in existing buildings. Amercian Council for an Energy-Efficient Economy, Washington, DC, USA, pp 1–46
62. Bhatt JG, Jani OK, Bhatt CB (2020) Automation based smart environment resource management in smart building of smart city. In: Smart environment for smart cities, pp 93–107

Chapter 10
Role of Pumped Hydro Storage to Mitigate Intermittency in Renewable Energy Systems

Satyajit Das Karmakar and Himadri Chattopadhyay

Abstract The ever-increasing demand for energy in the background of environmental concern is paving the way for replacement of fossil fuels with renewable energy sources. While hydropower is one of the most significant contributors in this sector for a long time, recent growth in renewables is coming from solar energy and wind energy. Despite their strong position of sustainability, a major problem of these sectors is the intermittent nature of energy supply. Hence, to suppress such fluctuations, energy storage is essential. Pumped hydro storage (PHS) in this context is one of the most attractive choices due to high efficiency, reliability and low cost. This paper discusses the use of PHS for removing the intermittency in supplies from solar and wind energy. The current state-of-art indicates that PHS shall be increasingly used in coming decades for energy generation in renewable and hybrid renewable sectors.

Keywords Renewable energy · Intermittency · Pumped hydro storage · Hybrid energy system · Energy droughts

10.1 Introduction

While the demand for energy is ever-increasing with the improving standard-of-living and increasing global population, the issues of global warming and environmental pollution are taking center stage, necessitating shift towards renewable energy generation. As such the major developments in energy field have shifted from high- to low-carbon energy sources, i.e., from conventional sources of energy to renewable sources of energy, which are sustainable forms of energy.

As of 2021, the global power capacity of renewable energy is about 3000 GW as shown in Fig. 10.1 [1, 2]. It is observed that hydropower contributes about one-third of this though the growth is marginal while solar and wind energy-based plants are

S. Das Karmakar · H. Chattopadhyay (✉)
Department of Mechanical Engineering, Jadavpur University, Kolkata 700032, India
e-mail: himadri.chattopadhyay@jadavpuruniversity.in

S. De et al. (eds.), *Challenges and Opportunities of Distributed Renewable Power*, Energy, Environment, and Sustainability, https://doi.org/10.1007/978-981-97-1406-3_10

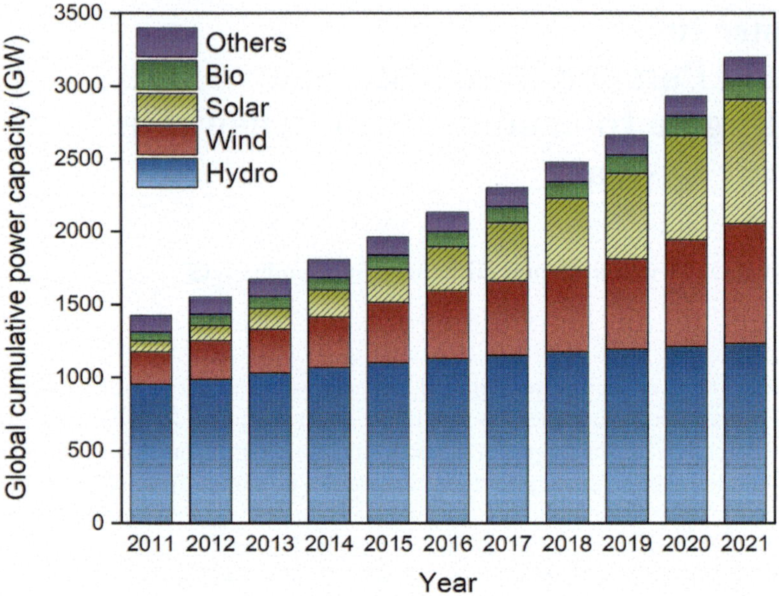

Fig. 10.1 Global renewable cumulative power capacity (2011–2021) [1, 2]

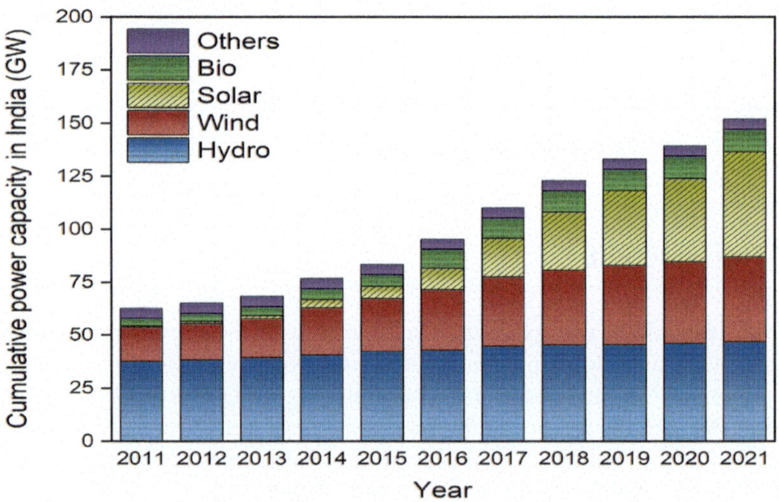

Fig. 10.2 Cumulative renewable power capacity in India (2011–2021) [1, 3, 4]

growing rapidly. A similar trend is also noted for India as in Fig. 10.2 [1, 3, 4]. The total installed capacity of hydropower is about 40 GW as on 2021 matched by solar energy. The next contributor is again wind sector.

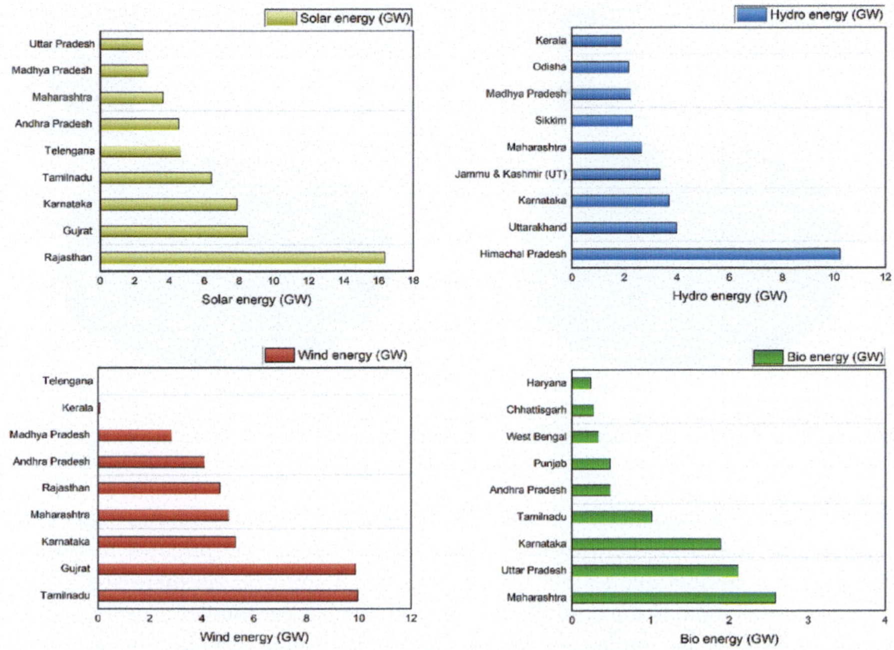

Fig. 10.3 State-wise cumulative installed capacity of solar, wind, hydro and bioenergy in India (in MW) [1, 3, 4]

Figure 10.3 [1, 3, 4] shows the state-wise cumulative installed capacity of solar, wind, hydro and bioenergy in India (in MW). Rajasthan emerges as an ideal location with immense future prospects for solar energy generation. Tamil Nadu and Gujarat stand at the forefront among states harnessing wind energy, while Maharashtra leads the way in the sector of bioenergy. Hilly terrains are characterized by diverse elevations and landscapes, an important factor in the context of PHS systems. Himachal Pradesh, Uttarakhand and other hilly regions provide excellent locations for PHS units due to their geography, water availability and potential role in maintaining safe and dependable energy systems. These areas can contribute to robust and sustainable energy infrastructure that will benefit the grid and the environment.

Globally the shares of hydro, wind and solar energies in total renewable energy are 38%, 26% and 27%, respectively, as seen in Fig. 10.4 [3, 4]. For India, the contribution of hydropower is 31%. The contributions of solar and wind sectors are 33% and 26%, respectively. Thus, in renewable energy share, India is above world average in solar, at par in wind and below average in hydropower. Of course, these proportions are governed by several geographical and socio-economic factors. In COP21 (United Nations Climate Change Conference), India has committed to reduce the dependence on fossil fuels and generate half of the energy requirement of the country from renewable sources by 2030 at a level of 500 GW.

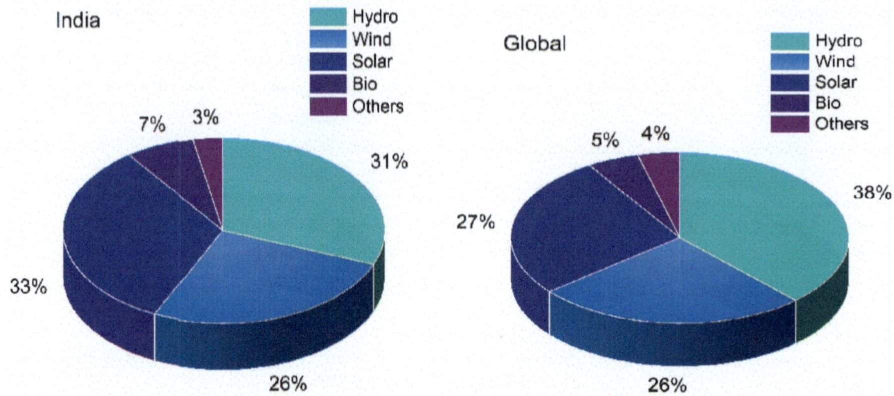

Fig. 10.4 Renewable energy shares (2021) in India (left) and all over the world [3, 4]

Transition to a mix-mode of renewable-conventional power or full renewable mode is a journey fraught with lot of challenges. Indeed, the price of renewable energy is now comparable to that from conventional sources, but there are several problems which include uncertainty of supply. In the background of ever-increasing demands of energy and emphasis on sustainability, this paper examines the problems of power generation from sources with intermittency. The role of energy storage especially of pumped hydro storage (PHS) in solving these issues is discussed.

10.2 Intermittency of Renewable Energy Sources and Drought

While renewable energy sources are occupying a larger space in the energy pie, a major concern about them is the intermittent stochastic nature of the supply. Solar intensity as well as wind speed at particular location varies through the day, months and years and is subjected to the vagaries of nature. As an example, stochastic nature of wind is depicted in Fig. 10.5 from data of four Indian cities: Imphal, Shilong, Guwahati and Kolkata [5, 6]. Throughout all months except winter, Kolkata consistently records the highest average wind speed, followed by Guwahati. However, during the winter season, Imphal experiences more significant variations in wind speed. With the exception of 2015–16, the average wind speed of Guwahati city has remained relatively stable over the past 5 years. This clearly establishes the fact that the wind speed varies from month to month and year to year leading to variation in energy generation. In fact, period of relatively lean winds is also quite possible. The available water at reservoirs is again subject to amount of rainfall and number of feeds from upstream source. To illustrate this point, in a case study for power generation in Romania, Spiru [7] showed that production from hydro-sources varied up to 60% between summer and spring. The variation for wind was 11% between

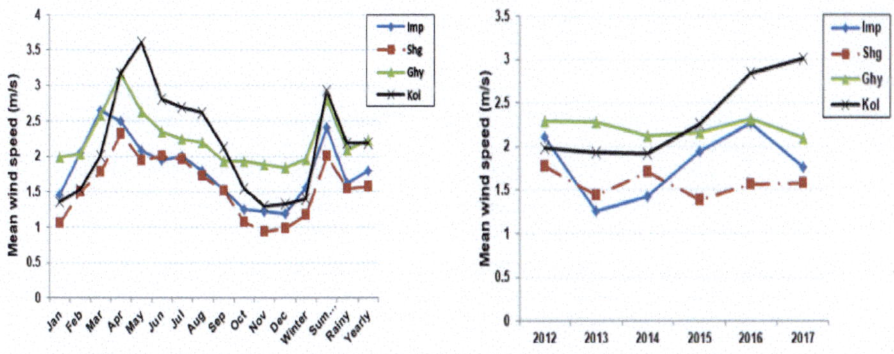

Fig. 10.5 Stochastic nature of wind month-wise average speed and year-wise average speed at Imphal, Shilong, Guwahati and Kolkata [5, 6]

summer and winter. The energy generated from solar power was drastically high at four times in summer as compared to that in winter.

Instead of relying on a particular source of energy, a modern trend is to go for hybrid power systems. Combinations of conventional power sources such as thermal with renewable energy sources are possible. Such hybrid systems, when properly optimized, can lead to reduced operating cost as well as minimization of emission. Panda et al. [8] demonstrated that a mix of four energy sources, i.e., thermal with solar, wind and hydro is generally offers the best solution in comparison with the combination of any three sources.

It is important here to discuss the concept of renewable energy drought [9]. The unpredictability of nature can often lead to drastic reduction or even total collapse in energy generation from hydro wind or solar power. It is not uncommon to have no generation days for wind or solar systems. Long absence of rain or even geopolitical disturbances can cause drying of hydro-sources. Multiple factors including Earth's orbit around the Sun, the angle of solar radiation with respect to Earth's surface, meteorological occurrences such as cloud cover and dust storms, variation of wind speed and direction due to atmospheric conditions collectively influence the variability of solar and wind energy [10]. Impact of droughts on various energy resources is shown in Table 10.1.

10.3 Energy Storage Technologies

To address these intermittencies and for smooth operation towards delivery on demand and store excess energy over demand, energy storage system is essential that can be later harnessed in systematic and regulated approach as needed [12]. Energy storage technology (EST) is going to be one of the most critical factors in deciding whether renewable energy can be used as base load for main grid dispatch strategy [13]. These are made to store extra energy and supply it when needed, depending

Table 10.1 Impact of droughts in energy resources [11]

Energy resources	Impacts of drought
Thermal power	A high temperature of cooling water reduces fuel availability and efficiency of power plant The energy produced by power plants is constrained due to scarcity of water
Hydropower	Rise in temperature increases the evaporation from water surface Shortage of water reduces the effective generating capacity
Biomass	Reduction in Net Primary Production of feedstock Increase in required area to grow feedstock Increase in pertinent costs
Wind energy	Changing wind patterns or less wind brought on by droughts may result in lower wind energy generation Increase in maintenance cost
Solar energy	Droughts cause a rise in airborne dust and particles, which stick to solar panels and decreases the efficiency of solar panels

on grid demands. ESTs can be classified into few groups based on their working principles such as mechanical, electrochemical, chemical and thermal ESTs [14–16]. Figure 10.6 illustrates the categorization of various storage systems based on their energy storage methods. Among these ESTs, mechanical energy storage (MES) excels in rapidly converting and releasing stored mechanical energy into electrical power. Electrical energy is stored as potential or kinetic energy during low-demand periods and during peak hours, it's transformed back into electrical power [17, 18]. Additionally, MES systems have three different categories: pumped hydro energy storage (PHES), compressed air energy storage (CAES) and flywheel energy storage (FES). For bulk energy storage over 100 MW, the two main options are pumped hydro storage (PHS) and compressed air energy storage (CAES). While 100 s of PHS plants are deployed worldwide with a total capacity around 130 GW, as per Javed et al. [13] only two large CAES plants are found in Germany and USA with capacity of 100 and 290 MW, respectively. The position of PHS vis-à-vis other available storage technologies can be realized by looking at Fig. 10.7. There are several attributes that make PHS the best choice. Capital investment and operating costs are 18 times higher in battery systems over PHS, which is capable of balancing large load variations. The carbon footprint of PHS is half of battery-based systems, where the issues of handling toxic materials are also involved.

The characteristics of bulk energy storage systems used in renewable energy sources are shown in Table 10.2. PHS takes the lead with largest energy range spanning from 100 to 5000 MW. PHS stands out due to its prolonged operational lifetime and boasts greater efficiency compared to other ESTs. Notably, PHS has lower energy capital costs than other storage strategies.

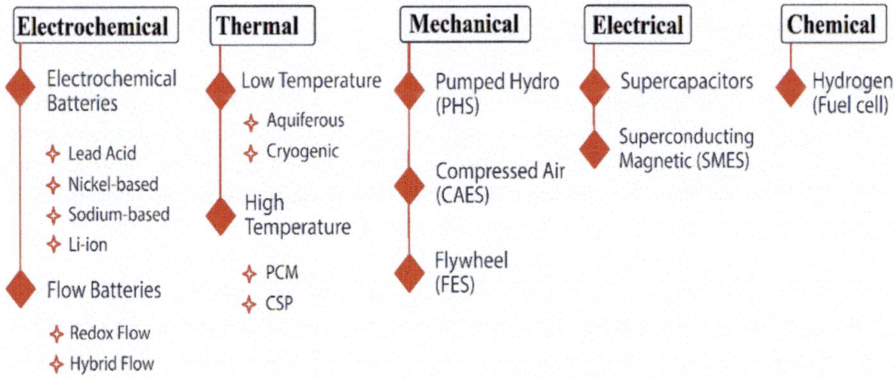

Fig. 10.6 Different types of energy storage systems [19]

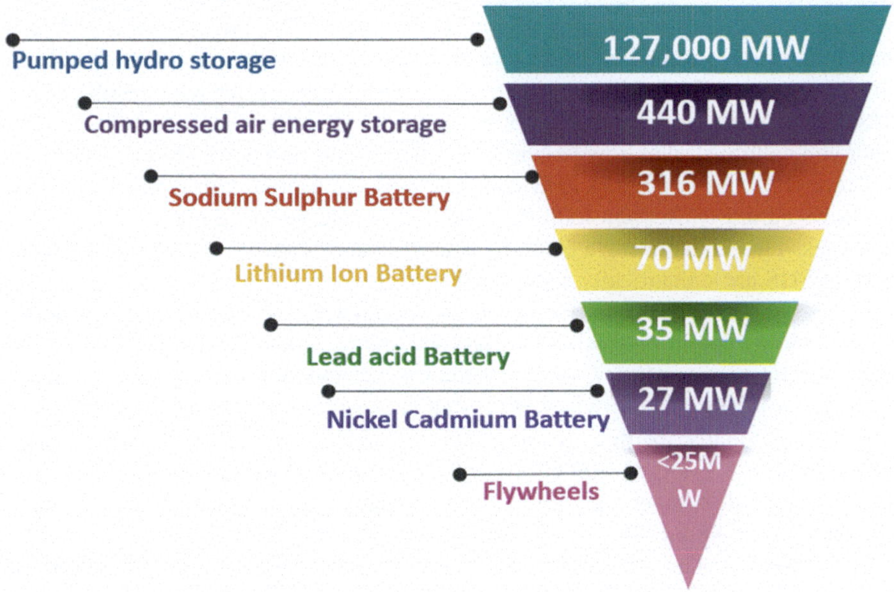

Fig. 10.7 Globally installed capacity of popular energy storage technologies [13]

Table 10.2 Characteristics of different ESTs [19–23]

Parameters	PHS	CAES	FES
Power rating (MW)	100–5000	5–1000	0.1–20
Efficiency (%)	65–87	50–89	85–95
Life time (years)	40–60	20–40	>15
Technical maturity	Mature	Commercial	Developed
Energy capital cost ($/kWh)	5–100	2–120	1000–14,000
Environmental issues	High (negative)	High (negative)	Low

10.4 Pumped Hydro Storage

While PHS as a technology had been known since the 1890s and was utilized in the region of and Italy and Switzerland, it came into renewed attention under the emergence of renewables [24]. Globally, the first commercial PHS system was the Pedreira Elevatory Plant, located in Cubató/SP, Brazil, and initiated its operations in 1939 [25].

PHS involves pumping and releasing of between higher and lower reservoir to store water and generate electricity. Water is pumped to higher reservoir from a lower reservoir by using electricity when demand low and then permits water to flow downhill through turbines during peak hours to generate electricity. Traditionally, PHS plants produce power during daylight hours and pumping during the night.

PHS systems can be classified into two main categories according to their operational design: (i) closed-loop PHS that is also known as pure PHS and (ii) open-loop or pump-back PHS [26]. In closed-loop PHSs or pure PHS, operational foundation entirely depends on pumped water that has been stored to an upper reservoir from a lower reservoir. Closed-loop systems are sometimes also referred to as off-stream PHS. On the other hand, open-loop or pump-back PHS systems utilize both pumped water and natural stream flow to generate power. Open-loop systems are sometimes also referred to as combined PHS and it always connected with a water resource that flows naturally. In open-loop PHS, the lower reservoir is generally established by harnessing and damming a naturally flowing water resource. These two different types of PHS are semantically shown in Fig. 10.8.

PHS's storage capacity is influenced by several factors, including the disparity in gravitational height between two reservoirs and the quantity of water that can be pumped. High energy density, low maintenance cost and long-life cycle are the major advantages of PHS. Region-wise PHS installed capacity till 2021 has been shown in Table 10.3. Globally, the three leading countries in PHS are China, Brazil and India with capacity of 391 GW, 109 GW and 51 GW, respectively, as on 2021 [1]. It is noted that PHS dominates the global energy storage capacity, boasting an impressive efficiency of 70–85% [28].

In the PHS system depicted in Fig. 10.9, solar and wind-based generation of electricity is being fed to grids. Now whenever there is less energy demand than the generation, excess electricity shall be used to pump water from lower reservoir to a reservoir situated at higher elevation. Water from the higher reservoir shall be fed to hydraulic turbine for power generation as and when demand arises. Thus, during a lean period of generation from solar or wind system, energy can be obtained from water flow between the reservoirs through the turbine. The working principle of PHS is quite suitable for handling intermittency. It is also important to ensure that overflow from the upper reservoir is directly fed to lower reservoir so as to conserve water. PHS is informally classified into four groups based on size: large (>10 MW), small (<10 MW), micro (<100 kW) and pico (<5 kW) as presented in Fig. 10.10 [24]. Significant advantages of micro and pico PHS include the ability to store solar and wind energy, which can then be used for small and remote communities

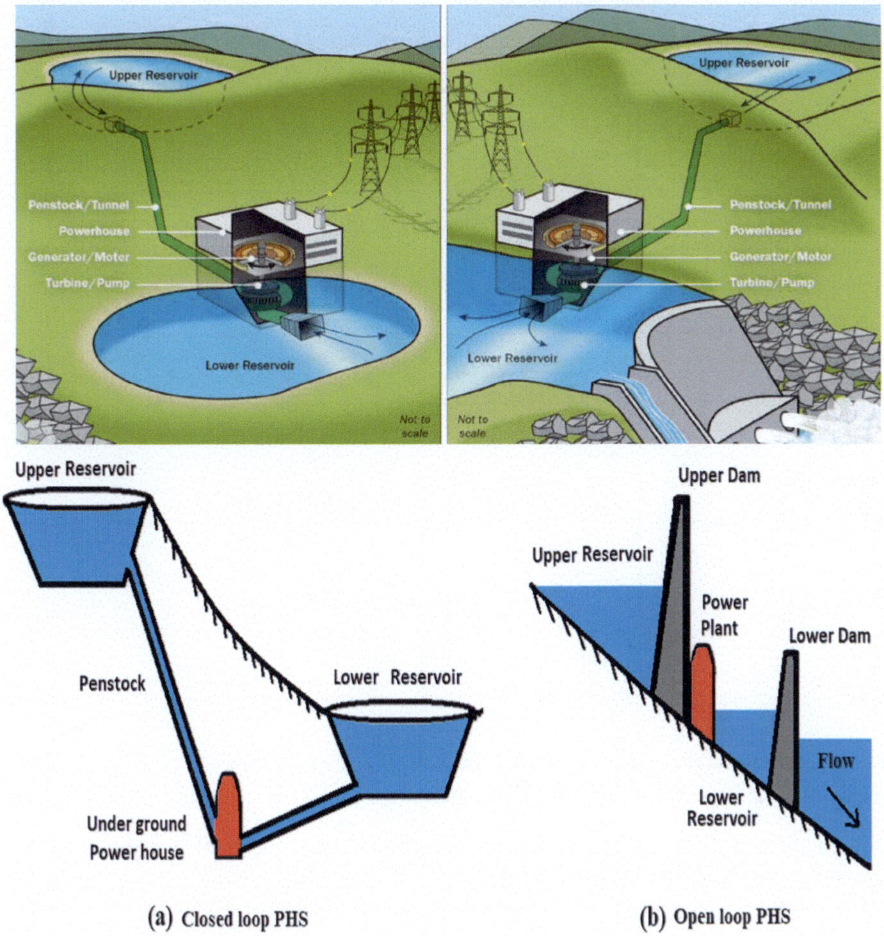

Fig. 10.8 **a** Closed-loop PHSs or pure PHS and **b** open-loop or pump-back PHS [26, 27]

Table 10.3 Region-wise PHS installed capacity in 2021 [1, 29]

Region	Total installed capacity including PHS (MW)	Pumped storage (MW)	Generation in 2021 (TWh)
East Asia and Pacific	522,891	75,540	1,639
Europe	254,901	55,050	689
North and Central America	205,058	22,089	702
South America	176,773	994	614
South and Central Asia	162,313	7,711	538
Africa	38,469	3,377	146

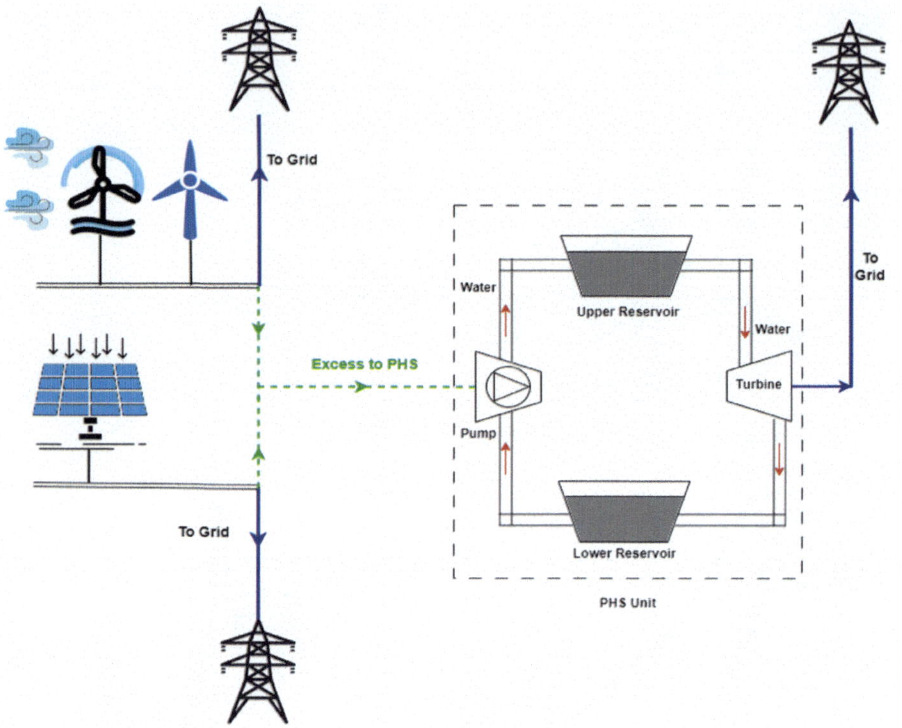

Fig. 10.9 Schematic diagram of solar-wind generation system with PHS

with exceptionally low power demands. Across all sizes, pumped storage efficiency typically ranges from 70 to 85% [30].

As the operating principles of pumps and turbines are complementary in nature, often a single pump-turbine unit is used in PHS systems, which offer the functionalities of both pump and turbine [31]. These units are generally a variation of Francis turbine or Darrieus turbines and can operate up to a head of 700–1200 m. The efficiency of such systems as pump and turbine is up to 90% and 95% and thus as a combined unit the efficiency level is about 85%.

The key features and benefits of PHS include adaptation to load change with quick response, voltage stability and frequency modulation, relatively lower operating cost. Deployment of PHS shall lead to higher penetration of renewable energy. Panda and Mishra [32] in a very recent work compared wind-solar-thermal, wind-hydro-thermal and wind-solar-thermal-PHS systems. Results showed that combined system with PHS came up with substantial reduction in cost, emission as well as enhanced voltage security. Apart from wind and solar, PHS systems have been analyzed in conjunction with generation using diesel, fuel cell and nuclear power. Intelligent scheduling of plant operation in tandem with PHS is the way for achieving best performance.

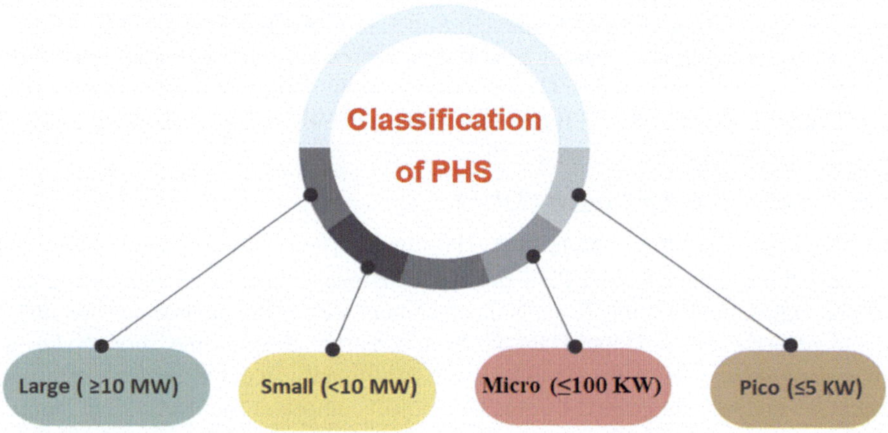

Fig. 10.10 Classification of PHS [24]

A study by Stocks et al. [33] has identified over 600,000 sites from publicly available geographical information systems, which could be used for closed-loop PHS energy generation. Intelligent utilization of such sites could overcome the barrier of land acquisition-related problems and add to energy generation in tandem with wind, solar or other energy sources. One of the important applications of PHS can be found in remote locations, which are not connected power grids. In such places, wind or solar-based energy generation is gaining ground. However, energy from both of them is stochastic and intermittent as already pointed out and thus a storage unit for energy is of immense benefit, see, e.g., Pali and Vadera [34] who have come with an idea of wind energy with PHS.

10.5 Examples of PHS Installed in West Bengal

Currently, there are two PHS projects commissioned in West Bengal, a state in India. They are known as Purulia Pumped Storage Project (PPSP) and Turga Pumped Storage Project (TPSP) both located in the district of Purulia [35, 36].

Purulia Pumped Storage Project (PPSP)

PPSP is located at Ajodhya hills under Baghmundi Block with total installed capacity of 900 MW. PPSP has been operational since 2007–2008 and its primary objective is to meet the system's peak load demand while harnessing surplus power during off-peak hours. This project features two Rockfill dams, namely, the Upper and Lower Dams, accompanied by two water conductors, an underground power house, and four reversible-type pump turbines. Central clay cores were employed in the construction of both upper and lower reservoirs. The project is linked to both the

Durgapur Sub-station and the Arambagh Sub-station through two 400 kV double-circuit transmission lines, enabling seamless power transmission and reception. PPSP operates in four distinct modes: Generation, Pumping, Synchronous Condenser and Line Charging (up to 400 kV Bus) mode. The generation voltage is maintained at 16.5 kV. Overall plant efficiency for PPSP is 77.80%

Turga Pumped Storage Project (TPSP)

TPSP, nestled along Turga Nala in the Purulia district of West Bengal, stands as an example of a closed-loop PHS project. This project envisions harnessing the abundant rainfall within the Turga Nala catchment area in the picturesque Ayodhya hills to generate power. Upper Dam has been constructed across the Turga Nala—a tributary of the majestic Subarnarekha River. Additionally, it entails water conductor system, an underground power house. The project comprises four Reversible Pump-Turbine units with 250 MW generating capacity, all housed within an underground power facility. Two of these units operate at a steady pace, whereas the other two runs at variable speeds, allowing them to harness intermittent solar and wind energy, ensuring the seamless integration of renewable energy into the grid.

10.6 Economics of PHS

The economics side of PHS system has two aspects. The first one introduces the direct cost of building a PHS and another a not so obvious side is the cost of intermittency. The electricity grid system faces a number of issues due to the renewable energy drought mainly for solar and wind resources that lead to the increase of cost of production [37, 38]. All of these intermittences lead to additional expenses that add to the overall cost of producing, transmitting and distributing power. The cost of intermittencies shows significant variation, yet in broad terms, they typically fall within the range of 1–12 euros per MWh [39]. A precise prediction of these intermittencies enhances the efficiency of the overall electrical system's control and operation.

The direct cost of PHS involves capital cost, operation and maintenance cost and levelized cost of electricity. The high cost of land acquisition, construction of dam reservoirs and installation of equipment such as pumps, turbines, generators and transformers are principally responsible for the high capital cost associated with PHS. So, the assessment of capital costs and investment requirements is essential in determining the economics and feasibility of such projects. It also depends on installed power and storage capacity of a particular PHS [40]. While the capital costs of PHS systems vary between 1000 and 4000 $/kW, their extended operational lifespan and large-scale energy storage capacity ultimately contribute to mitigating the initial capital cost over the course of time [26].

Operation and maintenance cost cover everything related to maintaining, repairing, and managing PHS systems that ensure their safe, secure and efficient operation throughout their prolonged operational lifespan. Regular maintenance of machineries like pumps, turbines and generators, any repairs or replacements required due to wear and tear are all included in this cost. Sustainability of PHS systems is heavily reliant on the operation and maintenance cost over the long term [41].

The levelized cost of electricity (LCOE) serves as a measure indicating the average cost associated with generating each unit of electricity across the lifespan of PHS. It is subject to a multitude of factors, encompassing reservoir size, site-dependent investment cost, operation and maintenance cost, fuel costs, construction and engineering costs, local energy market dynamics and other pertinent considerations. Indeed, LCOE stands as a vital indicator used to evaluate the long-term economic viability of PHS systems. The average LCOE of PHS was $ 0.048/kWh in 2010–2021 [42]. The LCOE associated with PHS is consistently among the most economical across various power generation techniques, positioning it as one of the most cost-effective options for electricity generation.

10.7 Hybrid PHS Systems

To fulfill the energy demands of a synchronized grid network, PHS can be seamlessly integrated with renewable energy systems, resulting in a highly efficient system. The surplus energy harnessed from renewable energy sources serves the purpose of elevating water to the upper reservoir within the PHS framework. During times of requirement, the stored water is utilized to drive a turbine, generating electricity to cater to load demands, especially when solar or wind energy failed to meet the energy demand.

Various integration approaches can be employed such as hybrid solar-PHS systems, hybrid wind-PHS systems, hybrid solar-wind-PHS systems. These integrated systems not only increase the efficiency of power grids but also contribute to the stabilization of energy prices. Upon optimizing a hybrid wind-PHS system to cater the power needs in remote regions of India, a minimal voltage drop of 3.5% was observed [34]. This noteworthy performance showcases PHS as a crucial component within a wind energy setup, capable of delivering consistent power to distant and remote locations.

In the context of hybrid solar-wind-PHS systems, PHS as storage is integrated with solar and wind energy systems owing to the inherent intermittency and reliance on seasonal fluctuations exhibited by solar and wind energy sources [43]. This combination ensures a more reliable, consistent power supply to remote populations and also mitigate the need for fossil fuel dependency, reduce the energy cost and greenhouse gas emissions [44].

Table 10.4 illustrates cost analysis of different hybrid PHS systems. Upon analysis, a noticeable pattern emerges where hybrid solar-wind-PHS systems exhibit the lowest energy cost contrast to hybrid solar-PHS systems, hybrid wind-PHS systems. This comparative assessment setups confirm that hybrid solar-wind-PHS systems stand out as the most financially efficient energy solution.

Table 10.4 Cost analysis of different hybrid PHS systems [45–49]

Type	Energy cost ($/kWh)
Hybrid solar-PHS systems	0.2–0.45
Hybrid wind-PHS systems	0.15–0.35
Hybrid solar-wind-PHS systems	0.072–0.27

10.8 Environmental Impact of PHS

PHS has advantages as well as potential negative effects on the environment, much like any other energy storage technique. It is crucial for balancing and stabilizing the electrical system, especially when including sporadic renewable energy sources like wind and solar.

The operation of PHS involves the mechanism of transferring water between two reservoirs. This creates an opportunity to exercise ecological control and optimize water utilization. Pumped hydro storage mitigates the intermittencies of renewable energy sources by storing extra energy during periods of high generation and releasing it during times of high demand or low power generation. It is an eco-friendly energy storage alternative because it emits very less greenhouse gases [50]. However, these gases can be emitted due to the decomposition of organic matter within reservoirs. As indicated by researchers, tropical climates could exhibit heightened emissions attributed to their warmer temperatures and high concentrations of organic waste [51]. The magnitude of emissions depends on the variables such as size of reservoir, water levels, temperature and the quantity of organic matter.

When constructing pumped hydro storage facilities like dam, reservoirs, substantial land areas may be required for infrastructure, potentially leading to soil erosion, and deforestation. These infrastructures can change the way that water naturally flows affecting aquatic environments. PHS also impacts the environment through the flooding of land inside reservoirs and these become worse if the reservoirs are situated in a river valley. Also, Wildlife habitats and biodiversity may be disturbed during the construction phase of PHS plants [27, 52].

PHS employs water for electricity storage, influencing nearby water supplies, aquatic habitats and downstream water availability. Water quality can be significantly impacted by the operation of PHS systems, especially in terms of nutrient concentrations and oxygen content. Water within reservoirs might undergo changes in temperature and oxygen concentration, potentially affecting downstream ecosystems. There may be the result of increased sedimentation and erosion due to discharge of water from reservoirs, which may have a negative impact on aquatic habitats and species [41]. Fish breeding habitats may become degraded as a result of sedimentation and the growth of aquatic organisms can be dramatically impacted by variations in water temperatures and oxygen levels [53]. PHS causes substantial water level variations, which may increase seismic activity in certain geological contexts [54, 55].

In socio economic point of view, the development of PHS has both positive and negative implications on the society. These systems can provide energy security, create jobs that lead to promote local economic growth. However, PHS projects carry the potential to uproot local communities, take away their land and possessions and can alter their way of life. Sometimes, entire towns may need to be relocated for the construction of reservoirs and other infrastructure. In the pursuit of sustainability and the alleviation of environmental impacts, it becomes imperative to include the affected communities in the decision-making processes and ensure they receive equitable compensation [56, 57].

10.9 Concluding Remarks

This paper discusses the growth in renewable energy sector where intermittency and variability of several renewable sources call for special strategy to smoothen the demand–supply gap. For continuous supply of power, energy storages are essential. Pumped hydro storage is found to be the preferred choice in this sector as it offers a low-cost long-life solution with high reliability and fast response. The development of PHS continues to face constraints related to suitable topography, efficient design outline and optimal environmental integration.

References

1. IRENA RES (2022) International renewable energy agency. Abu Dhabi
2. Dawn S, Tiwari PK, Goswami AK, Singh AK, Panda R (2019) Wind power: existing status, achievements and government's initiative towards renewable power dominating India. Energy Strateg Rev 23:178–199
3. MNRE (2022) Annual report of the ministry of new and renewable energy 2021–2022
4. IREDA (2022) Indian Renewable Energy Development Agency Limited
5. Rahman SM, Chattopadhyay H (2019) Statistical assessment of wind energy potential for power generation at Imphal, Manipur (India). Energy Sources, Part A Recover Util Environ Eff 1–13
6. Rahman SM, Chattopadhyay H (2020) A new approach to estimate the Weibull parameters for wind energy assessment: case studies with four cities from the Northeast and East India. Int Trans Electr Energy Syst 30:e12574
7. Spiru P (2023) Assessment of renewable energy generated by a hybrid system based on wind, hydro, solar, and biomass sources for decarbonizing the energy sector and achieving a sustainable energy transition. Energy Rep 9:167–174
8. Panda A, Mishra U, Tseng M-L, Ali MH (2020) Hybrid power systems with emission minimization: multi-objective optimal operation. J Clean Prod 268:121418
9. Gangopadhyay A, Seshadri AK, Sparks NJ, Toumi R (2022) The role of wind-solar hybrid plants in mitigating renewable energy-droughts. Renew Energy 194:926–937
10. Hay JE (1984) An assessment of the mesoscale variability of solar radiation at the earth's surface. Sol Energy 32:425–434
11. Shadman F, Sadeghipour S, Moghavvemi M, Saidur R (2016) Drought and energy security in key ASEAN countries. Renew Sustain Energy Rev 53:50–58
12. Thiyagarajan SR, Emadi H, Hussain A, Patange P, Watson M (2022) A comprehensive review of the mechanisms and efficiency of underground hydrogen storage. J Energy Storage 51:104490

13. Javed MS, Ma T, Jurasz J, Amin MY (2020) Solar and wind power generation systems with pumped hydro storage: review and future perspectives. Renew Energy 148:176–192
14. Gür TM (2018) Review of electrical energy storage technologies, materials and systems: challenges and prospects for large-scale grid storage. Energy Environ Sci 11:2696–2767
15. García-Pereira H, Blanco M, Martínez-Lucas G, Pérez-Díaz JI, Sarasúa J-I (2022) Comparison and influence of flywheels energy storage system control schemes in the frequency regulation of isolated power systems. IEEE Access 10:37892–37911
16. Gao T, Jiang L, Liu K, Xiong D, Lin Z, Bu W et al (2021) Field exploration and analysis of power grid side battery energy storage system. IEEE Access 9:63213–63218
17. Yao L, Yang B, Cui H, Zhuang J, Ye J, Xue J (2016) Challenges and progresses of energy storage technology and its application in power systems. J Mod Power Syst Clean Energy 4:519–528
18. Maletin YA, Stryzhakova NG, Zelinskyi SO, Chernukhin SI (2021) Energy storage technologies based on electrochemical double layer capacitors: a review. Theor Exp Chem 57:311–324
19. Argyrou MC, Christodoulides P, Kalogirou SA (2018) Energy storage for electricity generation and related processes: technologies appraisal and grid scale applications. Renew Sustain Energy Rev 94:804–821
20. Chatzivasileiadi A, Ampatzi E, Knight I (2013) Characteristics of electrical energy storage technologies and their applications in buildings. Renew Sustain Energy Rev 25:814–830
21. Chen H, Cong TN, Yang W, Tan C, Li Y, Ding Y (2009) Progress in electrical energy storage system: a critical review. Prog Nat Sci 19:291–312
22. Das CK, Bass O, Kothapalli G, Mahmoud TS, Habibi D (2018) Overview of energy storage systems in distribution networks: placement, sizing, operation, and power quality. Renew Sustain Energy Rev 91:1205–1230
23. Gallo AB, Simões-Moreira JR, Costa HKM, Santos MM, Dos Santos EM (2016) Energy storage in the energy transition context: a technology review. Renew Sustain Energy Rev 65:800–822
24. Rehman S, Al-Hadhrami LM, Alam MM (2015) Pumped hydro energy storage system: a technological review. Renew Sustain Energy Rev 44:586–598
25. Vilanova MRN, Flores AT, Balestieri JAP (2020) Pumped hydro storage plants: a review. J Brazilian Soc Mech Sci Eng 42:415
26. Deane JP, Gallachóir BPÓ, McKeogh EJ (2010) Techno-economic review of existing and new pumped hydro energy storage plant. Renew Sustain Energy Rev 14:1293–1302
27. Saulsbury JW (2020) A comparison of the environmental effects of open-loop and closed-loop pumped storage hydropower. Pacific Northwest National Lab. (PNNL). Richland, WA (United States)
28. Ibrahim H, Ilinca A, Perron J (2008) Energy storage systems—characteristics and comparisons. Renew Sustain Energy Rev 12:1221–1250
29. Association IH (2022) Hydropower status report; international hydropower association. London, UK
30. Ma T, Yang H, Lu L, Peng J (2015) Optimal design of an autonomous solar–wind-pumped storage power supply system. Appl Energy 160:728–736
31. Breeze P (2019) Power generation technologies. Newnes
32. Panda A, Mishra U (2023) An environmental optimal power flow framework of hybrid power systems with pumped hydro storage. J Clean Prod 391:136087
33. Stocks M, Stocks R, Lu B, Cheng C, Blakers A (2021) Global atlas of closed-loop pumped hydro energy storage. Joule 5:270–284
34. Pali BS, Vadhera S (2018) A novel pumped hydro-energy storage scheme with wind energy for power generation at constant voltage in rural areas. Renew Energy 127:802–810
35. Welcome to WBSEDCL. n.d. https://www.wbsedcl.in/irj/go/km/docs/internet/new_website/PPSP.html. Accessed 13 Sept 2023
36. Welcome to WBSEDCL. n.d. https://www.wbsedcl.in/irj/go/km/docs/internet/new_website/TPSP.html. Accessed 13 September 2023
37. Delucchi MA, Jacobson MZ (2011) Providing all global energy with wind, water, and solar power, Part II: Reliability, system and transmission costs, and policies. Energy Policy 39:1170–1190

38. Lew D, Brinkman G, Ibanez E, Florita A, Heaney M, Hodge B-M, et al (2013) Western wind and solar integration study phase 2. National Renewable Energy Lab. (NREL). Golden, CO (United States)
39. Notton G, Nivet M-L, Voyant C, Paoli C, Darras C, Motte F et al (2018) Intermittent and stochastic character of renewable energy sources: consequences, cost of intermittence and benefit of forecasting. Renew Sustain Energy Rev 87:96–105
40. Zakeri B, Syri S (2015) Electrical energy storage systems: a comparative life cycle cost analysis. Renew Sustain Energy Rev 42:569–596
41. Nikolaos PC, Marios F, Dimitris K (2023) A review of pumped hydro storage systems. Energies 16:4516
42. IRENA (2023) The changing role of hydropower: Challenges and opportunities. Int Renew Energy Agency, Abu Dhabi
43. Das BK, Zaman F (2019) Performance analysis of a PV/Diesel hybrid system for a remote area in Bangladesh: effects of dispatch strategies, batteries, and generator selection. Energy 169:263–276
44. Gupta SK, Purohit P (2013) Renewable energy certificate mechanism in India: a preliminary assessment. Renew Sustain Energy Rev 22:380–392
45. Duchaud J-L, Notton G, Darras C, Voyant C (2019) Multi-objective particle swarm optimal sizing of a renewable hybrid power plant with storage. Renew Energy 131:1156–1167
46. Papaefthymiou SV, Papathanassiou SA (2014) Optimum sizing of wind-pumped-storage hybrid power stations in island systems. Renew Energy 64:187–196
47. Bhayo BA, Al-Kayiem HH, Gilani SIU, Ismail FB (2020) Power management optimization of hybrid solar photovoltaic-battery integrated with pumped-hydro-storage system for standalone electricity generation. Energy Convers Manag 215:112942
48. Xu X, Hu W, Cao D, Huang Q, Chen C, Chen Z (2020) Optimized sizing of a standalone PV-wind-hydropower station with pumped-storage installation hybrid energy system. Renew Energy 147:1418–1431
49. Nyeche EN, Diemuodeke EO (2020) Modelling and optimisation of a hybrid PV-wind turbine-pumped hydro storage energy system for mini-grid application in coastline communities. J Clean Prod 250:119578
50. Yang C-J, Jackson RB (2011) Opportunities and barriers to pumped-hydro energy storage in the United States. Renew Sustain Energy Rev 15:839–844
51. Barros N, Cole JJ, Tranvik LJ, Prairie YT, Bastviken D, Huszar VLM et al (2011) Carbon emission from hydroelectric reservoirs linked to reservoir age and latitude. Nat Geosci 4:593–596
52. Patocka F (2014) Environmental impacts of pumped storage hydro power plants
53. Poff NL, Zimmerman JKH (2010) Ecological responses to altered flow regimes: a literature review to inform the science and management of environmental flows. Freshw Biol 55:194–205
54. Blakers A, Stocks M, Lu B, Cheng C (2021) A review of pumped hydro energy storage. Prog Energy 3:22003
55. Immendoerfer A, Tietze I, Hottenroth H, Viere T (2017) Life-cycle impacts of pumped hydropower storage and battery storage. Int J Energy Environ Eng 8:231–245
56. Sovacool BK, Bulan LC (2011) Behind an ambitious megaproject in Asia: the history and implications of the Bakun hydroelectric dam in Borneo. Energy Policy 39:4842–4859
57. Denholm P, Kulcinski GL (2004) Life cycle energy requirements and greenhouse gas emissions from large scale energy storage systems. Energy Convers Manag 45:2153–2172

Chapter 11
Harnessing Ocean Wave Energy—Flow Rectifications

Doddamani Hithaish and Abdus Samad

Abstract India can harness 40 GW of ocean energy from its 7500 km long coast. There are several devices to harness such power. One commonly researched device is an oscillating water column (OWC). One end of the duct faces the water, while the other is open to the atmosphere. When the wave enters the duct, the air is pushed out; when the wave recedes, the air gets sucked in. It uses an air duct, which contains a single turbine or two turbines. The conventional turbines face difficulty in harnessing this pneumatic energy as air direction changes with wave action in OWC. A bidirectional turbine can work in this situation without additional units. However, because of their symmetrical aerofoil geometry construction, these turbines have poorer efficiency than conventional air turbines. A pair of conventional turbines can be used with OWC such that one works as a power producer and another as a flow blocker. This turbine action reverses for the opposite wave cycle. However, the flow blocker turbine cannot block the airflow completely, causing poor performance at the power producer turbine. Mechanical valves can be placed along with the turbine to improve flow blockage, but this may be unreliable considering the ocean's dynamic nature. A fluidic diode (FD), similar to an electrical diode, offers variable resistance based on flow direction. This can be used with OWC to improve the performance of the air turbines. This article discusses various FDs used for engineering applications and a detailed numerical study on the FD used for wave energy applications.

Keywords Wave energy · Oscillating water column · Unidirectional turbine · Turbine duo · Fluidic diode

Abbreviations

BD	Bluff body
FD	Fluidic diode

D. Hithaish · A. Samad (✉)
Department of Ocean Engineering, Indian Institute of Technology Madras, Chennai, Tamil Nadu 600036, India
e-mail: samad@iitm.ac.in

S. De et al. (eds.), *Challenges and Opportunities of Distributed Renewable Power*, Energy, Environment, and Sustainability, https://doi.org/10.1007/978-981-97-1406-3_11

GCI	Grid convergence index
GV	Guide vane
OWC	Oscillating water column
TD	Turbine duo
UDT	Unidirectional turbine
WEC	Wave energy converter

Symbols

D	Diameter of the duct [m]
L_c	Length of the computational domain [m]
L_N	Normalized length of the nozzle [22]
k	Turbulent energy production (m^2/s^3)
Q	Flow rate [m^3/s]
R	Normalized radius [−]
T	Time [s]
t'	Normalized time [−]
U	Velocity of fluid inside the duct [m/s]
UA	Inlet axial velocity [m/s]
γ	Nozzle angle (°)
Δp	Pressure drop [Pa]
Δt	Time step [s]
ε	Turbulent energy dissipation (m^2/s^3)
Ψ	Diodicity [−]
ω	Angular speed [rad/s]

Subscripts

1, 2, 3	Index number
B	Bluff body
To	Toroidal

11.1 Introduction

Overdependence on conventional energy sources (fossil fuels) negatively impacts the environment as their usage releases harmful chemicals and gases responsible for climate change and global warming. Hence, an alternative source is essential to reduce the environmental impact and replace conventional energy.

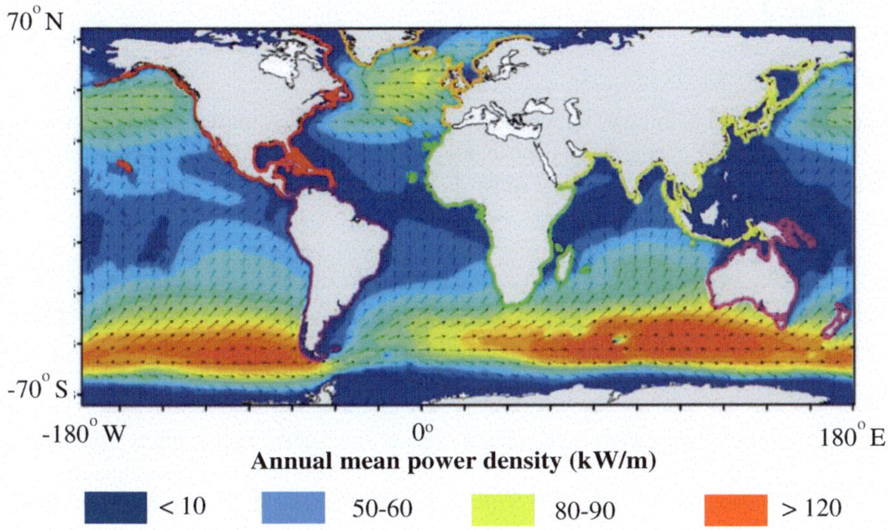

Fig. 11.1 Global wave energy resource distribution [1]

Ocean energy is a form of renewable energy source; unlike other sources, it is unaffected by seasonal variations and has little environmental impact. The waves are produced by energy exchange between the wind and the ocean surface. It can occur at different time scales and heights; they travel from deep sea to coastline and dissipate energy by breaking. If tapped well, this energy could serve a certain portion of world energy needs. Figure 11.1 shows the average yearly wave power for 10 years of databases worldwide. The power is represented in kW/m, corresponding to power per width of the wavefront.

However, developing a financially viable device or method for extracting wave energy requires meticulous research and development. Many wave energy enthusiasts and researchers have proposed various ways (devices) to extract wave energy into usable power.

11.1.1 Wave Energy Converter Devices

More than a thousand wave energy converter (WEC) concepts are proposed for harnessing wave power. However, they can be classified mostly based on site, type, and function.

WECs based on location

Shoreline devices

These are installed near cost, which makes them endure severe damage from storms. However, the power extraction by these devices is poor as the significant wave power gets attenuated as the waves travel toward shallow water.

Nearshore devices

These devices are often anchored to the seafloor in the shallow water area, which increases their stability. These units have poor energy extraction because of the lower wave power in the shallow water region.

Offshore devices

These devices are placed in the deep ocean, where waves have higher energy potential. They must operate in a higher energy potential environment, so their design must withstand these extreme conditions. This increases their installation and maintenance costs.

WECs based on the type

There are more variations of the WEC present under this category. However, majorly, that can be clubbed into the following groups (Fig. 11.2).

The attenuator device is a multi-segment floating structure that lies parallel to the wave direction, and varying sea height along its length causes flexing, which in turn drives the hydraulic unit. The principal axis of the terminator is parallel to the wave crest and terminates the wave. An oscillating water column is also a type of terminator. Based on sea-level variations, the point absorber's floating structure heaves up and down. This heaving motion can be converted into suitable power using the power take-off device (PTO). Some of these devices are used for specific applications like collecting data and as navigational buoys. The terminator device is

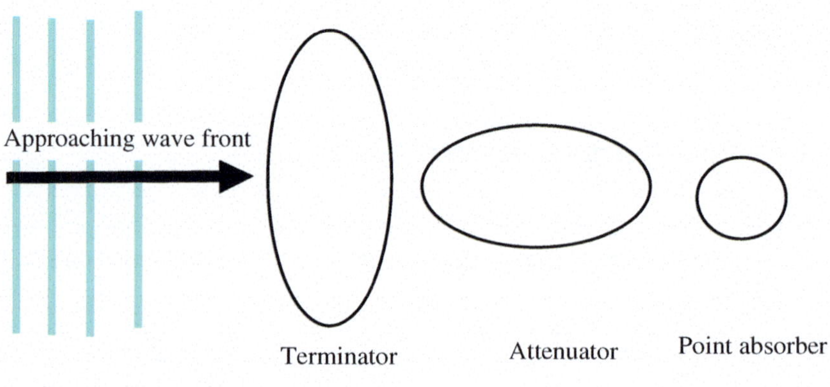

Fig. 11.2 WECs based on the type

installed at a right angle to the wave-approaching path. Salter Duck is an example of a device developed by the University of Edinburgh.

The orientation of these devices for the wave direction must be controlled according to the sea state for the time; otherwise, a lengthened converter might be forced to switch between terminating and attenuating waves at the same location under various sea states. The proper mooring configuration designs are necessary to manage the WEC orientation for specific sea states. For the same devices, a converter operating in attenuator mode often experiences lower mooring forces than one operating in terminator mode. When the WEC functions as a point absorber, the energy is collected from the wave field over an area with no significant orientation to the wave direction.

Mode of operations

Oscillating water column

Figure 11.3 shows the schematic of the oscillating water column (OWC) device. It has a submerged water column, air chamber, and duct. When a wave approaches the OWC, it raises the water level in the chamber, forcing air out of the OWC (Exhalation).

The receding wave draws water from the column, sucking air into the OWC from the atmosphere (Inhalation). This cyclic operation gives rise to oscillatory pneumatic energy. A PTO, typically an air turbine, converts this pneumatic energy into useful mechanical energy.

Submerged pressure differential

They comprise a movable cylinder and an air-filled cylindrical chamber similar to a submerged point absorber (Fig. 11.4). When the wave crest approaches, it pushes the movable part down, compressing the air. Similarly, the pressure on the movable part rises when the trough arrives. These devices do not experience any slamming force caused by the waves. However, these are submerged, which poses a significant challenge for maintenance.

Overtopping device

Excess water gets collected in the reservoir constructed above sea level in the event of a wave. The extra water passes through the turbine outlet and propels the turbine. Figure 11.5 shows a schematic of such a device (wave dragon).

Oscillating wave surge converter

These devices have a hinged deflector placed at a right angle to the wave path, as shown in Fig. 11.6. The oscillatory motion of the wave causes the deflector plate to move back and forth. This compresses or decompresses the liquid in the piston cylinder arrangement that can be used for pumping water or compressing the air, which can be later used for other applications.

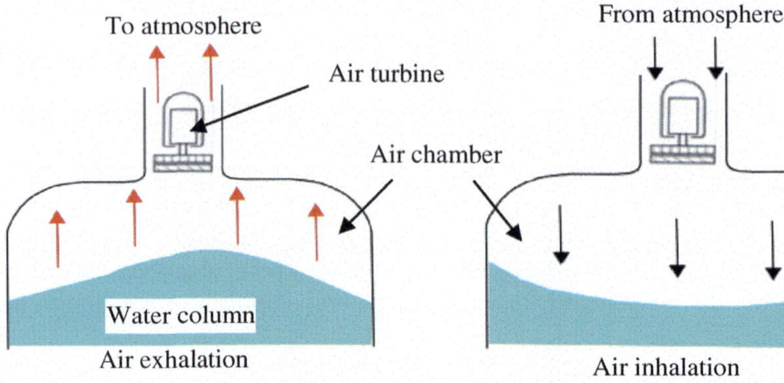

Fig. 11.3 Schematic of OWC [2]

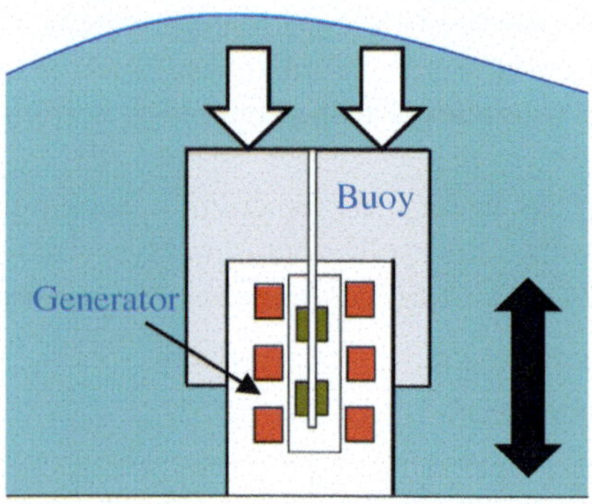

Fig. 11.4 Submerged pressure differential

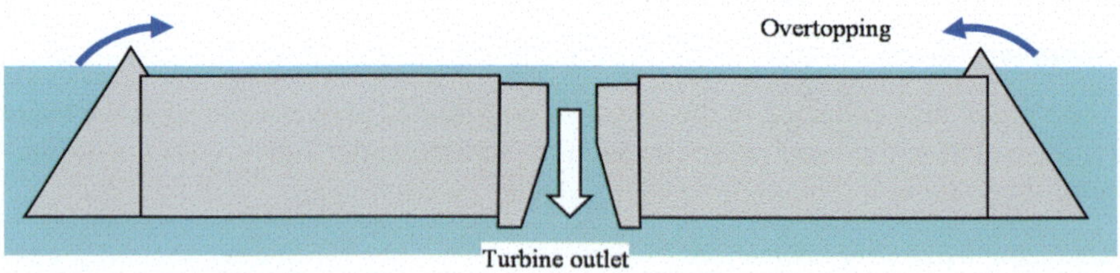

Fig. 11.5 Overtopping device

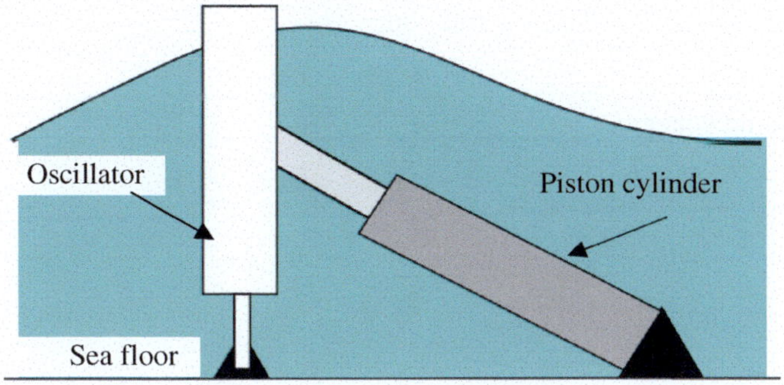

Fig. 11.6 Oscillating wave surge converter

11.1.2 Air Turbines

Reaction turbine

It has symmetrical airfoil blades located around the turbine hub. These are extensively used with OWC for their simplicity in design, fabrication, and efficient operation. Figure 11.7 shows the schematic of the turbine, also known as the Wells turbine.

A Wells turbine can have guide vanes (GVs) (Fig. 11.8a) and can be a mono-plane or bi-plane configuration. Mono-plane has a single row of turbine blades, whereas, in a bi-plane, a pair of rotor blades are attached staggered over the hub (Fig. 11.8b, c), and it can also be a counter-rotating type. The Wells turbine has a high peak efficiency of about 72%, but its operation is limited to a smaller range of flow co-efficient. It performs poorly at higher flow co-efficient because of the stall phenomenon, where its performance drops steeply.

Impulse turbine

An impulse turbine extracts power through the impulsive action of the airflow. The GVs are required to direct the airflow onto the blade. Based on the mode of power extraction, it can be axial or radial flow turbines and, further, bi-directional or unidirectional turbines.

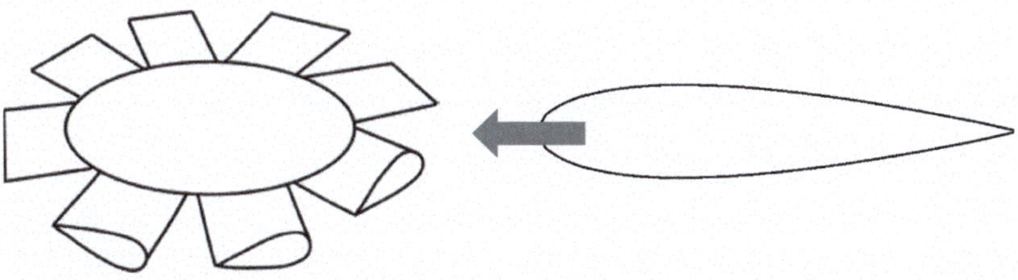

Fig. 11.7 Wells turbine

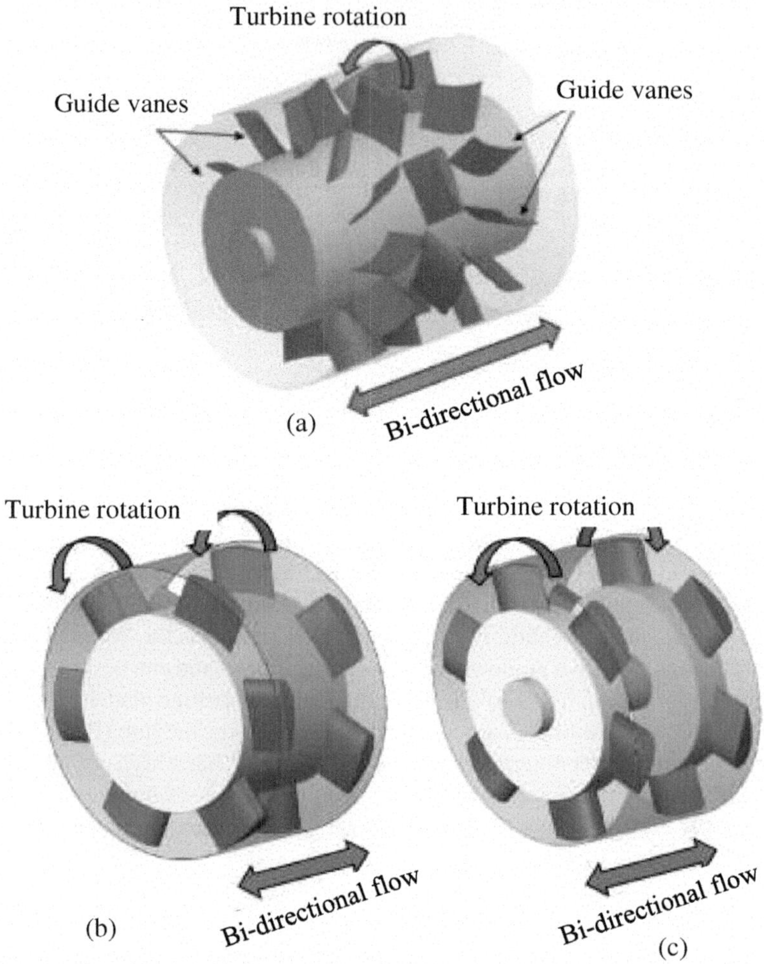

Fig. 11.8 Different Wells turbine **a** with GVs **b** bi-plane **c** counter-rotating [3]

Bi-directional impulse turbine

The rotor of this turbine has a symmetrical aerofoil section and a pair of GV rows, and they can be axial or radial turbines based on the airflow path onto the blade (Fig. 11.9). The symmetrical aerofoil and the GVs enable it to extract power from both airflow directions (exhalation or inhalation).

The GVs of the bi-directional impulse turbine can be pivoted to change the airflow direction to provide an aerodynamic moment called a pitch-controlled GV turbine (Fig. 11.10a). Drawing inspiration from the pitch-controlled GV is similar to a pitch-controlled blade proposed where the blade is oriented to airflow direction

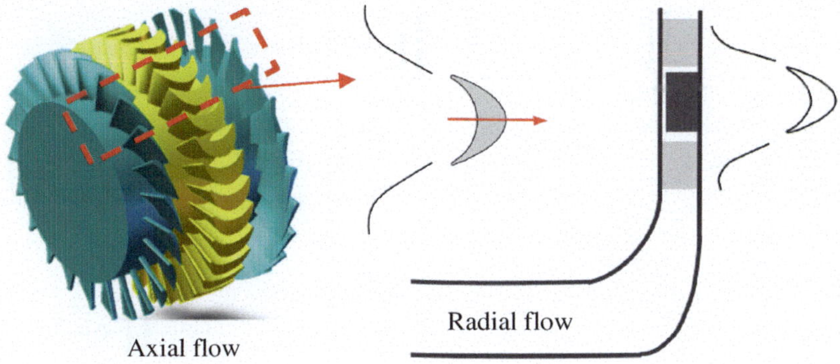

Fig. 11.9 Bi-directional impulse turbine [4]

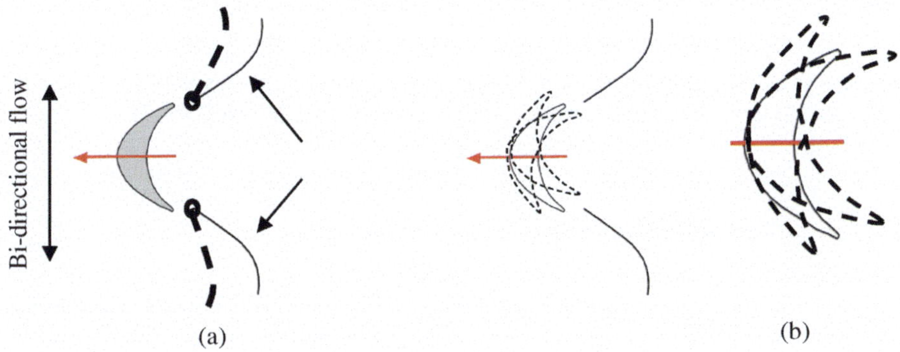

Fig. 11.10 Bidirectional turbine with **a** active pitch-controlled guide vane **b** active pitch-controlled blade [4]

(Fig. 11.10b). For a given wave cycle, blade arrangements around 5° yield better performance.

Unidirectional impulse turbine

They have a single row of rotors and GVs (Fig. 11.11). When air leaves the GV in a particular direction, it creates an impulse force on the rotor blade, similar to conventional steam turbine blades (Unidirectional axial flow). Unidirectional radial turbines can be either inward or outward radial turbines. An outward radial turbine experiences a centrifugal force from the air that leaves the turbine into the atmosphere. In contrast, an inward flow turbine feels the centripetal force when air enters the turbine from the atmosphere.

Wells and bi-directional turbines are self-rectifying turbines that rotate unidirectionally, irrespective of the airflow direction. Compared to them, a unidirectional axial turbine (UDT) has a wider operating range of flow coefficient with better peak

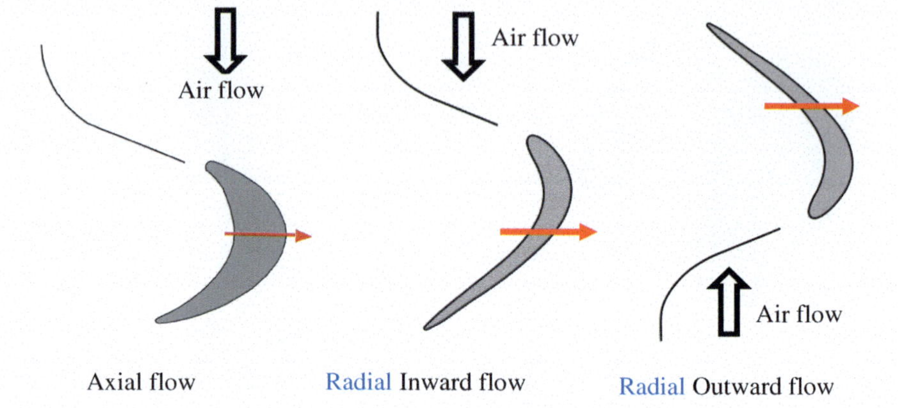

Fig. 11.11 Unidirectional impulse turbine

efficiency. Since these turbines are designed to function exclusively in a specific airflow direction, flow rectification is required to operate in OWC [5–7].

11.1.3 Flow Rectification

Flow rectification converts the bi-directional flow of the OWC into a unidirectional flow, enabling the UDT to operate in the OWC. In the early years, mechanical valves were used to rectify the flow, showing a schematic of such a unit, popularly known as the Masuda buoy (Fig. 11.12) [8]. The use of mechanical flow rectification has the following disadvantages,

I. The valve size must be large to handle the significant mass flow rate.
II. The response time of valves needs to be matched with the rapid change in the wave period.
III. Inertia concerns for the valves at lower flow rates.

The limitations of the mechanical valve-based flow rectification units hamper the performance of the UDT for WEC applications. Researchers have proposed using a pair of UDTs to overcome these issues, and this arrangement is called the turbine duo (TD).

11.1.4 Turbine Duo

The OWC duct is adjusted to accommodate the placement of two turbines, where one turbine harnesses power during a specific cycle of OWC while the other acts as a

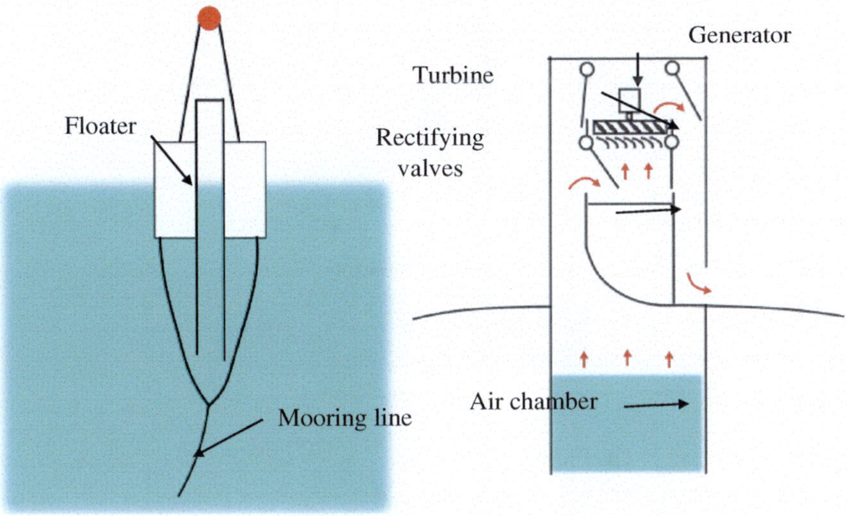

Fig. 11.12 Navigational buoys with flow rectification [4]

flow blocker, compelling more fluid to pass through the former turbine (Fig. 11.13). Likewise, the turbine's functionality is reversed for the subsequent cycle, facilitating flow rectification in both cycles of OWC without the need for mechanical valves [9].

From the literature, it can be concluded that the performance of TD is lesser than the individual UDT. This may be because the flow blocker turbine cannot choke the airflow entirely, leading to some flow leakage, which affects the performance of the power-extracting turbine. These units' performance can be enhanced by improving and preventing flow blockage. A fluidic diode (FD) can be used for this purpose.

11.1.5 Fluidic Diode

An FD has a variable geometrical shape for a given flow path; thus, fluid experiences variable resistance. It prevents flow reversal in applications such as biomedical, chemical processing, and nuclear reactors [10].

Figure 11.14 shows a simple nozzle-diffuser type FD, where the fluid experiences lesser resistance in forward flow (due to nozzle effect) than in reverse. Its performance is given by diodicity (Ψ) (Eq. 11.1).

$$\text{Diodicity}\,(\Psi) = \frac{\Delta p \text{ in reverse direction}}{\Delta p \text{ in forward direction}} = \frac{\Delta p_{2-1}}{\Delta p_{2-1}} \tag{11.1}$$

Various factors, such as the nature of the problems, fluid flow properties, and mechanical constraints, influence the design of fluidic diodes. Consequently,

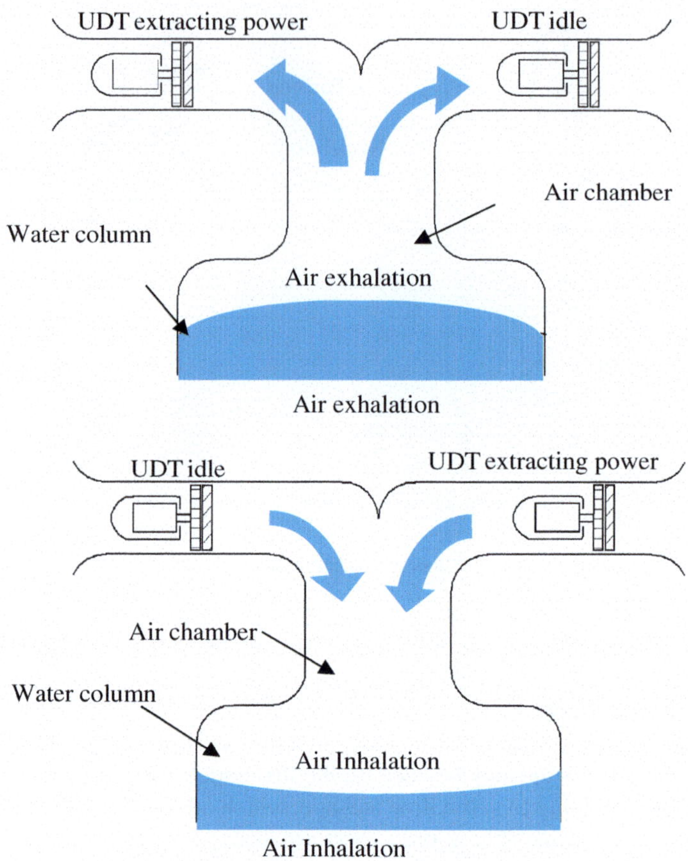

Fig. 11.13 Schematic of an OWC with UDTs (turbine duo)

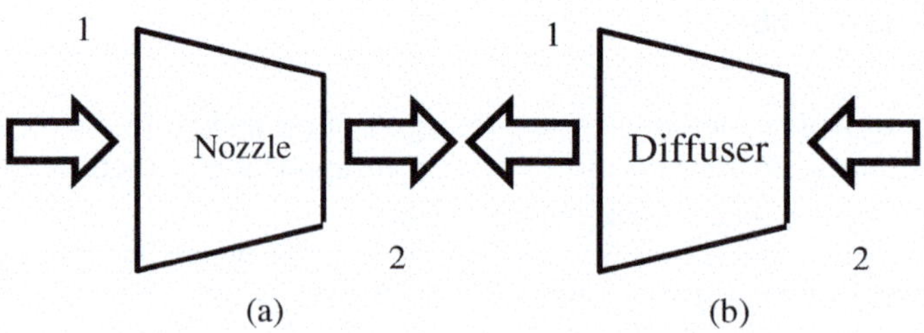

Fig. 11.14 Working of FD model **a** Forward direction **b** Reverse direction

researchers have developed new FD designs or adapted existing ones to suit specific applications. No distinct classification is available on FD; however, based on

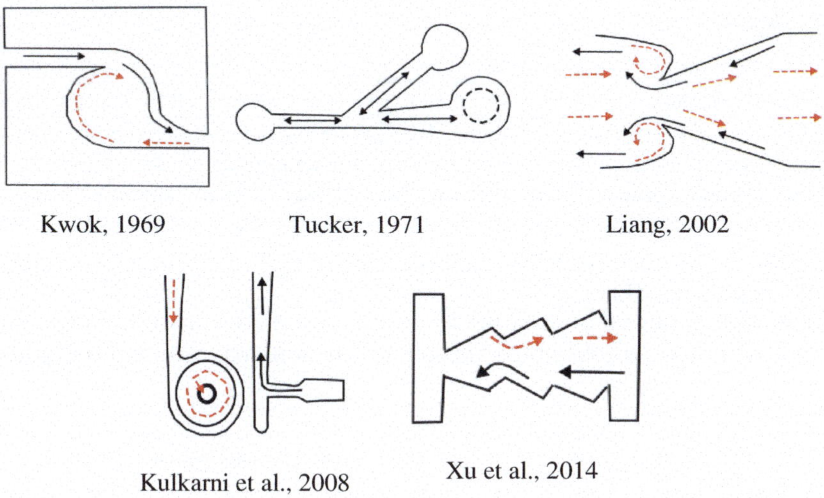

Fig. 11.15 Channel type FD found in the literature [13]

construction, they can be grouped into channel and bluff body types for explanation purposes. In a channel-type FD, only a flow path is modified to offer variable resistance to the flow. In the bluff body type, the variable resistance to the flow is provided by placing a bluff body inside the channel type. The solid line indicates the fluid flow in the forward direction, whereas the dotted line indicates the reverse flow.

Channel type

Figure 11.15 shows the channel type FD found in the open literature. Kwok proposed FD that offers resistance to the fluid flowing reversely using a vortex formed in the cavity region [11]. A series arrangement of this geometry could improve the unit's overall performance. Tucker presented a parametric study on the shape of the FD by varying the channel width ratio and intersection angle. They reported that an intersection angle of 30° yielded higher performance than the base case [12].

Ventrimeter-type FDs were also proposed; these operate on the Bernoulli principle. A vortex-type FD used in the nuclear reactor's cooling circuit detailed study on its performance is presented by researchers to improve its performance by modifying the entry and exit area of this diode [14–16]. They can be placed in series in transferring fluid with different pressure, and these enable preventing backflow. They are provided with the parametric study of the same. Xu et al. (2014) constructed a jig saw-shaped type FD, which can be easily fabricated. The experimental investigation is performed to understand its performance. A numerical parametric study was conducted to understand the performance of the saw angle and length on the performance. It was reported that the 7° saw angle yielded better performance [17].

Bluff body type

Figure 11.16 presents the bluff body type FD found in the literature. Telsa was the first one to introduce the concept of an FD [18]. It is constructed by keeping a bluff body placed inside the curved channel. This geometry is set in a series of repetitive geometry that offers smoother flow in one direction and higher resistance in the other. Telsa diode can be coupled with different diode designs like nozzle type to improve its performance [19]. Parametric studies on the Tesla diode also reported arriving at the number of repetitive geometry required to enhance the performance and response time of the unit [20, 21].

Fripp et al. presented a novel bluff body in a nozzle section. The shape offers variable resistance to the flow [22]. Hampton et al. proposed a split-type FD that can

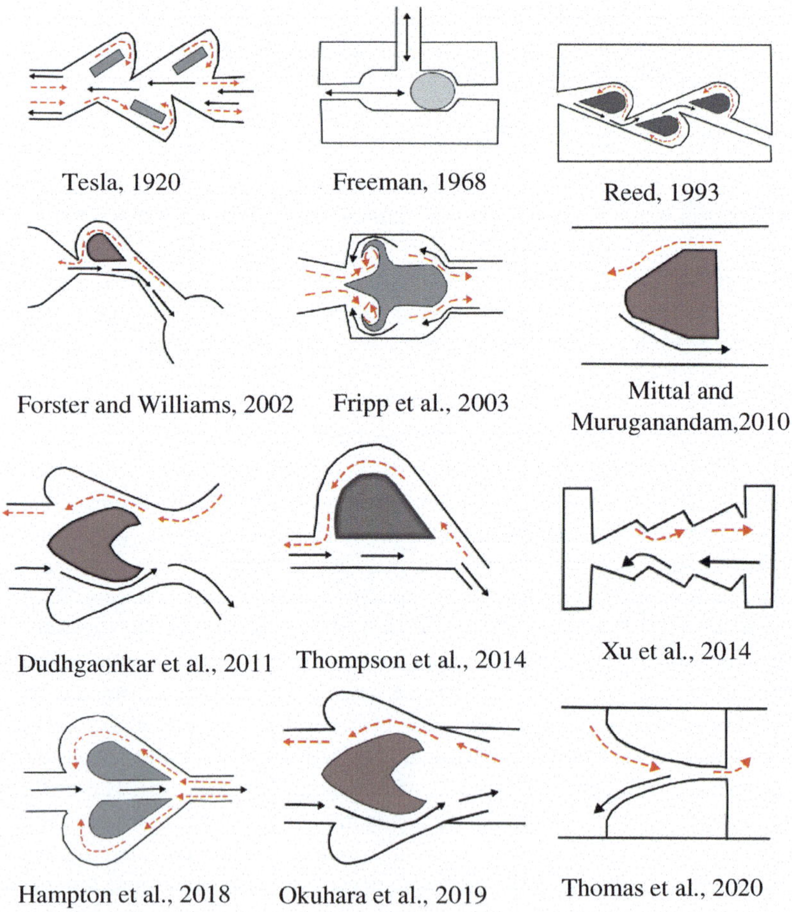

Fig. 11.16 Bluff body type FD found in the literature [13]

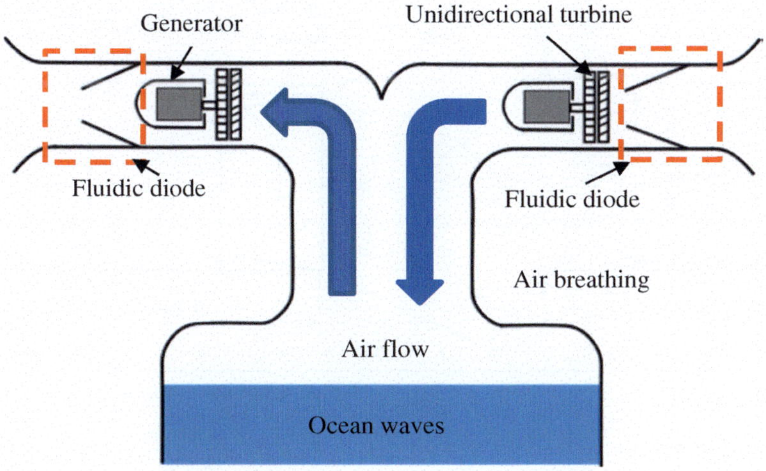

Fig. 11.17 Concept of the TD with FD [13]

provide smoother flow in the forward direction [23]. Dudhgaonkar et al. experimentally and numerically studied the FD for initial sizing for the wave energy system [24]. Okuhara et al. proposed a cascade-type FD for wave energy units [25]. An active method of FD is proposed by Freeman, which has a circular bluff body placed in the groove based on airflow; this pill can choke airflow based on the flow direction [26].

Figure 11.17 demonstrates the functionality of a fluidic diode with a turbine (TD). The fluidic diode either allows or resists the airflow to the turbine, depending on the direction of the flow. This variable resistance, combined with the turbine, enhances the flow blockage capabilities of the TD. Consequently, it directs the maximum airflow through the power-extracting turbine during each cycle of the OWC.

11.2 Numerical Modeling

11.2.1 Geometry of the FD Model

Figure 11.18 illustrates an FD designed for the OWC-WEC application, which is the focus of the current study. The FD features a tubular structure with a crescent-shaped bluff body (BD) positioned at its center. Surrounding the BD is a toroidal cup section, followed by a nozzle section (Fig. 11.18a). The linear dimensions are normalized based on the diameter of the duct (Table 11.1), and a nozzle angle (γ) of 20° is chosen for validation.

In forward flow, the fluid smoothly passes over the concave surface of the BD within the FD and exits through the nozzle section. In the case of reverse flow, the

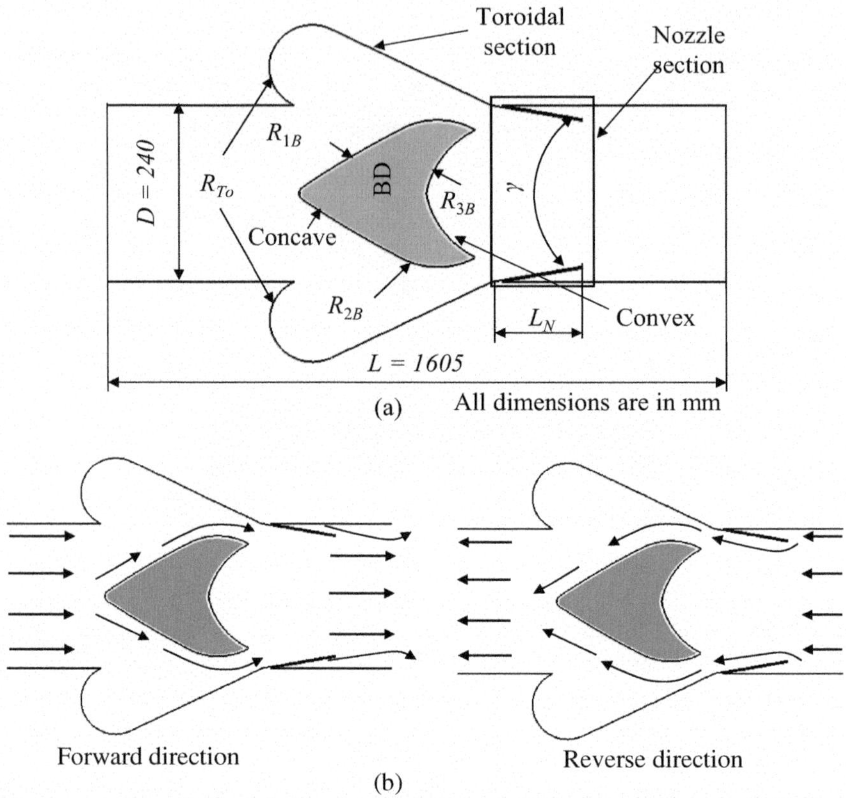

Fig. 11.18 Schematic of the FD model **a** geometrical details **b** working principle [13]

Table 11.1 Geometric details of the diode [25]

Geometry	Normalized dimension
R_{1B}	4.83
R_{2B}	0.42
R_{3B}	0.50
R_{To}	0.48
L_N	0.50

fluid enters through the nozzle section and undergoes diffusion. Subsequently, the diffused fluid interacts with the convex surface of the BD, creating a recirculation zone around it. This interaction introduces greater resistance to the incoming fluid (Fig. 11.18b).

11.2.2 Computational Domain and Setup

Figure 11.19 displays the computational domain used in the present study, which includes the fluidic diode (FD) and the duct. To discretize the computational domain, unstructured tetrahedral cells were employed, utilizing the ICEM CFD (V16.1) tool. To capture the details of sharp corners and the bluff body (BD), denser mesh elements were created in those regions (Fig. 11.20). Additionally, five prism layers were generated on the solid surface with an initial height of 0.2 mm and an exponential growth rate of 1.15.

The provided Fig. 11.21 illustrates the boundary conditions. In the case of forward flow, fluid enters the system through the inlet at a consistent flow rate and exits through an outlet maintained at atmospheric pressure. However, during reverse flow, these boundary conditions are reversed. The static pressure between the inlet and outlet is measured for the given flow direction. Experimental results were used to validate the numerical model.

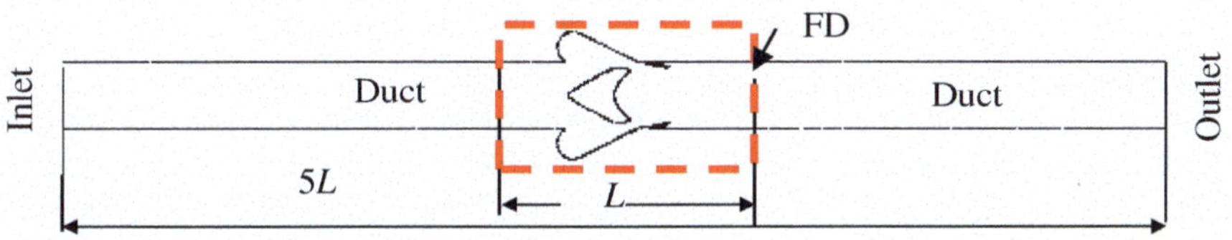

Fig. 11.19 Computational domain of the FD [13]

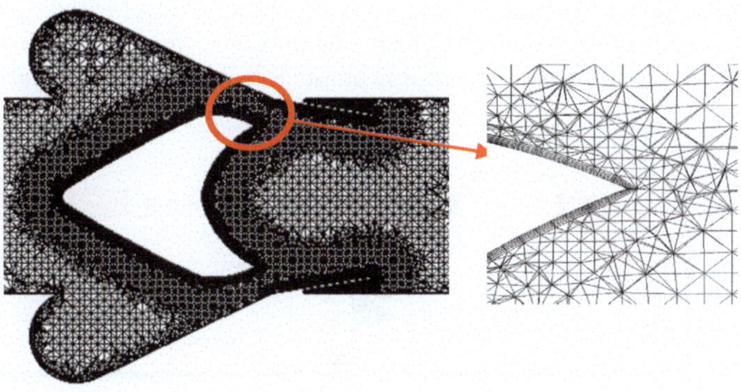

Fig. 11.20 Mesh around the BD [13]

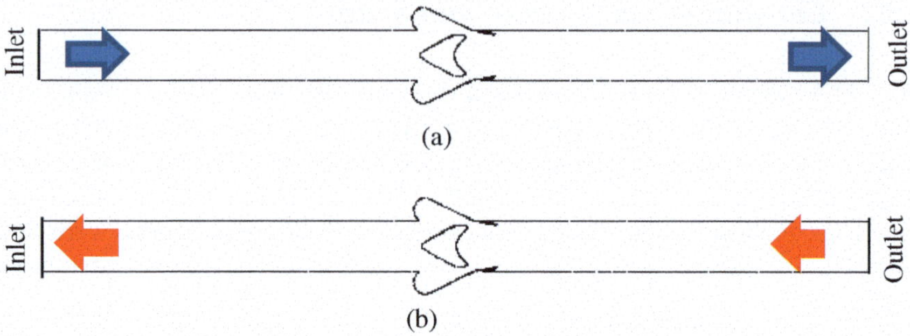

Fig. 11.21 Boundary condition **a** forward direction **b** reverse direction

11.2.3 Transient Study

The OWC output results in an oscillatory airflow that continuously fluctuates over time. This airflow variation can be approximated as a pseudo-sinusoidal profile (Fig. 11.22).

A wave approaching the OWC raises the water level within the water column. As a result, the air is gradually forced out of the OWC, representing the acceleration phase. Once the water column reaches its peak level, the rise stabilizes, and the water level slowly drops, signifying the deceleration phase. This cycle of acceleration and deceleration in the positive phase of the flow profile corresponds to the exhalation cycle of the OWC.

Similarly, as the wave departs from the OWC, the water level in the water column decreases, drawing air from the atmosphere into the OWC. Initially, the drop-in water level is rapid, representing the acceleration phase. Eventually, the drop stabilizes, and

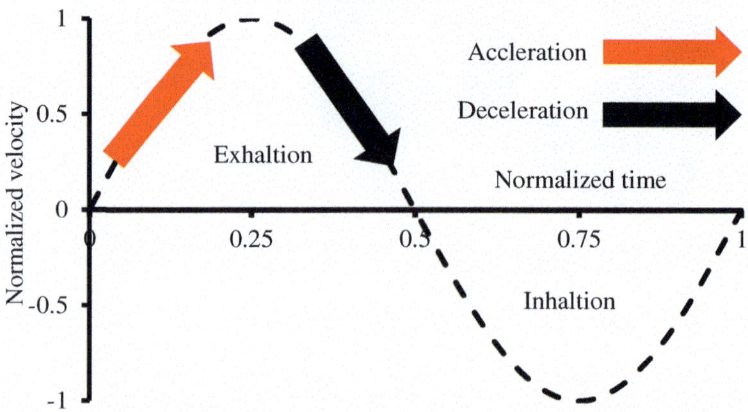

Fig. 11.22 Oscillatory airflow profile

the process slows down, indicating the deceleration phase. This cycle of acceleration and deceleration in the negative phase of the flow profile corresponds to the inhalation cycle of the OWC.

While the actual cyclic action of the OWC is not entirely smooth, the flow profile is commonly approximated as a pseudo-sinusoidal profile for analysis purposes. The axial velocity (UA) profile can be expressed using the following equation (Eq. 11.2).

$$\mathrm{UA} = \mathrm{UA}_{\mathrm{initial}} + \mathrm{UA}_{\mathrm{max}}\left(\sin\left(\frac{2\pi\,\Delta t}{T}\right)\right) \tag{11.2}$$

In practical situations, the air turbine and the diode assembly are utilized within an OWC system and are exposed to oscillatory airflow. In this oscillatory flow, the velocity magnitude undergoes continuous changes over time, resulting in the transition of the flow regime between laminar and turbulent states.

Due to the unsteady nature of the flow, the behavior of the fluid analyzed using a steady flow approach deviates significantly. The quasi-steady method can provide some insight into the unsteady performance but does not capture the complete transient characteristics of the system. Therefore, studying FD in the context of oscillatory flow becomes crucial for understanding the FD's transient flow behavior and performance when integrated with an air turbine.

To simulate the effect of natural sea waves, a pseudo-sinusoidal velocity profile is employed at the inlet of the numerical model using a user-defined function. This allows for the inclusion of wave profiles within the simulation, enabling a more realistic representation of the oscillatory flow conditions.

11.3 Case Study

11.3.1 Grid Sensitivity Study and Validation

The computational model's discretization error was assessed using the grid convergence index (GCI) method [27]. The GCI involves statistical calculations to determine the optimal number of mesh elements. The calculation results of the numerical study are presented in Table 11.2, indicating GCI values of 0.90% for the coarse grid and 0.17% for the fine grid. Since the GCI between consecutive grids is below 3%, it can be concluded that spatial convergence of the grid has been achieved, and further refinement is unlikely to impact the solution significantly. Therefore, the study maintains the medium grid, balancing computational time and accuracy.

Figure 11.23 shows the validation process of the numerical model by comparing it to experimental results. It also showcases the performance of various turbulent models employed. The pressure drop exhibits an upward trend with increasing flow rate in both flow directions, with a steeper rise observed during reverse flow compared to forward flow.

Table 11.2 Estimation of uncertainty due to discretization for the present numerical model

Parameter		Values
Number of cells (10^6)	$N_1/N_2/N_3$	6.063, 2.599, 1.151
Grid refinement factor	r_{21}, r_{32}	1.31, 1.33
Performance parameter (ϕ)	ϕ_1, ϕ_2, ϕ_3	183.15, 184.24, 190.51
Apparent order	H	6.18
Extrapolated values	$\phi_{\text{ext}}^{21}, \phi_{\text{ext}}^{32}$	182.89, 182.91
Approximate relative error	e_a^{21}, e_a^{32}	0.005951, 0.03403
Extrapolated relative error	$e_{\text{ext}}^{21}, e_{\text{ext}}^{32}$	0.00137, 0.00725
Grid convergence index	$\text{GCI}_{\text{fine}}^{21}, \text{GCI}_{\text{coarse}}^{32}$	0.170%, 0.90%

Three turbulence models were employed in the forward direction (Fig. 11.23a) to simulate the experimental results. All three models demonstrate closer agreement with the experimental data at lower flow rates (<0.4). With an increase in flow rate, the standard k-ε over-estimates the experimental results, with a root mean square error value of 52.04%. At the same time, the $k - \omega$ SST underpredicts the experimental results with a root mean square error value of 40.58%. In contrast, the realizable $k - \varepsilon$ model predicted the experimental results closely with a root mean square error value of 12.36%. Hence, the realizable $k - \varepsilon$ model was used for further investigations, and this turbulent model is also preferred for other applications with similar problem setups [28, 29]. The y + values obtained from simulations had a maximum value of 13, and the highest values were observed at the nozzle section.

11.3.2 Time Step Validation

The model's grid discretization was discussed in detail in the preceding section. To ensure the accuracy of the numerical results for transient simulations, it is important to validate the impact of time step discretization. Therefore, the total flow time was divided into three steps: $\Delta t_1 = 0.05$, $\Delta t_2 = 0.025$, and $\Delta t_3 = 0.0025$.

Figure 11.24 illustrates the pressure drop variation over the flow time for different time steps in the base model. The pressure drop during the inhalation cycle was found to be steeper compared to the exhalation cycle. There was a slight variation in the pressure drop pattern among the different time steps over the entire cycle. Based on this observation, it was assumed that further refinement in the time step would not significantly impact the results. Therefore, the fine time step (Δt_3) was selected for further processing.

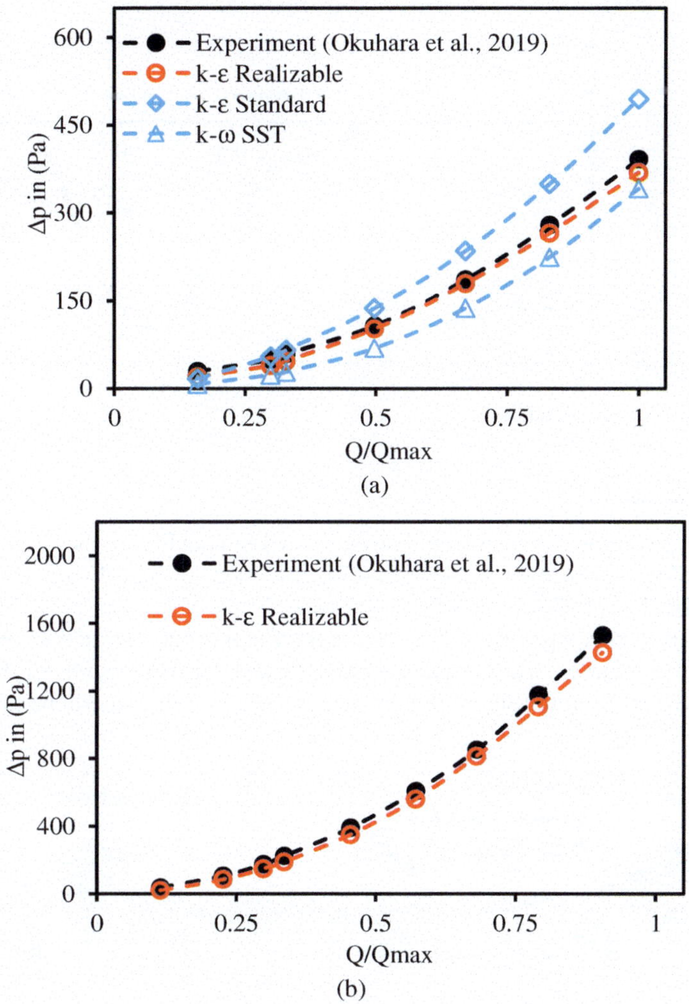

Fig. 11.23 Validation of the numerical model with experimental results **a** forward direction **b** reverse direction

11.3.3 Steady Flow Analysis

The internal fluid flow behavior of the fluidic diode is analyzed by obtaining the velocity contours and streamline distribution. These contours are obtained along the length of the middle section of the FD. It was observed that the flow pattern is symmetric along the axis. For simplicity, only a half section is considered.

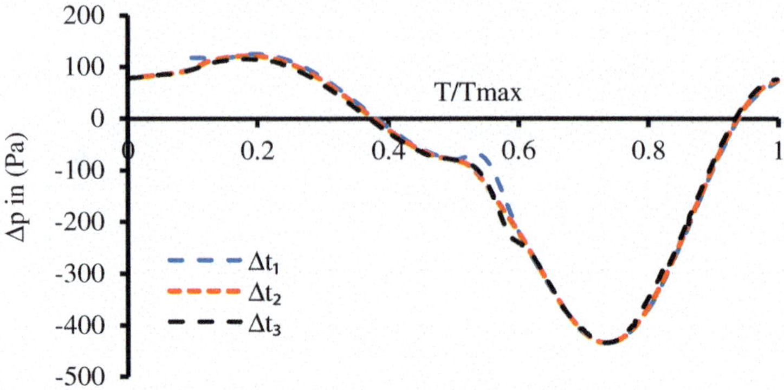

Fig. 11.24 Time step validation

Forward direction

Figure 11.25a presents the velocity distribution within the FD for the given flow rate. Normalized using the inlet velocity, and the direction of fluid flow is indicated by arrowheads.

As the fully developed fluid flow enters the FD through the duct in the forward direction, it encounters the BD and smoothly flows over its streamlined shape. This can be observed from the streamlined patterns (Fig. 11.25b), where the fluid moves

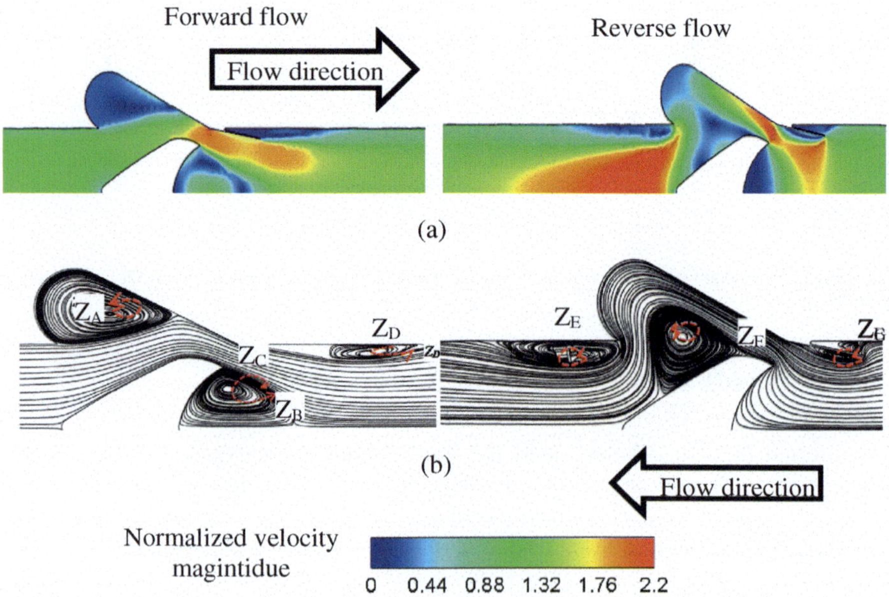

Fig. 11.25 Internal flow distribution **a** velocity contours **b** streamline distribution

smoothly over BD. However, the fluid following over the BD profile detaches from its surface as the concave surface abruptly ends, forming a wake behind the BD (Z_B).

Due to the smaller passage of the BD and the toroidal shape (Z_C) shown in Fig. 11.25a, the fluid experiences acceleration. The accelerated fluid (Z_C) tries to draw fluid in the cavity section along the flow. This differential velocity induces a circulation motion (Z_A).

Subsequently, fluid leaves the nozzle section, where the fluid swiftly transitions from a smaller area to a larger one. Due to the steepness of the nozzle, the exiting fluid experiences an adverse pressure gradient, causing backflow (Z_D).

Reverse flow

Let's consider fluid flow in the reverse direction (Fig. 11.25). The flow enters through the nozzle section, where a vena contract narrows the flow area, causing the fluid to accelerate within the nozzle unit. As it progresses, the fluid experiences additional acceleration as it passes through the smaller gap between the BD and the cup.

Due to the nozzle, the fluid impinges on the wall at the entry section, forming a recirculation zone that occupies the entire nozzle length (Z_E). As the fluid exits the nozzle, it encounters the convex shape of the BD, acting like a vane that deflects the flow from the nozzle into the toroidal area. However, the fluid, following the BD shape, experiences a sudden change in shape and is thus unable to attach to the surface, causing it to separate through the wall of the BD.

Finally, as the fluid gains circular motion following through the toroidal section, it cannot align with the duct flow at the end of the FD. These phenomena lead to flow separation on the walls of the duct and toroidal section, giving rise to the formation of a vortex (Z_G). The centerline velocity of the fluid leaving the FD is approximately twice the input velocity.

Pressure distribution

Figure 11.26 presents the static pressure distribution of the FD. As the flow rate increases, the pressure drop across also rises. When fluid flows through the FD, it experiences a lower resistance (pressure drop) in the forward direction than in the reverse. In forward flow, the fluid experiences higher pressure drop around the BD, as it obstructs the fluid flow, creating a stagnation zone. This leads to a decrease in velocity and an increase in pressure. The accelerating fluid passing through the narrow region of the BD and cup section overcomes resistance. Fluid flowing further through the nozzle gains kinetic energy, resulting in a further reduction in pressure.

During reverse flow, the fluid enters through the nozzle section, experiencing diffusion, increasing the pressure energy, thus experiencing higher resistance in this section. Subsequently, the fluid stream encounters a concave shape resembling a shallow cup, leading to increased drag resistance and higher pressure in that region.

Dynamic analysis

Figure 11.27 provides an overview of the streamline patterns within the FD for fluid flow in the forward path for various normalized time steps (t'). The flow regime transitions fall in the laminar region at smaller time steps. Fluid particles follow the

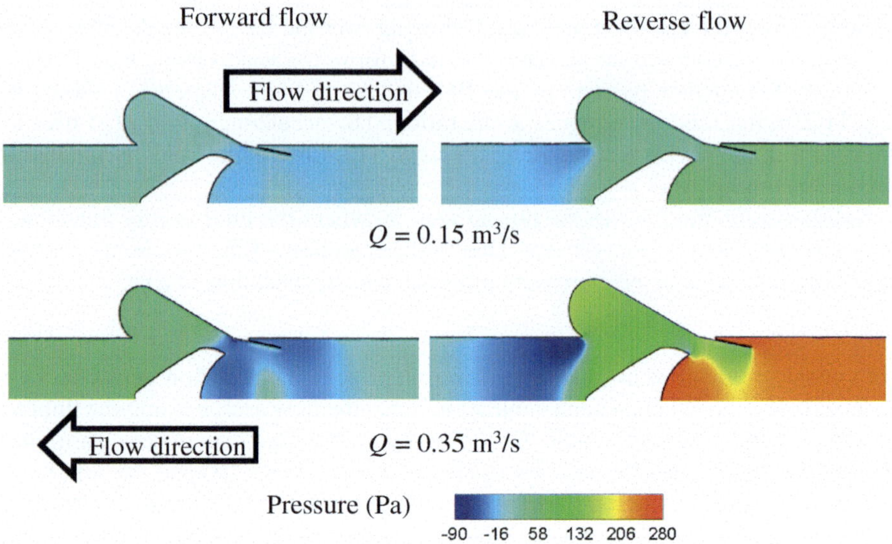

Fig. 11.26 Pressure distribution across the FD for different flow rates

orderly motion and stick to the walls. However, it detaches at sharp corners (Z_A, Z_B, and Z_C at $t' = 0.010$).

The saddle point, found on the wall of the cup section, travels along with the flow as the vortex dimension increases with longer flow times ($t' = 0.010$ to 0.066). However, it is movement restricted by the geometric constraints of the cup. This flow separation gradually develops with time and transforms into a corner vortex (Z_A). In the case of a single-objective optimized model, the vortex develops more rapidly and occupies the entire cup section earlier than in other cases. The fully developed flow comes into contact with the BD, known as the node of attachment, and subsequently alters its path.

The fluid flowing in the smaller gap between the BD and cup experiences acceleration when it exits through the nozzle. This causes a sudden expansion in the flow, thus causing separation from the nozzle walls due to this momentum difference. This separation evolves into vortex motion, intensifying with flow time and stabilizing at $t' = 0.066$.

Figure 11.28 presents a streamline distribution for different time steps in the reverse path. Here, the flow enters the nozzle section and exits through the FD. As the fluid flow gets reversed, the fluid in the forward path experiences resistance from the fluid entry through the outlet. This dynamic action creates a circulation zone (Z_G) at the end of the nozzle, leading to increased resistance at the beginning of the reverse direction, as shown in the streamline ($t' = 0.686$). With a further increase in flow time, the rotational direction of the vortex aligns with the flow path, and it merges into the mainstream, disappearing after $t' = 0.706$.

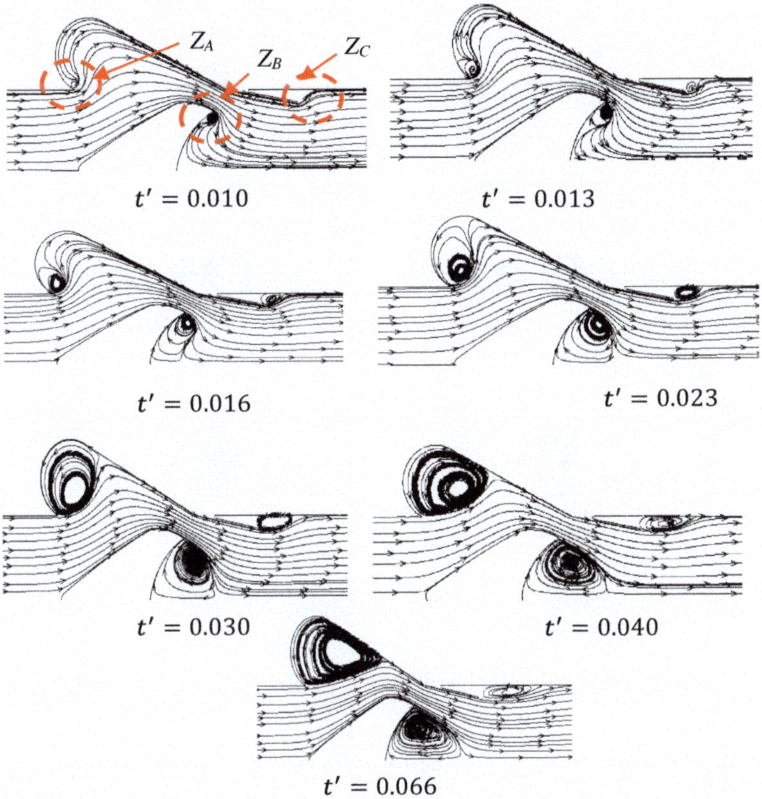

Fig. 11.27 Variation of the streamline for different flow times in the forward direction

Fluid traveling through the nozzle experiences an adverse pressure gradient due to the steepness of the nozzle wall. As a result, it detaches from its surface (Z_D). The saddle point travel along the flow, and thus vortex formed continues to grow and occupies the whole length of the nozzle at $t' = 0.706$. Due to the adverse pressure gradient created by the surface of the BD, the fluid downstream of the BD experiences separation. Simultaneously, the fluid in the cup section gains higher momentum due to its curvature, keeping the vortex attached to the BD surface (Z_E).

With increasing flow time, the flow velocity in the smaller gap between the BD and cup rises, drawing the vortex on the BD surface into the cup section along with the flow. This phenomenon resembles flow inside a cavity where a region of trapped fluid experiences relative movement. The toroidal shape of the section initiates a secondary vortex, as observed in the streamline contour (Z_E) ($t' = 0.866$). It also grows in size with increasing flow time.

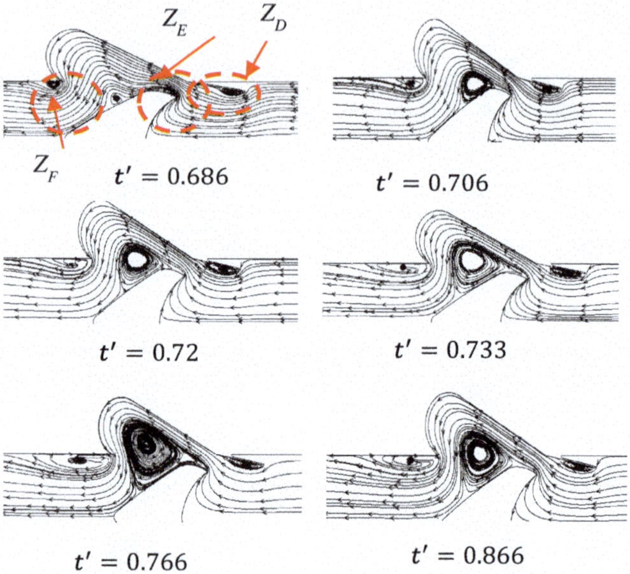

Fig. 11.28 Variation of the streamline for different flow times in the reverse direction

The flow exiting the toroidal cup section with larger momentum cannot align with a straight duct. This leads to the formation of a recirculation zone (Z_F) that expands with flow time. The rotational motion of this zone opposes the incoming flow path, resulting in resistance against the flow and an overall increase in fluid resistance.

11.4 Conclusion

A pair unidirectional turbine can be used in an oscillating water column type wave energy device. This kind of arrangement eliminates mechanical flow rectification. However, due to inadequate flow blockage, it has poor performance. A fluidic diode (FD) produces a variable resistance to the fluid depending upon the flow direction. It can be used with the turbine to enhance flow blockage and improve the unit's overall performance.

The didocity governs its performance, the ratio of pressure drops in forward to reverse direction. This article presents the numerical investigation of the FD used for wave energy application. Initially, steady state analysis is presented, followed by transient analysis. The numerical model studies the transient characteristics by incorporating oscillatory airflow conditions (natural sea wave profile). The FD works under periodic flow and experiences higher reverse flow resistance than a steady flow.

Due to the dynamic nature of the flow, the vortex intensities are more severe in the reverse direction compared to a steady flow. The shape of FD plays an important role in determining its performance.

References

1. Gunn K, Stock WC (2012) Quantifying the global wave power resource. Renew Energy 44:296–304. https://doi.org/10.1016/J.RENENE.2012.01.101
2. Doddamani, H, Samad, A.: Dynamic performance of a fluidic diode subjected to periodic flow. Ocean Eng. 268 (2023). https://doi.org/10.1016/j.oceaneng.2022.113381
3. Kumar PM, Halder P, Husain A, Samad A (2019) Performance enhancement of Wells turbine: combined radiused edge blade tip, static extended trailing edge, and variable thickness modifications. Ocean Eng 185:47–58. https://doi.org/10.1016/j.oceaneng.2019.05.041
4. Falcão AFO, Henriques JCC (2016) Oscillating-water-column wave energy converters and air turbines: a review. Renew Energy 85:1391–1424. https://doi.org/10.1016/j.renene.2015.07.086
5. Ansarifard, N., Fleming, A., Henderson, A., Kianejad, S.S., Chai, S.: Design optimisation of a unidirectional centrifugal radial-air-turbine for application in OWC wave energy converters. Energies (Basel) **12**(2791) (2019). https://doi.org/10.3390/en12142791
6. Takao M, Itakura K, Setoguchi T, Kim TH, Kaneko K, Thakker A (2002) Performance of a radial turbine for wave power conversion 3:62–67
7. Takao M, Setoguchi T, Kaneko K, Kim TH, Maeda H, Inoue M (2002) Impulse turbine for wave power conversion with air flow rectification system. Int. J. Offshore Polar Eng. 12:142–146. https://doi.org/10.1299/kikaib.66.646_1421
8. Masuda, Y., Yamazaki, T., Outa, Y., McCormick, M.: Study of backward bent duct buoy. In: Oceans Conference Record, 28 September to 01 October, 1987. Halifax, Canada (1987). https://doi.org/10.1109/oceans.1987.1160750
9. Jayashankar V, Anand S, Geetha T, Santhakumar S, Jagadeesh KV, Ravindran M, Setoguchi T, Takao M, Toyota K, Nagata S (2009) A twin unidirectional impulse turbine topology for OWC based wave energy plants. Renew. Energy 34:692–698. https://doi.org/10.1016/j.renene.2008.05.028
10. Thomas SK, Muruganandam TM (2018) A review of acoustic compressors and pumps from fluidics perspective. Sens. Actuators A Phys. 283:42–53. https://doi.org/10.1016/j.sna.2018.09.031
11. Kwok, C.C.K.: Vortex vent diode, US3461897 (1969)
12. Tucker, H.G.: Fluidic diode, US3604442 (1971)
13. Hithaish, D., Siddique, M.H., Samad, A.: A pareto optimal front of fluidic diode for a wave energy harnessing device. Ocean Eng. **260**, (2022). https://doi.org/10.1016/j.oceaneng.2022.111821
14. Liang, Z.: Autonomous fluid control system having a fluid diode, US9404349B2 (2002)
15. Suchezky, M.E., Drobnis, J.B.: Valvular conduit exhaust Manifold. US2019/0162104A1 (2019)
16. Kulkarni AA, Ranade VV, Rajeev R, Koganti SB (2008) CFD simulation of flow in vortex diodes. AIChE J 54:1139–1152. https://doi.org/10.1002/aic.11439
17. Xu, Y., Yan, W., Cao, T., Guo, L.: Study on the valveless micropumps with saw-tooth microchannels. In: 11th IEEE International Conference on Control and Automation (ICCA). 18–20. Taichung, Taiwan (2014). https://doi.org/10.1109/ICCA.2014.6871102.
18. Tesla, N.: Valvular conduit patent, US1329559 (1920)
19. Forster, F.K., Williams, B.E.: Parametric design of fixed geometry microvalves the Tesser valve. In: ASME International Mechanical Engineering Congress and Exposition 17–22 November, 2002. pp. 431–437. New Orleans, Louisiana (2002)
20. Reed, J.L.: Fluidic rectifier. US5265636 (1993)

21. Thompson SM, Ma HB, Wilson C (2011) Investigation of a flat-plate oscillating heat pipe with Tesla-type check valves. Exp. Therm. Fluid Sci. 35:1265–1273. https://doi.org/10.1016/j.expthermflusci.2011.04.014
22. Fripp, M.L., Dykstra, J.D., Dejesus, O.: Bidirectional downhole fluid flow control system and method (2003)
23. Hampton, K., Fletcher, D.E., Graichen, B.M., Gilmer, M.C., James, M.H., Niedert, A.D.: Fluidic diode check valve (2018)
24. Dudhgaonkar, P. V., Jayashankar, V., Jalihal, P., Kedarnath, S., Setoguchi, T., Takao, M., Nagata, S., Toyota, K.: Fluidic components for oscillating water column based wave energy plants. In: ASME-JSME-KSME 2011 Joint Fluids Engineering Conference. Hamamatsu, Japan (2011). https://doi.org/10.1115/AJK2011-07035
25. Okuhara, S., Ashraful, A.M.M., Takao, M., Kinoue, Y.: Performance of fluidic diode for a twin unidirectional impulse turbine. In: Earth and Environmental Science. Kyoto, Japan (2019). https://doi.org/10.1088/1755-1315/240/5/052011
26. Freeman, P.A.: Pill type fluidic devices. US3362633 (1968)
27. Celik IB, Ghia U, Roache PJ, Freitas CJ, Coleman H, Raad PE (2008) Procedure for estimation and reporting of uncertainty due to discretization in CFD applications. J. Fluids Eng. Trans. ASME 130:1–14. https://doi.org/10.1115/1.2960953
28. Hua, M.W., Xiao, K.Q., Lin, C., Lu, Q., Qiao, R.W.: Numerical study on energy-converging efficiency of the ducts of vertical axis tidal current turbine in restricted water. Ocean Eng. **210**(107320) (2020). https://doi.org/10.1016/j.oceaneng.2020.107320
29. Hollingshead CL, Johnson MC, Barfuss SL, Spall RE (2011) Discharge coefficient performance of venturi, standard concentric orifice plate, v-cone and wedge flow meters at low Reynolds numbers. J Pet Sci Eng 78:559–566. https://doi.org/10.1016/j.petrol.2011.08.008

Part IV
Modeling and Simulations

Chapter 12
Mathematical Modelling of Reactors Used for Syngas Fermentation—Contemporary Practices and Challenges

Dinabandhu Manna, Soumitra Pati, Sudipta De, and Ranjana Chowdhury

Abstract Biofuels can successfully replace fossil fuels as they emit less amount of greenhouse gas during combustion. Syngas fermentation is one of the sustainable routes of producing bioethanol, methane, etc., particularly from ligno-cellulosic biomass (LCB) rich in lignin. In this process, syngas generated via gasification of LCB is converted to liquid fuels with the aid of suitable microorganism (*Clostridium carboxidivorans*, *Clostridium ljungdahlii*, etc.) via acetyl-CoA pathway (Wood-Ljungdahl pathway). Syngas can also be converted to methane through microbial route. The present article discusses about the microbial principles, reactions and different reactors used for both bioethanol and methane production from syngas. The chapter aims to revisit the mathematical modelling practices for different reactors used for syngas fermentation for bioethanol formation along with identification of loopholes and possible recommendations.

Keywords Syngas fermentation · Biofuels · Biochemical reactions · Bioreactors · Mathematical models

Nomenclature

R_i	Mass transfer rate of any gaseous reactant i
C	Concentration of species
C_l	Bulk concentration of species
k_l	Mass transfer coefficient

D. Manna · R. Chowdhury (✉)
Department of Chemical Engineering, Jadavpur University, 188, Raja S.C Mallick Road, Kolkata 700032, West Bengal, India
e-mail: ranjana.chowdhury@jadavpuruniversity.in

S. Pati · S. De
Department of Mechanical Engineering, Jadavpur University, 188, Raja S.C Mallick Road, Kolkata 700032, West Bengal, India

S. De et al. (eds.), *Challenges and Opportunities of Distributed Renewable Power*, Energy, Environment, and Sustainability, https://doi.org/10.1007/978-981-97-1406-3_12

a	Surface area per unit volume
p_i	Partial pressure of ith species
R_m	Rate of assimilation of m
r_g	Microbial growth rate
Y	Yield coefficient
μ_{max}	Maximum specific growth rate
μ	Specific growth rate
X	Concentration of biomass
k_{Im}	Inhibition constant of substrate (or reactant) m
k_{sm}	Saturation constant for substrate m
V_g	Gas phase volume
t	Time
V_i	Inlet volume
V_o	Outlet volume
H_i	Henry's constant of ith component
N_{Gout}	Molar flow rate of outlet gas
Q_{out}	Volumetric flow rate of outlet gas
C^*	Equilibrium concentration
r_i	Reaction rate of ith component
v_i	Uptake rate of ith substrate
$v_{max,i}$	Maximum uptake rate of ith substrate
S_i	Dissolved concentration (mmol/L) of i-th gaseous substrate
ε_l	Liquid phase volume fraction
u_l	Liquid phase velocity
u_g	Gas phase velocity
D_A	Axial dispersion coefficient
M	Molecular weight
P	Pressure
T	Temperature
g	Gravitational constant
R	Universal gas constant
L	Length
ρ	Density
D	Dilution rate
r_b	Average radius of gas bubbles
A_C	Interfacial area
V_m	Working volume

Abbreviations

CNG	Compressed natural gas
LCB	Ligno-cellulosic biomass
ATR	Autothermal reforming

SMR	Steam methane reforming
CODH	Carbon monoxide dehydrogenase
NADH	Nicotinamide adenine dinucleotide
NADPH	Nicotinamide adenine dinucleotide phosphate
ADP	Adenosine diphosphate
ATP	Adenosine triphosphate
CSTR	Continuous stirred tank reactor
TBR	Trickle bed reactor
PBR	Packed bed reactor
HFMBR	Hollow-fibre membrane reactor
MBFR	Membrane biofilm reactor

12.1 Introduction

India along with the whole world is facing the grave environmental issues caused by the use of fossil fuels. Ever increasing price of fossil fuel is another serious concern. The fossil fuels cause serious damage to the environment due to their emission characteristics while being used in various industrial and automobile application. Also, the refinery process of these fuels is immensely energy intense causing further damage to the carbon neutral frame that the world is trying to achieve. In the light of this problem, new technologies and fuels need to be developed in order to find more environment friendly and cheap alternatives to fossil fuels. Biofuels in this regard have been making an impression in the global energy scenario for a long time now. The biofuels like bioethanol and methane are mostly carbon neutral and are produced through less energy intense processes, which can contribute significantly in reducing the global emission burden. The main limitation of using fossil fuel biofuels can be successfully used in substituting petrol and diesel in automobile and industrial applications. Methanol, ethanol, butanol and other higher alcohols can be used as an alternative to petrol fuels and long chain fatty acid esters commonly known as biodiesel can be used as an alternative to diesel fuels. Similarly, in place of compressed natural gas (CNG), or compressed biogas and methane can also act as an alternative to gasoline. The gasoline alternatives are divided into three categories according to the feedstock used, namely, first, s, and third generation biofuels. The first-generation biofuel is directly produced from food crops, i.e., edible biomass. The direct fermentation of rice, wheat, sugarcane juice, etc. in the presence of suitable enzymes is the process of first-generation biofuel production. In the second-generation biofuel production technology, the feedstocks used are the non-edible biomass resources like rice straw, wheat straw, sugarcane bagasse, corn stalk, mustard stalk, jute stick, etc. and other woody and non woody biomass that can be generally categorized as ligno-cellulosic biomass (LCB). The third-generation biofuels are produced from microalgae, which are cultured as dedicated energy crops having huge potential to produce energy in the form of fuels [1]. There are two major techniques of biofuel production in this second-generation pathway that is conducted by the biochemical route. First technique involves the removal of lignin from LCB via different pre-treatment methods followed by enzymatic hydrolysis and fermentation. This is the most common pathway to convert lingo-cellulose to biofuels. The second technique is the gasification of biomass followed by syngas fermentation in the presence of suitable acetogenic microorganisms. This process does not require lignin removal, hence can utilize the LCB in totality. The process of gasification involves thermal conversion of biomass in the presence of suitable gasifying agents like air, oxygen, steam, or CO_2 in general. The supply of gasifying agent is kept sub-stoichiometric and further production of gases with substantial heating value is achieved. The produced syngas is then fermented in the presence of suitable micro-organisms to convert the syngas to biofuels like ethanol, methane etc. Among the various renewable fuel options, ethanol and methane produced from syngas fermentation have emerged as promising candidates due to their application in automobile sector ensuring low carbon footprint. Steel plant off-gases composed mainly of CO, H_2 and CO_2, can also be converted via syngas fermentation [2]. The syngas fermentation is hybridization of thermochemical processes (namely pyrolysis and gasification) and biochemical process (namely fermentation of syngas) to generate liquid and gaseous biofuels. Syngas is a mixture of gases having main components as carbon monoxide (CO), hydrogen (H_2) and small amount of carbon dioxide (CO_2). In this process, the biomass is first converted either through gasification or pyrolysis to syngas, which in turn is converted to either alcohols or methane via anaerobic fermentation. However, the major bottleneck of this technology is associated with mass transfer limitation of the substrates ($CO + H_2$) from the gas phase to the liquid phase and inhibitory effects of CO on the fermentation process. For the syngas fermentation, various reactors like continuous stirred tank, bubble column, packed bed, trickle bed and membrane reactors are in practice. Mathematical modelling is an important aspect of developing any technology. The modelling of syngas fermentation is a critical task because both gas to liquid mass transfer and liquid-phase microbial kinetics play important roles in this process. The modelling procedure varies with varying reactor configuration.

In this chapter, emphasis has been made on the mathematical modelling of syngas fermentation considering different reactor configurations.

12.2 General Overview of Syngas Production

Synthesis gas, commonly referred to as syngas, is a versatile and crucial intermediate in the field of chemical engineering and energy production. It is a mixture primarily composed of carbon monoxide (CO) and hydrogen (H_2), carbon dioxide (CO_2) and other trace components. Syngas is the key building block for the synthesis of various chemicals and fuels, making it an essential component of modern industrial processes. This note delves into the methods, applications and significance of synthesis gas production.

12.2.1 Steam Methane Reforming (SMR)

This method involves the reaction of methane (CH_4) with steam (H_2O) over a catalyst, typically nickel-based, at high temperatures (700–1000 °C) and moderate pressures [3]. The endothermic nature of the reaction necessitates a continuous supply of heat to drive the process.

12.2.2 Autothermal Reforming (ATR)

ATR is a combination of SMR and partial oxidation, allowing for the production of syngas with an intermediate composition of hydrogen, carbon monoxide and carbon dioxide [4].

12.2.3 Biomass Gasification

Biomass, such as wood, agricultural residues or organic waste, can be converted into syngas through gasification. In this process, the biomass is reacted with a controlled amount of air, oxygen or steam in a high-temperature environment, resulting in the production of syngas along with char and ash residues. In this process, the biomass is fed to the gasifier and it passes through four different stages namely drying, pyrolysis, oxidation and reduction. The moisture is removed from the biomass in drying. In pyrolysis, the biomass is broken down to its elemental composition which is further oxidized in the presence of suitable oxidizing agent like O_2 and steam. The reduction reactions occur due to the insufficiency of the oxidizing agents and the syngas is produced. Biomass gasification is the most common procedure of producing syngas and has been extensively researched [5, 6].

12.2.4 Coal Gasification

Coal can similarly be gasified to produce syngas. Coal gasification involves reacting coal with oxygen and steam at high temperatures, generating syngas and leaving behind ash and other impurities [7]. This process reduces the pollutant emission due to the burning of coal without compromising with the heating value.

12.3 Fundamental Bioengineering of Syngas Fermentation

12.3.1 Microbial Reactions for Syngas Fermentation to Methane

Methanation of syngas can be done either catalytically or through microbial route. While catalytic conversion requires high pressure and temperature, the fermentative process can be conducted at atmospheric pressure and at temperature in the range of 35–75 °C. Due to the involvement of much milder operating condition, the biological methanation of syngas is gaining popularity. The reactions are as follows:

$$4CO + 2H_2O \rightarrow CH_4 + 3CO_2 \tag{12.1}$$

$$4H_2 + CO_2 \rightarrow CH_4 + 2H_2O \tag{12.2}$$

Biomethanation usually involves syntrophic microorganisms either in pure or mixed culture. In the fermentation, syngas serves both as carbon and energy source. According to current literature review, mainly carboxydotrophic hydrogenogenesis $(CO+H_2O \rightarrow H_2+CO_2)$, homoacetogenesis $(4H_2+CO_2 \leftrightarrow CH_3COOH+2H_2O)$, carboxydotrophic acetogenesis $(4CO + 2H_2O \rightarrow CH_3COOH + 2CO_2)$, carboxydotrophic methanogenesis $(4CO + 2H_2O \rightarrow CH_4 + 3CO_2)$, acetoclastic methanogenesis $(CH_3COOH \rightarrow CH_4 + 3CO_2)$ and hydrogenotrophic methanogenesis $(4H_2 + CO_2 \rightarrow CH_4 + 2H_2O)$ occur during syngas methanation. The mechanism varies with the operating temperature [8–10].

Traditionally carbon monoxide can be converted to acetate in the presence of *Peptostreptococcus productus* [11, 12] *Acetobacterium woodii* [13], *Clostridium thermoaceticum* [14, 15] and *Eubacterium woodii* [16] Anaerobic Bacteria like *Peptostreptococcus productus* can ferment hydrogen and carbon di oxide to acetic acid [17–19]. Two species of purple non-sulphur bacteria, *Rhodopseudomonas gelatinosa* [20, 21] and *Rhodospirillum rubrum* can convert CO and H_2O to CO_2 and H_2 through water gas shift reaction [22]. Methanogenic bacteria like *Methanospirillum hungatii, Methanobacterium formicicum, Methanobrevibacter smithii, Methanosarcina barkeri* can convert CO_2 and H_2 to methane [23–25]. Acetic acid can be converted to methane and CO_2 by *Methanothrix soehngenii* and *Methanosarcina sp.*, such as *Methanosarcina barkeri* [26].

12.3.2 Microbial Reactions for Syngas Fermentation to Ethanol

The syngas fermentation process is conducted by converting the mixture of CO, CO_2 and H_2 to alcohols and organic acids in the presence of acetogenic bacteria. *Moorellathermoacetica* (formerly called *C. thermoaceticum*) has been the oldest acetogen to be studied in lab scale for syngas fermentation. Other important acetogens used in this regard are *C. Ljungdahlii, C. aceticum, C. carboxidivorans, Acetobacteriumwoodii, C. ragsdalei* and *C. autoethanogenum.* Till now, there are more than 100 species of acetogens that have been developed by different researchers for the purpose of syngas fermentation. They can be collected from different sources like land, sludge, sediments, intestinal tracts of animals, etc. These microorganisms are mesophilic in nature and experiments suggest that 30–40 °C is the favorable temperature range for these microorganisms to facilitate gas–liquid mass transfer.

Harland Goff Wood and Lars Gerhard Ljungdahl first used *Moorellathermoacetica* to determine the acetyl-Coenzyme A(Acetyl- CoA) pathway enzymology in their laboratories. The bioconversion of the syngas following the acetyl-CoA pathway is therefore also known as the Wood- Ljungdahl pathway. This pathway is found to be a linear pathway that occurs in both oxidative and reductive direction. The reductive direction is the conversion of CO_2 to acetate and the oxidative direction is the conversion of acetate to CO_2. Two branches namely methyl and carbonyl branches are involved in the production of acetyl- CoA from CO and CO_2 respectively, which is further converted to alcohol and acetate having acetaldehyde and acetyl phosphate as their respective precursors. There are various literatures that describe the bio-chemistry involved in this process in detail [27, 28]. Acetyl-CoA is considered to be a metabolic intermediate in acetogens and is utilized to produce ethanol, butanol, hexanol, acetate, butyrate, hexanoate and cell mass. The conversion process from Acetyl-CoA to ethanol is known as solventogenesis and the conversion of Acetyl-CoA to acetate is known as acetogenesis. The chemical equations for the production of bio-ethanol are as follows.

$$6CO + H_2O \rightarrow C_2H_5OH + 4CO_2 \tag{12.3}$$

$$4CO + 2H_2O \rightarrow CH_3COOH + 2CO_2 \tag{12.4}$$

$$6H_2 + 2CO_2 \rightarrow C_2H_5OH + 3H_2O \tag{12.5}$$

$$4H_2 + 2CO_2 \rightarrow CH_3COOH + 2H_2O \tag{12.6}$$

However, these reactions do not take place naturally as phase change and gas-liquid mass transfer is involved in this process. In the acetyl-CoA pathway, the acetogens are responsible for metabolising the H_2, CO and CO_2 and produce ethanol and acetate. The required electrons for the reduction process are generated by oxidising

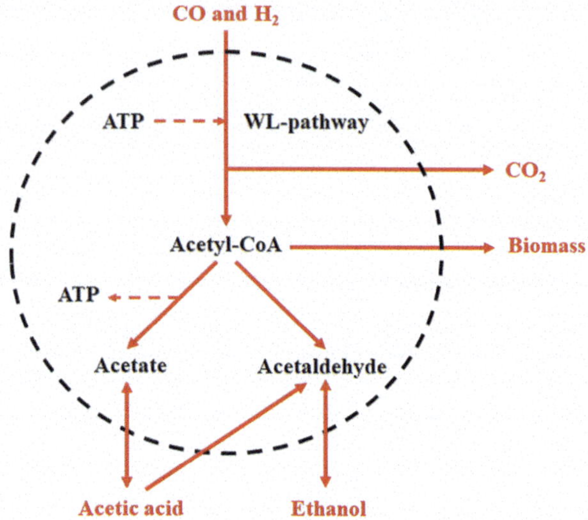

Fig. 12.1 Simplified pathway for syngas fermentation

the H_2 by hydrogenase and also by the oxidation of CO by carbon monoxide dehydrogenase (CODH). This donated reducing power is in the form of electrons and is carried by electron carrier pairs $NADH/NAD^+$, $NADPH/NADP^+$ or ferredoxin. Adenosine triphosphate (ATP) is responsible for the transport of chemical energy required for metabolism in the cells. The ATP is converted to adenosine diphosphate (ADP) by the energy released in hydrolysis of phosphate bonds. Generation of ATP for cell growth necessitates production of acetate from Acetyl-CoA. The reduced products like acetaldehyde, ethanol etc. are produced by the reduction of acetate. The pathway of fermentative production of ethanol from syngas is represented in Fig. 12.1 [29–31].

During formation of acetyl-CoA, some amount of carbon di-oxide is produced. Acetyl-CoA is then either integrated in cellular biomass or converted to metabolic products (acetate and acetaldehyde). Ethanol and acetic acid are produced from acetate and acetaldehyde, respectively. All possible reactions for syngas fermentation have been reported in Table 12.1.

12.3.3 Microbial Reactions for Higher Alcohols and Medium Chain Fatty Acids

Syngas platform can also be utilized for the production of C4 to C8 compounds [36]. Among this *Clostridium carboxidivorans* can directly convert syngas to C4 to C6 compounds. Two stage fermentation strategy—first stage for syngas fermentation to ethanol and acetate and second stage for chain elongation using *Clostridium kluyveri*, can also be followed to produce C4 to C6 compounds. Mixed culture strategy can

Table 12.1 All reactions involved in syngas fermentation [29–35]

Product	Biochemical reaction	Equation	ΔG° (kJ/mol)
Ethanol	$6H_2 + 2CO_2 \rightarrow C_2H_5OH + 3H_2O$	[A]	−96.0
	$6CO + 3H_2O \rightarrow C_2H_5OH + 4CO_2$	[B]	−220.6
	$5CO + H_2 + 2H_2O \rightarrow C_2H_5OH + 3CO_2$	[C]	−197.3
	$4CO + 2H_2 + H_2O \rightarrow C_2H_5OH + 2CO_2$	[D]	−177.3
	$3CO + 3H_2 \rightarrow C_2H_5OH + CO_2$	[E]	−157.2
	$2CO + 4H_2 \rightarrow C_2H_5OH + H_2O$	[F]	−137.1
Acetate	$4H_2 + 2CO_2 \rightarrow C_2H_3O_2^- + H^+ + 2H_2O$	[G]	−74.4
	$4H_2 + 2HCO_3^- + H^+ \rightarrow C_2H_3O_2^- + 4H_2O$	[H]	−87.8
	$4CO + 2H_2O \rightarrow C_2H_3O_2^- + H^+ + 3CO_2$	[I]	−172.2
Acetic acid	$C_2H_3O_2^- + C_2H_5OH \rightarrow C_4H_7O_2^- + H_2O$	[J]	−38.5
	$C_2H_3O_2^- + 6H_2 + 2CO_2 \rightarrow C_4H_7O_2^- + H_2O$	[K]	−143.0
Water shift reaction	$CO + H_2O \rightarrow CO_2 + H_2$	[L]	−20.1

also be utilized for the production of C4 to C8 compounds. The reactions occurring to produce C4 to C8 compounds, in the presence of *Clostridium kluyveri* and *Clostridium autoethanogenum*, are as follows:

Chain elongation using *Clostridium kluyveri*

$$CH_3COO^- + CH_3CH_2OH \rightarrow CH_3CH_2CH_2COO^- + H_2O \tag{12.7}$$

$$CH_3CH_2CH_2COO^- + CH_3CH_2OH \rightarrow CH_3CH_2CH_2CH_2CH_2COO^- + H_2O \tag{12.8}$$

Production of higher alcohols using *Clostridium Autoethanogenum*

$$CH_3CH_2CH_2COOH + CO + H_2 \rightarrow CH_3CH_2CH_2CH_2OH + CO_2 \tag{12.9}$$

$$CH_3CH_2CH_2CH_2CH_2COOH + CO + H_2 \rightarrow CH_3CH_2CH_2CH_2CH_2CH_2OH + CO_2 \tag{12.10}$$

12.4 Bioreactors for Syngas Fermentation

Continuous stirred tank bioreactor (CSTBR); **bubble column or air lift bioreactor;** hollow-fibre membrane biofilm reactor (HFMBR); rotating packed bed biofilm reactor (RPBR), trickle bed reactor (TBR) as represented in Figs. 12.2, 12.3, 12.4, 12.5, 12.6 are used for syngas fermentation. Monolithic biofilm reactors (MBRs), represented in Fig. 12.7, are also used in syngas fermentation. The working principle and configurations of the reactors can be described as follows:

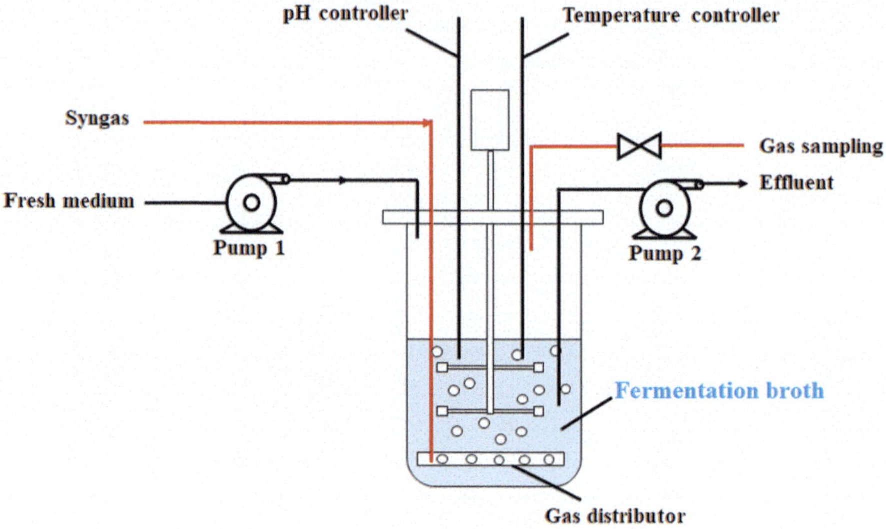

Fig. 12.2 CSTR used for syngas fermentation

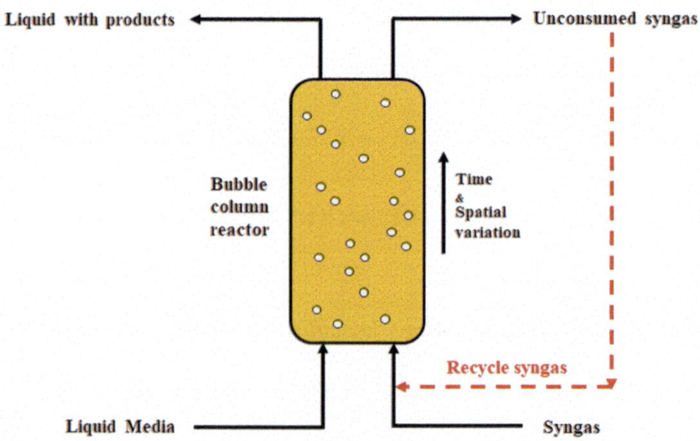

Fig. 12.3 Schematic diagram of bubble column reactor for syngas fermentation

The continuous stirred tank reactors (CSTBRs) are very often used in syngas fermentation [37, 38]. Unlike classical homogeneous CSTRs, this is a multi-phase reactor in which cells (solid) are suspended in stirred liquid medium through which syngas is sparged. Agitator speed is increased to increase the gas–liquid interfacial surface area of mass transfer. However, the high impeller speed demands high energy.

The trickle bed reactors are packed with solid matrix having high specific surface area [38–41]. The film is developed on the packing. The liquid usually trickles down through the bed and gas is passed either in counter-current or co-current direction.

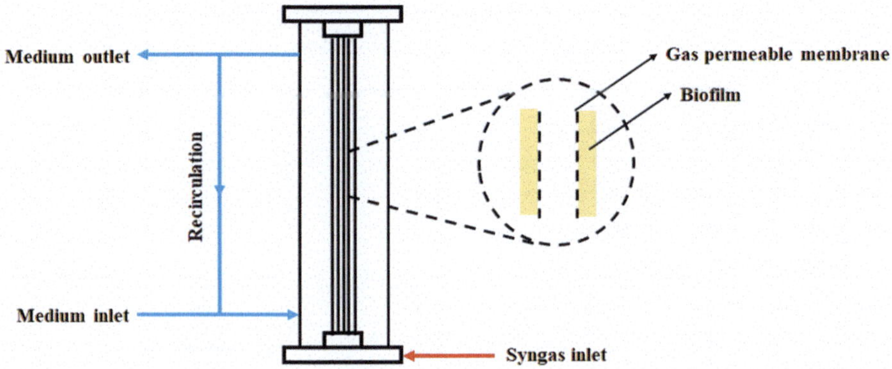

Fig. 12.4 Schematic diagram of hollow-fibre membrane biofilm reactor for syngas fermentation

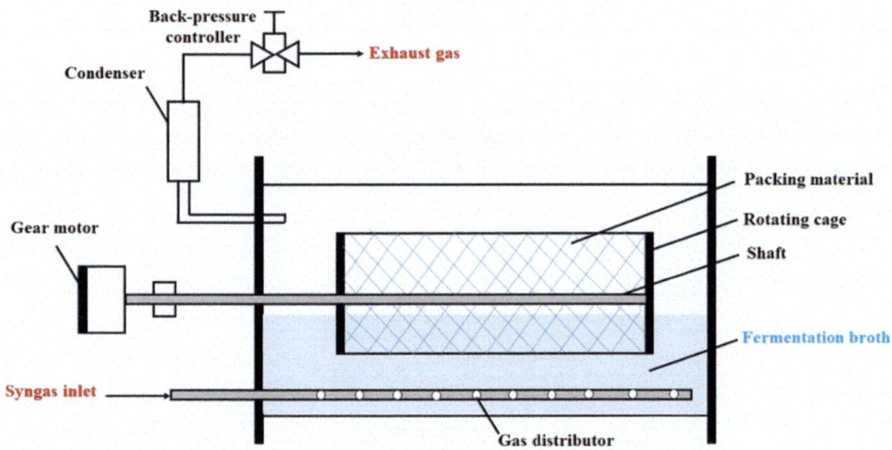

Fig. 12.5 Schematic diagram of horizontal rotating packed bed biofilm reactor

The velocity of both gas and liquid phases should be in trickle flow regime, i.e., very low in any trickle bed reactor [42]. Due to enhanced value of $k_l a$ and large residence time, the mass transfer is very effective.

In bubble column or air lift bioreactor, cells are suspended in liquid medium and gas is passed through the column in bubbles. In BCBRs, there is no mechanical agitation and hence there is no energy investment for this purpose. The value of kla depends on the bubble diameter, gas hold up and gas phase superficial velocity [43].

The rotating packed bed biofilm reactors (RPBRs) are usually cylindrical in shape and another concentric cylindrical cage, packed with solid matrix carrying fixed biofilm of microbial culture, is placed inside the mail body [44–48]. While the outer cylinder is fixed, the inner cylindrical cage is rotated around horizontal axis. The outer cylinder is half-filled with liquid medium. The RPBRs are particularly effective for

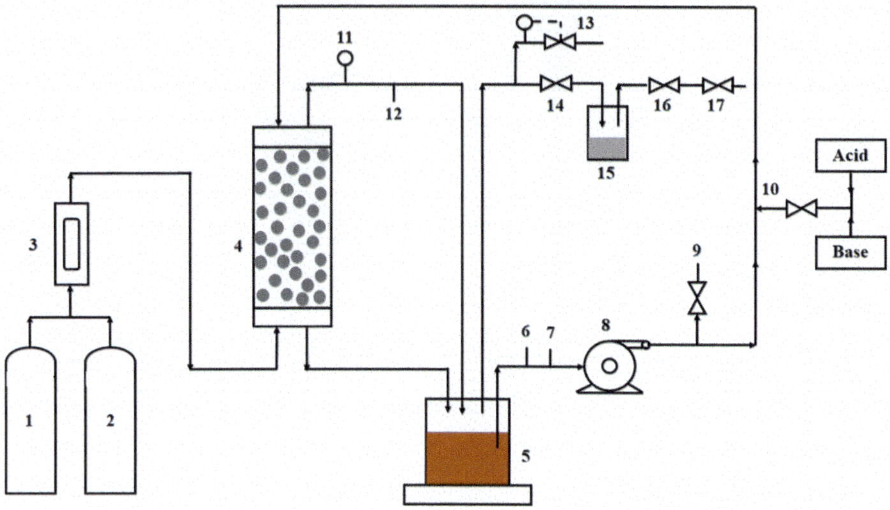

Fig. 12.6 Trickle bed reactor semi-continuous syngas fermentation setup. (1) Nitrogen cylinder, (2) syngas cylinder, (3) rotameter, (4) trickle bed reactor, (5) medium sump, (6) oxidation reduction potential (ORP) probe, (7) pH probe, (8) master-flux pump, (9) liquid sampling port, (10) acid/base addition, (11) pressure gauge, (12) gas sampling port, (13) by pass line with solenoid valve and pressure switch, (14) back pressure regulator, (15) gas bubbler, (16) 108 kPa check valve and (17) ball valve

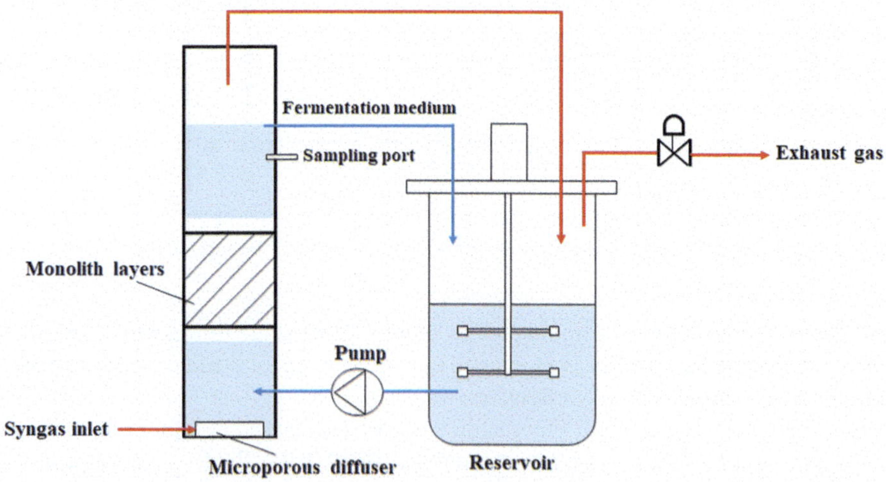

Fig. 12.7 Monolith biofilm reactor setup for syngas fermentation

mass transfer of substrate from gas phase to growth medium and then to cells. Unlike continuous stirred tank reactors, the rotational speed of the inner vessel is also kept very low and is about 5 rpm. The mass transfer of syngas is particularly enhanced on the side of cage exposed to overhead gas during each rotation. The rectors do not face the problems of clogging and pressure drop.

In hollow fibre membrane bioreactor (HFMBR), cells are attached to hollow and porous fibre membrane through which gas is passed [49–51]. As the cells are in close vicinity with the cells, the mass transfer resistance is low.

In monolithic biofilm reactor (MBR), the cells are attached in the inner surface of monolithic channels through which gas is passed in bubbles [49]. The mass transfer rate is enhanced due to reduction in boundary layer thickness between gas phase and microbial cells.

As reported in the literature, trickle bed reactors are mostly used for biomethanation of syngas [52–56]. In some research studies, floating membrane reactor [57], reverse membrane bioreactor [58], CSTR [59] and free cell bioreactor have been used (Table 12.2).

Table 12.2 Comparative studies for methane production from syngas using different reactor configurations

Reactor type	P (Bar)	V (L)	T (°C)	Syngascomposition	Volumetric methane production rate (mmol/(L_R.h))	Refs.
Pressurized agitated column	4	10	55	$CO:H_2:CO_2$ (40: 40: 20)	6.8	[10]
Floating membrane reactor	–	1.7	55	$CO:H_2:CO_2:N_2$ (55: 20: 10: 15)	1.43	[57]
CSTR	1	0.75	65	$CO:H_2$ (33.3: 66.6)	7.44	[59]
Reverse membrane bioreactor	–	0.6	55	$CO:H_2:CO_2:N_2$ (55: 20: 10: 15)	0.2	[58]
Free cell bioreactor	–	0.6	55	$CO:H_2:CO_2:N_2$ (55: 20: 10: 15)	0.2	[57]
TBR	1	5	60	$CO:H_2:CO_2:N_2$ (20: 45: 25: 10)	10.6	[54]
TBR	1	1.05	37	$CO:H_2:CO_2:Ar$ (55.62: 19.68: 9.67: 14.82)	3.00	[56]
TBR	1	26.0	37	$CO:H_2:CO_2:Ar$ (54.42: 21.11: 9.72: 14.74)	0.12	[60]

(P: Pressure, V: working volume, T: Temperature)

Bioreactors like continuous stirred tank reactor (CSTR), trickle bed reactor (TBR), packed bed reactor (PBR), membrane-based reactor and bubble column reactor are used for syngas fermentation for ethanol production. Tables 12.3 and 12.4 represent the values of $k_l a$ and performance of different bioreactors used ethanol production through syngas fermentation.

Table 12.3 Mass transfer coefficient, $k_l a$ for several bioreactor configurations

Reactor configuration	Gas composition	Working volume (L)	$k_l a$ (h^{-1})	Refs.
HFMBR	50% CO, 30% H_2, 20% CO_2	0.13 (HFM), 2.4 (Reservoir)	385 CO	[50]
Composition hollow fibre membrane (CHF module)	99.99% CO	3	946	[45]
Column with column diffuser	>99.99% CO	3	129.6 CO, 97.2 H_2	[61]
Air-lift combined with single point gas entry	99.99% CO	3	45	[62]
TBR	Air	1 (packed bed volume)	171	[63]
Hollow fibre	Air	5 (water holding tank)	55	[52]
CSTR	Air	3 (glass vessel)	114	[52]

Table 12.4 Comparative studies for ethanol and acetate production from syngas fermentation using different reactor configurations

Reactor configuration	Gas composition	Working volume	Micro-organisms	Temp (°C)	Products (g/L)	Refs.
RPBR	$CO:H_2:CO_2:N_2$ (20:5:15:60)	3.3 L	*Clostridium carboxidivorans P7*	37	Ethanol: 7.0	[64]
MBR	$CO:H_2:CO_2:N_2$ (20:5:15:60)	8 L	*Clostridium carboxidivorans P7*	37	Ethanol: 4.89 Acetate: 3.05	[49]
HFMBR	$CO:H_2$ (40:60)	390 mL	Mixed culture	35	Acetate: 4.22	[30]
TBR	$CO:H_2: CO_2:N_2$ (38:28.5:28.5:5)	500 mL	*Clostridium ragsdalei*	37	Ethanol: 5.7 Acetic acid: 12.3	[39]
CSTR	$CO:H_2: CO_2:Ar$ (56.1:15.5:10:18.4)	350 mL	Mixed culture	37	Ethanol: 2.8 Acetate: 6.0	[65]

RPBR: rotating packed bed biofilm reactor, MBR: monolithic biofilm reactor, HFMBR: hollow-fibre membrane reactor, TBR: trickle bed reactor, CSTR: continuous stirred tank reactor

12.5 Mathematical Modelling of Syngas Fermentation

For the production of either methane or ethanol simultaneous, mass transfer and microbial growth are involved. It is established that for gaseous reactants with low solubility, the reactor size is controlled by mass transfer [66, 67]. A suitable reactor should ensure (I) high mass transfer rate and (II) sufficient cell density. Similar to multiphase heterogeneous catalytic reactors, the following steps for the gaseous reactants are principally present [68]:

1. Mass transfer from the bulk gas phase to the gas–liquid interface;
2. Mass transfer from the gas–liquid interface to the bulk liquid phase;
3. Mass transfer from the bulk liquid to the surface of cells;
4. Internal mass transfer across the cell membrane;
5. Intracellular microbial reaction for growth and product generation.

In all syngas fermentation, either for methane or ethanol, the mass transfer of gaseous reactant to the liquid medium is the main challenge. The mass transfer rate of any gaseous reactant, i, can be calculated as follows:

$$R_i = k_{lm}a(C_{im} - C_{lm}) \tag{12.11}$$

where, C_i = Concentration of gaseous reactant, m, at the gas–liquid interface

C_{lm} = Bulk concentration of gaseous reactant, m (CO or H_2 or CO_2), in the liquid phase.

k_{lm} = Volumetric mass transfer coefficient of reactant, m

a = Surface area per unit volume

$$C_{im} = H_m p_{im} \tag{12.12}$$

where, H = Henry's constant and p_{im} = Partial pressure of gaseous reactant, m

For any reactor under steady state,

$$R_m = k_{lm}a\left(C_{im} - C^*_{lm}\right) = \text{Rate of assimilation of m for growth} = \frac{1}{Y_{\frac{X}{m}}} r_{gm} \tag{12.13}$$

$$r_{gm} = \text{microbial growth rate on substrate (or reactant) m}$$

$$Y_{\frac{X}{m}} = \text{yield coefficient of biomass with respect to substrate (or reactant) m}$$

Under uninhibited condition, growth kinetics on substrate (or reactant) *m* follow Monod model and can be represented as follows [69]:

$$r_{gm} = \frac{\nu_{max} C_{lm}}{K_{sm} + C_{lm}} X \tag{12.14}$$

Under inhibited condition, growth kinetics on substrate (or reactant)m follow Haldane model and can be represented as follows:

$$r_{gm} = \frac{\nu_{max} C_{lm}}{K_{sm} + C_{lm} + \frac{C_{lm}^2}{K_{lm}}} X \tag{12.15}$$

where, k_{Im} = inhibition constant of substrate (or reactant) m

μ_{max} = maximum specific growth rate; μ = Specific growth rate = $\frac{1}{X}\frac{dX}{dt}$

K_{sm} = saturation constant for substrate (or reactant) m

Particularly in case of CO, usually substrate inhibition cannot be neglected.

In this section, the mathematical models developed for bioreactors for ethanol production through syngas fermentation are focused. Thus, the models for CSTR, bubble reactor and membrane-based reactors are reviewed.

12.5.1 Continuous Stirred Tank Reactors

Most common type of bioreactor used for the syngas fermentation is the continuous stirred tank reactor. The fermentation process here is carried out by continuously agitating the reactor with the help of a motor and agitator. The agitation process helps to develop a constant temperature and concentration inside the reactor volume. The CSTR has a continuous gas supply into the liquid phase, while agitator controls the gas–liquid mass transfer. Higher agitation speeds lead to a higher mass transfer rate between the substrate gases and the microbes. However, in industrial-scale fermenters, higher agitation speeds increase the agitator's power consumption, thus increasing the operational cost. Transfer of sparingly soluble gases to suspended growth cultures in STRs requires substantial amount of energy (1–10 kW/m^3). High agitation speed breaks up the large bubbles into smaller bubbles, which have less rise velocity, smaller surface/volume ratios and longer liquid contact time. Other limitations of CSTRs include possible damage to shear-sensitive microorganisms, low lucrative cell concentrations due to biomass wash out at short hydraulic retention times, and poor mixing in stirred tanks with large diameters.

Peptostreptococcus productus was one of the first bacteria ever studied for the fermentation of syngas compounds to liquid fuels. 4.8 mmol.L_{RWV}^{-1} h^{-1} was the maximum measured productivity of acetate with an agitation speed of 400 rpm and gas flow rate of 19.54 mL.$min^{-1}L_{RWV}^{-1}$ [70]. In another work, syngas fermentation was conducted in a 2 L CSTR using *Clostridium ljungdahlii* the acetogenic micro-organism with a gas flow of 14 mL/min with 500 rpm rotational speed at

37 °C temperature. Results suggested that the ethanol production achieved was 1.7 mmol.$h^{-1}L_{RWV}{}^{-1}$ and the acetate production was 1.1 mL.$h^{-1}L_{RWV}{}^{-1}$ [71].

The modelling of syngas fermentation in CSTR reactor is based on the following assumptions:

i. CO is the rate limiting substrate.
ii. Other nutrients were supplied in the culture media.

Following these basic assumptions, mass balance equations have been developed.

In gas phase.

CO balance,

$$V_g\frac{dC_{CO,g}}{dt} = V_i\frac{P}{TR} - V_o C_{CO,g} - V_l(k_l a_g)_{CO}\left[\left(\frac{P_{CO,g}}{H_{CO}}\right) - C_{CO,l}\right] \quad (12.16)$$

CO_2 balance,

$$V_g\frac{dC_{CO_2,g}}{dt} = -V_o C_{CO_2,g} - V_l(k_l a_g)_{CO_2}\left[\left(\frac{p_{CO_2,g}}{H_{CO_2}}\right) - C_{CO_2,l}\right] \quad (12.17)$$

In liquid phase,

CO balance,

$$\frac{dC_{CO,l}}{dt} = (k_l a)_{CO}\left[\left(\frac{C_{CO,g}}{H_{co}}\right) - C_{CO,l}\right] - \left[\frac{\mu X}{Y_{\frac{X}{CO}}}\right] \quad (12.18)$$

CO_2 balance,

$$\frac{dC_{CO_2,l}}{dt} = (k_l a)_{CO_2}\left[\left(\frac{C_{CO_2,g}}{H_{CO_2}}\right) - C_{CO_2,l}\right] - \left[\frac{\mu X}{Y_{\frac{X}{CO_2}}}\right] \quad (12.19)$$

where,

$Y_{\frac{X}{CO}} = \frac{\text{biomass produced(g/L)}}{\text{CO consumed(moles/L)}}$ and $Y_{\frac{X}{CO_2}} = \frac{\text{biomass produced (g/L)}}{\text{CO}_2\text{consumed(moles/L)}}$.

Represent the yield of biomass per unit consumption of the substrates.

The product balance has been provided as,

$$Y_{\frac{P}{X}}\mu = \frac{1}{X}\frac{dp}{dt} \quad (12.20)$$

where $Y_{\frac{P}{X}} = \frac{\text{product formed (moles/L)}}{\text{biomass generated(g/L)}}$ represents the corresponding yield of the product.

For acetic acid,

$$\frac{dAA}{dt} = X\left(\mu Y_{\frac{AA}{X}}\right) \quad (12.21)$$

For ethanol,

$$\frac{dE}{dt} = X\left(\mu Y_{\frac{E}{X}}\right) \tag{12.22}$$

Molar flow rate of the outlet gas

$$N_{Gout} = \frac{V_i P}{RT} - V_l(k_l a_g)_{CO}\left[\left(\frac{p_{CO,g}}{H_{CO}}\right) - C_{CO,l}\right] - V_l(k_l a_g)_{CO_2}\left[\left(\frac{p_{CO_2,g}}{H_{CO_2}}\right) - C_{CO_2,l}\right] \tag{12.23}$$

Now,

$$Q_{out} = \frac{N_{Gout} RT}{P} \tag{12.24}$$

The overall conversion rate of CO and production rate of CO_2 is now is given as

$$r_{CO} = \left(\frac{\mu}{Y_{\frac{X}{CO}}} + Y_{CO/E}\right)X \tag{12.25}$$

$$r_{CO_2} = \left(\mu Y_{CO_2/X} + Y_{CO_2/E}\, r_{ET}^{AA}\right)X \tag{12.26}$$

Now solving the differential equations, Eqs. (12.16)–(12.22), the concentration of ethanol and other components can be calculated at any given instance of time.

12.5.2 Bubble Column Bioreactor

In the bubble column reactor, mass transfer of CO and H_2 from gaseous phase to liquid phase can be enhanced due to increase in average mass transfer driving force triggered by: (i) the gas–liquid contact time and (ii) favourable gas composition spatial profiles. The schematic diagram of bubble column reactor for syngas fermentation has been shown in the Fig. 12.3. From Fig. 12.3, it is clear that syngas and liquid media (with inoculum) are introduced to the bubble column reactor from the bottom. Due to cellular consumption, the concentration of CO and H_2 decreases as the gas flow up the column. Therefore, the dissolved concentration of CO and H_2 is being varied at different positions of column. High dissolved CO and H_2 concentration is desired. Unconsumed syngas has been recycled to enhance the efficiency for conversion of syngas to products. Liquid products mainly ethanol and acetic acid are collected from the top of the column. A very comprehensive modelling of bubble column reactor was developed by Chen et al. (2015) [72]. According to them, the uptake rake rate of gaseous reactant including inhibition is as follows:

$$v_{CO} = \frac{v_{max,\,CO} C_{CO}}{K_{m,CO} + C_{CO} + \frac{C_{CO}^2}{K_{I,CO}}} \times \frac{1}{1 + \frac{E+A}{K_I}} \tag{12.27}$$

$$v_{H_2} = \frac{v_{max,\,H_2} C_{CO}}{K_{m,H_2} + C_{H_2,l}} \times \frac{1}{1 + \frac{E+A}{K_I}} \times \frac{1}{1 + \frac{C_{CO}}{K_{I,H_2}}} \tag{12.28}$$

where, $K_{I,CO}$ and K_{I,H_2} are parameters that account for CO inhibition of CO uptake and H_2 uptake, respectively.

Uptake kinetics were specified for the dissolved gaseous substrates CO and H_2 as well as for CO_2 that can be consumed again. Uptake kinetics were assumed to follow inhibition Monod expression form,

$$v_i = \frac{v_{max,\,i} S_i}{K_{m,I} + S_i} \times \frac{1}{1 + \frac{E+A}{K_I}} \tag{12.29}$$

where v_i = uptake rate (mmol/gDW/h) of the i-th substrate,

S_i = dissolved concentration (mmol/L) of i-th gaseous substrate,

v_{max} = maximum uptake rate,

$K_{m,I}$ = saturation constant and K_I is an inhibition constant.

A combined term consisting both the concentrations of ethanol (E) and acetate (A) was used to account for the known inhibitory effects of these two products on microorganism growth. The substrate uptake kinetics and genome-scale reconstruction of intracellular metabolism were combined with convection–dispersion reaction type equations for the bubble column transport processes with the following assumptions:

1. Ideal plug flow for the vapour phase and plug flow plus axial dispersion for liquid phase. This is reasonable at gas superficial velocities <5 cm/s, i.e., in the bubbly flow regime and liquid velocities < 0.02 cm/s used in the study.
2. Convection and dispersion were assumed to occur only in the axial direction

The mass balance of *Clostridium ljungdahlii* biomass was as follows:

$$\frac{\partial X(z,t)}{\partial t} = \mu X - \frac{u_l}{\varepsilon_l}\frac{\partial X}{\partial z} + D_A \frac{\partial^2 X}{\partial z^2} \tag{12.30}$$

$$u_L X(0,t) - \varepsilon_L D_A \frac{\partial X(0,t)}{\partial z} = 0; \quad \frac{\partial X(L,t)}{\partial z} = 0; \quad X(z,0) = X_0; \tag{12.31}$$

where X is the biomass concentration (g/L), μ is cellular growth rates (h^{-1}) obtained from the flux balance calculation, u_l is the liquid phase velocity, ε_l is the liquid phase volume fraction and D_A is the axial dispersion coefficient of the liquid phase. A typical Danckwert's boundary condition was imposed at the reactor entrance (z = 0), while a zero slope boundary condition was applied at reactor exit (z = L). A uniform biomass concentration profile within the reactor was used as the initial condition.

Mass balances of the dissolved gaseous substrates had the form,

$$\frac{\partial C_{CO,l}(z,t)}{\partial t} = v_{CO}X + \frac{k_{m,C}}{\varepsilon_l}\left(C^*_{CO} - C_{CO,l}\right) - \frac{u_l}{\varepsilon_l}\frac{\partial C_{CO,l}}{\partial z} + D_A\frac{\partial^2 C_{CO,l}}{\partial z^2} \tag{12.32}$$

$$u_l C_{CO,l}(0,t) - \varepsilon_l D_A \frac{\partial C_{CO,l}(0,t)}{\partial z} = 0; \quad \frac{\partial C_{CO,l}(L,t)}{\partial z} = 0; \; C_{CO,l}(z,0) = C_{CO,l0}; \tag{12.33}$$

$$\frac{\partial C_{H_2,l}(z,t)}{\partial t} = v_{H_2}X + \frac{k_{m,H_2}}{\varepsilon_l}\left(C^*_{H_2} - C_{H_2,l}\right) - \frac{u_l}{\varepsilon_l}\frac{\partial C_{H_2,l}}{\partial z} + D_A\frac{\partial^2 C_{H_2,l}}{\partial z^2} \tag{12.34}$$

$$u_l C_{H_2,l}(0,t) - \varepsilon_l D_A \frac{\partial C_{H_2,l}(0,t)}{\partial z} = 0; \quad \frac{\partial C_{H_2,l}(L,t)}{\partial z} = 0; \; C_{H_2,l}(z,0) = C_{H_2,l0}; \tag{12.35}$$

where $C_{CO,l}$ and $C_{H_2,l}$ are the liquid phase CO and H_2 concentrations (mmol/L).

v_{CO} and v_{H_2} are the CO and H_2 uptake rates (mmol/gDW/h) obtained from the flux balance calculation,

$k_{l,CO}$ and k_{l,H_2} the corresponding gas–liquid mass transfer co-efficient,

C^*_{CO} and $C^*_{H_2}$ are the saturated liquid concentrations (mmol/L) calculated from the corresponding gas phase concentrations using Henry's law at specified temperature and pressure.

Mass balances of the two substrates in the gas phase had the form,

$$\frac{\partial C_{CO,g}(z,t)}{\partial t} = -\frac{k_{l,CO}}{\varepsilon_g}\left(C^*_{CO} - C_{CO,l}\right) - \frac{u_g}{\varepsilon_g}\frac{\partial C_{CO,g}}{\partial z};$$
$$C_{CO,g}(0,t) = C_{GF}; \; C_{CO,g}(z,0) = C_{CO,g0}; \tag{12.36}$$

$$\frac{\partial C_{H_2,g}(z,t)}{\partial t} = -\frac{k_{l,H_2}}{\varepsilon_g}\left(C^*_{H_2} - C_{H_2,l}\right) - \frac{u_g}{\varepsilon_g}\frac{\partial C_{H_2,g}}{\partial z};$$
$$C_{H_2,g}(0,t) = C_{H_2,gF}; \; C_{H_2,g}(z,0) = C_{H_2,g0}; \tag{12.37}$$

where $C_{CO,g}$ and $C_{H_2,g}$ are the gas phase CO and H_2 concentrations (mmol/L), $\varepsilon_g = 1 - \varepsilon_l$ is the gas phase volume fraction and u_g is the superficial gas velocity.

The gas concentrations at the reactor entrance $C_{CO,gF}$ and $C_{H_2,gF}$ were calculated from the partial pressures of the feed gas using ideal gas law. Uniformly initial concentrations were specified by setting $C_{CO,g0} = C_{CO,gF}$ and $C_{H_2,g0} = C_{H_2,gF}$.

Mass balances on the two major metabolic products ethanol and acetate had the form,

$$\frac{\partial E_l(z,t)}{\partial t} = M_E v_E X - \frac{u_L}{\varepsilon_L}\frac{\partial E_l}{\partial z} + D_A\frac{\partial^2 E_l}{\partial z^2} \tag{12.38}$$

$$u_L E_l(0,t) - \varepsilon_L D_A \frac{\partial E_l(0,t)}{\partial z} = 0; \frac{\partial E_l(L,t)}{\partial z} = 0; E_l(z,0) = E_{l0}; \quad (12.39)$$

$$\frac{\partial A_l(z,t)}{\partial t} = M_A v_A X - \frac{u_L}{\varepsilon_L}\frac{\partial A_l}{\partial z} + D_A \frac{\partial^2 A_L}{\partial z^2} \quad (12.40)$$

$$u_L A_L(0,t) - \varepsilon_L D_A \frac{\partial A_L(0,t)}{\partial z} = 0; \frac{\partial A_L(L,t)}{\partial z} = 0; A_L(z,0) = A_{L0}; \quad (12.41)$$

where E_L and A_L are the concentrations of liquid phase ethanol (g/L) and acetate (g/L), v_E and v_A are the corresponding fluxes (mmol/L) calculated from the flux balance model, and M_E and M_A are the corresponding molecular weight (g/mmol). Gas phase balances on the ethanol and acetate were omitted under the assumption of low volatility at the column conditions.

Mass balances on the liquid and gas phase carbon dioxide had the form

$$\frac{\partial C_{CO_2,l}(z,t)}{\partial t} = v_{CO_2} X + \frac{k_{l,CO_2}}{\varepsilon_L}\left(C^*_{CO_2} - C_{CO_2,l}\right) - \frac{u_L}{\varepsilon_L}\frac{\partial C_{CO_2,l}}{\partial z} + D_A \frac{\partial^2 C_{CO_2,l}}{\partial z^2} \quad (12.42)$$

$$u_L C_{CO_2,l}(0,t) - \varepsilon_L D_A \frac{\partial C_{CO_2,l}(0,t)}{\partial z} = 0; \frac{\partial C_{CO_2,l}(L,t)}{\partial z} = 0; \quad C_{CO_2,l}(z,0) = C_{CO_2,l0}; \quad (12.43)$$

$$\frac{\partial C_{CO_2,g}(z,t)}{\partial t} = -\frac{k_{l,CO_2}}{\varepsilon_G}\left(C^*_{CO_2} - C_{CO_2,l}\right) - \frac{u_G}{\varepsilon_G}\frac{\partial C_{CO_2,l}}{\partial z}; \quad C_{CO_2,g}(0,t) = C_{CO_2,gF}; \; C_{CO_2,g}(z,0) = C_{CO_2,g0}; \quad (12.44)$$

where $C_{CO_2,l}$ and $C_{CO_2,g}$ are the concentration of liquid phase CO_2 (mmol/L) and gas phase CO_2 (mmol/L), v_{CO_2} is the CO_2 flux (mmol/L) calculated from the flux balance model, k_{l,CO_2} is the CO_2 gas–liquid mass transfer coefficient and $C^*_{CO_2}$ is the saturated liquid CO_2 concentration (mmol/L) calculated from the corresponding gas phase concentration using Henry's law.

The reactor was assumed to be isothermal, while the pressure profile was calculated from liquid head as

$$P(z) = P_L + \rho_L g(L - z) \quad (12.45)$$

where L is the length of the column, ρ is the liquid phase density assumed to be equal to the density of water and P_L is the pressure (Pa) at the top of the column, which was assumed to be atmospheric pressure.

Accordingly, gaseous substrate was modelled to dissolve more readily in the lower portion of the column. Calculation of gas and liquid volume fractions in bubble column reactors in notoriously difficult, as the volume fractions are known to depend on a number of operating parameters. The effect of gas flow rate is known to be particularly important. Therefore, we fit gas flow rate versus volume fraction data to a simple model to drive following relationship:

$$\varepsilon_g = \frac{\varepsilon_{g,max} u_g}{K_g + u_g} \tag{12.46}$$

12.5.3 Membrane Biofilm Reactor

Production of both short-chain and medium-chain fatty acids from the H_2 and CO_2 by mixed microbial culture in a hollow-fibre membrane biofilm reactor (MBFR) has been demonstrated experimentally. In the MBFR, H_2 and CO_2 were supplied through gas-permeable membranes, which also served as by biofilm support [73]. Based on the experimental observations and the known metabolisms revealed the production of acetate, butyrate and caproate from syngas containing 60% H_2 and 40% CO_2 could be described as

$$4H_2 + 2CO_2 \rightarrow C_2H_3O_2^- + H^+ + 2H_2O \ \left(\text{at pH 7.0, } \Delta G^0 = -95\,\text{kJmol}^{-1}\right) \tag{12.47}$$

$$C_2H_3O_2^- + 6H_2 + 2CO_2 \rightarrow C_4H_7O_2^- + 4H_2O \ \left(\text{at pH 7.0, } \Delta G^0 = -143\ \text{kJmol}^{-1}\right) \tag{12.48}$$

$$3C_2H_3O_2^- + 2H^+ + 4H_2 \rightarrow C_6H_{11}O_2^- + 4H_2O \ \left(\text{at pH 7.0, } \Delta G^0 = -97\ \text{kJmol}^{-1}\right) \tag{12.49}$$

$$C_4H_7O_2^- + 6H_2 + 2CO_2 \rightarrow C_6H_{11}O_2^- + 4H_2O \ \left(\text{at pH 7.0, } \Delta G^0 = -143\,\text{kJmol}^{-1}\right) \tag{12.50}$$

Monod equation was applied to describe the rate of each reaction, with non-competitive inhibition functions incorporated to depict the inhibition effect of non-substrate fatty acids. Equations (12.51–12.53) show the Monod-based process rate equations for acetate (r_A), butyrate (r_{Bu}) and caproate (r_{Ca}) production, respectively. The stoichiometric relationships between substrates and productions were formulated based on the characteristics of each reaction, with continuity checking performed to ensure correct mass balance of all components. The develop model integrates the relationships among four soluble species, i.e., hydrogen (H_2), acetate (S_A), butyrate (S_{Bu}) and caproate (S_{Ca}) and two particulate species, i.e., functional biomass (X_B) and inert biomass (X_I).

$$r_A = \nu_A \frac{S_{H2}}{K_{H2} + S_{H2}} \frac{K_A}{K_{Ac} + S_{Ac}} \frac{K_{Bu}}{K_{Bu} + S_{Bu}} X_B \tag{12.51}$$

$$r_{Bu} = \nu_{Bu} \frac{S_{H2}}{K_{H2} + S_{H2}} \frac{S_A}{K_A + S_{Ac}} \frac{K_{Bu}}{K_{Bu} + S_{Bu}} X_B \tag{12.52}$$

$$r_{Ca} = \nu_{Ca} \frac{S_{H_2}}{K_{H_2} + S_{H_2}} \frac{S_{Bu}}{K_{Bu} + S_{Bu}} \frac{K_A}{K_A + S_A} X_B \tag{12.53}$$

where ν_A, ν_{Bu} and ν_{Ca} are the maximum uptake rates for acetate, butyrate and caproate production, respectively. K_{H_2}, K_A and K_{Bu} are the hydrogen, acetate and butyrate affinity constants, respectively.

The flux of syngas L from the gas to biofilm matrix compartment through the membrane was modelled using the following equation

$$L = k\left(\frac{S_g}{H} - S_l\right) \tag{12.54}$$

where S_g and S_l are the syngas concentrations in the gas and biofilm matrix components (gm^{-3}), respectively, k is the overall mass transfer coefficient ($m\ d^{-1}$) and H is the Henry's law constant ($molm^{-3}$ gas/$molm^{-3}$ liquid).

12.5.4 *Hollow-Fibre Membrane Reactor (HFMBR)*

The hollow-fibre membrane bioreactor (HFMBR) is used for syngas fermentation. A HFMBR having a membrane filtration unit attached to bubble column reactor is shown in Fig. 12.4[29]. The microorganism is being attached on the surface on the membrane and the nutrients in the medium have been supplied. Final products like ethanol, acetic acid, etc. are diffused through the membrane into the bulk liquid. Due to large specific surface area of gas permeable membrane, the mass transfer rate between the gas and liquid is very high. The following equation was used to model the mass transfer rate in the kinetic simulation [74].

$$\frac{dC_{CO}}{dt} = k_l a\left(C_{CO}^* - C_{CO}\right) \tag{12.55}$$

where C_{CO} is the dissolved concentration of CO (mg/L). Equation 12.52 could be integrated to obtain Eq. (12.56). A plot of C_{CO} using the linear least squares technique was used to determine k_La. The mean values of k_La were obtained from three tests:

$$\ln\left(\frac{C_{CO}^* - C_{CO}}{C_{CO}^* - C_{CO,0}}\right) = -k_l a_t \tag{12.56}$$

The maximum gas–liquid mass transfer rate (R_{max}) was calculated using k_la and C_{CO}^*, assuming $C_{CO} = 0$ in Eq. (12.55).

$$r_{max} = k_l a C_{CO}^* \tag{12.57}$$

Equation (12.58) was used to calculate k_la and R during the reactor operation. The values of q_{CO} were assumed to be 962.2 mg/g cell/h.

$$q_{CO}X = k_l a C_{CO}^* \tag{12.58}$$

where q_{CO} is the specific CO consumption rate (mg/g cell/h), i.e., $\frac{\text{Consumption rate}}{X}$ and X is the cell concentration (g/L).

In a previous study, using hollow fibre membrane diffuser for syngas fermentation, the following correlation was obtained [50]

$$\ln\left(\frac{C_{CO}^{*}}{C_{CO}^{*} - C_{CO}}\right) = \frac{Q}{V}\left[1 - \exp\left(k_{l}a\frac{L}{V_{l}}\right)\right]t \quad (12.59)$$

where, C_{CO}^{*} is the aqueous phase CO concentration in equilibrium with the gas phase (mg/L); C_{CO} is the measured dissolved CO concentration (mg/L).

A typical dependence of $k_l a$ on membrane specific surface area, water recirculation rate and volumetric flow rate of gas has been shown in Table 12.5 [50].

A kinetic simulation was conducted to predict the fermentation performance by the change in $K_l a$ and C*. Model equations were developed on the process stoichiometry and reaction kinetics. Following equations were used to gaseous molar fraction (y) by using simulated values of cumulative moles of CO ($N_{CO,cum}$) and CO_2 ($N_{CO_2,cum}$) in the head space.

$$y_{CO} = \frac{N_{CO,cum}}{N_{CO,cum} + N_{CO_2,cum}} \quad (12.60)$$

$$y_{CO_2} = \frac{N_{CO_2,cum}}{N_{CO,cum} + N_{CO_2,cum}} \quad (12.61)$$

The partial pressure (P_{CO}) and saturated concentration ($C_{CO,\,sat}$) of CO could be expressed by following equations

$$P_{CO} = P_t y_{CO} \quad (12.62)$$

$$C_{CO,sat} = H_{CO} P_{CO} \quad (12.63)$$

Table 12.5 Experimental runs using hollow fibre membrane diffuser for syngas fermentation

Fibre numbers/membrane specific surface area per liquid working volume (A/v, 1/cm)	Water recirculation rate (Q_L, mL/min)/water velocity (V_L, cm/s)	Gas supply rate (Q_g, mL/min)/specific gas flow rate (V_g, 1/min)	Mass transfer coefficient for CO ($k_L a$)
100/1.4	310/1.02	35/0.26	21.0
200/0.28	670/2.20	140/1.02	90.0
250/0.35	670/2.20	140/1.02	283.0
350/0.49	670/2.20	70/0.51	170.0
450/0.56	670/2.20	105/0.77	252.0

where P_t is total gas pressure (atm), and H is the Henry's constant (mol/L/atm).

q_{CO} and the specific growth rate (μ) were expressed as a function of C_{CO}. Inhibition constant (K_I) was introduced to integrate the substrate inhibition kinetics. The acetate and ethanol concentration terms were added to show the product inhibition such as uncoupler effect and alcohol toxicity. The kinetics reflect the increasingly reducing microbial activity (q_{CO} and μ) proportional to the product concentration. The simulated CO, ethanol and acetate concentration were used as variables.

$$q_{CO} = \left(\frac{q_{CO}^{max} C_{CO}}{k_S + C_{CO} + \frac{C_{CO}^2}{K_I}}\right)\left(1 - \frac{C_E}{C_{E,cri}}\right)^{0.8}\left(1 - \frac{C_A}{C_{A,cri}}\right)^{0.8} \quad (12.64)$$

$$\mu = \left(\frac{\mu_{CO}^{max} C_{CO}}{k_S + C_{CO} + \frac{C_{CO}^2}{K_I}}\right)\left(1 - \frac{C_E}{C_{E,cri}}\right)^{0.8}\left(1 - \frac{C_A}{C_{A,cri}}\right)^{0.8} \quad (12.65)$$

where, q_{CO}^{max} is the maximum specific CO consumption rate (mol/ g cell/h), k_S is the half-saturation constant (mol/L), $C_{E,cri}$ is the critical concentration of ethanol (g/L), $C_{A,cri}$ is the critical concentration of acetate.

The microbial and gas–liquid equilibrium reactions occurred simultaneously in the reactor. The gas phase was maintained at 1 atm. The time course molar fraction of the gas phase and output gas flow rates varied and could be calculated from the pressure difference created by excessive virtual gases in the head space.

$$\frac{dN_{CO,cum}}{dt} = N_{g,in} y_{CO.in} - k_L a\left(C_{CO,sat} - C_{CO}\right) V_l \quad (12.66)$$

$$\frac{dN_{CO_2,cum}}{dt} = N_{g,in} y_{CO_2.in} + q_{CO} X V_l Y_{CO_2} \quad (12.67)$$

where, $N_{g,in}$ is the molar flow rate of the input gas (mole/h), $y_{CO.in}$ is the molar fraction of CO in the input flow, $y_{CO_2.in}$ is the molar fraction of CO_2 in the input flow, V_l is the liquid phase volume, X is the cell concentration (g/L) and Y_{CO_2} is the yield of coefficient of CO_2 from CO. Considering the cell-recycled operation, dilution rate (D) was incorporated.

$$\frac{dC_{CO}}{dt} = DC_{CO,in} - DC_{CO} + k_l a\left(C_{CO,sat} - C_{CO}\right) - q_{CO} X \quad (12.68)$$

The biomass, ethanol and acetate concentrations were simulated using Eqs. (12.63), (12.64) and (12.65), respectively. In case of Eq. (12.63), the effect of biomass degradation was ignored due to the product inhibition term in Eq. (12.59).

$$\frac{dX}{dt} = \mu X \quad (12.69)$$

$$\frac{dC_E}{dt} = DC_{E,in} - DC_E + q_{CO}XY_E \tag{12.70}$$

$$\frac{dC_A}{dt} = DC_{A,in} - DC_A + q_{CO}XY_A \tag{12.71}$$

where D is the dilution rate (hr^{-1}), Y_E is the yield coefficient of ethanol from CO and Y_A is the yield of acetate from CO.

The gas–liquid interfacial area is calculated by Eq. (12.45).

$$A = \frac{k_l a V_l}{k_l} = 4\pi r_b^2 \times \text{No. of bubbles} \tag{12.72}$$

where k_l is the specific gas–liquid mass transfer coefficient (cm/hr), A is gas–liquid interface area (cm^2) and r_b is the average radius of gas bubbles (cm).

Based on the experimental data, k_l value of 27.02 cm/h was calculated using following equation

$$k_l = \frac{k_l a}{\left(\frac{A_C}{V_m}\right)} \tag{12.73}$$

where A_C and V_m are the interfacial area (cm^2) and working volume (cm^3) of the medium inside the vial.

Total volume of the gas bubble contacting with the liquid phase is determining using change in the height of water before and after gas sparging in reactor as follows

$$V_g = \pi r_b^2 (h_{g+l} - h_l) = \frac{4}{3}\pi r_b^3 \times \text{No.of bubbles} \tag{12.74}$$

All the model equations were solved numerically.

12.6 Challenges Involved in Modelling

The primary purpose of having a mathematical model of any physical process is to accurately replicate the process in mathematical terms. However, it is quite evident that there will be some limitation or physical challenge in the process that the model will not be able to incorporate. The main challenges or limitations in the mathematical model of syngas fermentation are the following.

i. The kinetics of the syngas fermentation process is mostly empirical based on smaller number of experimental runs which can vary as the bacterial strain varies.
ii. There exist certain mass transfer limitations as the process is dependent on gas–liquid mass transfer, which are not accounted for in the mathematical models.

iii. Accounting for the unpredicted behaviour of the culture media is another important challenge in syngas fermentation modelling.
iv. There are very few examples where the syngas fermentation technology has been implemented in large industrial production. So, the changes due to scaling up of the process are yet to be incorporated in the mathematical models.

To overcome these challenges, more rigorous experiments need to be done and dynamic process models need to be developed by mathematically interpreting the various physical phenomena of syngas fermentation.

12.7 Recommendation

From the above review, it is clear that mathematical models are available for the most popular bioreactors, e.g., CSTR, bubble column and membrane reactors, already used for syngas fermentation. The researchers have also shown that the predictions of the model are in good agreement with the experimental results. However, all the models are based on some assumptions. More attempts should be made to develop models by avoiding simplistic assumptions. More attempts should be made to integrate the mathematical models with process models using ASPEN Plus so that realistic prediction about large-scale syngas fermentation units can be made using the lab-scale data. The integrated models will be helpful for the assessment of the possibility of implementation of commercial units of Syngas fermentation for the production of ethanol from lignocellulosic biomass and low-grade coal.

12.8 Conclusion

The article focuses on the state-of-the art review on the mathematical modelling of syngas fermentation from the perspective of methane and ethanol production. Different reactors used for this purpose have been described with the salient reactions. The article also discusses the concepts of the microbial pathways with reference to the specific microorganisms involved in the process. Since the mass transfer of the gaseous substrates to the liquid medium is one of the controlling steps in this process, equal weightage has been given on this. Different mathematical models for various reactors have been detailed along with their assumptions. The challenges of the models have also been discussed with some recommendations. Since lignin content of many of the dry agricultural wastes available in India and other countries are high, their conversion to ethanol or biogas through conventional fermentative routes is only possible after energy-intensive pre-treatment. Through gasification, all carbohydrates, namely, cellulose, hemicelluloses and lignin in the agricultural feedstocks can be converted to syngas and in turn can be converted to either ethanol or methane. During gasification, heat required for the endothermic reactions is supplied from the exothermic heat of partial combustion, and hence the process is almost self-driven from the perspective of energy requirement. The mathematical models are particularly helpful for the prediction of reactor performance. The integration of mathematical models with process modelling software like ASPEN Plus is expected to facilitate the implementation of this technology in commercial scale.

References

1. Mofijur M, Rasul MG, Hassan NMS, Nabi MN (2019) Recent development in the production of third generation biodiesel from microalgae. Energy Procedia 156:53–58
2. Kleinekorte J, Leitl M, Zibunas C, Bardow A (2022) What Shall We Do with Steel Mill Off-Gas: Polygeneration Systems Minimizing Greenhouse Gas Emissions. Environ Sci Technol 56:13294–13304
3. Mosayebi A, Nasabi M (2020) Steam methane reforming on LaNiO3 perovskite-type oxide for syngas production, activity tests, and kinetic modeling. Int J Energy Res 44:5500–5515
4. Aasberg-Petersen K, Christensen TS, Nielsen CS, Dybkjær I (2003) Recent developments in autothermal reforming and pre-reforming for synthesis gas production in GTL applications. Fuel Process Technol 83:253–261
5. Pati S, De S (2021) Model development and thermodynamic analysis of biomass co-gasification using Aspen plus®. Indian Chem Eng 63:172–183. https://doi.org/10.1080/00194506.2021.1887770
6. Pati S, De S, Chowdhury R (2021) Process modelling and thermodynamic performance optimization of mixed Indian lignocellulosic waste co-gasification. Int J Energy Res 45:17175–17188. https://doi.org/10.1002/er.6052
7. Gür M, Canbaz ED (2020) Analysis of syngas production and reaction zones in hydrogen oriented underground coal gasification. Fuel 269:117331
8. Grimalt-Alemany A, Łężyk M, Kennes-Veiga DM et al (2020) Enrichment of mesophilic and thermophilic mixed microbial consortia for syngas biomethanation: the role of kinetic and thermodynamic competition. Waste Biomass Valoriz 11:465–481
9. Li C, Zhu X, Angelidaki I (2020) Carbon monoxide conversion and syngas biomethanation mediated by different microbial consortia. Bioresour Technol 314:123739
10. Figueras J, Benbelkacem H, Dumas C, Buffière P (2021) Biomethanation of syngas by enriched mixed anaerobic consortium in pressurized agitated column. Bioresour Technol 338:125548
11. Barik S, Corder RE, Clausen EC, Gaddy JL (1987) Biological conversion of coal synthesis gas to methane. Energy Prog (United States) 7
12. Lorowitz WH, Bryant MP (1984) Peptostreptococcus productus strain that grows rapidly with CO as the energy source. Appl Environ Microbiol 47:961–964
13. Kerby R, Niemczura W, Ziekus JG (1983) Single carbon catabolism in Acetobacterium woodii and Butyribacterium methylotrophicum by fermentation and 13C nuclear magnetic resonance measurement. J Bacteriol 155:1208–1218
14. Wood HG (1982) Studies with Clostridium thermoaceticum and the resolution of the pathway used by acetogenic bacteria that grow on carbon monoxide or carbon dioxide and hydrogen. In: Proceedings of the biochemistry symposium, Palo Alto, California. Annual Reviews Inc., pp 29–56
15. Kerby R, Zeikus JG (1983) Growth of Clostridium thermoaceticum on H_2/CO_2 or CO as energy source. Curr Microbiol 8:27–30
16. Genthner BRS, Bryant M (1982) Growth of Eubacterium limosum with carbon monoxide as the energy source. Appl Environ Microbiol 43:70–74
17. Mayer F, Lurz R, Schoberth S (1977) Electron microscopic investigation of the hydrogen-oxidizing acetate-forming anaerobic bacterium Acetobacterium woodii. Arch Microbiol 115:207–213

18. Balch WE, Schoberth S, Tanner RS, Wolfe RS (1977) Acetobacterium, a new genus of hydrogen-oxidizing, carbon dioxide-reducing, anaerobic bacteria. Int J Syst Evol Microbiol 27:355–361
19. Sleat R, Mah A, Robinson R (1985) Acetoanaerobium noterae: new genus, new species, an anaerobic bacterium that forms acetate from hydrogen and carbon dioxide. Int J Syst Bacteriol 35:10–15
20. Uffen RL (1976) Anaerobic growth of a Rhodopseudomonas species in the dark with carbon monoxide as sole carbon and energy substrate. Proc Natl Acad Sci 73:3298–3302
21. Dashekvicz MP, Uffen RL (1979) Identification of a carbon monoxide-metabolizing bacterium as a strain of Rhodopseudomonas gelatinosa (Molisch) van Niel. Int J Syst Evol Microbiol 29:145–148
22. Bergey DH (1994) Bergey's manual of determinative bacteriology. Lippincott Williams & Wilkins
23. Thauer RK, Jungermann K, Decker K (1977) Energy conservation in chemotrophic anaerobic bacteria. Bacteriol Rev 41:100–180
24. Balch WE, Fox GE, Magrum LJ et al (1979) Methanogens: reevaluation of a unique biological group. Microbiol Rev 43:260–296
25. Whittenbury R, Phillips KC, Wilkinson JF (1970) Enrichment, isolation and some properties of methane-utilizing bacteria. Microbiology 61:205–218
26. Jones WJ, Nagle DP Jr, Whitman WB (1987) Methanogens and the diversity of archaebacteria. Microbiol Rev 51:135–177
27. Henstra AM, Sipma J, Rinzema A, Stams AJ (2007) Microbiology of synthesis gas fermentation for biofuel production. Curr Opin Biotechnol 18:200–206. https://doi.org/10.1016/j.copbio.2007.03.008
28. Bengelsdorf FR, Straub M, Dürre P (2013) Bacterial synthesis gas (syngas) fermentation. Environ Technol (United Kingdom) 34:1639–1651. https://doi.org/10.1080/09593330.2013.827747
29. Gunes B (2021) A critical review on biofilm-based reactor systems for enhanced syngas fermentation processes. Renew Sustain Energy Rev 143:110950
30. Shen N, Dai K, Xia X-Y et al (2018) Conversion of syngas (CO and H2) to biochemicals by mixed culture fermentation in mesophilic and thermophilic hollow-fiber membrane biofilm reactors. J Clean Prod 202:536–542
31. Zhang F, Ding J, Zhang Y et al (2013) Fatty acids production from hydrogen and carbon dioxide by mixed culture in the membrane biofilm reactor. Water Res 47:6122–6129
32. Phillips JR, Huhnke RL, Atiyeh HK (2017) Syngas fermentation: a microbial conversion process of gaseous substrates to various products. Fermentation 3:28
33. Wang Y-Q, Yu S-J, Zhang F et al (2017) Enhancement of acetate productivity in a thermophilic (55 C) hollow-fiber membrane biofilm reactor with mixed culture syngas (H_2/CO_2) fermentation. Appl Microbiol Biotechnol 101:2619–2627
34. Wang H-J, Dai K, Xia X-Y et al (2018) Tunable production of ethanol and acetate from synthesis gas by mesophilic mixed culture fermentation in a hollow fiber membrane biofilm reactor. J Clean Prod 187:165–170
35. Zhang F, Ding J, Shen N et al (2013) In situ hydrogen utilization for high fraction acetate production in mixed culture hollow-fiber membrane biofilm reactor. Appl Microbiol Biotechnol 97:10233–10240
36. Stoll IK, Boukis N, Sauer J (2020) Syngas fermentation to alcohols: reactor technology and application perspective. Chemie Ing Tech 92:125–136
37. Ungerman AJ, Heindel TJ (2007) Carbon monoxide mass transfer for syngas fermentation in a stirred tank reactor with dual impeller configurations. Biotechnol Prog 23:613–620
38. Bredwell MD, Srivastava P, Worden RM (1999) Reactor design issues for synthesis-gas fermentations. Biotechnol Prog 15:834–844. https://doi.org/10.1021/bp990108m
39. Shen Y, Brown R, Wen Z (2014) Syngas fermentation of Clostridium carboxidivoran P7 in a hollow fiber membrane biofilm reactor: Evaluating the mass transfer coefficient and ethanol production performance. Biochem Eng J 85:21–29

40. Devarapalli M, Atiyeh HK, Phillips JR et al (2016) Ethanol production during semi-continuous syngas fermentation in a trickle bed reactor using Clostridium ragsdalei. Bioresour Technol 209:56–65. https://doi.org/10.1016/j.biortech.2016.02.086
41. Yasin M, Jeong Y, Park S et al (2015) Microbial synthesis gas utilization and ways to resolve kinetic and mass-transfer limitations. Bioresour Technol 177:361–374
42. Smith JM (1956) Chemical engineering kinetics. McGraw-Hill
43. Froment GF, Bischoff KB, De Wilde J (1990) Chemical reactor analysis and design. Wiley, New York
44. Schulte MJ, Wiltgen J, Ritter J et al (2016) A high gas fraction, reduced power, syngas bioprocessing method demonstrated with a Clostridium ljungdahlii OTA1 paper biocomposite. Biotechnol Bioeng 113:1913–1923
45. Munasinghe PC, Khanal SK (2012) Syngas fermentation to biofuel: evaluation of carbon monoxide mass transfer and analytical modeling using a composite hollow fiber (CHF) membrane bioreactor. Bioresour Technol 122:130–136
46. Picard C, Logette S, Schrotter J-C et al (2012) Mass transfer in a membrane aerated biofilm. Water Res 46:4761–4769
47. Mederos FS, Ancheyta J, Chen J (2009) Review on criteria to ensure ideal behaviors in trickle-bed reactors. Appl Catal A Gen 355:1–19
48. López LR, Bezerra T, Mora M et al (2016) Influence of trickling liquid velocity and flow pattern in the improvement of oxygen transport in aerobic biotrickling filters for biogas desulfurization. J Chem Technol Biotechnol 91:1031–1039
49. Shen Y, Brown R, Wen Z (2014) Enhancing mass transfer and ethanol production in syngas fermentation of Clostridium carboxidivorans P7 through a monolithic biofilm reactor. Appl Energy 136:68–76
50. Lee PH, Ni SQ, Chang SY et al (2012) Enhancement of carbon monoxide mass transfer using an innovative external hollow fiber membrane (HFM) diffuser for syngas fermentation: Experimental studies and model development. Chem Eng J 184:268–277
51. Yasin M, Park S, Jeong Y et al (2014) Effect of internal pressure and gas/liquid interface area on the CO mass transfer coefficient using hollow fibre membranes as a high mass transfer gas diffusing system for microbial syngas fermentation. Bioresour Technol 169:637–643
52. Asimakopoulos K, Gavala HN, Skiadas IV (2018) Reactor systems for syngas fermentation processes: a review. Chem Eng J 348:732–744
53. Asimakopoulos K, Łężyk M, Grimalt-Alemany A et al (2020) Temperature effects on syngas biomethanation performed in a trickle bed reactor. Chem Eng J 393:124739
54. Asimakopoulos K, Kaufmann-Elfang M, Lundholm-Høffner C et al (2021) Scale up study of a thermophilic trickle bed reactor performing syngas biomethanation. Appl Energy 290:116771
55. Klasson KT, Ackerson MD, Clausen EC, Gaddy JL (1992) Bioconversion of synthesis gas into liquid or gaseous fuels. Enzyme Microb Technol 14:602–608. https://doi.org/10.1016/0141-0229(92)90033-K
56. Kimmel DE, Klasson KT, Clausen EC, Gaddy JL (1991) Performance of trickle-bed bioreactors for converting synthesis gas to methane. Appl Biochem Biotechnol 28:457–469
57. Chandolias K, Pekgenc E, Taherzadeh MJ (2019) Floating membrane bioreactors with high gas hold-up for syngas-to-biomethane conversion. Energies 12:1046
58. Westman SY, Chandolias K, Taherzadeh MJ (2016) Syngas biomethanation in a semi-continuous reverse membrane bioreactor (RMBR). Fermentation 2:8
59. Diender M, Uhl PS, Bitter JH et al (2018) High rate biomethanation of carbon monoxide-rich gases via a thermophilic synthetic coculture. ACS Sustain Chem Eng 6:2169–2176
60. Klasson KT, Ackerson MD, Clausen EC, Gaddy JL (1991) Bioreactors for synthesis gas fermentations. Resour Conserv Recycl 5:145–165
61. Munasinghe PC, Khanal SK (2014) Evaluation of hydrogen and carbon monoxide mass transfer and a correlation between the myoglobin-protein bioassay and gas chromatography method for carbon monoxide determination. RSC Adv 4:37575–37581
62. Munasinghe PC, Khanal SK (2010) Syngas fermentation to biofuel: evaluation of carbon monoxide mass transfer coefficient (kLa) in different reactor configurations. Biotechnol Prog 26:1616–1621

63. Orgill JJ, Atiyeh HK, Devarapalli M et al (2013) A comparison of mass transfer coefficients between trickle-bed, hollow fiber membrane and stirred tank reactors. Bioresour Technol 133:340–346
64. Shen Y, Brown RC, Wen Z (2017) Syngas fermentation by Clostridium carboxidivorans P7 in a horizontal rotating packed bed biofilm reactor with enhanced ethanol production. Appl Energy 187:585–594
65. Vega JL, Prieto S, Elmore BB et al (1989) The biological production of ethanol from synthesis gas. Appl Biochem Biotechnol 20:781–797
66. Vega JL, Clausen EC, Gaddy JL (1989) Study of gaseous substrate fermentations: carbon monoxide conversion to acetate 1 batch culture. Biotechnol Bioeng 34:774–784
67. Pheil CG, Ordal ZJ (1967) Sporulation of the "thermophilic anaerobes." Appl Microbiol 15:893–898
68. Bailey JE, Ollis DF (2018) Biochemical engineering fundamentals. McGraw-Hill
69. Schuler M, Kargi F (2002) Bioprocess engineering basic concepts, 2nd edn. Prentice-Hall Inc., New Jersey
70. Vega JL, Antorrena GM, Clausen EC, Gaddy JL (1989) Study of gaseous substrate fermentations: carbon monoxide conversion to acetate 2 continuous culture. Biotechnol Bioeng 34:785–793
71. Mohammadi M, Younesi H, Najafpour G, Mohamed AR (2012) Sustainable ethanol fermentation from synthesis gas by Clostridium ljungdahlii in a continuous stirred tank bioreactor. J Chem Technol Biotechnol 87:837–843. https://doi.org/10.1002/jctb.3712
72. Chen J, Gomez JA, Höffner K et al (2015) Metabolic modeling of synthesis gas fermentation in bubble column reactors. Biotechnol Biofuels 8:1–12
73. Chen X, Ni B-J (2016) Anaerobic conversion of hydrogen and carbon dioxide to fatty acids production in a membrane biofilm reactor: A modeling approach. Chem Eng J 306:1092–1098
74. Jang N, Yasin M, Kang H et al (2018) Bubble coalescence suppression driven carbon monoxide (CO)-water mass transfer increase by electrolyte addition in a hollow fiber membrane bioreactor (HFMBR) for microbial CO conversion to ethanol. Bioresour Technol 263:375–384

Chapter 13
Hardware-In-Loop Approach of ELC with Battery Storage as Dump Load in Off-Grid SEIG

Shalini Sinha and Rajen Pudur

Abstract The Self-excited Induction Generator (SEIG) is widely employed as a significant power generation source in geographically isolated rural areas inside the nation. Nevertheless, the functionality of the machine is limited by the fluctuating loading circumstances at its output. The uninterrupted provision of power supply is crucial for the efficient functioning of the system, as it ensures the continuous delivery of electricity to meet the demands of the consumer load. The utilization of an Electronic Load Controller (ELC) is employed within a self-excited induction generator system with the objective of sustaining a consistent voltage and frequency at the bus bar. One of the benefits of this technology is its ability to store and utilize excess kinetic energy at the receiving end for various purposes in rural regions. The study utilized induction machines with a power rating of 2.2 kW and a current rating of 4.8A. These machines were designed to operate at a voltage of 415 V and a frequency of 50 Hz. The excitation capacitor of the machines was employed to supply power to a three-phase load. Excess power generated is stored in a dump load battery. The suggested system underwent testing in the MATLAB/SIMULINK environment and was afterwards verified using the OP4510 RCP/HIL FPGA-based real-time simulator.

Keywords SEIG · ELC · Battery storage

13.1 Introduction

Currently, India is making significant progress in overcoming the formidable obstacle of electricity scarcity, potentially resulting in reduced reliance on fossil fuels and more utilization of renewable energy sources [1]. India is currently the site of the

Shalini Sinha and Rajen Pudur are in NIT Arunachal Pradesh.

S. Sinha · R. Pudur (✉)
Department of Electrical Engineering, NIT Arunachal Pradesh, Jote 791113, Arunachal Pradesh, India
e-mail: rajenpudur@nitap.ac.in

S. De et al. (eds.), *Challenges and Opportunities of Distributed Renewable Power*, Energy, Environment, and Sustainability, https://doi.org/10.1007/978-981-97-1406-3_13

largest small gasifier program and the second largest biogas program globally. China is currently the dominant force in worldwide capacity expansion for solar photovoltaic, wind, bio-electricity, and hydro power. On the other hand, Spain and the United States have emerged as leaders in terms of net capacity increase specifically in geothermal power and solar-thermal power throughout the year 2019. In many instances, small-scale hydroelectric systems operate as "run-of-river" installations, devoid of any dam or water storage infrastructure. These systems are widely regarded as a highly cost-effective and ecologically friendly energy solution, particularly when contemplating the electrification of rural areas. The utilization of squirrel cage architecture in induction generators has emerged as a dominant method of power generation in distant areas, surpassing other types of generators [2]. These devices have several benefits, including their compact size and low maintenance costs, which contribute to their economic viability. Additionally, they demonstrate the capacity to effectively operate under challenging conditions, including load perturbations, rapid load changes, and three-phase faults. Unlike typical synchronous generators, they do not necessitate a distinct DC excitation mechanism for DC supply. These several elements collectively enable the successful implementation of small-scale hydropower plants, including Pico, micro, and mini power plants [3]. The categorization is determined by the power rating, with pico-hydro power plants falling within a rating of 10 kW, micro hydropower plants ranging from 5 to 100 kW, and mini hydropower plants falling within the range of 100 kW to 1 MW. The Pico-hydropower rated generators have demonstrated their utility in small, isolated villages where a modest amount of electrical power is needed. For instance, they may effectively supply electricity to approximately 50 households, enabling the operation of a few fluorescent light bulbs, as well as a television or radio. Applications for micro hydro power plants include small-scale enterprises, flood mitigation, irrigation, and the provision of drinking water. Rivers in urban areas should be managed to serve a variety of purposes, including the generation of renewable energy using small hydropower technology. The induction machine in MHPP operates with a relatively fixed input power of the turbine, and depending on the maximum power output and peak load demand, the power capacity of the generator is decided. There is always variation in load demand which causes voltage and frequency fluctuation at the PCC [4]. To mitigate these fluctuations, the induction generator control (IGC) method is used in SEIG. Here the balance between the input and output power is achieved with an additional load known as a dump load. After feeding to the consumer load, the excess generated power is diverted to the dump load. The direction of power flow between two parallel loads (consumer & dump load) is decided by the suitable approach used with the operation of the load controller. In MHPP, the application of power electronics devices with different topologies on the SEIG gives the evolution of the Electronic Load Controller (ELC) [5]. Here the on–off status of power electronics switches decides the amount and direction of generated power flow to keep PCC voltage and frequency constant. The use of a governor system in Pico-hydro is hindered by its high cost. However, the ELC offers an affordable and resilient alternative for power control. The present study is centered on the measurement of power at the consumer level, with the aim of efficiently managing power distribution

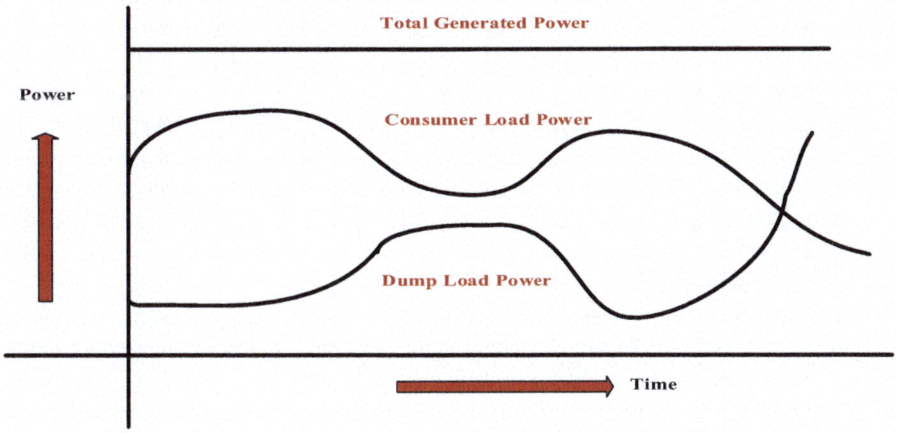

Fig. 13.1 The relationship between consumer power and dump load power in an electrical load control (ELC) system

throughout the dump load. When the amount of consumer power used is lower than the amount of power generated, the excess power is sent to the dump load through the utilization of the Electronic Load Controller (ELC). The notion is sufficiently explained in Fig. 13.1.

The concept of Electric Load Control (ELC) is depicted in Fig. 13.1, with the objective of maintaining a constant total power at the generator terminals. The aforementioned objective is accomplished through the adjustment of power allocation to the dump load in accordance with variations in power demand from the customer.

In literature, several methods are available that have used thyristor-based switches, controlled rectifiers, switched resistors using binary weighted techniques, and control of chopper-based dump load [6–9]. The specified last technique has several advantages from an economic point of view, reliable and no need for reactive power control, but only needs a dump load. However, there is the dissipation of heat energy in the form of a resistor which is used to be dumped. With the extensive generation of power in the case of self-excited induction generator-based MHPP, there is a wastage of power in the dump load when there is less load demand. This excessive unused power can be stored in batteries for better application in isolated areas of the country when insufficient generation is available to feed the consumer load [10–14]. The paper in Sen et al. [15] depicts voltage and frequency control in a single-phase self-excited induction generator feeding the linear variable load. The approach followed is based on the communication-less method in an off-grid system. It uses a single-phase inverter (VSI) with a battery energy storage system (BESS). It is connected with the primary winding of SEIG via a connecting cable, which is highly inductive and synchronized using a dynamically phase-shifted phase-locked-loop (PLL). The proposed technique increases the reliability of the overall system and is verified with MATLAB and Simulink. The paper [16] presents a microgrid system with a small hydro-photovoltaic (PV) array and a battery energy storage system. It signifies various power quality issues based on a multichannel structure using the Volterra filtered-X least mean square (VFXLM) technique. There is the implementation of the bidirectional converter, which is used to manage power flow between the generator and battery. This has been analyzed with the MATLAB/SIMULINK platform. In Meena et al. [17], power quality is improved with the help of the Leaky-Momentum Control Algorithm (LMA) in three-phase SEIG. This algorithm is used to control a voltage-source converter (VSC). It reduces the higher demand for reactive power, the unbalancing of loads, and harmonic distortion. It is a modification of the Leaky and Momentum algorithm.

Different observations have been performed in both steady-state and dynamic conditions in MATLAB/Simulink. There is the implementation of an energy-saving scheme in the grid-connected SEIG with an electronic load controller [18]. It has used different load types: dc load, three-phase ac load, and grid. It proved an alternative to the conventional SEIG system with a dump load. Both steady-state and dynamic conditions have been considered. It has been verified on the MATLAB platform and in an experimental environment. In Mishra et al. [19], an autonomous system based on wind-battery energy sources has been presented. It signifies a proper power management system and presents different protection schemes. A drift-free perturb and observe algorithm is used in wind energy systems. A bidirectional converter is used with a unified controller and a battery energy storage system to regulate DC-bus voltage. The proposed scheme is validated in MATLAB/SIMULINK and on the DS1103 digital platform. The method proved to be fast and seamless. In Gbadega and Sun [20], the author presented a hybrid system with a PV-wind-microturbine-storage energy system. They have used the JAYA optimization problem. The main purpose of the demonstration is to minimize the cost of power generation, increase the operating efficiency of microturbines, and reduce environmental pollution.

The presented paper, Chopper-based ELC has been applied in 2.2 kW rated SEIG to feed the three-phase load with a battery bank system as a dump load to maintain constant voltage and frequency at the PCC of the system. The paper's organization is as follows – section II depicts the proposed system configuration, section III shows the mathematical modeling of parallel connected SEIGs, ELC, and consumer load; section IV includes the result and discussion and section V gives the conclusion of the work.

13.2 Proposed System Configuration

The work being proposed involves the utilization of a 2.2 kW three-phase Self-Excited Induction Generator (SEIG) to supply power to a three-phase consumer load. The Electronic Load Controller (ELC) assists in maintaining a steady voltage and frequency of the PCC (Point of Common Coupling) despite fluctuations in the primary load. The consumer load's active power demand is met by the power generation capacity of the machine, while the required reactive power is supplied through the use of a delta-connected capacitor bank. The Energy Load Controller (ELC) is a system that utilizes a chopper-based mechanism to store surplus power generated as a dump load in the battery bank system. In instances where a self-excited induction generator (SEIG) produces an excess of power beyond the immediate load demand, the surplus energy is subsequently redirected to a dump load.

The power balance equation at the Point of Common Coupling (PCC) is expressed as follows.

$$P_{gen} = P_{load} + P_{dump} \tag{13.1}$$

The Eq. (13.1) represents the relationship between the generated power (P_{gen}), the power used by the load (P_{load}), and the power dissipated as waste (P_{dump}). The purpose of this device is to facilitate the process of recharging the battery. During the moment of highest demand, when consumers require electricity, the battery undergoes discharge in order to supply power. ELC is effectively implementing its strategies in this manner. The determination of the IGBT's switching status is contingent upon the command signal generated by the closed-loop control system. The voltage at the point of common coupling (PCC) is measured in response to fluctuations in load demand. The obtained measurement is then compared to a reference voltage of 415 V, which is provided in a three-phase configuration. The erroneous signal produced is minimized through the utilization of a proportional-integral (PI) controller, achieved by appropriately adjusting the values of the proportional gain (Kp) and integral gain (Ki). The error voltage produced is compared to a carrier wave that operates at a predetermined switching frequency. This comparison is then followed by a conditional block, which afterwards provides a switching instruction to turn the IGBT on or off. The power flow between the primary and dump loads is determined by this factor, influencing both the direction and magnitude of power transfer. The figures presented below in Fig. 13.2 demonstrate the various findings.

13.3 Mathematical Modeling of the Proposed System Configuration

A. Modeling of SEIG

Induction machine is widely used in hilly or remote areas of the country for power generation. When the rotor speed of the induction machine exceeds its synchronous machine, it is called an induction generator. The generating amount of the machine itself fulfills the requirement of the active power of the load. A bank of delta-connected capacitors is used to provide enough reactive power support to the load and machine. Thus, the induction machine is known as a self-excited induction machine (SEIG).

The rotating field set up by the stator current of three-phase induction motor rotates at synchronous speed (N_{sync}) given by,

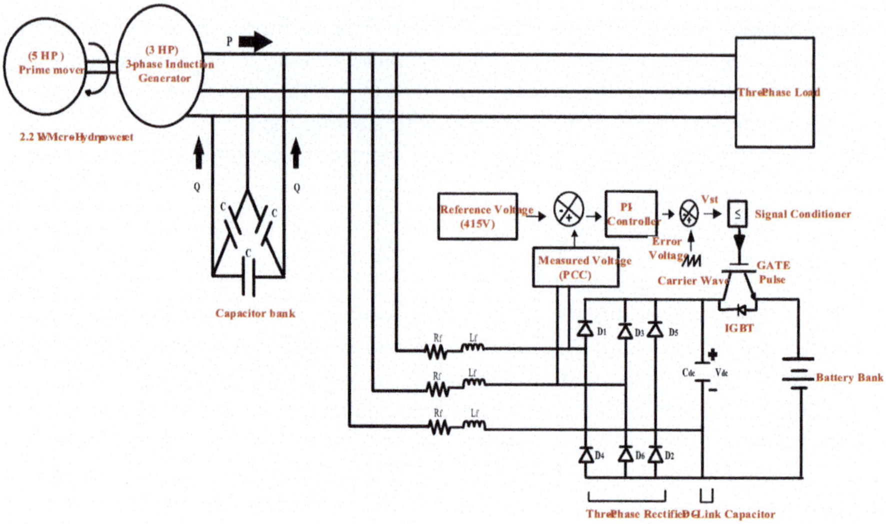

Fig. 13.2 Block diagram of the proposed configuration

$$N_{sync} = \frac{120 * f}{p} \tag{13.2}$$

where, f = supply frequency, p = stator windings poles pair.

Now, slip-speed is given as the difference of speed of rotating magnetic field (N_s) and rotor speed (N_r) given as,

$$\text{Slip} - \text{speed} = N_s - N_r \tag{13.3}$$

Representation of slip (s) as fraction is given as,

$$s = \frac{N_s - N_r}{N_s} \tag{13.4}$$

There are different values of slip (s) on which the working mode of induction machine is decided.

When, $(0 < s < 1) \leftrightarrow$ Positive slip, works as induction motor.

$(s < 0) \leftrightarrow$ Negative slip, works as induction generator.

$(s > 1) \leftrightarrow$ works in braking mode.

To analyze the unbalanced and transient operation of the induction machine, d-q reference frame theory has followed [15]. The electromagnetic developed torque is given as,

$$Te = \left(\frac{3P}{4}\right) L_m \left(i_{qs} i_{dr} - i_{ds} i_{qr} \right) \tag{13.5}$$

The balance equation of electromagnetic torque is given as,

$$T_{shaft} = T_e + j\left(\frac{2}{p}\right)pw_g \tag{13.6}$$

As shown in Fig. 13.2 for three-phase output power, three capacitors are connected as delta configuration in the bank.

Capacitor (C) rating is calculated as,

Apparent-power(S)

$$S = \sqrt{3} * V_L * I_L \tag{13.7}$$

Active-power(P)

$$P = S * \cos\theta \tag{13.8}$$

Absorbed reactive power (Q)

$$Q = S * \sin\theta \tag{13.9}$$

For minimum value of capacitance (C),

$$C = \frac{Q}{3} \tag{13.10}$$

Voltage per-phase (V_P)

$$V_p = \frac{V_L}{\sqrt{3}} \tag{13.11}$$

Capacitance current (I_C)

$$I_C = \frac{Q}{V_P} \tag{13.12}$$

Per phase capacitive reactance (X_c)

$$X_C = \frac{V_P}{I_C} \tag{13.13}$$

Minimum value of per-phase capacitance (C) is given as,

$$C = \frac{1}{2\pi f X_C} \tag{13.14}$$

B. Modeling of ELC

Consider V_{dc} = voltage across dump load resistance (r_{dc}).
P_{gen} = rated generated power of SEIG.
Output voltage (V_O) of ELC is given as,

$$V_O = \frac{3\sqrt{2}}{\pi} V_{dc} \tag{13.15}$$

Current through ELC (I_{dc}) is given as,

$$I_{dc} = \frac{P_{gen}}{V_{dc}} \tag{13.16}$$

Its peak value is given as,

$$V_P = \sqrt{2} * V_{LL} \tag{13.17}$$

Now, rating of current of switches and rectifier is given as

$$I_{AC} = \frac{P_{gen}}{\sqrt{3} * V_{LL}} \tag{13.18}$$

The value of dump resistance (R_d) is given by,

$$R_d = \frac{V_{dc}^2}{P_{gen}} \tag{13.19}$$

C. Control topology

The PCC voltage undergoes a drop or increase with each change in load, and this change in voltage is detected by the voltage sensor. Subsequently, an error signal is generated based on the detected voltage change. The occurrence of this error is reduced when it is inputted to the controller and the appropriate instruction signal, whether analogue or digital, is inputted to the plant. Closed-loop operation has been observed to exhibit higher levels of accuracy and stability. The block diagram illustrating the aforementioned statement is presented in Fig. 13.3.

The proposed control method consists of a comparison of actual and reference voltage values and generates the error command signal. The generated error signal is fed to the Proportional-Integral (PI) Controller. With proper fine tuning with Kp & Ki values, errors get minimized and input to the conditioning block. The signal conditioning block compared this input command with carrier wave having a certain switching frequency and generates pulse.

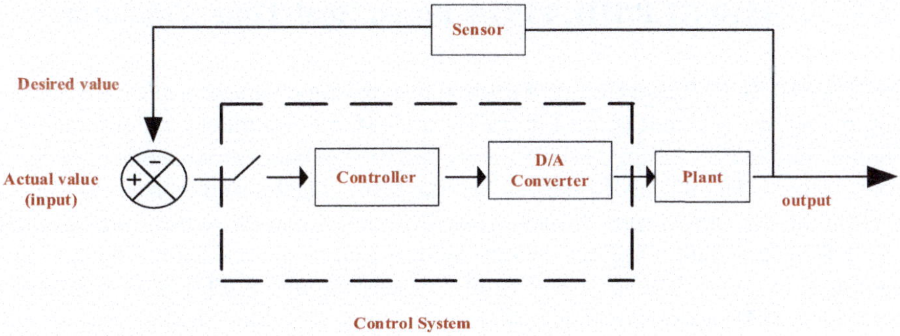

Fig. 13.3 Block diagram of control system

D. Three Phase resistive load

The equation of line current of three-phase resistive load is given as,

$$i_{aL} = i_{pa} - i_{pc} = \left(\frac{V_a}{R_{La}}\right) - \left(\frac{V_c}{R_{Lc}}\right) \tag{13.20}$$

$$i_{bL} = i_{pb} - i_{pa} = \left(\frac{V_b}{R_{Lb}}\right) - \left(\frac{V_a}{R_{La}}\right) \tag{13.21}$$

$$i_{cL} = i_{pc} - i_{pb} = \left(\frac{V_c}{R_{Lc}}\right) - \left(\frac{V_b}{R_{Lb}}\right) \tag{13.22}$$

where i_{aL}, i_{bL}, i_{cL} are the line currents of the load and V_a, V_b, V_c are the voltage of phase a, b, and c respectively.

E. Dump Load (Battery)

The extra generated power after feeding to the dump load is stored in the battery to charge it and some amount is wasted as losses. The stored energy in battery ($E_{battery}$) during t hours of charging process is given as,

$$E_{battery} = \int_0^t \left[P_{battery} - \left(P_{Lbattery}\right)\right]dt + E_O \tag{13.23}$$

where, $P_{Lbattery}$ = Power consumed in battery as dump load
E_o = Stored energy in the battery at the beginning of charging process.

13.4 OP4510 RCP/HIL FPGA-Based Real-Time Simulator

The OP4510 RCP (Rapid Control Prototyping) real-time simulator provides a range of applications for agent control in microgrid systems, laboratory experimentation, Uninterruptible Power Supply (UPS), and emulation of battery management systems, among others. The RT-LAB program facilitates the transfer of data from MATLAB or Simulink onto a real-time environment. The device provides pulse width modulation (PWM) generation and accurately captures it in its precise form. The system is comprised of an Intel Xenon E3 v5 central processing unit (CPU), a random-access memory (RAM) capacity of 16 gigabytes, and a solid-state drive (SSD) with a storage capacity of 128 gigabytes. The digital output of the system is designed to accommodate a user-provided voltage range of around 5 to 30 V. The system comprises a total of 32 channels, with the digital input range being changed to accommodate voltages ranging from 4 to 50 V. The device features a true differential analogue input with a range of ± 20 V, offering a resolution of 16 bits across 16 channels. Similarly, it provides an analogue output with a range of ± 16 V, a resolution of 16 bits, and 16 channels.

Figure 13.4 represents a block schematic of the real-time simulator based on FPGA.

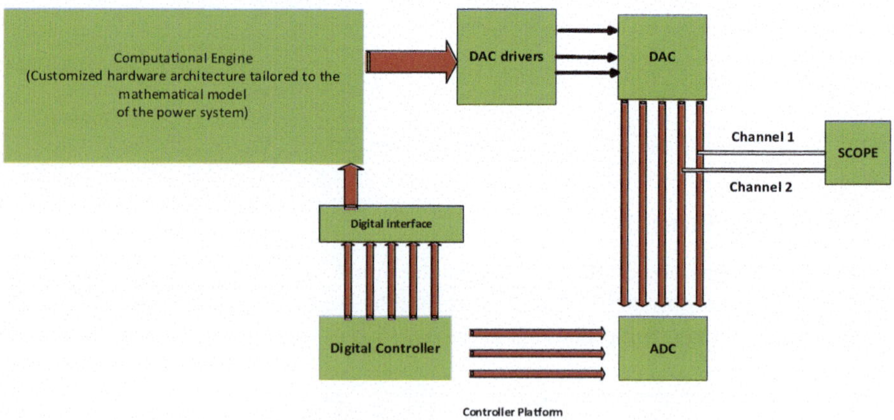

Fig. 13.4 Block diagram of real-time simulator

13.5 Result and Discussion

The proposed methodology was implemented using the MATLAB 2015b Simulink platform, and the simulation was conducted for a duration of 10 s. From the onset of 1 s, a three-phase voltage build-up of 415V occurs. In order to assess the flexibility of the proposed ELC, load switching experiments were conducted. The switching process of the three-phase load occurred in a sequential manner, with the first step taking place between 3 and 6 s, followed by a second step between 4 and 5 s, and finally a third step between 4 and 6 s. During a time, interval ranging from 1 to 3 s, the power that is produced is directed towards the dump load, specifically for the purpose of charging the battery until it reaches its maximum capacity. During a time period of 3–4 s, the consumer load is activated, resulting in a drop in the amount of power transmitted to the dump load. This decrease is observed when the PCC voltage decreases to a value of 390V and the frequency reaches 49.46 Hz, both of which remain within acceptable parameters. The load switch was completed within a time frame of 4–5 s. During this particular time frame, there is a peak in the demand for electricity. Consequently, the period in question is characterized by the transmission of maximum power, while a minimal amount of power is allocated towards the charging of the battery. During this specific time interval, the voltage experiences a decrease to a maximum value of 340V, while the frequency decreases to 48.80 Hz. Once again, within a time frame of 5–6 s, the requested power is decreased and the remaining power is redirected back to the battery. This redirection is accompanied by an increase in the voltage at the point of common coupling (PCC), which rises from 340 V, and a corresponding increase in frequency to 49.1 Hz. The aforementioned study is demonstrated in Figs. 13.5 and 13.6.

In the laboratory, a real-time simulator based on RCP/HIL FPGA technology was utilized in conjunction with a four-channel DSOX1204G oscilloscope for the purpose of capturing and presenting the output results. This is demonstrated in Fig. 13.7 depicted.

The hardware-in-loop observation was documented using a four-channel DSOX1204G. The four channels are allocated for monitoring the PCC terminal voltage, consumer load, dump load, and terminal frequency. It is assumed that.

CH0: Waveform of PCC Voltage, CH1: Waveform of consumer load.

CH2: waveform of dump load; CH1: Waveform of terminal frequency.

The analogue representation of the output PCC voltage is measured at 586 V peak-to-peak. The current system is not capable of capturing output from OPAL-RT within the specified range of its digital and analogue ports. Therefore, it is divided by an appropriate gain value of 20.75, resulting in a voltage of 28.24 V. The root mean square (rms) value of the given quantity is calculated as 28.24 divided by the square root of 2, resulting in a value of 19.9V. The output value of PCC Voltage can be obtained by the use of the RT simulator. The yellow color is depicted in the output result provided in Fig. 13.9. In a similar vein, the consumer load has been assigned a gain value of 90, the dump load a gain value of 1.75, and the PCC frequency a gain value of 2.5. The output waveforms of the consumer load, dump load, and PCC

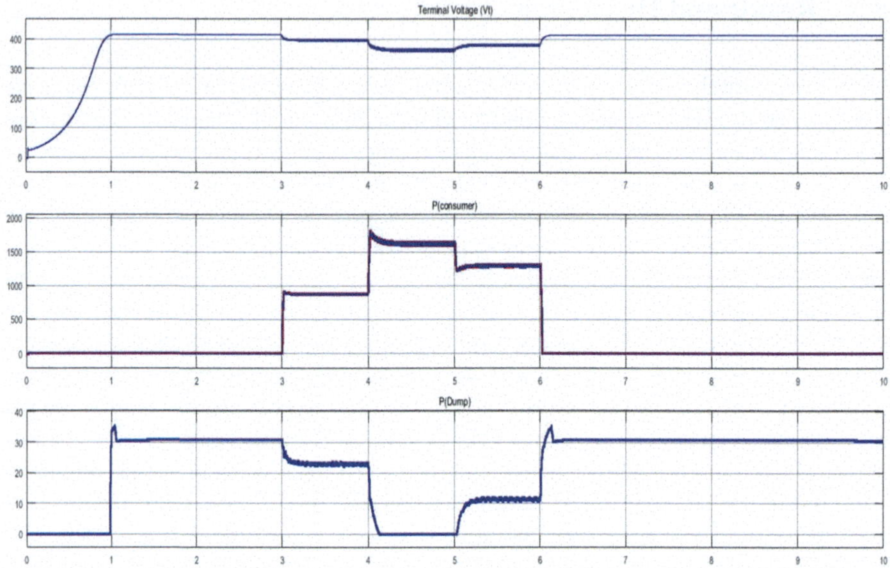

Fig. 13.5 Simulink result of terminal voltage, power consumed by the main load and dump load upon perturbation of load

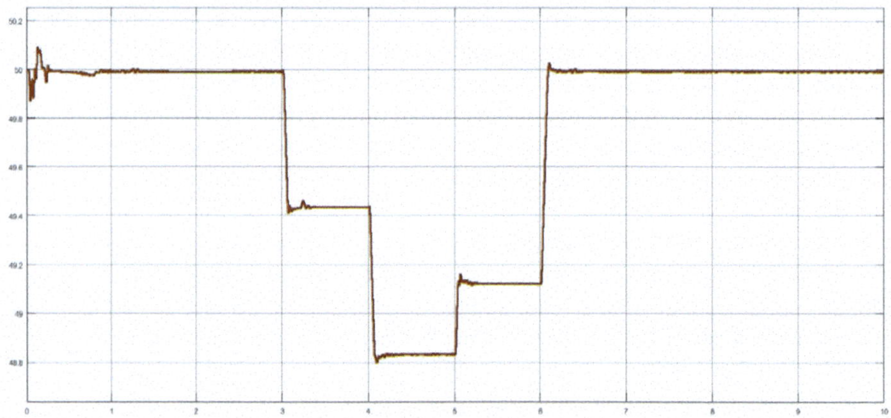

Fig. 13.6 Simulink result in PCC frequency upon perturbation of load

frequency are represented by the colors green, blue, and red, respectively. Figure 13.8 illustrates the impact of a MATLAB-based environment on a personal computer (PC) when interfaced with the OPAL-RT real-time simulator.

Fig. 13.7 Laboratory arrangement of real-time simulator

Fig. 13.8 Schematic design illustrating the integration of a real-time simulator with a personal computer (PC)

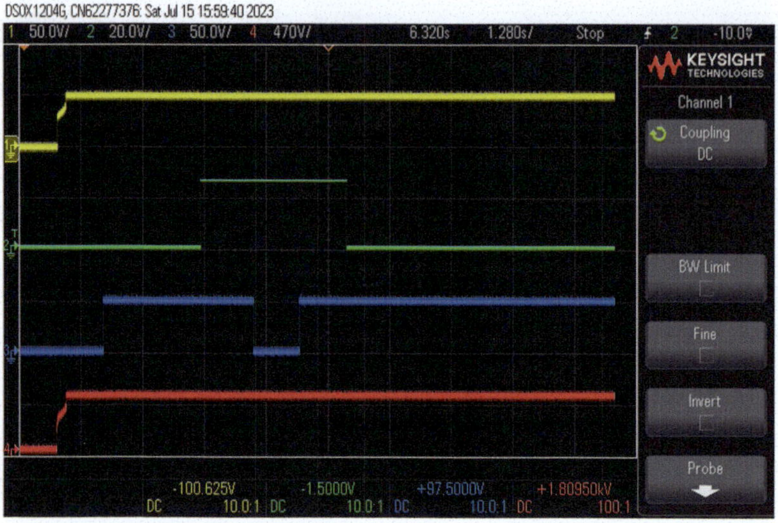

Fig. 13.9 Output waveform of PCC Voltage, consumer load, dump load, and PCC frequency during load switching recorded in four-channel DSOX1204G through real-time simulator OP4510 RCP

13.6 Conclusion

In this work, a successful design has been developed for the energy storage system on the battery, which effectively regulates the output voltage of the Self-Excited Induction Generator (SEIG) during load variation. This design incorporates the Pulse Width Modulation (PWM) method. The system under consideration is a configuration that involves the utilization of an Uncontrolled Rectifier in conjunction with a PWM DC Chopper for the purpose of controlling the charging of batteries. The process of supplying power to both the consumer load and the lead-acid battery in a closed-loop is performed using the SEIG. The battery's state of charge is consistently monitored, and the charging or discharging operation will cease once it hits the predetermined upper and lower thresholds, respectively. The study demonstrated the implementation of a control approach aimed at maintaining the ideal levels of bus voltage and system frequency. A model using MATLAB/Simulink is constructed in order to implement the aforementioned principle. The data indicates that as the primary load fluctuates between 300 and 500W, the terminal voltage exhibits a range of 340 to 390 V, and remains constant until reaching 415V. The frequency of the system exhibited a range of 48.80 to 49.1 Hz during load switching, which is within the acceptable limits. Following a load perturbation, the frequency stabilized at 50 Hz, aided by a 600V battery acting as a dump load. The aforementioned study was conducted on a real-time platform utilizing the OP4510 RCP/HIL FPGA-based real-time simulator. Hence, the suggested system utilizing an Electronic Load controller is deemed appropriate for the provision of power supply in distant places through small-scale micro hydro power generation, employing induction generators.

Acknowledgements The authors thank the DST-SERB (EEQ/2020/00076) project for providing all the hardware equipment required for these experiments/studies.

References

1. Bansal RC (2005) Three-phase self-excited induction generator: an overview. IEEE Trans Energy Convers 20(2):292–299
2. Rana K, Meena DC (2018) Self excited induction generator for isolated pico hydro station in remote areas. In: 2nd IEEE international conference on power electronics, intelligent control and energy systems (ICPEICES). IEEE
3. Sachdev HS, Akella AK, Kumar N (2015) Analysis and evaluation of small hydropower plants: a bibliographical survey. Renew Sustain Energy Rev 1(51):1013–1022
4. Singh B, Murthy SS, Gupta S (2006) Analysis and design of electronic load controller for self-excited induction generators. IEEE Trans Energy Convers 21(1):285–293
5. Kalla UK, Singh B, Murthy SS (2016) Modified electronic load controller for constant frequency operation with voltage regulation of small hydro-driven single-phase SEIG. IEEE Trans Ind Appl 52(4):2789–2800
6. Roodsari BN, Nowicki EP, Freere P (2014) A new electronic load controller for the self-excited induction generator to decrease stator winding stress. Energy Proc 1(57):1455–1464
7. Kumar P, Kalla UK, Bhati N, Agarwal KL (2022) Performance investigation of synchronized three-phase AC chopper-based controller for small hydrogeneration systems. IEEE Trans Ind Appl 58(2):2217–2228
8. Sofian Y, Iyas M (2011) Design of electronic load controller for a self excited induction generator using fuzzy logic method based microcontroller. In: Proceedings of the 2011 international conference on electrical engineering and informatics. IEEE, pp 1–6
9. Nazir R, Syafii S, Pawawoi A, Akbar F, Doriza A (2019) Differences in the impact of harmonic distortion due to the installation of electronic load controller in self-excited induction generator and synchronous generator. Int J Power Electron Drive Syst 10(1):104
10. Sathyakala M, Arutchelvi M (2012) Design of controller for wind driven self-excited induction generator with battery storage for stand-alone application. In: International conference on emerging trends in science, engineering and technology (INCOSET). IEEE, pp 409–415
11. Chilipi RR, Singh B, Murthy SS (2014) Performance of a self-excited induction generator with DSTATCOM-DTC drive-based voltage and frequency controller. IEEE Trans Energy Convers 29(3):545–557
12. Tariq M, Yuvarajan S (2013) Simulink based modeling, analysis and simulation of self-excited induction generator for use in remote areas. IU-J Electr Electron Eng 13(1):1623–1628
13. Barrado JA, Grino R, Valderrama-Blavi H (2010) Power-quality improvement of a stand-alone induction generator using a STATCOM with battery energy storage system. IEEE Trans Power Deliv 25(4):2734–2741
14. Nazir R, Pawawoi A, Rahmadhani F (2011) The energy storage system on the batteries with PWM method to regulate the output voltage of SEIG. In: Proceeding of international conference on electrical power and energy systems, Netherlands
15. Sen S, Singh SP, Choudhuri SG (2019) A communication less control of single phase two winding SEIG By VSI battery based system. In: IEEE region 10 symposium (TENSYMP). IEEE, pp 196–201
16. Kumar S, Singh B, Kalla U, Singh S, Mittal A (2021) Power quality control of small hydro-PV array and battery storage based microgrid for rural areas. In: International conference on sustainable energy and future electric transportation (SEFET). IEEE, pp 1–6
17. Meena DC, Singh M, Giri AK (2021) Leaky-momentum control algorithm for voltage and frequency control of three-phase SEIG feeding isolated load. J Eng Res ICARI Spec Issue 109:120

18. Krishnan A, Natarajan K, Dwivedi AC (2019) Energy saving scheme for grid connected pico-hydro energy conversion system supplying hybrid loads. In: 2019 National power electronics conference (NPEC). IEEE, pp 1–5
19. Mishra J, Behera PK, Pattnaik M, Samanta S (2022) An efficient supervisory power management scheme for a wind–battery-assisted hybrid autonomous system. IEEE Syst J 17(1):768–779
20. Gbadega PA, Sun YX (2023) JAYA algorithm-based energy management for a grid-connected micro-grid with PV-wind-microturbine-storage energy system. Int J Eng Res Afr 10(63):159–184

Chapter 14
Hardware-In-The-Loop Simulation Approach for Grid Integration of a 2.2 kW Self-Excited Induction Generator Using Synchronous Reference Frame Theory

Mrinal Kanti Rajak and Rajen Pudur

Abstract This paper explores the integration of a micro-hydro-based self-excited induction generator (SEIG) with the power grid to harness the advantages of renewable energy sources in the power sector. By incorporating renewable energy into the grid, significant enhancements can be achieved in terms of sustainability, economic viability, resilient capacity, and mitigation of power quality issues in generation and distribution. The study focuses on integrating a specific system comprising a 2.2 kW, 415-V, 50-Hz squirrel cage induction motor (SCIM) driving a three-phase SEIG. The MATLAB machine model is calibrated using machine parameters obtained through rigorous testing, including no-load characteristics, block rotor tests, magnetic saturation, and excitation capacitor calculations. A hydroelectric prime mover powers the SEIG realized with a 3.3 kW AC motor. Grid integration is facilitated through rectification and inversion operations employing an uncontrolled rectifier with a DC-link capacitor and a 6-pulse IGBT-based inverter. Switching operations utilize a voltage source converter (VSC) control topology based on synchronous reference frame theory. Simulations are conducted in the MATLAB/Simulink environment (R2018a version, 40,832,900), and the validity of the results is verified through real-time simulation using OPEL RT(OP4510). This investigation contributes to the understanding and implementation of micro-hydro-based self-excited induction generators for effective grid integration in the power sector.

Keywords Hardware-in-the-loop · SEIG · VSC · AC-DC-AC converter · Micro-hydro plant

M. K. Rajak · R. Pudur (✉)
Department of Electrical Engineering, National Institute of Technology Arunachal Pradesh, Jote 791113, India
e-mail: rajenpudur@nitap.ac.in

S. De et al. (eds.), *Challenges and Opportunities of Distributed Renewable Power*, Energy, Environment, and Sustainability, https://doi.org/10.1007/978-981-97-1406-3_14

14.1 Introduction

Energy shortages pose a significant challenge for many developing countries, primarily due to the depletion of local energy resources and reliance on imported fuels. Among the various renewable energy technologies, micro-hydropower is a viable and cost-effective solution, especially when compared to wind and solar power plants. In 2017, hydropower demonstrated the lowest weighted average cost of electricity worldwide, with a rate of ₹3.71/kWh, whereas industrial-scale photovoltaic projects exhibited the highest cost at ₹7.42/kWh. Regional variations were observed, with Asia and South America experiencing rates of ₹2.97/kWh, North America at ₹4.45/kWh, Africa, Eurasia, and the Middle East at ₹5.19/kWh, and Central America and the Caribbean at ₹7.42/kWh, as reported by the International Renewable Energy Agency (IRENA) in 2016. Micro-hydropower is an energy bridge in developing regions, meeting substantial energy demands in mountainous areas alongside other energy sources. Classification of hydropower plants is based on generation capacity, with three categories recognized: Small Hydro Power Plants (<25 MW), micro-hydro ($\leq$100 kW), mini-hydro (101 kW–2 MW), and small hydro (2 MW–25 MW). The hilly regions of several countries, including Uttarakhand, Chhattisgarh, Kerala, Himachal Pradesh, Maharashtra, and Arunachal Pradesh, possess significant potential for micro-hydropower plants. These installations play a vital role in electrifying remote villages situated in these areas.

As evidenced by numerous studies, self-excited induction generators (SEIGs) have proven effective in mini-micro-hydro plants for power generation [1–6]. However, these generators exhibit poor voltage and frequency regulation, necessitating the incorporation of power electronics devices. To meet the power demand on the consumer side, SEIGs can be operated in parallel or tandem fashion [7–9]. This allows for combining multiple SEIGs or their integration with other power sources, such as grid or renewable energy systems, to ensure a reliable and continuous power supply. There has been an increased focus on grid integration technologies to address the challenges arising from the limited voltage regulation capabilities of SEIGs. Grid connection enables the equalization of reactive power flow from the SEIG's excitation capacitor with the active power flow from the grid. This synchronization helps maintain a constant voltage at the point of standard coupling (PCC), the connection point between the SEIG and the grid. SEIGs can benefit from their stable and well-regulated voltage by integrating with the grid, improving their overall performance and power quality. The grid integration process involves regulating the reactive power flow from the SEIG to align with the active power flow from the grid, thereby balancing the power exchange and maintaining a stable voltage at the PCC.

Additionally, it is worth noting that the induction generator itself consumes reactive power from the grid. This reactive power is necessary to maintain the core magnetization of the generator, ensuring its proper operation. Overall, grid integration with SEIGs is crucial in mitigating the voltage regulation challenges associated with these generators. It facilitates a more stable and reliable operation, allowing SEIGs to effectively contribute to the power supply while maintaining a consistent voltage at the PCC. In off-grid operation, the reactive power absorbed by induction generators, especially those with squirrel cage rotors, must be compensated for using capacitors. Induction machines are widely favored for power generation in micro hydropower plants due to their numerous advantages over synchronous machines. While connecting induction generators to the grid is common, one challenge is the high inrush current that can potentially damage the machine. Although this can be mitigated using a suitable soft starter, the cost is often prohibitive for micro-level generation. To address the inrush current issue, an uncontrolled rectifier and an excitation capacitor are employed on the machine-side to handle reactive power within the induction machine. The primary requirement for grid connection is to ensure frequency and voltage matching. Frequency matching can be achieved by adjusting the speed of the prime mover, which, in turn, regulates the amount of water in the storage tank.

There are various methods available for grid integration with self-excited induction generators (SEIGs), including:

1. Direct-on-line method involves directly connecting the SEIG to the grid, which requires careful coordination to ensure proper synchronization.
2. Soft starter method: A soft starter helps manage the inrush current during grid connection, but it can be costly for micro-level generation applications.
3. AC-DC-AC method involves using a machine-side converter that is either controllable or uncontrollable. It enables effective control of the power flow and enhances grid integration capabilities.

Grid integration of micro hydropower plants (MHPPs) has significant advantages for renewable energy sources (RES). MHPPs offer a constant and more predictable electricity generation compared to other renewables. Additionally, their relatively low cost and high reliability make them attractive for remote or isolated areas where they can quickly and easily meet local energy demands. Consequently, the grid integration of MHPPs in the mountainous regions has gained considerable research interest in the power sector, given its potential economic benefits [10–12]. Significant progress has been made in this field, and the literature review presented below provides a detailed overview of the recent scientific advancements in grid integration of micro hydropower plants.

In [10], a novel technique for micro hydropower plant (MHP) control is introduced. The study presents a unique three-phase grid-connected MHPP control scheme, incorporating a maximum power point tracking (MPPT) algorithm for dynamic power generation and distribution system adjustment. In [13], the author presents a novel grid synchronization scheme based on the adaptive linear optimal filtering (ALOF) technique. The proposed scheme improves the performance of the phase-locked loop (PLL) in achieving accurate synchronization with the grid. The work presented in [14] discusses a maximum power point tracking (MPPT) control scheme specifically designed for the grid integration of self-excited induction generators (SEIGs). This scheme aims to optimize power extraction from the SEIG and enhance its performance during grid connection. In [15], a topology structure for micro hydropower plants (MHPPs) is proposed, providing a framework for designing and configuring MHPP systems. The study described in [16] focuses on connecting a squirrel cage induction generator (SCIG) wind turbine and an inverter to the grid. The research addresses the challenges associated with the nonlinear behavior of the system.

In [17], a new robust predictive power controller (RPCC) is introduced, which operates in the synchronous reference frame. This controller enhances the stability margins of conventional predictive controllers and improves the overall robustness of the system. The authors of [18] present an efficient topology for a grid-connected wind-solar power plant. The proposed topology optimizes the power conversion and control mechanisms to ensure effective integration with the grid. [19] discusses the design and control of a back-to-back converter for wind energy conversion systems. The paper outlines a step-by-step technique for the efficient and reliable operation of the converter. Finally, [20] proposes a control strategy for a grid-connected photovoltaic inverter with a two-stage operation. The strategy aims to optimize the photovoltaic system's overall performance and power generation during grid connection. These references highlight various control techniques, grid synchronization schemes, maximum power point tracking approaches, and system topologies to enhance the grid integration of micro hydropower plants, wind turbines, and photovoltaic systems.

This study analyzes the real-time integration of a 2.2 kW self-excited induction generator (SEIG) with the grid using a synchronous reference frame approach combined with a feed-forward decoupling strategy. Various machine tests were conducted to model the SEIG to determine the equivalent circuit parameters accurately. The performance of the SEIG model was then evaluated using real-time simulation on the OPEL RT(OP4510) platform. The system consists of two converters: machine-side converters and grid-side converters. The machine-side converters regulate the SEIG's power output and maintain the required voltage and frequency levels.

The study on real-time integration using advanced control strategies for a 2.2 kW self-excited induction generator SEIG with the network is essential for the following reasons.

1. **Renewable Energy Integration:** As the world transitions towards cleaner and more sustainable energy sources, integrating small-scale renewable energy generators like SEIGs becomes crucial. This study contributes to the knowledge and technology needed to harness renewable energy efficiently.
2. **Grid Stability:** Maintaining electrical grids' stability is crucial. When integrated well, SEIGs can function as distributed energy resources. It improves grid stability and resilience to know how to control them in real-time to match grid requirements.
3. The study's emphasis on SEIGs is consistent with the movement toward decentralized power generation. This research promotes the development of distributed energy systems, increasing energy security and lowering transmission losses by enabling SEIGs to connect seamlessly to the grid.

4. Better utilization of energy resources is possible through the effective integration of SEIGs. Energy can be produced and used more effectively by tuning these generators' power output, voltage, and frequency.
5. **Advanced Control Techniques:** Advanced control techniques, such as the synchronous reference frame approach and feed-forward decoupling, apply to SEIGs and can be adapted to other power generation systems. This research contributes to the broader field of power system control.
6. **Real-Time Simulation:** Real-time simulation platforms like the OPEL RT (OP4510) are valuable for testing and validating control strategies. It provides a safe and controlled environment to experiment with various scenarios before real-world implementation, reducing risks and costs.
7. **Modeling Accuracy:** Accurate modeling of SEIGs is essential for efficient control. The efforts made in this study to identify the equivalent circuit parameters improve our comprehension of SEIG behavior and serve as a guide for upcoming investigations and applications.
8. **Environmental Impact:** This research supports the integration of SEIGs, which are commonly used in renewable energy systems. This indirectly helps reduce greenhouse gas emissions and mitigate climate change.
9. **Technological Innovation:** Creating control systems and integration strategies for SEIGs is a technological innovation that could pave the way for progress in other power generation and distribution fields.
10. **Practical Application:** The insights gathered from this research can be put into practice in practical situations, allowing for using SEIGs and comparable systems in various applications such as producing power from a distance, microgrids, and installations for renewable energy.

Integrating renewable energy, improving grid stability, and using sustainable energy resources to achieve success. The study emphasizes that advanced control techniques and accurate modeling in power systems engineering play a significant role. The research findings may inspire innovative energy management and distribution strategies. Integrating renewable energy sources into the current grid infrastructure, can reduce our reliance on fossil fuels and decrease greenhouse gas emissions. This study highlights the importance of continued investment in advanced control techniques and precise modeling to ensure a smooth transition towards a sustainable and robust energy future.

On the other hand, the grid-side converters ensure the seamless integration of the SEIG with the grid, enabling efficient power exchange and adherence to grid requirements. The paper is structured as follows: Section II discusses SEIG parameter calculations, where the necessary parameters for accurate modeling of the SEIG are determined. Section III focuses on the calculations related to the DC-link capacitor, which plays a crucial role in maintaining stable operation and power exchange. In Section IV, the design of the LCL filter is presented, addressing the mitigation of harmonics and ensuring a clean power output. Section V delves into the control system design, outlining the strategies employed to regulate the SEIG and achieve reliable grid integration. Sinusoidal pulse width modulation (PWM) techniques are discussed in Section VI, elaborating on the methods used for controlling the converter switches. Both MATLAB and real-time simulations are performed to validate the proposed approach, and the results are analyzed and discussed in Section VII. Finally, in the concluding section, Section VIII, the essential findings and contributions of the study are summarized, and potential avenues for future research are outlined.

14.2 Proposed Technique

Efficient integration of a 2.2 kW self-excited induction generator (SEIG) with the electrical grid is crucial for renewable energy systems, backup power generation, and distributed energy sources. To achieve this, two advanced control strategies must be utilized: the synchronous reference frame approach and the feed-forward decoupling strategy. The former regulates voltage, current, and frequency while enhancing system stability. Accurate modeling is essential to determine SEIG's equivalent circuit parameters and responsive control algorithms adaptable to load and grid conditions. Real-time simulations validate system performance before real-world implementation.

The integrated system includes machine-side and grid-side converters, each playing a vital role. Machine-side converters regulate SEIG's power output to align it with the grid's requisites. Grid-side converters bridge the interface between SEIG and the electrical grid, facilitating power injection while maintaining voltage and frequency stability. Maintaining stable and synchronized operation between SEIG and the grid is critical, ensuring compliance with the grid's rigorous voltage and frequency requirements.

This study comprehensively explains integrating a self-excited induction generator with the electrical grid. It harnesses advanced control strategies, meticulous modeling, and real-time simulations to guarantee that SEIG effectively contributes power to the grid while upholding its stringent standards. Such research promotes renewable and distributed energy sources within the modern electrical grid framework.

This study examines the integration of a 2.2 kW self-excited induction generator (SEIG) with the grid in real-time. The main aim of this study is to create a smooth and effective connection between a 2.2 kW self-excited induction generator (SEIG) and the electrical grid. This connection is extremely important for renewable energy systems, backup power generation, and distributed energy sources. It is crucial to synchronize the SEIG's output with the grid to provide a reliable power supply.

In this study, two advanced control strategies were used: the synchronous reference frame approach and a feed-forward decoupling strategy. The synchronous reference frame approach uses a mathematical transformation to simplify the control of AC machines and ensure accurate regulation of voltage, current, and frequency. Additionally, the feed-forward decoupling strategy enhances control by minimizing interactions between different control loops, leading to improved stability of the system overall. Creating an accurate model of the SEIG is crucial for seamless integration.

This study conducts several machine tests to identify the equivalent circuit parameters of the SEIG. These parameters are vital for developing a precise mathematical representation of the SEIG's behavior, which is necessary for designing effective control algorithms that can adapt to fluctuations in load and grid conditions. In order to properly evaluate the SEIG model's efficiency and control strategies' effectiveness, conducting real-time simulations is crucial. The OPEL RT platform, with plate number OP4510, offers a controlled environment that closely mimics real-world operating conditions. This allows researchers to assess the SEIG's reaction to different scenarios and is an essential tool for adjusting control algorithms and verifying the system's performance prior to practical implementation. The converter systems have two important parts: machine-side converters and grid-side converters. The machine-side converters serve as a link between the SEIG and the control system, making sure that the power output meets the grid's needs. Meanwhile, the grid-side converters help the SEIG inject power into the electrical grid while keeping the voltage and frequency stable. It is crucial to achieve stable and synchronized operation between the SEIG and the grid. The machine-side converters are responsible for regulating the power output, voltage, and frequency levels of the SEIG to ensure that it generates power while meeting the grid's strict standards. This control is essential to maintain the required voltage and frequency levels, which are crucial for the reliability and safety of the grid.

To sum up, this in-depth investigation explores the complex procedure of connecting a self-excited induction generator to the electrical grid. Through the use of advanced control methods, thorough modeling, and dynamic simulations, it guarantees that the SEIG can efficiently supply power to the grid while complying with the grid's strict voltage and frequency regulations. This study is vital in promoting the use of renewable and decentralized energy sources in the contemporary electric grid.

The Synchronous Reference Frame (SRF) Theory has existed for a while. It involves a frame that rotates at a synchronous speed. However, to determine the speed at which this reference frame should rotate, need to understand the practical applications of this theory. SRF Theory is particularly useful in 3-phase control applications where a 3-phase quantity (voltage/current) rotates in space at a high speed. Instead of using individual controllers in all three phases, which can be expensive, SRF Theory allows eliminating one controller for each of the three phases while keeping two controllers (for P and Q control ability) for all three phases. In SRF Theory, consider the d and q axes, with the q axis leading the d axis by 90 degrees. This becomes the reference frame (dq axis), which rotates synchronously at the same speed as the system frequency. The rotating phasor in the image rotates at a speed "w", and so does the reference frame. Therefore, the relative speed between the rotating phasor and the frame is zero (0), leading to a DC quantity that can be resolved along the d and q axes. SRF can deal with DC quantities, which are easier to control in control applications and do not result in steady-state errors (Fig. 14.1).

The feed-forward decoupling strategy is a highly advanced control technique for grid-connected inverters to enhance their performance and stability. This approach aims to compensate for two critical factors affecting the inverter's operation: grid

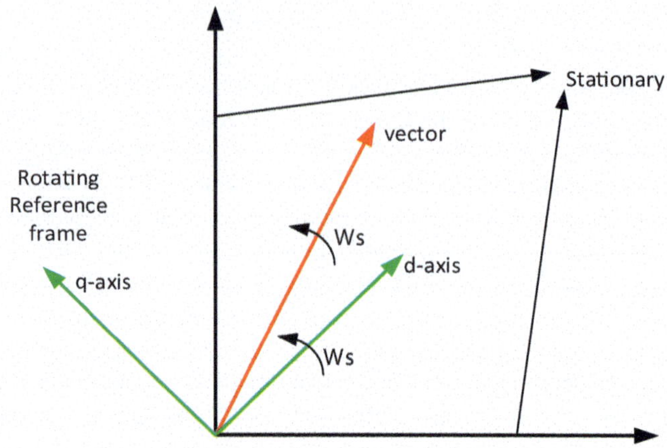

Fig. 14.1 Synchronous Reference Frame (SRF) Theory

voltage disturbances and changes in internal parameters. Grid voltage disturbances occur due to fluctuations or disruptions in the electrical grid, and they can potentially disrupt the inverter's output. Similarly, variations in the inverter's internal parameters, such as temperature changes or component aging, can also affect its performance. The feed-forward decoupling strategy utilizes predictive signals called feed-forward signals that provide advanced information about the expected changes in the grid voltage and internal parameters. The inverter can accurately adjust its reference currents, specifically within the d-q synchronous frame, to ensure that its output aligns optimally with the grid's requirements. This strategy allows the inverter to proactively respond to disturbances and parameter variations, improving its overall stability and performance in grid-connected applications.

14.3 SEIG Parameters Calculation

A 2.2 kW 50 Hz 4.8 Amps SCIM makes a self-excited induction generator. For MATLAB simulation machines, parameters have been calculated through no-load and block rotor tests. For the magnetic saturation test, first, the SCIM coupled with the prime mover. The direction of rotation was confirmed by driving the set by 3.3 kW, 50 Hz Variable frequency drive (VFD), and a three-phase autotransformer because to perform saturation test, the direction of rotation of prime mover fed by VFD and SCIM fed by auto-transformer should be same. Prime mover is rotated at its full rated speed, then the supply to the generator gradually increases using auto-transformer, and current and voltage are measured. The best around ten readings have plotted voltage and current in the MATLAB platform accordingly; values are shown in the appendix. The experiment setup for magnetic saturation can be seen in Fig. 14.2.

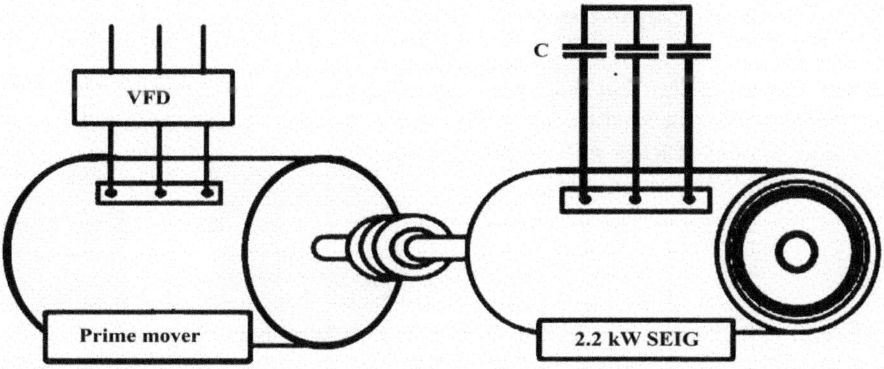

Fig. 14.2 SEIG-Prime mover setup for parameter calculation

14.3.1 Excitation Calculation

$$c = \frac{1}{2\pi f X_c} \tag{14.1}$$

14.3.2 Parameters Calculation

After the Block rotor and No-load test (Table 14.1).

$$L_m = \frac{X_m}{2\pi f}$$

After getting all the parameters, the induction machine block was set in MATLAB (Table 14.2).

Table 14.1 Machine rating

Voltage (V)	Current (A)	Power factor	Power (kW)	Capacitor (μF)
415	4.8	0.8	2.2	12.5

Table 14.2 Machine parameters after NL & BR test

Rated power (kW)	R_s (Ω)	R_r (Ω)	$X_{ls} = X_{lr}$ (Ω)	Base Impedance	X_m (Ω)
2.2	3.81	2.87	4.45	86.113	87.010

14.4 DC-Link Capacitors

The DC-Link capacitor balances the fluctuating instantaneous power and stabilizes the "ripple" generated by high-frequency circuits. It is selected based on the maximum amount of RMS AC and DC/voltage a capacitor can withstand before it fails. Its equation is given as [25].

$$C = \frac{S}{4\pi f v_{dC}^{*} \Delta v_{dC}} \tag{14.2}$$

Here, S is the apparent power of the converter, f is nominal frequency, Δv_{dC} is the peak-to-peak voltage ripple of the DC-link capacitor in a steady-state. Its value has been taken as 2% of v_{dC}^{*}.

14.5 LCL Filter Design

The combination of an inductor and capacitor forms an LCL filter. The inverter and grid connect two inductors, L1 and L2. The capacitor (C) is connected in a star configuration between the two inductors. This design of the filter has parameters like inverter voltage (Vi), grid Voltage (Vg), and Grid current (Ig), as shown in Fig. 14.3 [19–26].

After converting, the circuit is drawn in S-domain, as shown in Fig. 14.4.

$$\frac{I_g}{v_i} = \frac{1}{sL\left(1 + s^2 C L_P\right)}$$

The final equation for the transfer function to calculate the minimum inductor values and the resonance frequency can be expressed as follows.

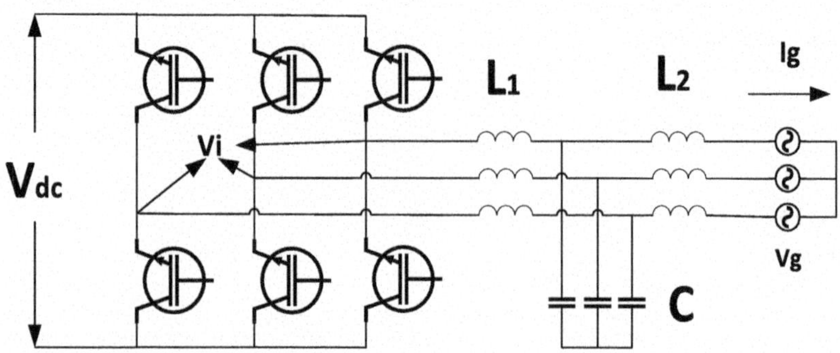

Fig. 14.3 Three-phase grid-connected Inverter with LCL filter

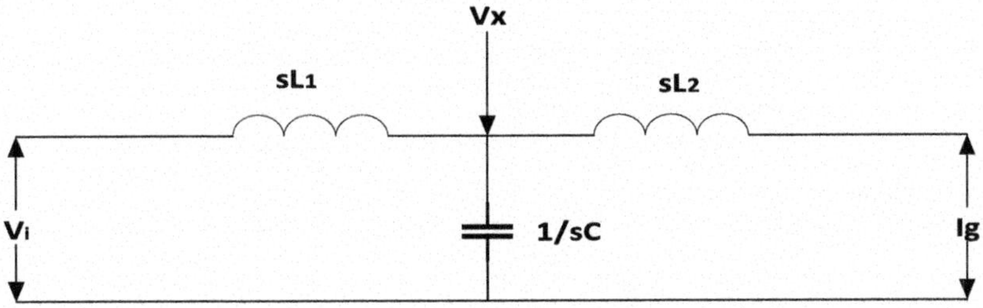

Fig. 14.4 Filter circuit in S-domain

$$L = \left| \frac{1}{\omega_{SW} \frac{Ig(Sw)}{v_i(Sw)} \left(1 - \frac{\omega_{Sw}^2}{\omega_{res}^2}\right)} \right| \tag{14.3}$$

The value of a capacitor is determined by how much reactive power the capacitor needs.

Reactive power = 5% of the rated power (S)

$$c = \frac{0.05 \times s}{v^2 \times 2 \times \pi \times f} \tag{14.4}$$

14.6 Control System

An inverter, which controls the quality of electricity fed into the grid, is a critical component of many grid-connected distributed generation systems. Inverter control design is primarily concerned with synchronization with the utility grid since the grid frequency, and voltage must be regulated. Power can only be supplied if the inverter power matches the local grid connection requirements [27]. The control algorithms used in this work are based on synchronous reference frame theory. Assuming that the voltage in a three-phase electrical system is balanced, the following equations apply [20].

$$\begin{aligned} V_{g_a} &= V_m \sin \omega_0 t \\ V_{g_b} &= V_m \sin\left(\omega_0 t - \frac{2\pi}{3}\right) \\ V_{g_c} &= V_m \sin\left(\omega_0 t + \frac{2\pi}{3}\right) \end{aligned} \tag{14.5}$$

The electric grid's maximum voltage, denoted V_m, and its angular frequency, denoted ω_0, are both specified. The fundamental equation of an inverter is shown in Fig. 14.5.

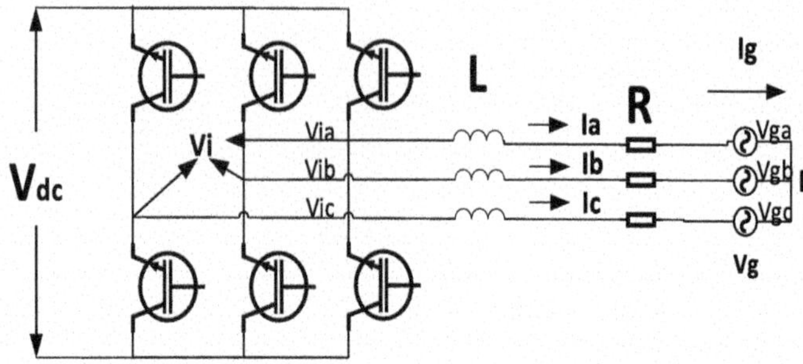

Fig. 14.5 Representation of a three-phase connected inverter

$$\begin{bmatrix} \frac{di_a}{dt} \\ \frac{di_b}{dt} \\ \frac{di_c}{dt} \end{bmatrix} = \begin{bmatrix} -\frac{R}{L} & 0 & 0 \\ 0 & -\frac{R}{L} & 0 \\ 0 & 0 & -\frac{R}{L} \end{bmatrix} \begin{bmatrix} I_a \\ I_b \\ I_c \end{bmatrix} + \frac{1}{L} \begin{bmatrix} V_{\text{ia}} - V_{\text{ga}} \\ V_{\text{ib}} - V_{\text{gb}} \\ V_{\text{ic}} - V_{\text{gc}} \end{bmatrix}$$

synchronize the transition from a three-phase rotating ABC reference frame to a two-phase synchronous dq reference frame.

$$\begin{bmatrix} \frac{di_d}{dt} \\ \frac{di_q}{dt} \end{bmatrix} = \begin{bmatrix} -R & \omega L \\ -\omega L & R \end{bmatrix} \begin{bmatrix} i_d \\ i_q \end{bmatrix} - L \begin{bmatrix} V_{gd} \\ V_{gq} \end{bmatrix} + \frac{1}{L} \begin{bmatrix} V_{i_d} \\ V_{i_q} \end{bmatrix}$$

Three-phase grid-connected inverter has d-axis and q-axis components of the output current represented as id and iq, Vgd and Vgq are d-axis and q-axis components of the three-phase voltage, respectively and the output current of the three-phase grid-connected inverter has d-axis and q-axis component as V_{id} and $V_{iq.}$

The above Equation is expressed as.

$$\begin{aligned} V_{i_d} &= L\frac{di_d}{dt} + Ri_d - \omega Li_q + V_{\text{gd}} \\ V_{i_q} &= L\frac{di_q}{dt} + Ri_q - \omega Li_d + V_{\text{gq}} \end{aligned}$$

The approach of decoupling using feed-forward technique is introduced here. For the closed-loop operation of the system, there is an addition of PI controller with the existing d and q connection to increase the overall system's stability.

The control equation is given as.

$$\begin{aligned} V_{i_d} &= \left(k_p + \frac{k_i}{s}\right)\left(i_d^* - i_d\right) - \omega Li_q + V_{\text{gd}} \\ V_{i_q} &= \left(k_p + \frac{k_i}{s}\right)\left(i_q^* - i_q\right) + \omega Li_q + V_{\text{gq}} \end{aligned} \quad (14.6)$$

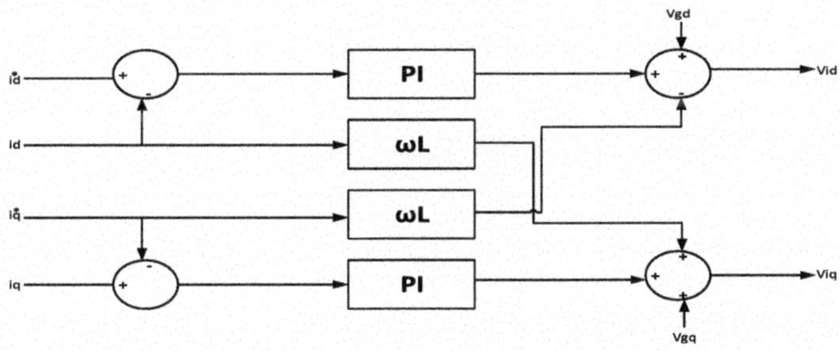

Fig. 14.6 Representation of inner-loop current grid controller

In three-phase grid-connected inverter, there is a need for proper power flow of active and reactive power. For this, the feed-forward decoupling technique is incorporated here. The schematic diagram is shown below in Fig. 14.6.

PLLs are responsible for maintaining coherence between the input (reference) signal frequency (fi) and the output frequency (fo). A PLL has been used to determine the phase angle (ωt) (Fig. 14.7).

For sine PWM the relation between modulation index and inverter voltage is given by.

$$V_{i_d} = \frac{m_d \times v_{dc}}{2} \tag{14.7}$$

$$V_{i_q} = \frac{m_q \times v_{dc}}{2} \tag{14.8}$$

Therefore

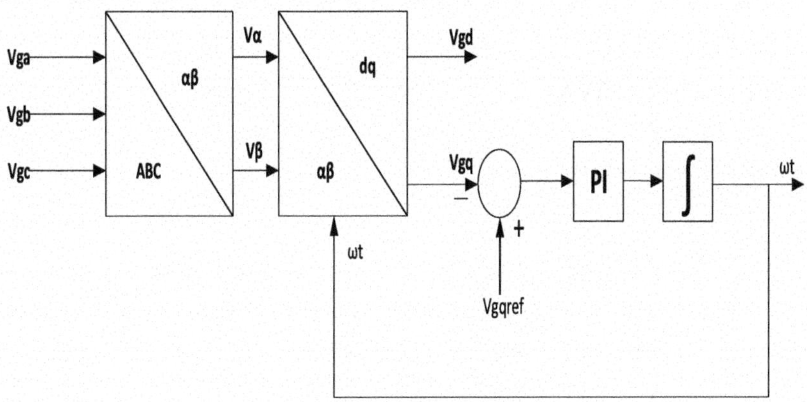

Fig. 14.7 PLL block diagram

$$V_{gd} = V_{i_d}\left(\frac{2}{v_{dc}}\right) \tag{14.9}$$

$$V_{gq} = V_{i_q}\left(\frac{2}{v_{dc}}\right) \tag{14.10}$$

The voltage is subsequently converted to ABC voltage, which serves as a reference for PWM production.

A three-phase grid-connected inverter's instantaneous active and reactive power utilizing d-axis and q-axis voltages and currents may be calculated using the following formulae.

$$\begin{aligned} P &= \frac{3}{2}\left(v_{g_d} i_d + v_{gq} i_q\right) \\ Q &= \frac{3}{2}\left(v_{g_q} i_d - v_{gd} i_q\right) \end{aligned} \tag{14.11}$$

The above equation may be stated as follows if the d-axis is positioned as the grid voltage synthesized vector, i.e., $v_{g_q} = 0$.

$$\begin{aligned} P &= \frac{3}{2}\left(v_{g_d} i_d\right) \\ Q &= -\frac{3}{2}\left(v_{g_d} i_q\right) \end{aligned} \tag{14.12}$$

According to Eq. (14.12), there is a linear relation between the active power (P) and reactive power (Q) with the d and q axis components of current. Thus, in three-phase grid-connected inverter, any change in d or q component of current can affect power directly (Fig. 14.8).

14.7 Sinusoidal PWM

The modulation index (m) is

$$m = \frac{v_p}{v_T} \tag{14.13}$$

Vp is the modulating wave's peak, and VT is the carrier wave's peak. The modulation index can be changed between 0 and 1, resulting in a linear connection between the magnitudes of the modulating and output waves. It is a linear amplifier when modulating/reference signals are applied to the inverter's output voltage (Fig. 14.9).

A sinusoidal signal is taken as a reference signal and has frequency as an output signal. Generally, the carrier signal is usually a triangular or sawtooth wave with a higher frequency than the reference signal. Therefore, there is a comparator that

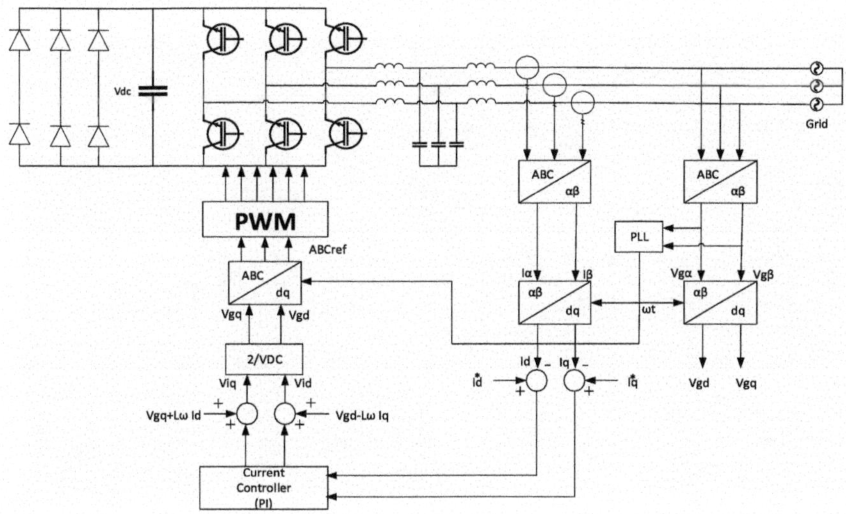

Fig. 14.8 Grid side converter control

compares these two signals. Its output has two states. In one state, either the reference signal exceeds the carrier signal, or in another, the carrier signal exceeds the reference signal, and accordingly, the output is decided [28].

14.8 MATLAB and Real-Time Simulation Results

To provide the appropriate voltage and frequency for grid integration, a 2.2 kW, 415 V, 50-Hz, 4.8A self-excited induction generator is utilized with a calculated range of delta connected capacitor driven by a prime mover 3.7 kW, 415 V, 50-Hz, 7.5A squirrel cage induction motor, but in practical Variable Frequency Drive (VFD) used.

The analogue output of the real-time simulator is limited to 16 V. To represent the Vpk (peak-to-peak voltage output of SEIG), a gain block is used to divide the voltage (1/58) (Fig. 14.10).

An uncontrolled diode rectifier performs the rectification action (AC-DC) with a DC-link capacitor, and the inversion action is performed by a 6-pulse IGBT-based inverter (DC-AC).

The capacitor builds up voltage Vdc in real-time, as shown in Fig. 14.11. The gain block was used to divide the scale (1/60) to show the full Vdc voltage. The Vdc voltage is 600V. After converting the AC output of the SEIG, a 2500 μF DC-link capacitor is used.

To obtain the pure sinusoidal voltage and reduced harmonics, an LCL filter is used. It is an interface between the inverter connected in grid-connected mode and

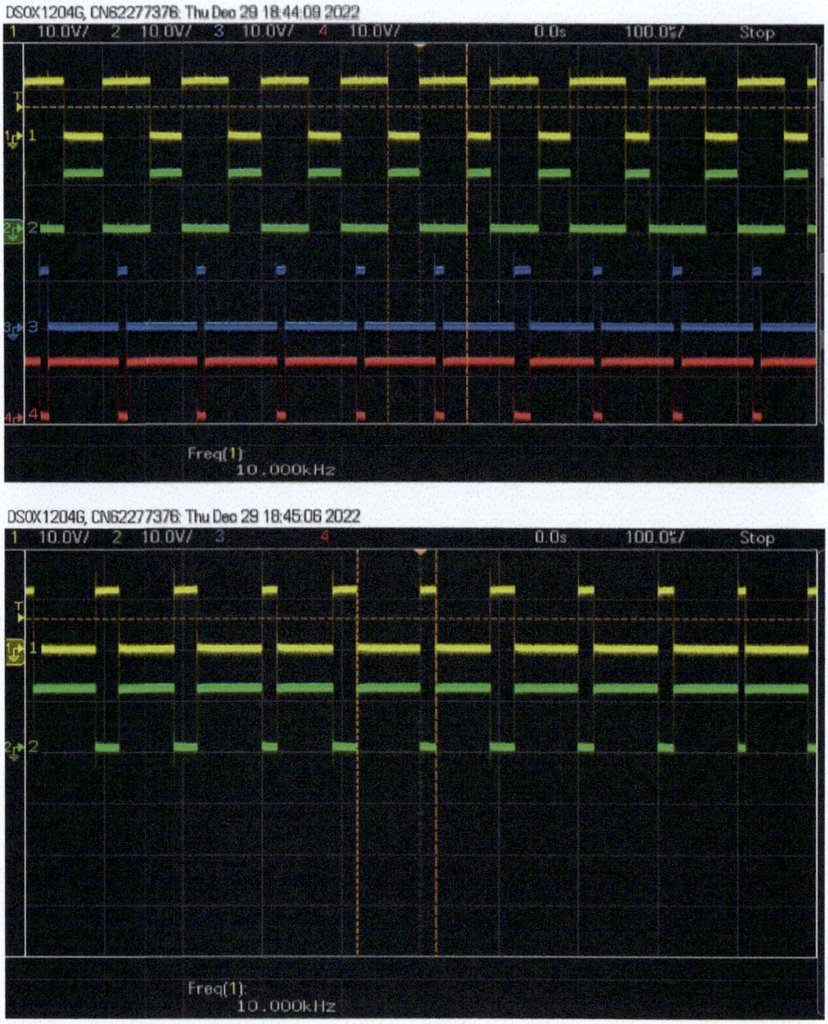

Fig. 14.9 Sinusoidal PWM generation from real-time

the grid. It proves to dampen switching frequency harmonics cost-effectively, which is present in grid current compared to L-filter (Figs 14.12 and 14.13).

The equivalent circuit of a three-phase grid-connected inverter, where V_{a0}, $V_{b0,}$ and V_{c0} are the output voltages of the three inverter bridges concerning the center point O. ia1, ib1, and ic1 are the inverter-side inductor current. V_{cx} (x = a,b,c) and i_{cx} are the voltage and current of the filter capacitor, respectively. i_{2x} is the grid-side inductor current [29].

From Fig. 14.14 V_{ao}, $V_{bo,}$ and V_{co} can be expressed as.

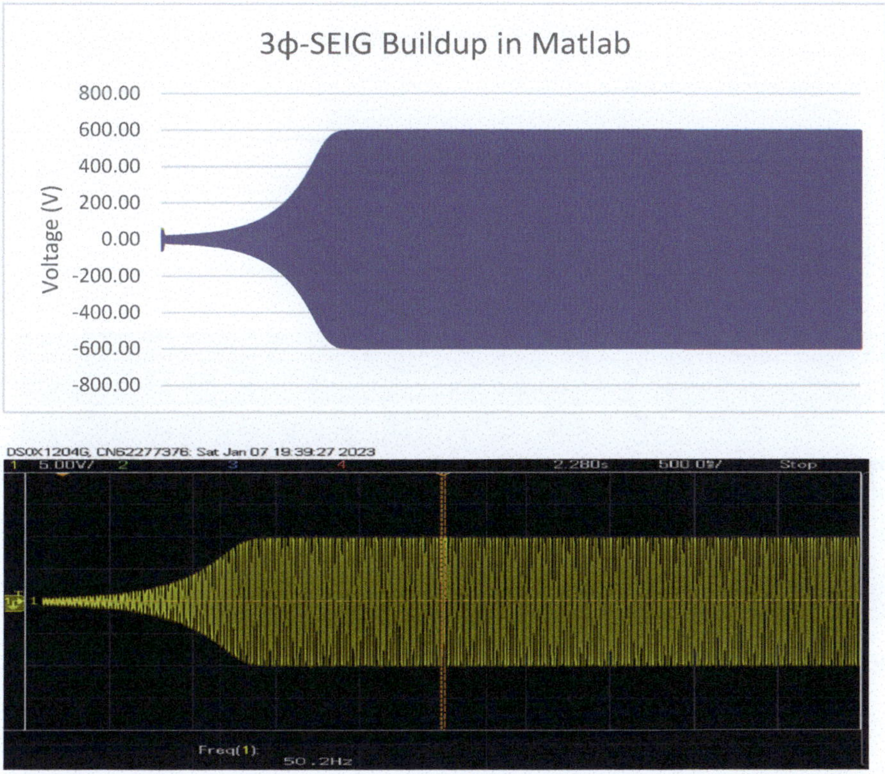

Fig. 14.10 SEIG build-up Voltage in MATLAB & SEIG build-up voltage in real SIMULATION time

$$
\begin{aligned}
V_{\text{ao}} &= j\omega L_1 \cdot i_{1a} + V_{\text{ca}} + V_{\text{NO}} \\
V_{\text{bo}} &= j\omega L_1 \cdot i_{1b} + V_{\text{cb}} + V_{\text{NO}} \\
V_{\text{co}} &= j\omega L_1 \cdot i_{1c} + V_{\text{cc}} + V_{\text{NO}}
\end{aligned}
\tag{14.14}
$$

where $V_{\text{NO is}}$ the voltage across N and O.

Three-phase filter capacitor voltage are.

$$
V_{\text{Ca}} = \frac{i_{\text{ca}}}{jwC}
$$

$$
V_{\text{Cb}} = \frac{i_{\text{cb}}}{jwC}
$$

$$
V_{\text{Cc}} = \frac{i_{\text{cb}}}{jwC}
$$

For a three-phase system, $i_{1a} + i_{1b} + i_{1c} = 0$, $i_{ca} + i_{cb} + i_{cc} = 0$, according to Eq. (14.14), the zero-sequence component V_{NO} is.

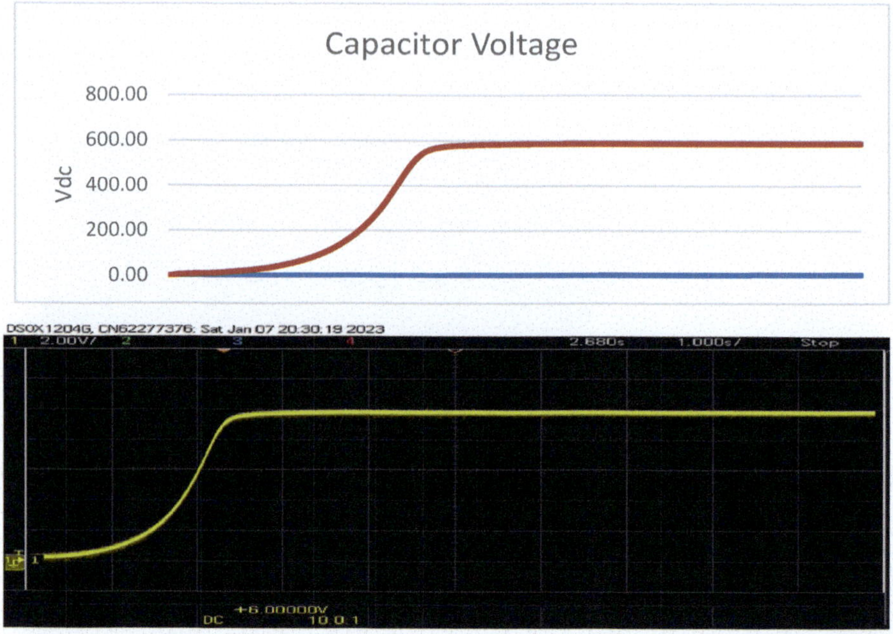

Fig. 14.11 Vdc output from MATLAB & Vdc output from real-time simulation. Simulator

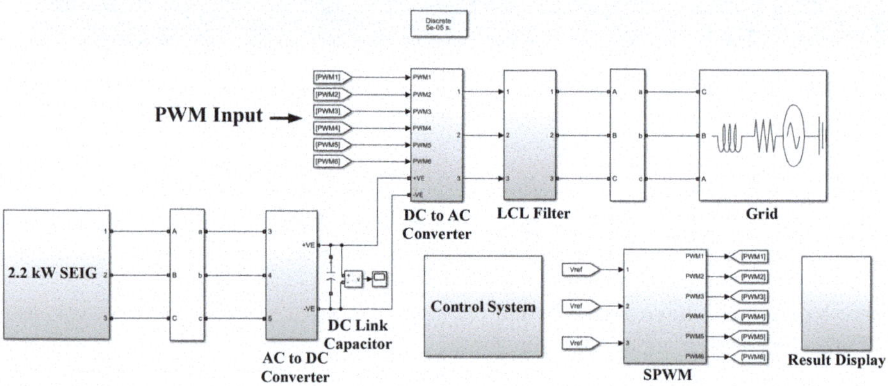

Fig. 14.12 MATLAB Simulink model

$$V_{\text{NO}} = \frac{v_{a0} + v_{b0} + v_{c0}}{3} \tag{14.15}$$

And $V_{NN'}$, the voltage across points N and N' can be expressed as.

$$V_{NN'} = \frac{v_{\text{ga}} + v_{\text{gb}} + v_{\text{gc}}}{3} \tag{14.16}$$

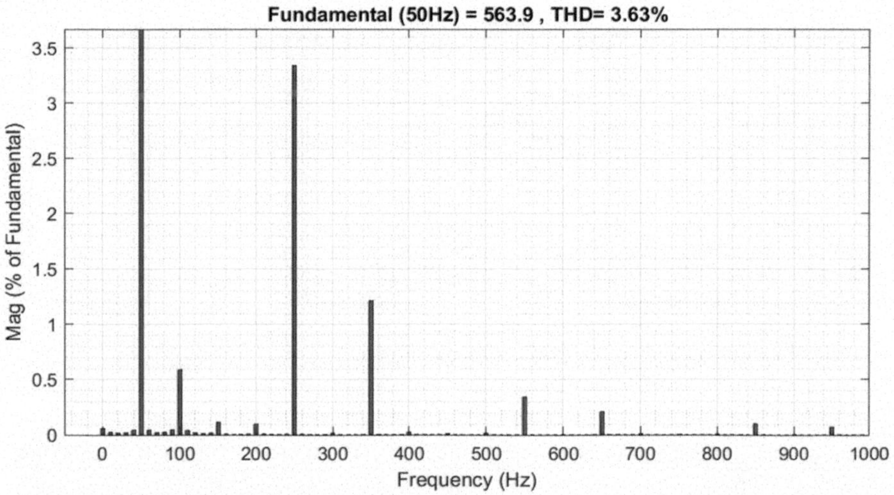

Fig. 14.13 Inverter current THD

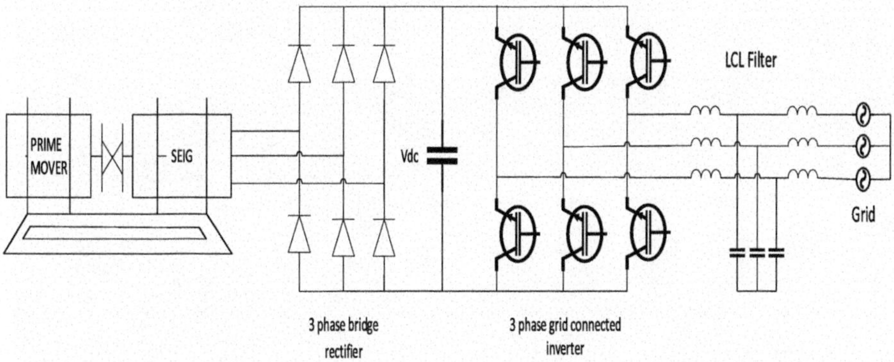

Fig. 14.14 SEIG with grid

And with PWM control $v_{a0} + v_{b0} + v_{c0} \neq 0$.

According to filter capacitor equation, V_{NO} is not equal to zero, and the potential of N and O are not equal, when the three-phase grid voltage is balanced (Figs. 14.15 and 14.16).

When the three-phase grid voltage is balanced, then $v_{\text{ga}} + v_{\text{gb}} + v_{\text{gc}} = 0$ and the potential of N and N' are equal to Eq. (14.15).

The output voltage range of the real-time simulator is 16 V, so the voltage is reduced by using a gain block in MATLAB simulation, as shown in Fig. 14.17.

The output of the mains voltage and the inverter current is shown in Fig. 14.18. The output range of the real-time simulator is 16 V, so a gain block is used to divide the voltage and current (1/40 for the mains voltage and 1/20 for the inverter current),

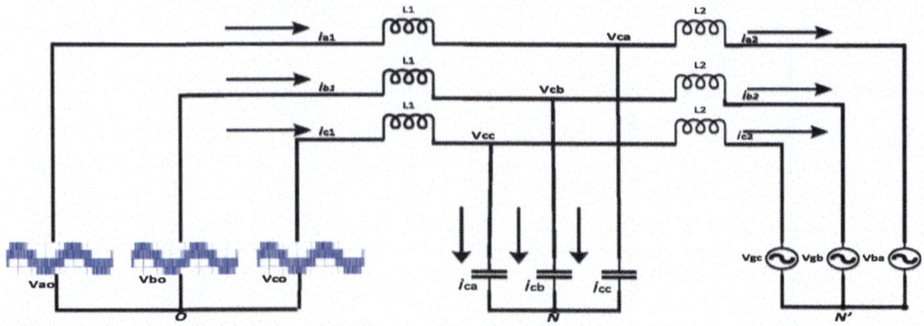

Fig. 14.15 Equivalent circuit of the grid-connected inverter

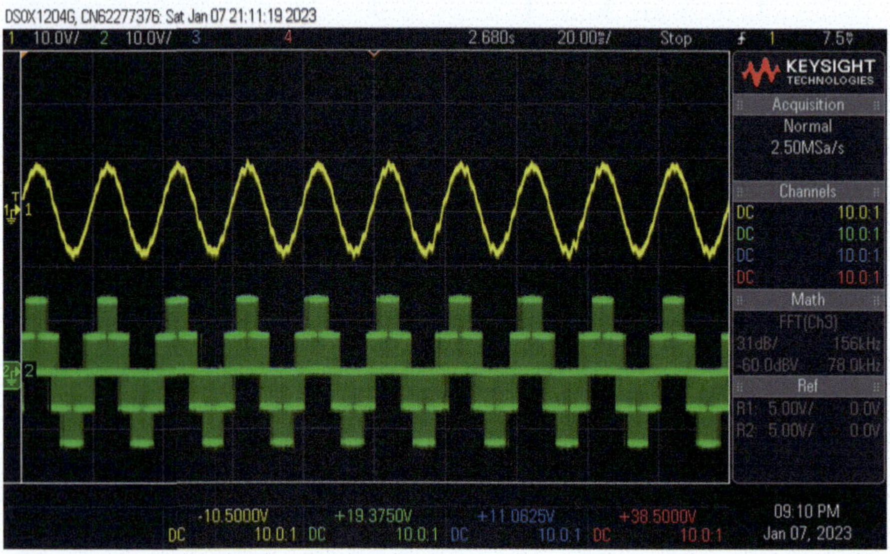

Fig. 14.16 Voltage across NN' and NO in real-time

an active current of 100 A is set in the control system to transfer the active current to the mains. The condition for active power transfer into the grid is as follows.

To send the active power to the grid, the inverter voltage leads the grid voltage with some angle, and the current must be in phase with the grid voltage (Fig. 14.19).

When applying KVL

$$V_{\text{in}} = V_g + V_L \tag{14.17}$$

When sending active power, the current and voltage are in phase and inductor case voltage leads by current with an angle of 90°.

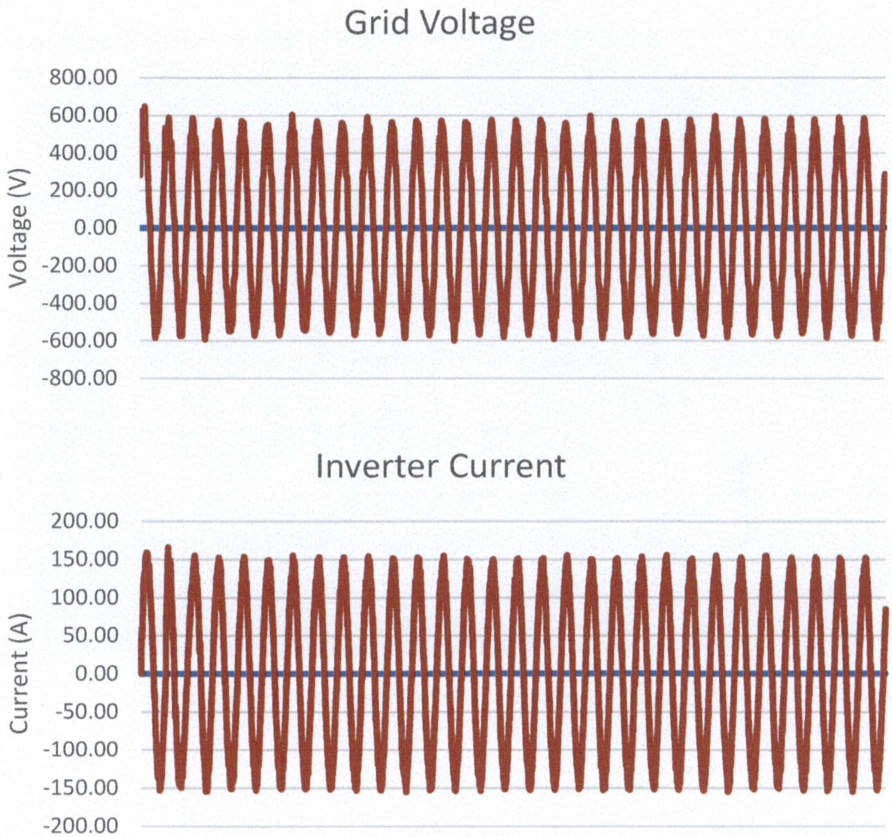

Fig. 14.17 Grid voltage and Inverter current in MATLAB simulation

From Eq. (14.17) the inverter voltage is the vector sum of grid voltage and voltage drops across the inductor.

Figure 14.20 shows that the voltage drop across the filter inductor is led by 90º from grid voltage. Therefore, the active power is feeding to the grid.

The active and reactive power are shown in Fig. 14.21, the active power from the inverter is sent to the grid, and the system is successfully connected.

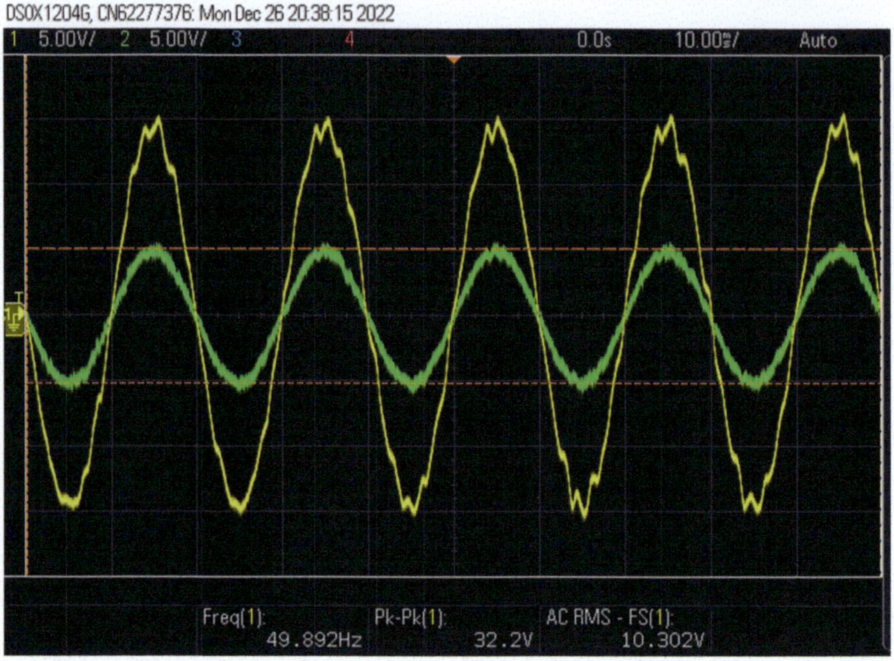

Fig. 14.18 Grid voltage and Inverter current in the real-time simulator

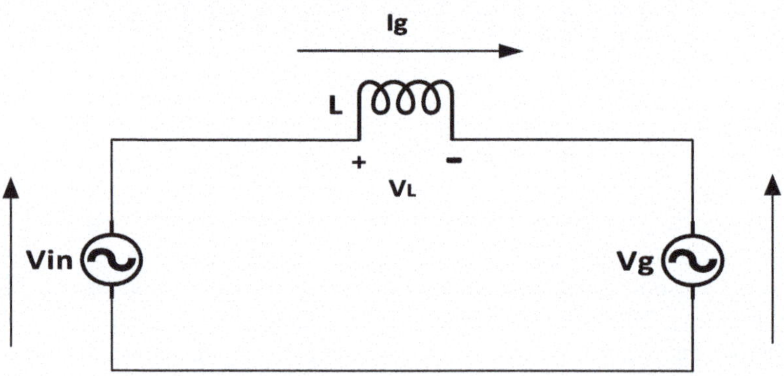

Fig. 14.19 Circuit diagram of grid-connected inverter

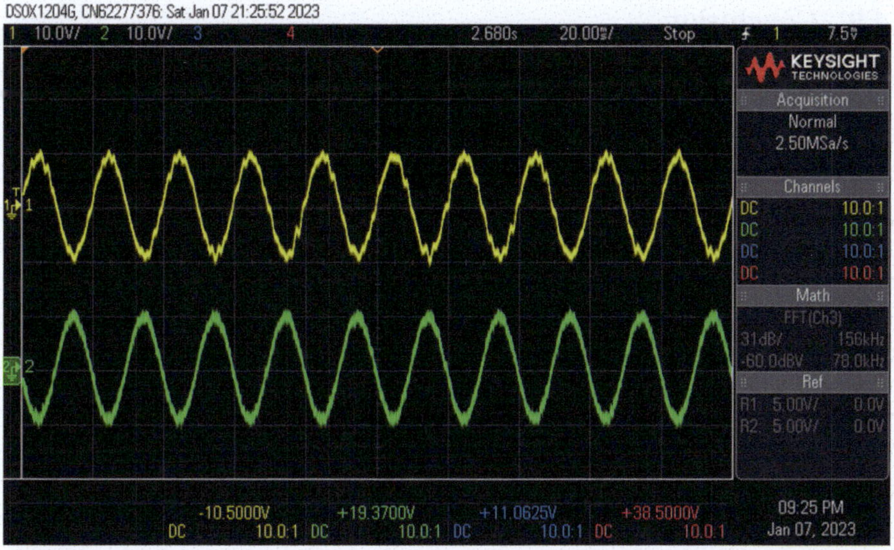

Fig. 14.20 Voltage drop across the inductor, grid voltage in real-time

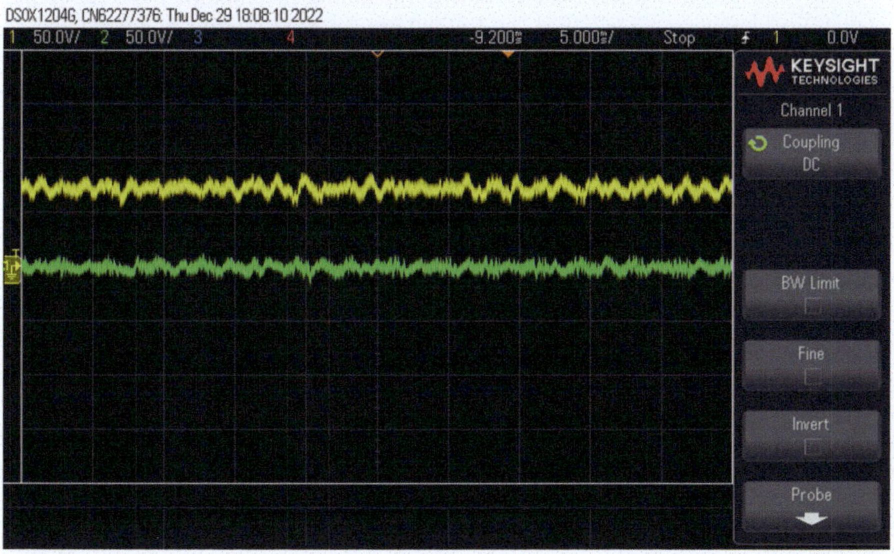

Fig. 14.21 Active and reactive power in real-time

14.9 Conclusion

This study aimed to develop a small-scale hydroelectric power system that utilized a Self-Excited Induction Generator (SEIG) and determine its compatibility with the electrical grid. The SEIG had a capacity of 2.2 kilowatts, 415 V, 50 Hz, and 4.8 amps. Thorough simulations were conducted to ensure that the SEIG's voltage and frequency matched the grid requirements, and necessary adjustments were made before connections were made. An LCL filter was also introduced to minimize the harmonics produced by the inverter output. The simulation results displayed successful integration of the SEIG system with the grid using a real-time Hardware-in-the-loop (HIL) simulation method, which allowed for a seamless connection without any issues. The results confirmed that the mains voltage and current were perfectly aligned when the active current reached 100 amps and the reactive current was zero, indicating that the system was robust and could be used in practical applications. The following points are concluded as follows:

1. The paper delves into integrating a micro-hydro-based self-excited induction generator (SEIG) with the power grid, unlocking the potential of renewable energy resources within the power sector. By seamlessly infusing renewable energy into the grid, many benefits emerge, spanning sustainability and economic viability, bolstering resilience, and alleviating power quality concerns across generation and distribution networks.
2. The central focal point of this study revolves around integrating a meticulously designed system featuring a 2.2 kW, 415-V, 50-Hz squirrel cage induction motor (SCIM) propelling a three-phase SEIG. The underpinning MATLAB machine model stands fortified and meticulously calibrated via extensive tests encompassing no-load characteristics, block rotor assessments, magnetic saturation analyses, and meticulous excitation capacitor calculations. The dynamic energy source, driven by hydroelectric prime movers, breathes life into the SEIG, instantiated through a 3.3 kW AC motor.
3. The research focuses on effectively integrating the Self-Excited Induction Generator (SEIG) with the grid, ensuring that the voltage and frequency match grid specifications. It is crucial for the stable and efficient operation of such systems.
4. Using an LCL filter to reduce harmonics from the inverter output highlights the study's commitment to maintaining the quality of power injected into the grid, preventing disturbances, and ensuring a reliable power supply.
5. To connect the SEIG to the grid successfully, it must synchronize with its voltage and frequency. Synchronization is crucial to ensure that the SEIG generates power in phase with the grid, preventing issues such as power surges or instability. SEIGs, particularly those designed for grid integration, are equipped with controls that facilitate synchronization.
6. The research voyage traverses the vast expanse of simulations, meticulously executed within the MATLAB/Simulink environment (R2018a version, 40,832,900). The magnificence of the results is not left unattended, meticulously scrutinized through the lens of real-time simulation harnessed by OPEL

RT(OP4510). The HIL simulation enabled a comprehensive test of the integration of the SEIG with the grid. This involved emulating various grid conditions, such as voltage and frequency fluctuations, to ensure that the SEIG's performance remained stable and within grid specifications. It allowed for rigorous evaluation in a wide range of scenarios.

Acknowledgements The authors thank the DST-SERB (EEQ/2020/00076) project for providing all the hardware equipment required for these experiments.

Appendix

Sl. no	Parameters	Value
1	Machine excitation capacitor	12.5 μF
2	DC-Link capacitor	2500 μF
3	L_1	50 mH
4	L_2	50 mH
5	C	30 μF
6	K_p	5
7	K_I	5000

References

1. Pudur R, Gao S (2016) Savonius rotor-based grid-connected hydrokinetic power generation scheme. Sustain Energy Grids Netw 5:148–155
2. Chanu LK, Pudur R (2022) Voltage and frequency regulation of micro hydro-based renewable energy generation. J Optoelectr Laser 41(7):1271–1289
3. Chakraborty S, Pudur R (2021) Supply of single-phase power for rural area using three-phase self-excited induction generator. In: Asian conference on innovation in technology (ASIANCON), pp 1–6
4. Murthy SS, Malik OP, Tandon AK (1982) Analysis of self-excited induction generators. IEEE Proc C (Gener Transm Distrib) 129(6). IET Digital Library
5. de Oliveira JDA et al (2023) Synchronverter-based frequency control technique applied in wind energy conversion systems based on the doubly-fed induction generator. Electric Power Syst Res 214:108820
6. Singh GK (2017) Self-excited induction generator for renewable applications:239–256
7. Rajak MK, Samanta J, Pudur R (2022) A hardware-based novel approach for parallel operation of two differently rated SEIGs. Results Eng:100825
8. Al-Bahrani AH, Malik NH (1994) Parallel operation of self-excited induction generators. J King Saud Univ Eng Sci 6(1):79–96

9. Al-Senaidi S, Alolah A, Alkanhal M (2022) Parallel operation of three-phase self-excited induction generators with different numbers of poles. Eng Sci Technol Int J 25:100988
10. Márquez JL, Molina MG, Pacas JM (2010) Dynamic modelling, simulation and control design of an advanced micro-hydro power plant for distributed generation applications. Int J Hydrogen Energy 35(11):5772–5777
11. Nair BV et al (2014) Design of micro-hydro power plant using an induction motor as a generator. In: 2014 International conference on green computing communication and electrical engineering (ICGCCEE). IEEE
12. Pudur R, Kumari P, Rajak MK (2020) Mitigation of power fluctuations in UPFC-based network. In: Electronic systems and intelligent computing. Springer, Singapore,
13. Han Y et al (2009) A novel synchronisation scheme for grid-connected converters using adaptive linear optimal filter based PLL (ALOF–PLL). Simul Model Pract Theory 17(7):1299–1345
14. El-Sousy Fayez FM, Mohamed O, Hatem G (2006) Maximum power point tracking control scheme for grid connected variable speed wind driven self-excited induction generator. J Power Electr 6(1):52–66
15. Nan X et al (2017) Research of micro hydro power control based on self-excited induction generator. Recent Pat Comput Sci 10(4):340–346
16. Konstantopoulos GC, Alexandridis AT (2014) Full-scale modeling, control, and analysis of grid-connected wind turbine induction generators with back-to-back AC/DC/AC converters. IEEE J Emerging Sel Topics Power Electr 2(4):739–748
17. Huerta JME et al (2009) A synchronous reference frame robust predictive current control for three-phase grid-connected inverters. IEEE Trans Ind Electr 57(3):954–962
18. Radwan AAA, Mohamed YARI (2019) Grid-connected wind-solar cogeneration using back-to-back voltage-source converters. IEEE Trans Sustain Energy 11(1):315–325
19. Quéval L, Ohsaki H (2012) Back-to-back converter design and control for synchronous generator-based wind turbines. In: International conference on renewable energy research and applications (ICRERA). IEEE
20. Huang T et al (2013) Three-phase photovoltaic grid-connected inverter based on feedforward decoupling control. In: 2013 International conference on materials for renewable energy and environment, vol 2, pp 476–480
21. Zhou D, Pietkiewicz A, Cuk S (1999) A three-switch high-voltage converter. IEEE Trans Power Electron 14(1):177–183
22. Lee JS, Lee KB, Blaabjerg F (2014) Open-switch fault detection method of a back-to-back converter using NPC topology for wind turbine systems. IEEE Trans Ind Appl 51(1):325–335
23. Guo X, Xu D, Guerrero JM, Wu B (2015) Space vector modulation for dc-link current ripple reduction in back-to-back current-source converters for microgrid applications. IEEE Trans Industr Electron 62(10):6008–6013
24. Dai J, Xu D, Wu B (2009) A novel control scheme for current-source-converter-based PMSG wind energy conversion systems. IEEE Trans Power Electron 24(4):963–972
25. Liserre M, Blaabjerg F, Hansen S (2005) Design and control of an LCL-filter-based three-phase active rectifier. IEEE Trans Ind Appl 41(5):1281–1291
26. Dursun M, DÖŞOĞLU MK (2018) LCL filter design for grid connected three-phase inverter. In 2018 2nd International symposium on multidisciplinary studies and innovative technologies (ISMSIT). IEEE,
27. Mohan N, Undeland TM, Robbins WP (2003) Power electronics: converters, applications, and design. John Wiley & Sons
28. Doucet J, Eggleston D, Shaw J (2007) DC/AC pure sine wave inverter. PFC Worcester Polytecnic Institute
29. Ruan X, Wang X, Pan D, Yang D, Li W, Bao C (2018) Control techniques for LCL-type grid-connected inverters. Springer, Singapore

Chapter 15
Performance Enhancement of Three-Phase SEIG to Feed Single-Phase Load in Micro-hydro Systems Using a Novel Capacitor Excitation Topology

Samrat Chakraborty and Rajen Pudur

Abstract Three-phase self-excited induction generator (SEIG) plays a vital role in micro-hydro systems (MHSs) to generate off-grid single-phase power for hilly terrain-based remote areas. In general, various capacitor excitation topologies (CETs) are available for supplying the requisite reactive power to the three-phase IG while producing single-phase power from the system. Generation of single-phase power from a three-phase SEIG driven by the MHS indicates a worst case of imbalance in the system in terms of phase voltages and stator currents of the generator. This chapter aims to implement a new CET to generate single-phase power from an MHS-driven 2.2 kW three-phase SEIG, whose stator is connected in delta, and the excitation cum reactive power is supplied from a star connected capacitor bank. The proposed CET-IG system is simulated in MATLAB/Simulink and analyzed in steady state for supplying single-phase load under fixed SEIG speed with 55% loading. The result indicates that the recommended CET can effectually supply the single-phase load having voltage regulation of 1.26% and exhibiting a perfect balance in the phase voltages and stator currents of the three-phase SEIG. Hence, in order to generate clean-cum-sustainable power practically in a cheap way for the remote hilly areas from hydro resources at small scale level (i.e., MHS), the proposed CET-IG scheme seems to be an effective one. The recommended CET therefore can be seen to supply the single-phase load in an efficient way, keeping the phase voltages and stator currents of the SEIG balanced while supporting a loading of 55%.

Keywords Capacitor excitation topology · Induction generator · Micro-hydro · Off-grid · Remote loads · Symmetrical component analysis · Two-port network

---

S. Chakraborty · R. Pudur (✉)
Department of Electrical Engineering, National Institute of Technology Arunachal Pradesh, Jote, Arunachal Pradesh 791113, India
e-mail: rajenpudur@nitap.ac.in

S. De et al. (eds.), *Challenges and Opportunities of Distributed Renewable Power*, Energy, Environment, and Sustainability, https://doi.org/10.1007/978-981-97-1406-3_15

15.1 Introduction

15.1.1 Renewable Energy Scenario

The last ten years have seen a surge in the production of electricity from renewable sources due to the progressive decaying in fossil fuel accessibility in nature along with the accompanying trend towards the global accomplishment of net zero [1]. Therefore, the global energy mix as of 2021 is found to be 71.7% from non-renewable sources (including nuclear power) and 28.3% from renewable sources, which was 20.4% from renewable sources and 79.6% from non-renewable sources (including nuclear power) one decade before i.e., in the year 2011 [2] as shown in Fig. 15.1. At present in terms of global scenario, hydro-power contributes the largest share for renewable-based energy generation which accounts to 15% of the global energy generation [3]. Figure 15.2 represents the current global share of power generation from the available renewable sources. It has been aimed by the International Energy Agency that by the end of 2040, there will be more generation of energy from renewables, up to 4,550 GW, and a decline in carbon-based emission to net zero by the end of 2050 [2]. In India, the current scenario showcases an energy mix of 59% from non-renewable (including nuclear power) and 41% from renewable sources, respectively [4].

Among them, the contribution of hydro-power as a renewable-based energy generator is the second highest (12.6%) in terms of Indian context. The percentage of energy generated from renewables in the Indian market is shown in Fig. 15.3. Thus, the generation of renewable-based energy from hydro-power has a huge impact globally and in India.

Micro-hydro systems (MHSs) are largely regarded as a potential option for green energy production with sustainable capability [5]. In comparison to large-scale hydro systems, the environmental impact of these MHSs is quite negligible. The north-eastern and northern Indian provinces in the foothill regions of Himalayas have enormous prospects of generating energy from hydro resources [6]. The hilly Indian

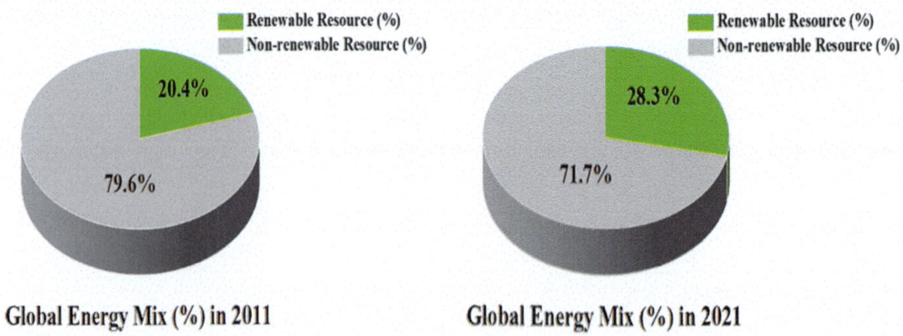

Fig. 15.1 Global energy mix (%) 2011 and 2021 [2]

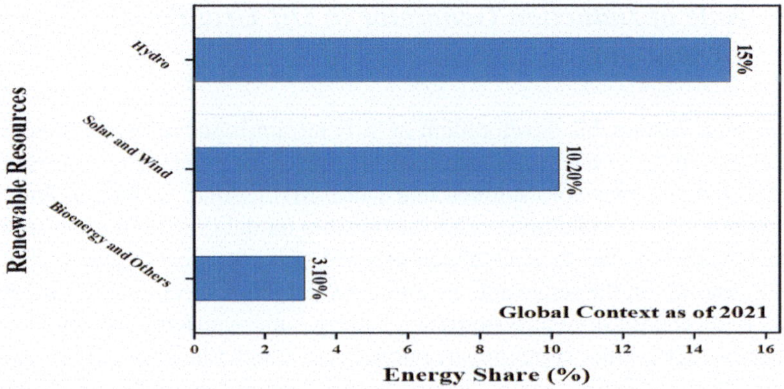

Fig. 15.2 Global energy share from renewable resources as of 2021 [3]

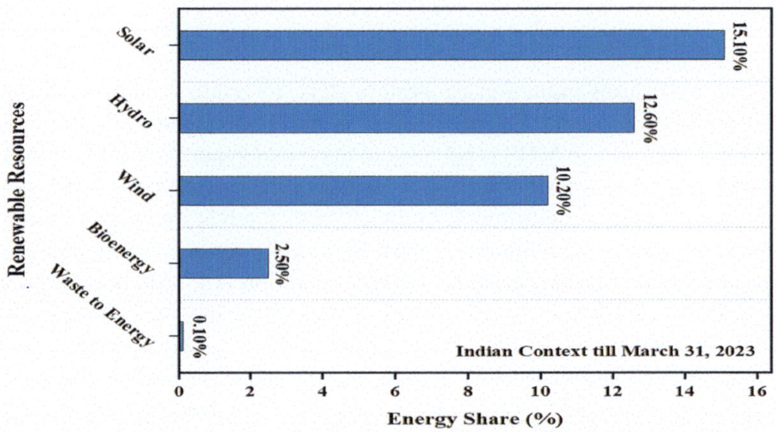

Fig. 15.3 Energy share from renewable resources till March 31 2023, in the Indian market [3]

province of Arunachal Pradesh has the most considerable hydro-power capability of 60,000 MW in the country [7]. Integrating central grid for fulfilling huge electricity requirements in these hilly fur-flung regions becomes difficult because of the challenging geographical terrain and other technical & engineering factors. A report showcases that around 77% of India's unpowered villages are from Arunachal Pradesh [8]. Hence, MHS can play a pivotal role in supplying off-grid power as well as developing sustainable growth and producing green-cum-clean energy for these remote villages.

15.1.2 Role of Induction Generator in Off-Grid Power Generation

For smart-cum-sustainable development of hilly terrain-based remote villages, electrifying those areas by means of localized off-grid systems plays a significant key factor. The use of renewable sources like wind, MHS, solar, or biomass/biogas is highly essential for establishing such a sustainable model [9]. In order to optimally and successfully execute such localized off-grid system, selecting a suitable technology depending on the renewable resource availability as well as determining the necessity of an energy storage system (i.e., battery, super-capacitor, flywheel, etc.) is very important [10]. Thus, supplying a remote area through a localized off-grid micro-level generating mode like MHS can easily overcome issues such as transmission-cum-distribution losses during grid connection and geographical challenges [11]. A traditional MHS can have limitations in generating power when there is a varying water-flow rate. Therefore, an MHS should be capable enough to generate power and perform better under such a scenario when there is a variation in water-flow rate [12].

Presently, MHS uses induction generators (IGs) to produce single-phase power for the off-grid remote locations [13]. The crucial and advantageous reasons for using such types of generators in MHS is because such type of generators are very economical, possess gratifying transient response, and have the capability of self-protection from any sort of adverse conditions [14]. In addition, the IGs are easily found within the marketplace, easily maintainable, robust-natured, have rotors without any brush, and don't need any sort of separate DC-based excitation [15, 16]. To operate in off-grid mode, the IG requires reactive power to magnetize its core, which is usually provided by a capacitor bank [17]. In such cases, an IG is generally termed as a self-excited induction generator (SEIG). Hence, the frequency as well as the generated voltage of the machine relies on the excitation capacitor bank, machine's rotation speed, and also the load's impedance [18], making the study of a SEIG more complex. In general, the remote hilly region load comprises domestic bulbs, fans, water heating equipment, and motors for pumping water, which is of single-phase nature. Therefore, a two-winding based single-phase SEIG is capable of easily supplying these single-phase loads [19]. These single-phase SEIGs are normally deployed where the output power requirement is very low [20]. However, for high power requirements, a three-phase SEIG is generally preferred in place of a single-phase SEIG since they have promising efficiency, are compact in architecture, and are low cost [21]. On the other hand, when loading a three-phase SEIG machine with single-phase load, the phase properties (currents and voltages) of the machine become extremely unbalanced [22]. This unbalanced nature causes extreme pulsation in the machine's torque and thereby enhances the rotor loss because of the backward-flux factor. Hence, to maintain the machine's temperature under an alarming limit, de-loading of the SEIG is performed [23].

15.1.3 Existing Capacitor Excitation Topologies (CETs)

Various concepts on CETs for a three-phase SEIG to supply single-phase load are being presented by different researchers from the late 1980s to 2004 in [24–29] as an initial stage of research. The C–2C excitation scheme with the steady state analysis was first proposed in [24], where a delta connected three-phase SEIG is being excited by two capacitors (C). The capacitor C and 2C are connected parallel separately with two phases of the machine, while the capacitor excitation for the other phase is kept null. The single-phase resistive load is being supplied across the C-connected phase. A steady state study to supply single-phase resistive load from delta connected three-phase SEIG using a single capacitor scheme has been shown in [25]. A single excitation capacitor was used to provide the necessary excitation to the SEIG, and the single-phase resistive load is being supplied from across the excitation capacitor. The proposed scheme was also compared with Steinmetz—I & II connections for supplying single-phase resistive load from three-phase SEIG, which outperforms the proposed scheme [25] in terms of efficiency, voltage unbalance, and power output if the single excitation-feeding capacitor is connected across the lagging-phase. The steady state analysis-based three-capacitor excitation scheme to feed single-phase resistive load from a self-regulated star connected three-phase SEIG has been presented in [26]. Here, two series capacitors are connected across a phase, and another capacitor is connected across to the load and another phase. The proposed scheme claimed to have better voltage regulation, minimal vibration, and low noise under a single-phase resistive loaded scenario. The two excitation capacitors-based Steinmetz topology for a delta connected three-phase SEIG is shown in [27], where one capacitor is connected across a phase, and another capacitor is connected in series with the single-phase resistive load across the other phase. The steady state study of this scheme showcases improved voltage regulation and is a low-cost technique for autonomous energy generation. A modified three-capacitors-based Smith connection for supplying single-phase resistive load from three-phase SEIG is proposed in [28]. The proposed scheme for the first time tried to make an attempt to minimize the stator current unbalance under the variable speed of the machine. In [29], a new two capacitor-based topology is introduced for delta connected three-phase SEIG to supply single-phase resistive load. One of the capacitors is connected across a phase, while the other capacitor is connected in parallel with the load to another phase. Result showcases that the value of the capacitor across the load is nearly half the other one.

In the last ten years, from 2012 to 2021, few more CETs [18, 30–34] were reported for feeding single-phase load from a three-phase SEIG. In [18], delta connected excitation-feeding three-capacitor-based topology was reported for a delta connected three-phase SEIG to supply single-phase resistive-inductive load. The steady state analysis of the circuit was done. Results showcase that an improved current-cum-voltage unbalance factor was achieved with the topology under varying SEIG speeds. A new three excitation capacitors-based topology was proposed in [30]. Here, two of the capacitors are connected individually across each phase of a delta connected three-phase SEIG and the third capacitor is connected in series with the single-phase resistive load across the remaining phase. On steady state analysis of the topology under variable SEIG speed, it is investigated that there is improvement in power output of the machine as well as in the voltage regulation. In [31], a modification in C–2C topology has been reported, where the prime mover of the three-phase SEIG machine is driven by a biogas-based engine for feeding single-phase resistive-inductive load. Both the dynamic and steady state study is done and the result indicates that only for the single-phase resistive load, the topology exhibits balanced phase voltages and stator currents at a particular point. A novel three-capacitor-based excitation scheme was proposed in [32], where a star connected three-phase SEIG is excited by star connected capacitor bank to feed single-phase resistive load. The steady state analysis of the scheme under a fixed SEIG speed showcases a very high percentage of current-cum-voltage unbalance factor. For the smart selection of capacitors, which largely depends on load and SEIG's speed, a novel switched excitation capacitor-based scheme has been proposed in [33] on the topology as reported by [18], for supplying single-phase resistive load from a three-phase SEIG. The proposed control scheme uses Arduino-based microcontroller kit and the steady state analysis-based result showcases an improved balancing in stator voltages and currents. In [34], a new two capacitors-based topology has been reported where a star connected three-phase SEIG is excited by two series capacitors connected across two phases for feeding single-phase resistive load. The reported scheme showcases commendable voltage regulation improvement of the SEIG because of the two series capacitors, making it a cost-effective approach.

15.1.4 Challenges Towards Existing CETs

After studying several literatures, it is observed that the existing CETs, although can supply single-phase load, face some drawbacks on the machine side. These are:

1. The phase voltages of the three-phase SEIG remain imbalanced.
2. The stator current of the three-phase SEIG remains imbalanced.

This unbalanced nature causes extreme pulsation in the machine's torque and thereby enhances the rotor loss because of backward-flux factor. Therefore, to maintain the machine's temperature under an alarming limit, de-loading of the SEIG is performed. Hence, to avoid all these complexities, there is a necessity for low-cost novel CET, which can address the drawbacks and supply single-phase load from three-phase SEIG.

15.1.5 Technology Towards Proposed Novel Capacitor Excitation Topology, Contributions, and Chapter Organization

This chapter reflects a novel CET for a 2.2 kW three-phase SEIG whose stator is delta connected, and it is being excited using a star configured bank of capacitors. The proposed CET is intended to supply single-phase resistive load. In this chapter, the steady state examination of the recommended novel CET has been carried out, implementing two port-based circuital theory along with the application of symmetrical component analysis in order to irradicate zero sequence component. Among the three capacitors which are configured in star topology, one of the capacitors remained to be fixed as it acts as a source for constant reactive power supply. The values for the other two are being computed depending on the load that the three-phase SEIG is serving along with the SEIG's rotational speed for keeping the phase voltages and its stator currents totally balanced. The mathematical-based investigation of the overall proposed circuit fetches all together two equations having non-linear nature. Therefore, in solving these two equations two unknown capacitor values are found. The proposed topology has been tested to supply single-phase resistive load under normal operational speed (1500 rpm). Thus, the following are the contributions of this chapter:

a. Framing a new CET that can efficiently supply single-phase load from a MHS-driven three-phase SEIG.
b. The proposed topology has been tested for 55% loading of the considered 2.2 kW three-phase SEIG.
c. The proposed topology has been examined to completely balance the stator currents and phase voltages of the three-phase SEIG.
d. The terminal voltage regulation of the three-phase SEIG has been found to be very satisfactory.

The remaining chapter is organized as follows: Sect. 15.2 deals with the mathematical modeling of the proposed CET, Sect. 15.3 showcases the simulation model, and Sect. 15.4 reflects the simulation results achieved for the topology. The conclusion, along with future directions of the chapter has been mentioned in Sect. 15.5.

15.2 Mathematical Framing of the Novel CET

The novel CET for the three-phase SEIG driven by MHS has been shown in Fig. 15.4, in which a capacitor bank of star topology excites the three-phase IG having stator of delta connection. This proposed CET-IG system generates single-phase power to feed the single-phase load. Here, the excitation providing C_{fixed}, C_{var2} and C_{var1} capacitors are connected to phases (*pha*, *phb*, and *phc*) in series. The single-phase resistive load impedance is considered to be Z_{rl}. The value of C_{fixed} is kept constant

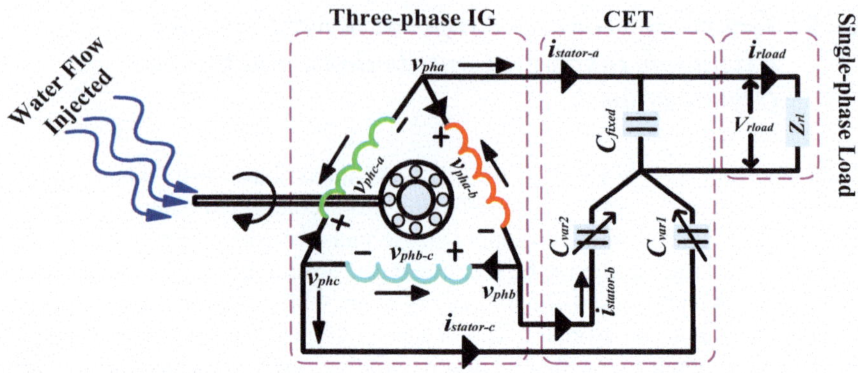

Fig. 15.4 Proposed CET-IG scheme for a MHS-driven three-phase SEIG for feeding single-phase load

as it supplies reactive power to the IG, while the value of C_{var2} and C_{var1} varies with the Z_{rl} and turbine speed.

The parallel connected circuit of C_{fixed} and Z_{rl} comprises of a corresponding impedance of Z_{fixed}. The impedances across C_{var2} and C_{var1} are considered to be Z_{var2} and Z_{var1} respectively as expressed by Eqs. (15.1–15.3).

$$Z_{fixed} = \frac{Z_{rl} \times \frac{1}{i\omega C_{fixed}}}{Z_{rl} + \frac{1}{i\omega C_{fixed}}} \tag{15.1}$$

$$Z_{var2} = \frac{1}{i\omega C_{var2}} \tag{15.2}$$

$$Z_{var1} = \frac{1}{i\omega C_{var1}} \tag{15.3}$$

For a convenient computation, the star connected CET is mathematically converted into its corresponding delta configuration by applying star-to-delta conversion formulae. Thus, the equivalent delta configured CET has impedances of $Z_{fixed}{'}$, $Z_{var2}{'}$ and $Z_{var1}{'}$ respectively and admittances of Y_{fixed}, Y_{var2} and Y_{var1} respectively as expressed by eqs. (15.4–15.6). This star-to-delta conversion technique helps to make the system balanced such that a three-phase SEIG supplies a balanced three-phase load.

$$Z'_{fixed} = \frac{(Z_{fixed} \times Z_{var2}) + (Z_{var1} \times Z_{var2}) + (Z_{fixed} \times Z_{var1})}{Z_{fixed}} = \frac{1}{Y_{fixed}} \tag{15.4}$$

$$Z'_{var2} = \frac{(Z_{fixed} \times Z_{var2}) + (Z_{var1} \times Z_{var2}) + (Z_{fixed} \times Z_{var1})}{Z_{var2}} = \frac{1}{Y_{var2}} \tag{15.5}$$

$$Z'_{var1} = \frac{(Z_{fixed} \times Z_{var2}) + (Z_{var1} \times Z_{var2}) + (Z_{fixed} \times Z_{var1})}{Z_{var1}} = \frac{1}{Y_{var1}} \tag{15.6}$$

The voltage of the SEIG phases in Fig. 15.4 is represented by v_{pha}, v_{phb} and v_{phc} respectively with stator currents of $i_{stator-a}$, $i_{stator-b}$ and $i_{stator-c}$. Equation (15.7) indicates a relation among the stator currents and phase voltages as computed by applying nodal analysis.

$$\begin{bmatrix} i_{stator-a} \\ i_{stator-b} \\ i_{stator-c} \end{bmatrix} = \begin{bmatrix} (Y_{var1} + Y_{var2}) & -Y_{var2} & -Y_{var1} \\ -Y_{var2} & (Y_{fixed} + Y_{var2}) & -Y_{fixed} \\ -Y_{var1} & -Y_{fixed} & (Y_{fixed} + Y_{var1}) \end{bmatrix} \begin{bmatrix} v_{pha} \\ v_{phb} \\ v_{phc} \end{bmatrix} \tag{15.7}$$

Now, applying symmetrical component-based techniques on Eq. (15.7), a per phase relationship among positive and negative sequences with zero sequence currents ($i_{positive}$, $i_{negative}$ and i_{zero}) and voltages ($v_{positive}$, $v_{negative}$ and v_{zero}) are acquired from Eq. (15.8).

$$\begin{bmatrix} i_{positive} \\ i_{negative} \\ i_{zero} \end{bmatrix} = \frac{1}{3} \begin{bmatrix} 1 & a & a^2 \\ 1 & a^2 & a \\ 1 & 1 & 1 \end{bmatrix} \begin{bmatrix} (Y_{var1} + Y_{var2}) & -Y_{var2} & -Y_{var1} \\ -Y_{var2} & (Y_{fixed} + Y_{var2}) & -Y_{fixed} \\ -Y_{var1} & -Y_{fixed} & (Y_{fixed} + Y_{var1}) \end{bmatrix} \begin{bmatrix} 1 & 1 & 1 \\ a^2 & a & 1 \\ a & a^2 & 1 \end{bmatrix} \begin{bmatrix} v_{positive} \\ v_{negative} \\ v_{zero} \end{bmatrix} \tag{15.8}$$

The operator 'a' as used in Eq. (15.8) can be defined as $a = e^{j\frac{2\pi}{3}}$. Further, simplification of the above equation nullifies the zero sequence part as shown in Eq. (15.9).

$$\begin{bmatrix} i_{positive} \\ i_{negative} \end{bmatrix} = \begin{bmatrix} Y_{total} & Y_{total_1} \\ Y_{total_2} & Y_{total} \end{bmatrix} \begin{bmatrix} v_{positive} \\ v_{negative} \end{bmatrix} \tag{15.9}$$

Now, Y_{total}, Y_{total_1} and Y_{total_2} are expressed in Eqs. (15.10–15.12).

$$Y_{total_1} = -aY_{fixed} - Y_{var2} - a^2 Y_{var1} \tag{15.10}$$

$$Y_{total_2} = -Y_{fixed} - a^2 Y_{var2} - aY_{var1} \tag{15.11}$$

$$Y_{total} = Y_{fixed} + Y_{var2} + Y_{var1} \tag{15.12}$$

It can be concluded from Eqs. (15.10 and 15.11) that when the loading is balanced, $Y_{total_1} = Y_{total_2} = 0$ for $Y_{fixed} = Y_{var2} = Y_{var1}$. It is observed that $v_{positive}$, $v_{negative}$, Y_{total}, Y_{total_1} and Y_{total_2} all are the functions of frequency per unit (F). Thus, utilizing Eqs. (15.9–15.12), the two port-based representations of the proposed scheme can be obtained as per [35]. Now, in the two port-based representation, the resistance and reactance for the stator and rotor are represented as R_{rotor} & R_{stator} and X_{rotor} & X_{stator} respectively, with the speed in per unit being τ. . Therefore, the net admittance for the negative sequenced side (Y_{net}) of the two port-based systems can be expressed by Eq. (15.13) [36].

$$Y_{net} = Y_{total} + [\frac{R_{stator}}{F} + \frac{R_{rotor}}{F+\tau} + j(X_{stator} + X_{rotor})]^{-1} \tag{15.13}$$

Again, to balance the phase voltages as well as the stator currents, the voltage unbalanced factor (UF_v) and current unbalanced factor (UF_i), for the total proposed scheme as be expressed as per Eqs. (15.14 and 15.15) [18].

$$UF_v = \left|\frac{Y_{total_2}}{Y_{net}}\right| \tag{15.14}$$

$$\mathrm{UF}_i = \left|\frac{(Y_{total_2} Y_{net}) - (Y_{total_2} Y_{total})}{(Y_{total} Y_{net}) - (Y_{total_2} Y_{total_1})}\right| \tag{15.15}$$

Now, for a perfect balance in the operation, $UF_v = UF_i = 0$. This implies that if the right side numerator of part of the Eq. (15.14) i.e., Y_{total_2} is 0, then the whole equation becomes 0. Therefore, the real-cum-imaginary part of Y_{total_2} will be separately 0, which is shown in Eq. (15.16 and 15.17).

$$Rl(-Y_{fixed} - a^2 Y_{var2} - aY_{var1}) = 0 \tag{15.16}$$

$$Img(-Y_{fixed} - a^2 Y_{var2} - aY_{var1}) = 0 \tag{15.17}$$

Solving Eqs. (15.16 and 15.17), the two unknown variable capacitors C_{var1} and C_{var2} acting as the phase balancers are obtained when τ, C_{fixed} and Z_{rl} are known along with other 2.2 kW three-phase IG parameters are known. These parameters are shown in *appendix*.

15.3 Simulation Model

The simulation model for the proposed CET has been designed and run in MATLAB/Simulink (ver. 2015b) platform. The overall simulation model has been showcased in Fig. 15.5. Here, an MHS is designed, where a three-phase/415 V/50 Hz/4.8 A/1435 rpm, delta connected SEIG of 2.2 kW rating supplies single-phase resistive load. A star connected capacitor bank supplies the required excitation to IG

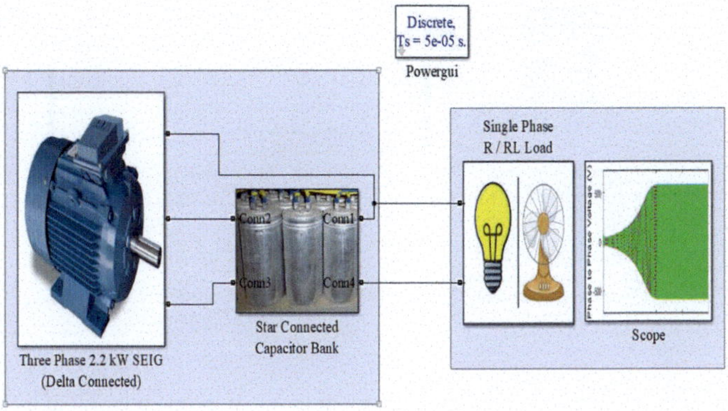

Fig. 15.5 Simulation model of the proposed CET-IG scheme for an MHS-driven three-phase SEIG for feeding single-phase load

Fig. 15.6 Star connected capacitor bank

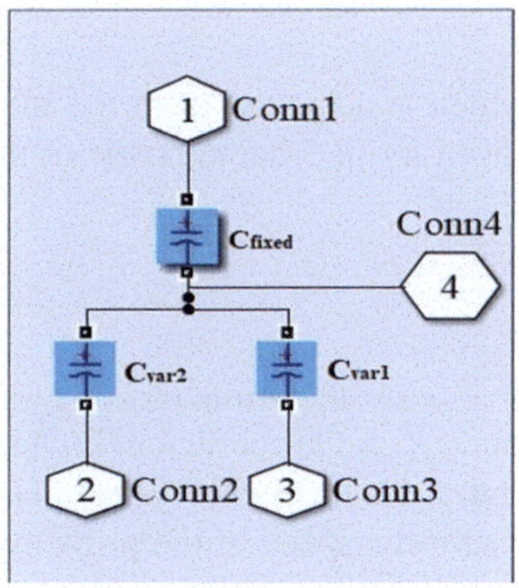

and also helps in balancing the phases, whose arrangement is shown in Fig. 15.6. The single-phase resistive load is connected across the terminal A or phase A of the three-phase SEIG and the common point of connection of the capacitors i.e., Conn4. The terminal Conn1 of the capacitor C_{fixed} is also connected to the terminal A of the machine. The other two terminals Conn2 and Conn3 of the capacitor C_{var2} and C_{var1} are connected to the terminals B and C respectively of the machine. The value of the capacitor C_{fixed} is taken fixed at 36 μF. While, the value of C_{var2} and C_{var1} depends on the load impedance and speed of the turbine.

The details behind fixing the of C_{fixed} value, has been described in Sect. 15.4. Using Eqs. (15.16 and 15.17), the mathematically computed values of C_{var2} and C_{var1} are achieved, which are tested on the simulation model. The simulation model for the three-phase 2.2 kW IG is shown in Fig. 15.7. Here, a three-phase asynchronous machine is made to function as a three-phase/415 V/50 Hz delta connected IG by using the original machine parameters (shown in *appendix*) and saturation data,

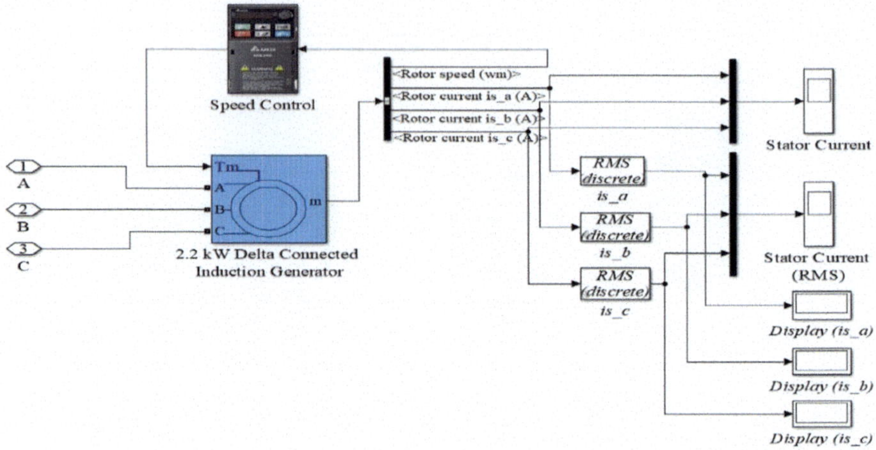

Fig. 15.7 Simulation model for the 2.2 kW induction generator

which is achieved from the machine's no-load test (discussed in Sect. 15.4). The speed-torque characteristic of the prime mover—turbine is defined by Eq. (15.18).

$$T_{mech} = T_0 - K_1\omega_{rot} \tag{15.18}$$

T_{mech} is the prime mover's torque supplied to the machine, T_0 is the torque coefficient, K_1 is the coefficient for angular speed, which varies according to the water flow and ω_{rot} is the rotational speed of the prime mover in rpm. The tuned value of T_0 and K_1 for the speed of the prime mover (1500 rpm) is shown in *appendix*. Figure 15.8 shows the simulation model for prime mover's speed control. All the results of the simulation are being advocated from the scope block.

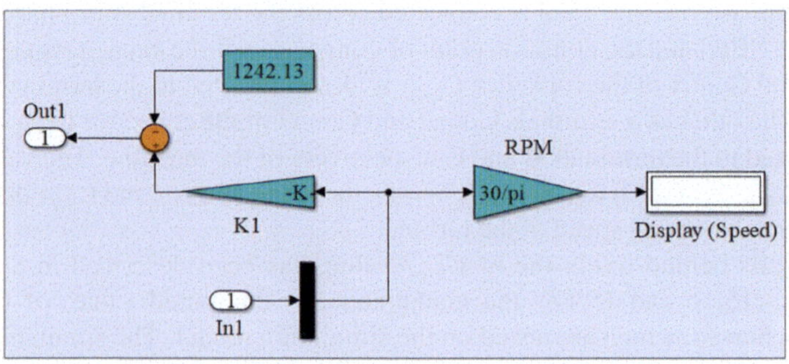

Fig. 15.8 Simulation model for prime mover's speed control of the 2.2 kW SEIG

15.4 Results

For simulating the proposed scheme, a perfect saturation data of the considered 2.2 kW machine has to be known. The data regarding the saturation of the considered machine has been taken from [22]. Following the methods as described in [18, 37, 38], the value of C_{fixed} is found to be 36 μF. This capacitor C_{fixed} supplies a continuous reactive power to the three-phase IG. In this work, a single-phase resistive load of 100 Ω has been considered when the speed of the SEIG is 1500 rpm. Now, using the Eqs. (15.16 and 15.17), the unknown values of C_{var2} and C_{var1} are calculated. The obtained values of C_{var2} and C_{var1} are then used to simulate the proposed scheme in MATLAB environment. It is noticed that there is a slight difference in the values of C_{var2} and C_{var1} achieved from calculation while performing the simulation. Table 15.1 shows the calculated and simulated values of C_{var2} and C_{var1}.

The overall simulations have been run for $t = 2\,sec$. At $t = 0.9\,sec$, the breaker connects the single-phase resistive load with the CET-IG scheme and at $t = 1.4\,sec$, the breaker gets disconnected. Figure 15.9 showcases the balanced phase-to-phase voltages and stator currents when the single-phase load is connected to the generator.

It is observed from Fig. 15.9, that from $t = 0\,sec$ to $t = 0.7\,sec$, at no load condition, the voltage buildup of the SEIG occurs and the rms voltage during that time is about 425 V. With the application of load from $t = 0.9\,sec$ to $t = 1.4\,sec$, the voltage drops to 412 V maintaining a balance in the phase-to-phase voltage. From

Table 15.1 Calculated and simulated values of C_{var1} and C_{var2} when $C_{fixed} = 36\mu F, Z_{rl} = 100\Omega$ and speed = 1500 rpm

Variable capacitors	Calculated	Simulated
C_{var1}	78.3	78.4
C_{var2}	32.4	32.5

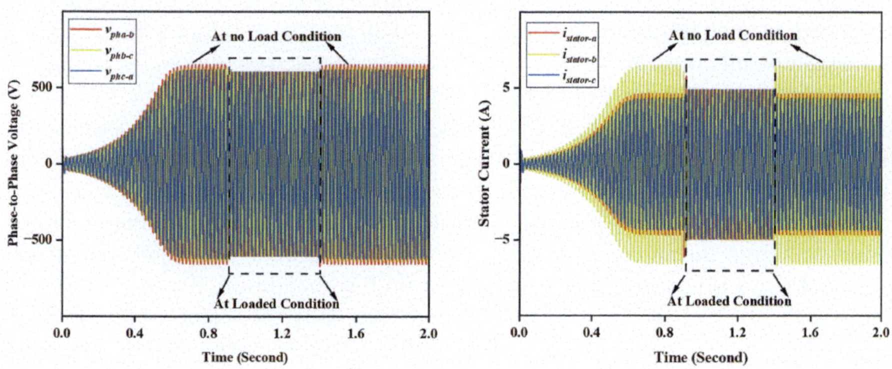

Fig. 15.9 Balanced phase-to-phase voltage and stator current of 2.2 kW three-phase SEIG while supplying single-phase load

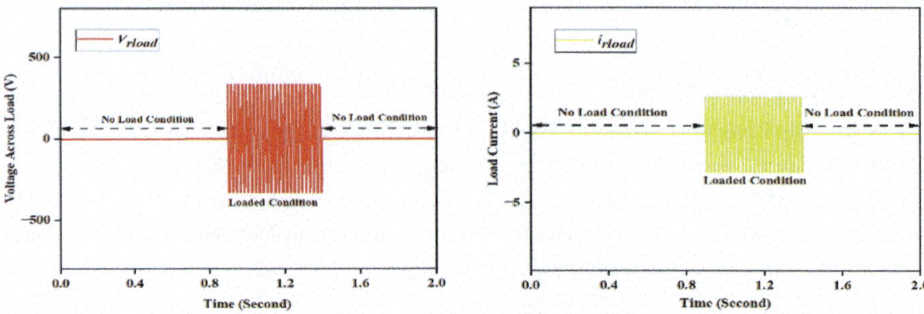

Fig. 15.10 Load terminal voltage and current of 2.2 kW three-phase SEIG supplying single-phase load

$t = 1.41\,sec$ to $t = 2\,sec$, the voltage again rises. Similarly, during the time interval while application of the load at $t = 0.9\,sec$ to $t = 1.4\,sec$, the stator currents are totally balanced with rms current value being 3.34 A. The stator currents during the unloaded condition ($t = 0\,sec$ to $t = 0.89\,sec$ and $t = 1.41\,sec$ to $t = 2\,sec$) are found to be highly unbalanced. Figure 15.10 showcases the load terminal voltage along with the load current consumed by the single-phase resistive load.

It is observed from Fig. 15.10, that from $t = 0.9\,sec$ to $t = 1.4\,sec$, with the application of load, the terminal voltage of 237 V (rms) is obtained. While, from $t = 0\,sec$ to $t = 0.89\,sec$ and from $t = 1.41\,sec$ to $t = 2\,sec$, at no load condition, since the breaker is kept off, therefore the voltage across the load is found to be 0 V. Similarly, in case of load current, from $t = 0.9\,sec$ to $t = 1.4\,sec$, it is found that, the single-phase load draws a current of 2.64 A (rms). Therefore, it can be said that with the proposed scheme, the terminal voltage regulation is calculated to be 1.26%. It is also observed that with $Z_{rl} = 100\Omega$, the i_{rload} is found to be 2.64 A. This indicates that the proposed scheme enables the generator to withstand 55% of the loading.

From Fig. 15.11 it is observed that, the load consumes an active power of about 625.68 W. From $t = 0\,sec$ to $t = 0.89\,sec$, i.e., at no load condition, the frequency of the system is maintained at 50.15 Hz. At $t = 0.89\,sec$ to $t = 0.9\,sec$, the frequency of the system falls from 50.15 Hz to 49.35 Hz and at loaded condition, the frequency is maintained to 49.35 Hz from $t = 0.9\,sec$ to $t = 1.4\,sec$ while from $t = 1.41\,sec$ to $t = 2\,sec$ i.e., at no load, the frequency of the system is again regained and maintained at 50.15 Hz. All the results achieved from the simulation study are shown in Table 15.2.

Therefore, with the application of the proposed CET scheme to supply single-phase load from three-phase SEIG, it is observed that the two variable capacitors C_{var1} and C_{var2} acts as a phase balancer. At no load condition, it can be observed from the results that the voltage buildup and eventually the stator current buildup of the SEIG occurs due to the presence of the constant reactive power supplying capacitor C_{fixed}. Under this scenario, the current and voltage profile is unbalanced. With the introduction of the single-phase load in the system, drop in the phase voltage and stator current can be observed. Capacitors C_{var1} and C_{var2} makes the

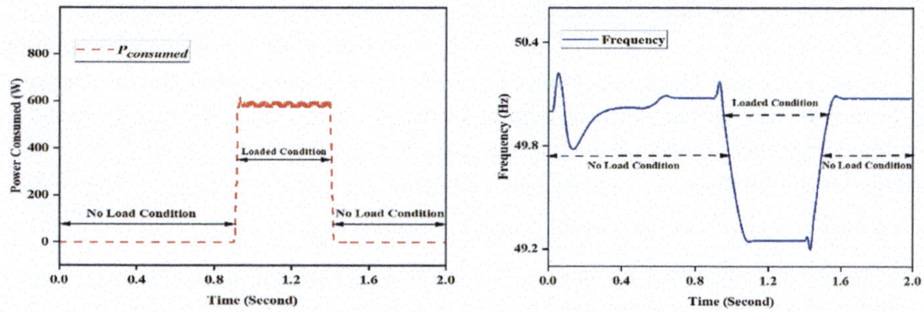

Fig. 15.11 Active power consumed by the single-phase load along with the frequency of the total system

Table 15.2 Results obtained from simulation study when $C_{fixed} = 36\ \mu F$, $C_{var1} = 78.4\ \mu F$, $C_{var2} = 32.5\ \mu F$, $Z_{rl} = 100\ \Omega$ and speed = 1500 rpm

Simulated parameters	Result obtained
v_{pha-b}	412 V (rms)
v_{phb-c}	412 V (rms)
v_{phc-a}	412 V (rms)
$i_{stator-a}$	3.34 A (rms)
$i_{stator-b}$	3.34 A (rms)
$i_{stator-c}$	3.34 A (rms)
v_{rload}	237 V (rms)
i_{rload}	2.64 A (rms)
Power consumed	625.68 W
Frequency (F)	49.35 Hz (0.987 p.u.)
Terminal voltage regulation	1.26%
Loading	55%

phase voltage and stator current of the three-phase SEIG totally balanced, thereby supplying tangible power to the single-phase load. A drop in the frequency is also observed due to the introduction of load. Again, when the load is disconnected, the phase voltages, stator currents, and frequency are restored to their actual state. Thus, the effect of the capacitor C_{fixed} in the proposed scheme makes a perfect buildup of the machine while C_{var1} and C_{var2} balances the system under loaded condition, enhancing the performance of the SEIG.

15.5 Conclusion

The use of three-phase SEIG in an MHS to generate off-grid power for remote single-phase loads has gained a great research prospect with the utilization of renewable-based energy around the globe. The proposed CET has proved its effectiveness in supplying power to remote areas. Simulation result indicates that under the proposed topology, both the phase voltages and stator currents are obtained to be balanced with a promising terminal voltage regulation of 1.26%. It is also observed that, with the proposed topology, the SEIG can withstand a loading of 55% while maintaining the system balanced. The current work can be extended by implementing and testing the proposed CET scheme under a laboratory environment and investigating the total system. The effectiveness of the scheme can be compared with simulation results obtained from the present research. Since, the load does not always remain constant, therefore studying the transient-based effect on the SEIG due to sudden change in load for the proposed scheme can be another area of research. Thus, employing a three-phase SEIG in MHS for generating off-grid power in remote areas has multi-faceted significance, encompassing social, economic, environmental, and technological aspects. It not only addresses energy access challenges but also promotes sustainability, economic development, and community empowerment, ultimately improving the overall well-being of remote communities.

Acknowledgements The authors thank the DST-SERB (EEQ/2020/00076) project for providing all the hardware equipment required for this experiment/study.

Appendix

2.2 KW IG parameter values: $R_{rotor} = 2.87\Omega$, $R_{stator} = 3.81\Omega$, $X_{rotor} = X_{stator} = 4.46\Omega$, $L_{rotor} = L_{stator} = 14.196mH$, pole number = 4 and moment of inertia (J) = 0.00759 kgm^2.

Speed-torque characteristic of the prime mover at 1500 rpm: $T_0 = 1242.13$, $K_1 = 7.899$ and $\omega_{rot} = 9.54$.

References

1. Chakraborty S, Mukherjee D, Guchhait PK, Bhattacharjee S, Abdelaziz AY, El-Shahat A (2023) Optimum design of a renewable-based integrated energy system in autonomous mode for a remote hilly location in northeastern India. Energies 16(4):1588
2. IEA, Reports World Energy Outlook (2022). https://www.iea.org/, last accessed 2023/03/31
3. Renewables 2022 Global Status Reports. https://www.ren21.net/, last accessed 2023/03/31
4. Power Sector at a Glance ALL INDIA. https://powermin.gov.in/en/content/power-sector-glance-all-india, last accessed 2023/03/31
5. Pudur R, Gao S (2016) Savonius rotor based grid connected hydrokinetic power generation scheme. Sust Energy Grids Netw 5:148–155
6. Singh PD, Gao S (2022) Clean and sustainable micro hydro power generation using parallel variable asynchronous generators controlled by AC-DC-AC converter and fuzzy PSO/PI D-STATCOM for remote areas. Sustain Cities Soc 77:103527
7. Industry Scenario. https://www.investindia.gov.in/, last accessed on 2023/03/31
8. Rural Electrification: Arunachal a Major Hurdle in Achieving May 1 Deadline. https://indianexpress.com/article/business/budget/union-budget-rural-electrification-arunachal-a-major-hurdle-in-achieving-may-1-deadline-5042809/ last accessed on 2023/03/31.

9. Ahmad F, Alam MS (2018) Economic and ecological aspects for microgrids deployment in India. Sustain Cities Soc 37:407–419
10. Ahmad F, Alam MS (2017) Feasibility study, design and implementation of smart polygeneration microgrid at AMU. Sustain Cities Soc 35:309–322
11. Robert FC, Gopalan S (2018) Low cost, highly reliable rural electrification through a combination of grid extension and local renewable energy generation. Sustain Cities Soc 42:344–354
12. Nababan S, Muljadi E, Blaabjerg F (2012) An overview of power topologies for micro-hydro turbines. In: 2012 3rd IEEE International Symposium on Power Electronics for Distributed Generation Systems (PEDG), pp 737–744. IEEE, Aalborg, Denmark
13. Bansal RC (2005) Three-phase self-excited induction generators: an overview. IEEE Trans Energy Convers 20(2):292–299
14. Simoes MG, Farret FA (2008) Alternative energy systems – design and analysis with induction generators, 2nd edn. CRC Press, Boca Raton, FL
15. Blaabjerg F, Ma K (2013) Future on power electronics for wind turbine systems. IEEE J Emer Sel Top Pow Electr 1(3):139–152
16. Kiselychnyk O, Bodson M, Wang J (2017) Comparison of two magnetic saturation models of induction machines and experimental validation. IEEE Trans Industr Electron 64(1):81–90
17. Bae J, Kim Y, Yoo C (2016) Torque control topologies of a self-excited induction generator as an electric retarder. IEEE Trans Energy Convers 31(2):557–565
18. Wang YJ, Lee MH (2012) A method for balancing a single-phase loaded three-phase induction generator. Energies 5(9):3534–3549
19. Kalla UK, Singh B, Murthy SS (2016) Intelligent neural network-based controller for single-phase wind energy conversion system using two winding self-excited induction generator. IEEE Trans Industr Inf 12(6):1986–1997
20. Murthy SS, Singh B, Sandeep V (2011) A novel and comprehensive performance analysis of a single-phase two-winding self-excited induction generator. IEEE Trans Energy Convers 27(1):117–127
21. Singh B, Murthy SS, Chilipi RSR (2014) STATCOM-based controller for a three-phase SEIG feeding single-phase loads. IEEE Trans Energy Conver 29(2):320–331
22. Chakraborty S, Pudur R (2021) Supply of single-phase power for rural area using three-phase self-excited induction generator. In: 2021 Asian Conference on Innovation in Technology (ASIANCON), pp 1–6. IEEE, Pune, India
23. Murthy SS, Bhuvaneswari G, Gao S, Ahuja RK (2010) Self excited induction generator for renewable energy applications to supply single-phase loads in remote locations. In: 2010 IEEE International Conference on Sustainable Energy Technologies (ICSET), pp 1–8. IEEE, Kandy, Sri Lanka
24. Bhattacharya JL, Woodward JL (1988) Excitation balancing of a self-excited induction generator for maximum power output. IEE Proc C (Generation, Transmission and Distribution) 135(2):88–97
25. Chan TF (1999) Performance analysis of a three-phase induction generator self-excited with a single capacitance. IEEE Trans Energy Convers 14(4):894–900
26. Fukami T, Kaburaki Y, Kawahara S, Miyamoto T (1999) Performance analysis of a self-regulated self-excited single-phase induction generator using a three-phase machine. IEEE Trans Energy Convers 14(3):622–627
27. Chan TF, Lai LL (2001) A novel single phase self-regulated self-excited induction generator using a three-phase machine. IEEE Trans Energy Convers 16(2):204–208
28. Chan TF, Lai LL (2002) Single-phase operation of a three-phase induction generator with the Smith connection. IEEE Trans Energy Convers 17(1):47–54
29. Murthy SS, Singh B, Gupta S, Gulati BM (2003) General steady-state analysis of three-phase self-excited induction generator feeding three-phase unbalanced load/single-phase load for stand-alone applications. IEE Proc Gener Transm Distrib 150(1):49–55
30. Anagreh YN, Iqteit NA, Mohammad SF (2013) Performance analysis of a new configuration of three-phase self-excited induction generator feeding a single-phase load. Intern J Power Energy Conv 4(2):167–181

31. Iqteit NA, Daud AK (2019) A new model of self-excited induction generator to feed a single phase load with an application in lighting animal farm. Intern J Power Energy Conv 10(1):32–50
32. Benhacine TZE, Nesba A, Mekhtoub S, Ibtiouen R (2019) A new approach for steady state analysis of three-phase SEIG feeding single-phase load. COMPEL–Intern J Comput Mathem Elect Electr Eng 38(1):46–67
33. Benhacine TZE, Nesba A, Mekhtoub S, Ibtiouen R (2021) Low-cost three-phase self-excited induction generator for supplying isolated single-phase loads. Intern J Dig Sign Smart Syst 5(1):80–102
34. Ray S, Chatterjee HS, Mahato SN, Roy NK (2021) Analysis of a novel three-phase asynchronous generator based on graph theory for supplying single-phase load. IETE J Res, 1–18
35. Chakraborty S, Pudur R (2022) Hardware investigation of a new phase balancing topology for supplying single-phase loads using three-phase SEIG. In: 2022 IEEE Sustainable Power and Energy Conference (iSPEC), pp 1–5. IEEE, Perth, Australia
36. Chakraborty S, Pudur R (2022) A novel balancing operation topology for single-phase supply to remote locations from ThreePhase SEIG. In: 2022 IEEE 19th India Council International Conference (INDICON), pp 1–6. IEEE, Kochi, India
37. Rajak MK, Samanta J, Pudur R (2023) A hardware-based novel approach for parallel operation of two differently rated SEIGs. Resul Eng 17:100825
38. Krishna VBM, Sandeep V (2020) Experimental study on different modes of self excited induction generator operation. In: 2020 2nd PhD Colloquium on Ethically Driven Innovation and Technology for Society (PhD EDITS), pp 1–2. IEEE, Bangalore, India

Chapter 16
Modified P-Q Control Algorithm of Shunt Active Power Filter to Reduce Harmonics, Voltage Regulation, Power Factor Correction and Unbalanced Load for Hydro Electric Power Plant

Jotirmoy Samanta, Ralli Sangno, and Rajen Pudur

Abstract Power system harmonics, poor power factor, and unbalanced load are different and prominent problems in modern power networks. Also, for the standalone mode, voltage regulation is the major problem to tackle. It is always challenging to perform all corrective operations at once, but since these problems are theoretically correlated, one new algorithm is possible. Many techniques are available to solve these issues, but the complexity and lack of a single technique to solve these problems simultaneously have motivated this present work. An attempt is made to mitigate harmonics, correct the Power Factor (PF), and improve voltage regulation under balanced and unbalanced load conditions with a single and simple control technique using Shunt Active Power Filter (SAPF) based on p-q theory. Details of Self-Excited Induction Generator (SEIG) are discussed in this chapter. 2.2 kW SEIG has been considered for Hydro- electric power plant. The mathematical formulation is done and tested in the MATLAB environment, and the results are convincing and satisfactory.

Keywords Power factor · Power quality · p-q theory · Shunt active power Filter (SAPF) · Total harmonics distortion (THD) · Hydro power

16.1 Introduction

Renewable energy is the energy collected from natural resources like wind, sunlight, flowing water from river, tides, heat energy from earth, etc. These energies can be transferred to other forms of energies. For example, to convert solar energy to electrical energy solar panels can be used, for converting wind energy to electrical

J. Samanta · R. Sangno · R. Pudur (✉)
Department of Electrical Engineering, National Institute of Technology, Jote, Arunachal Pradesh, India
e-mail: rajenpudur@nitap.ac.in

S. De et al. (eds.), *Challenges and Opportunities of Distributed Renewable Power*, Energy, Environment, and Sustainability, https://doi.org/10.1007/978-981-97-1406-3_16

energy wind mills can be used, and to convert the flowing water of river to electrical energy hydro power plants are used. Non-renewable energy is limited and it emits carbon on burning. According to U.S. Energy Information Administration in the year of 2020, 60% of total electrical energy is produced from fossil fuels. From European Environmental Agency survey, 34% of electricity is generated by using renewable energy sources which has doubled since the year 2005 in Europe. Then also 38% of European electrical energy is generated from fossil fuels. According to IQ Air report, India is in 5th position in pollution in the year 2021. Recent report of 31 March 2022 says India is producing 236,109 MW from thermal power plants and 156,607 MW from renewable energy resources. Where 46,723 MW is from hydro power plants and 109,885 MW from other renewable energy resources including solar, wind, and geothermal energy sources. India had set a target of producing electricity by using renewable energy resources (175 GW by the year 2022 and 500 GW by 2030). 70% fossil fuels are used to produce electrical energy in India, and 30% energy is from renewable energy resources. India is currently undergoing a shortage of electrical energy due to shortage of coal since October 2021. Promoting renewable energy is the only option to fulfill the demand of electrical energy in future.

India has different geographical locations and different climatic regions. There are several rivers including Ganga, Yamuna, Brahmaputra, Kaveri, and Godavari. Every state in India is having river so Hydroelectric power plants can be promoted to generate electrical energy. Hydro Power Plants are planned constructed power plants over rivers, large reservoirs, and water falls. Koyna Hydro power plant is India's largest hydro power plant. The dam is established in 1964 over Koyana River, Maharashtra. This hydro power plant has a total capacity of 1930 MW. Globally India is in 5th position for its total capacity of hydro power plants. Hydro Power Plants are planned constructed power plants over rivers, large reservoir, and water falls. This type of plant converts the Kinetic energy of flowing water to electrical energy. HPP can be classified into three main types of HPP Mini Hydro (up to 100 kW), Small Hydro (101 kW to 2 MW), and Large hydro (2.1 MW to 25 MW). Among these Mini hydro and small hydro power plants are renewable and sustainable. Large hydro power plants require large reservoirs and high installation cost so they cannot be categorized as renewable energy.

Generating power is not that much difficult than correcting its power quality issues. Whole world including India is also facing power quality issues nowadays. Some common power quality issues in power systems are voltage sag and swell, unbalanced voltage, harmonics distortion, regular breakdown, and load shedding. So, it is very essential to solve the power quality issues. The prevailing primary challenges within electrical systems involve voltage sags (dips), voltage swells, voltage fluctuations, harmonic distortions, and voltage imbalances. Non-linear loads like SMPS, computers, TV, etc. are widely used nowadays. Non-linear loads are the leading cause of power quality issues and are solely responsible for lousy power. Due to these loads, there will be power quality issues such as harmonics, voltage distortion, and noise in the power system. Good power will provide 50 Hz frequency, sinusoidal voltage waveform, and voltage magnitude of 230 V (±5%) for single phase system and 415 (±5%) for three-phase system. The main aim is to reduce the harmonics distortion, improve the power factor in the system, and to provide good power quality from the system.

SAPFs are used to reduce harmonics from a power system. There are different control techniques to control SAPF [1, 2]. It includes current reference generation using the time domain method, frequency domain method, and soft computing method. S. Kumaresan et al. proposed a method to calculate the DC-link voltage and capacitor values for controlling the SAPF [3]. If the inductance value is large enough, then proper compensating current can be obtained from SAPF [4]. Control systems utilizing the p-q theory in the time domain are frequently used to control SAPF [5–8]. SAPF control using the FFT algorithm is given in [9], with THD reduced to 6.67%. Poor PF is another problem that is to be solved by power engineers. The correlation between power factor and THD is explained by L. Cividino in 1992 and W. M. Grady in 1993 [10, 11]. DC-link capacitor voltage regulation and the peak overshoot problem are discussed in [12]. An Improved control algorithm to control SAPF using DSP is discussed in [13]. In addition, it was observed that voltage regulation and power factor correction could not be simultaneously achieved. The slide mode control method has been proposed by I. Ullah et al. to control SAPF using the p-q theory [14]. Reactive power control methods are discussed in [15, 16]. The relations given by W.M./ Gray can be used to create a new technique to control the filter for a polluted power system. However, the conventional p-q technique is a complex process, with the limitation of correcting one problem at a time. An attempt is made to harmonic mitigation and power factor correction simultaneously, keeping good DC-link capacitor voltage regulation and under balanced and unbalanced loaded conditions.

Induction generators are widely used to generate electricity because it has less cost as compared to other motors. It is also rugged, brushless, has self-protection from short circuits, and needs low maintenance cost. SEIG is normally an induction machine except the capacitor is connected to it for excitation. It is generally used in isolated areas and standalone systems where grid is not available for supplying reactive power to the machine [17–19]. S. Chakraborty et al. proposed a scheme to convert three-phase supply to single phase in SEIG [21]. A novel technique is also proposed by MK. Rajak et al. to connect two differently rated SEIG to grid [20]. The details of 2.2 kW machine is given in [22].

The main objective of this chapter is to model Shunt Active Power Filter for Hydro-electric power plant and its control system to irradicate multiple power quality issues at the same time. SAPF has to be built for mini hydro power plant (up to 100 kW) and simulate the model in MATLAB/Simulink. This chapter will also help to promote SEIG based hydro electric power plant as it is having various advantages than synchronous generator. According to IEEE 519 and IEC 1000–3, THD% should be below 5%. Also for good power quality, the range of power factor should be more than 0.95. So different combinations of non-linear loads (i.e., balanced and unbalanced loads) are to be connected to the system and the results are to be tested in MATLAB/Simulink platform.

This chapter is composed of 6 sections. The introduction and literature survey are given in Sect. 16.1. In Sects. 16.2 and 16.3, the designing and proposed control system of SAPF are given. Section 16.4 discusses about SEIG. Section 16.5 shows the simulation and its result, and finally, the conclusion is given in Sect. 16.6.

16.2 SAPF and Its Design

Active Power filters are categorized into two types, they are Series Active Power Filter and Shunt Active Power Filter. VSI is used to model SAPF whereas Current Source Inverter is used to model Series Active Power Filter. Active Power filter is mainly used to reduce the harmonics from a system. SAPF introduces compensating current to the power system to rectify the power factor as well as reduce harmonics from the power system. Generally, the working principle of SAPF is to inject an equal amount of current harmonics but opposite to the produced harmonics due to load in the system. SAPF is realized by voltage source inverter, DC-link capacitor, and coupling inductor as shown in the schematic diagram in Fig. 16.1. In Fig. 16.1 I_{S*}, I_{L*}, and I_{C*} indicate source currents, load currents, and compensating currents. The value of coupling inductance plays a vital role here. The compensation currents required for mitigation entirely depend on the current limiting reactor, as explained in [4]. The calculation of inductance and DC capacitor is done as per (16.1), (16.2), and (16.3).

$$V_{DC} = \frac{2\sqrt{2} \times V_{line}}{\sqrt{3}m} \tag{16.1}$$

$$\frac{V_{DC}}{8f_s(Level - 1)\Delta I_r} \leq L_f \leq \frac{\delta V_{DC}}{r\omega I_c} \tag{16.2}$$

$$C_f = \frac{P \times T}{0.5 \times V_{dc}{}^2} \tag{16.3}$$

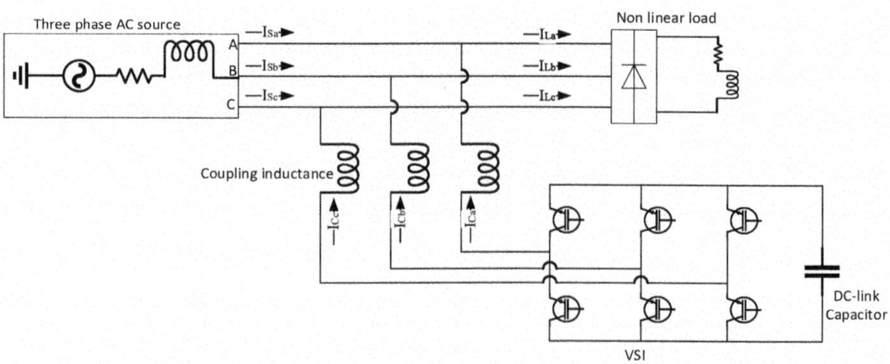

Fig. 16.1 Schematic diagram system

Keeping the switching frequency at 10 kHz, with a 0.7 modulation index Vdc is set to 650 V, and using the above equations, the DC capacitor's value and coupling inductance are calculated as 100 μF and 20 mH, respectively.

16.3 Control Scheme of SAPF

Generating reference current for SAPF has divided into three categories one is in time domain and the other is in frequency domain and soft computing method on which time domain is simple and dependent on circuit analysis and algebraic transformation whereas frequency domain and soft computing method is complex and requires large memory. Time domain is further categorized into active and reactive power theory which is instantaneous in nature (i.e., p-q theory) and synchronous reference frame theory (d-q theory).

A. **Power factor and harmonics:**

The cosine angle between voltage and current of an AC circuit is called power factor. It can also be defined as the actual electrical power dissipation ratio to the product of RMS value of voltage and current in an AC circuit. The range of good power factor is 0.95 to 1. There are two types of power factors leading and lagging depending upon the property of loads. If the load is highly inductive then it will be lagging and if the load is capacitive in nature, then it will be leading power factor. Harmonics is defined as the multiple of integers in fundamental frequency of voltage and current waveform. If harmonics is present then the waveform may or may not be sinusoidal but it may be continuous. THD is the total amount of distortion in voltage or current waveform produced due to the presence of harmonics in it. The harmonics distortion in the network is inversely proportional to the power factor. PF can be represented as (16.4) and displacement factor as (16.5).

$$Power\ factor = \frac{P_{avg}}{(V_{rms})(I_{rms})} \tag{16.4}$$

$$Displacement\ factor = \cos(\theta_v - \theta_i) \tag{16.5}$$

where P_{avg} is average power, V_{rms} is RMS voltage, I_{rms} is RMS current, θ_v is the phase of the voltage and θ_i is the phase of the current. These two equations are only valid if the current and voltage profile is sinusoidal.

When power system is polluted due to the presence of non-linear loads then current will be periodic in nature but not sinusoidal. Let us take steady state polluted voltage and current as (16.6, 16.7) then its root mean square can be (16.8, 16.9).

$$V(t) = \sum_{K=1}^{\infty} V_K \sin(K\omega_0 t + \theta_{vk}) \tag{16.6}$$

$$I(t)=\sum_{K=1}^{\infty} I_K \sin(K\omega_0 t+\theta_{ik}) \tag{16.7}$$

$$V_{rms}=\sqrt{\sum_{K=1}^{\infty} V_{Krms}^2} \tag{16.8}$$

$$I_{rms}=\sqrt{\sum_{K=1}^{\infty} I_{Krms}^2} \tag{16.9}$$

Therefore, by using the above four equations the average power can be predicted as (16.10) and total harmonics distortion in terms of polluted rms current can be written as (16.11).

$$P_{avg}=\sum_{K=1}^{\infty} V_{Krms}\times I_{Krms}\times \cos(\theta_v-\theta_i) \tag{16.10}$$

$$\text{THD}=\frac{\sqrt{\sum_{K=2}^{\infty} I_{Krms}^2}}{I_{1rms}} \tag{16.11}$$

Solving Eqs. (16.10) and (16.11), Eq. (16.12) can be achieved. Neglecting voltage THD as it is generally below 10% power factor equation can be derived to equation number (16.13).

$$I_{rms}=I_{1rms}\sqrt{\frac{1}{1+THD^2}} \tag{16.12}$$

$$PF=\frac{P_{avg}}{(V_{1rms})(I_{1rms})}\times\sqrt{\frac{1}{1+THD^2}} \tag{16.13}$$

And distortion factor is given by (16.14).

$$\text{Distortion factor}=\sqrt{\frac{1}{1+THD^2}} \tag{16.14}$$

Now, the PF can be termed as the product of the distortion and displacement factors (16.15). The distortion factor is due to the harmonics present in the current waveform. Using the PF and distortion factor, the power factor can be generated as shown in Fig. 16.2.

$$\textit{Power factor}\ (\cos\varnothing)=\cos\ (\theta_v-\theta_i)\times\sqrt{\frac{1}{1+\text{THD}^2}} \tag{16.15}$$

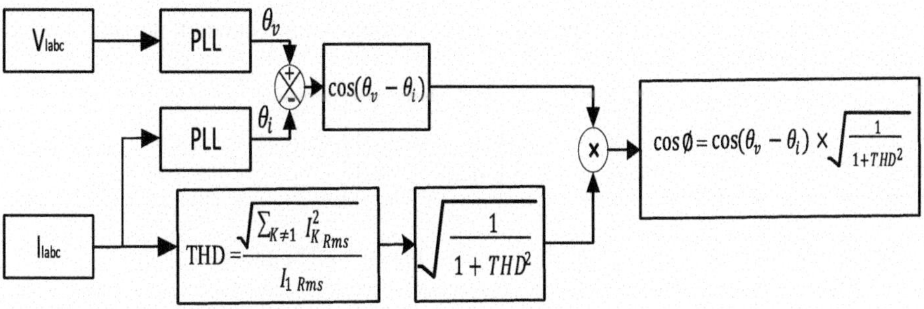

Fig. 16.2 Power factor generation block diagram

Power can be classified into three parts: active, reactive, and apparent. Active power is the true power consumed by the load given in (16.17). Reactive power is the complex power and its wastage of power retrieved from the load represented in (16.18).

$$Powerfactor\ (\cos\varnothing) = \cos\ (\theta_v - \theta_i) \times \sqrt{\frac{1}{1 + THD^2}} \tag{16.16}$$

$$Active\ power\ (P) = VI\ \cos\varnothing \tag{16.17}$$

$$Reactive\ power(Q) = VI\ \sin\varnothing \tag{16.18}$$

2 Conventional p-q theory:

The concept was initially introduced in 1982 by H. Akagi et al. [8]. p-q theory is generally active and reactive power which is instantaneous in nature. Instantaneous active power is the power which flows continuously from source to load and Instantaneous active power represents the continuous power flow from a source to a load, while instantaneous reactive power denotes the bidirectional flow of power between a source and a load. This theory finds its application in three-phase systems, with or without the presence of a neutral wire. Importantly, there are no restrictions on the waveform of voltage and current. In AC signal voltage and current are converted from *abc* reference to αβ0 using Clark's transformation, as shown below in (16.19). Conversely, the αβ0 coordinates are converted back to the abc reference frame through the inverse Clark's transformation, as indicated in (16.20).

$$\begin{bmatrix} 0 \\ \alpha \\ \beta \end{bmatrix} = \sqrt{\frac{2}{3}} \begin{bmatrix} \frac{1}{\sqrt{2}} & \frac{1}{\sqrt{2}} & \frac{1}{\sqrt{2}} \\ 1 & \frac{-1}{2} & \frac{-1}{2} \\ 0 & \frac{\sqrt{3}}{2} & \frac{\sqrt{3}}{2} \end{bmatrix} \begin{bmatrix} a \\ b \\ c \end{bmatrix} \tag{16.19}$$

$$\begin{bmatrix} a \\ b \\ c \end{bmatrix} = \sqrt{\frac{2}{3}} \begin{bmatrix} \frac{1}{\sqrt{2}} & 1 & 0 \\ \frac{1}{\sqrt{2}} & \frac{-1}{2} & \frac{\sqrt{3}}{2} \\ \frac{1}{\sqrt{2}} & \frac{-1}{2} & \frac{-\sqrt{3}}{2} \end{bmatrix} \begin{bmatrix} 0 \\ \alpha \\ \beta \end{bmatrix} \tag{16.20}$$

For balanced system, the zero sequence voltages and current are eliminated thereby giving the (16.21)

$$\begin{bmatrix}\alpha\\ \beta\end{bmatrix}=\sqrt{\frac{2}{3}}\begin{bmatrix}1 & \frac{-1}{2} & \frac{-1}{2}\\ 0 & \frac{\sqrt{3}}{2} & \frac{-\sqrt{3}}{2}\end{bmatrix}\begin{bmatrix}a\\ b\\ c\end{bmatrix} \tag{16.21}$$

Similarly for power equation, this method uses (16.22) and (16.23).

$$\begin{bmatrix}p_0\\ p\\ q\end{bmatrix}=\begin{bmatrix}V_0 & 0 & 0\\ 0 & V_\alpha & V_\beta\\ 0 & -V_\beta & V_\alpha\end{bmatrix}\begin{bmatrix}i_0\\ i_\alpha\\ i_\beta\end{bmatrix} \tag{16.22}$$

$$\begin{bmatrix}p\\ q\end{bmatrix}=\begin{bmatrix}V_\alpha & V_\beta\\ -V_\beta & V_\alpha\end{bmatrix}\begin{bmatrix}i_\alpha\\ i_\beta\end{bmatrix} \tag{16.23}$$

The conventional method involves multiple conversions and different matrices for balanced and unbalanced systems, which eventually gives complexity. In the proposed control system, the complexity is eliminated, and a wide range of control can be achieved simultaneously.

3 Proposed control scheme:

Using (16.17) and (16.18), load power and harmonics are detected, thereby achieving the power factor and active and reactive powers as shown in Figs. 16.2 and 16.3. Unlike conventional theory, during steady state, the use of active filter for filtering p-q signals has been eliminated. Elaborating (16.23) Eqs. (16.24), and (16.25) can be obtained.

$$P = V_\alpha i_\alpha + V_\beta i_\beta \tag{16.24}$$

$$Q = -V_\beta i_\alpha + V_\alpha i_\beta \tag{16.25}$$

$$P = \left(K_p \Delta V_{DC} + K_i \int_0^t \Delta V_{DC} dt\right) - VI\cos\varnothing \tag{16.26}$$

$$Q = K_p\left(Q_{ref} - VI\sin\varnothing\right) = K_p \Delta Q \tag{16.27}$$

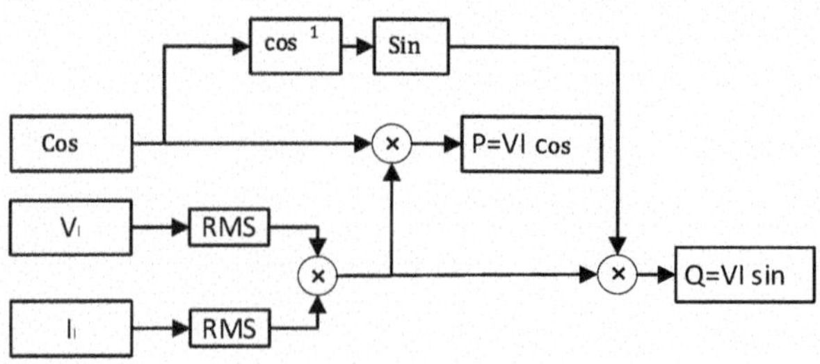

Fig. 16.3 Active and reactive power block diagram

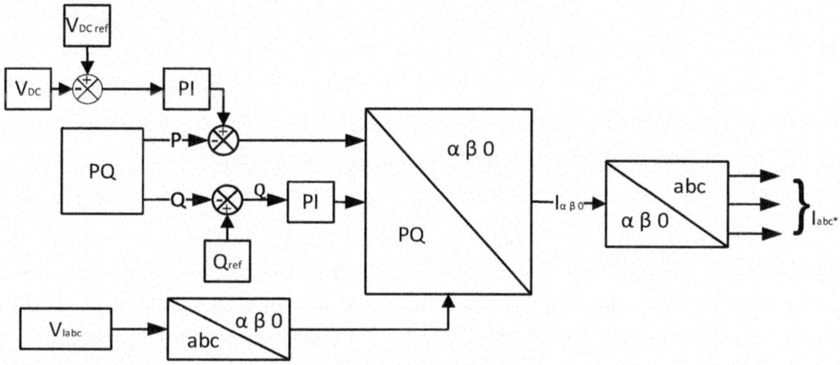

Fig. 16.4 Reference current generation block diagram

Regulating the voltage across the DC capacitor and using the PI controller (16.26) can be written [5]. Reactive power can also be regulated using Eq. (16.27), which is very helpful for controlling reactive power. Comparing the actual Q with reference Q (i.e., $Q_{ref} = 0$) reactive power can be minimized. The reference current generation block is given in Fig. 16.4.

After sensing both active and reactive power from the system, it is controlled with two closed-loop systems, one for active power and another for reactive power, as shown in Fig. 16.4. Gate pulses are generated using a hysteresis current controller comparing reference currents and source currents. Then, the αβ0 to p-q conversion is done using (16.22).

Gate pulse for switching the VSI can be generated in two ways, PWM career control scheme and Hysteresis Current Control (HCC). The easiest way to produce gate pulse from reference current is HCC. As the generated current reference and the current drawn from source are nearly the same, minimal hysteresis band produces the phase current. Higher switching frequency can be obtained from smaller hysteresis band. HCC is basically a feedback current control method of PWM which is instantaneous in nature, where the current generated as reference continually tracks the current drawn from source within a hysteresis band (Table 16.1).

In Fig. 16.7, I_{abc*} and I_{abc} are the generated reference current after transformation techniques and the actual current drawn from the source. The output current after comparing both the reference current and actual current is passed through hysteresis band to generate switching pulse for VSI of the SAPF. In this method, the source current is compared with the reference current generated. The hysteresis

Table 16.1 Switching using HCC

Switch	Hysteresis band (HB)
Lower VSI switches ON	$i_{aa} - i_a < -HB$
Upper VSI switches ON	$i_{aa} - i_a > HB$

current controller produces pulses that ensure the source current tracks the generated reference current effectively.

16.4 Self-Excited Induction Generator (SEIG)

In SEIG for excitation purpose, a bank of capacitor should be connected to the stator terminal to provide reactive power to the generator as well as to the load. A normal induction motor cannot be self-excited induction generator, so it uses capacitor to generate magnetizing current and reactive power to develop the rotating magnetic field. When the prime mover rotates the rotor then a minimal amount of voltage is produced in the stator terminal due to the existence of residual magnetism in rotor. The frequency of the voltage is depending upon the rotor speed. The over speed variations in SEIG are very small (1 to 2%), therefore, for all purposes an induction generator is considered as a "constant speed machine". Its mechanical simplicity and robust construction made the SEIG quite popular in a considerable amount of time. The interesting fact about SEIG is that, it was originally an induction motor and its rotor adapts the number of poles as in the stator automatically by itself. The same rotor can therefore be used with a variety of pole pairs. The term "motor" is used when an induction machine operates at speeds below its synchronous speed, while the term "generator" is employed when the machine runs at speeds exceeding its synchronous speed. Another interesting mechanical property is that whenever the torque varies, the induction generator will change its speed automatically in accordance with torque, thereby resulting in less tear and wear of gear boxes (if any), which is the most important reason for using induction generator rather than synchronous generator when connected directly to grid. This is a single-fed induction generator as only stator windings are being connected to loads or grid. The two main functions of SEIG stator windings are (a) it provides excitation or magnetisation, and (b) it carries the generated current and supplies it to load. The speed of SEIG is determined by the pole-pairs, grid frequency, and the gear box ratio (if any). The SEIGs are not self-exciting; it depends on an external voltage source to produce a rotating magnetic field by stator. The rotor of an induction motor rotates in proximity to the synchronous speed (N_S), although the actual speed is not identical to the synchronous speed. The variance between the synchronous speed and the rotor speed (N_r) is termed as slip speed, represented as ($N_s - N_r$). The proportion of the slip speed to the synchronous speed is referred to as slip, as indicated in Eq. (16.28).

$$S = (N_s - N_r)/N_s \tag{16.28}$$

Slip of an induction machine will be unity if it operates as a motor during starting. While the slip will be zero at its synchronous speed. During motoring mode, the slip will be positive say about + 1% and it will be negative during generating mode which is about -1 to -2%. The rotor frequency is given by (16.29).

$$f_r = sf \tag{16.29}$$

where f_r is the frequency of rotor current.

An induction generator either can be operated in isolated self-excited mode or in grid connected mode. In isolated mode, the terminal voltage and frequency depend on so many factors. An induction machine requires external current supply for excitation and it is not self-starting in nature. Because the field of rotor lags the field of stator. An induction may operate as a generator or a motor but it always consumes reactive power. The current for exciting the stator (i.e., reactive power) of SEIG can be supplied from the utility grid or, a bank of capacitors when it operates in isolated mode. Therefore, induction generator with external excitation is best suited for small or mini hydro power plants, for injecting additional reactive power capacitor bank is adopted in this work. The capacitor values can be determined using the following formulas (16.30–16.36).

$$S = \sqrt{3}V_L I_L \tag{16.30}$$

$$P_A = S\cos\theta \tag{16.31}$$

$$Q = \sqrt{P^2 + S^2} \tag{16.32}$$

$$V_P = \frac{V_L}{\sqrt{3}} \tag{16.33}$$

$$I_c = \frac{Q}{V_P} \tag{16.34}$$

$$X_C = \frac{V_P}{I_C} \tag{16.35}$$

$$C = \frac{1}{\omega X_C} \tag{16.36}$$

where, S is Apparent power, V_L is line voltage, and I_L is line current. P_A is Active power of SEIG and $\cos\theta$ is power factor. Q is Reactive power to be absorbed by the SEIG. V_P is phase voltage of SEIG, I_c is capacitor current and X_C is capacitor reactance. Required capacitor value can be obtained using (16.36). Therefore, minimum value of capacitor is obtained to be 10.62 μF theoretically, whereas in simulation for tangible output the capacitor value is selected as 12.5 μF.

2.2 kW induction motor of ABB (3GBA 102 520-ADDIN) is considered for modeling of SEIG in MATLAB [22]. For MATLAB simulation, circuit model parameters are to be calculated hence Block rotor and No-load test are required. The data of the tests are obtained from datasheet and the values are given in Table 16.2.

Table 16.2 Test parameters details

Test parameter	Values
P_o	118 W
V_o	415 V
I_o	0.51 A
P_{sc}	281 W
V_{sc}	125 V
I_{sc}	4.8 A

$$R_{eq} = \frac{P_{SC}}{3I_{SC}^2} \tag{16.37}$$

$$Z_{eq} = \frac{\frac{V_{SC}}{\sqrt{3}}}{I_{SC}} \tag{16.38}$$

$$X_{eq} = \sqrt{Z_{eq}^2 - R_{eq}^2} \tag{16.39}$$

$$X_s = X_r = \frac{X_{eq}}{2} \tag{16.40}$$

$$L_s = L_r = \frac{X_s}{2\pi f} \tag{16.41}$$

$$\cos\theta_o = \frac{P_o}{\sqrt{3}V_o I_o} \tag{16.42}$$

$$I_m = I_0 \sin\theta_0 \tag{16.43}$$

$$X_m = \frac{\frac{V_o}{\sqrt{3}}}{I_o} \tag{16.44}$$

In the above Eqs. (16.37, 16.38, 16.42) P_{sc}, V_{sc}, and I_{sc} are short circuit power, voltage, and current and P_o, V_o, and I_o are no load power, voltage, and current. The stator resistance and rotor resistance can be obtained using Eq. (16.40), stator and rotor inductance can be calculated from Eq. (16.41) and magnetic inductance can be obtained from Eq. (16.44). The calculated results are given in Table 16.3.

Table 16.3 Calculated equivalent circuit parameters

Obtained parameters	Values
$X_s = X_r$	7.2376Ω
$L_{lr} = L_{ls}$	0.0230 H
L_m	1.5793 H

16.5 Simulation and Results

Based on the analysis of the above equations, the simulation model is prepared in a matrix laboratory (MATLAB) environment (R2022a (8.6.0.267246) version, 40,832,900). Sample time for simulation of 50 μ sec has been used. Two types of loads have been tested in the system: a balanced non-linear load (150 Ω and 40 mH) and an unbalanced non-linear load (250Ω, 150Ω, and 40 mH). SAPF is realized with an IGBT-based Voltage Source Inverter (VSI). In simulation, an isolator has been connected to the system which has been switched ON at 0.5 s. SEIG and SAPF were simulated separately. Simulation diagram of SAPF is shown in Fig. 16.5, and simulation diagram of SEIG is given in Fig. In the results obtained in all cases I_S, I_C, I_L, and V_{DC} in the y-axis represent source current, compensating current, load current, and DC voltage across the DC capacitor.

K_p and K_i values for voltage regulation have been chosen as 0.0008 and 13. The K_p value for reactive power control is selected as 0.1. The proposed control system is tested on unbalanced and balanced load systems, and the power factor correction, harmonics elimination, and source voltage regulation have been achieved.

Figure 16.6 shows the extraction of active and reactive power in simulation. In this figure Eqs. (16.16–16.18) is modeled in simulation. Hysteresis Current control simulation has been shown in Fig. 16.7. Hysteresis band is considered as -0.05 to 0.05. In Fig. 16.8 asynchronous machine from MATLAB library is used to model SEIG. Resistive loads of 100 Ω were connected to the system to verify the results.

A. **Under balanced loaded conditions:**

With a balanced non-linear load, the THD was reduced to 3.02% from 25.09%, as shown in Figs. 16.9 and 16.13. A peak voltage of 1200 V is observed in the DC-link

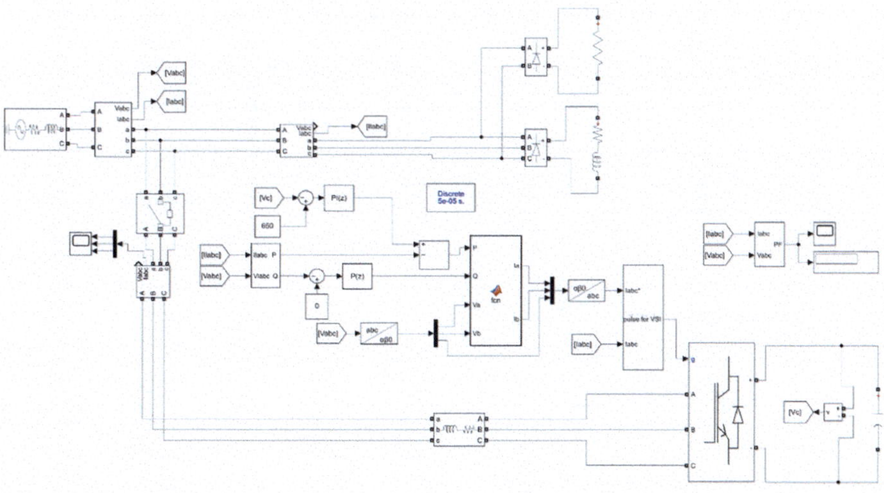

Fig. 16.5 SAPF simulation diagram

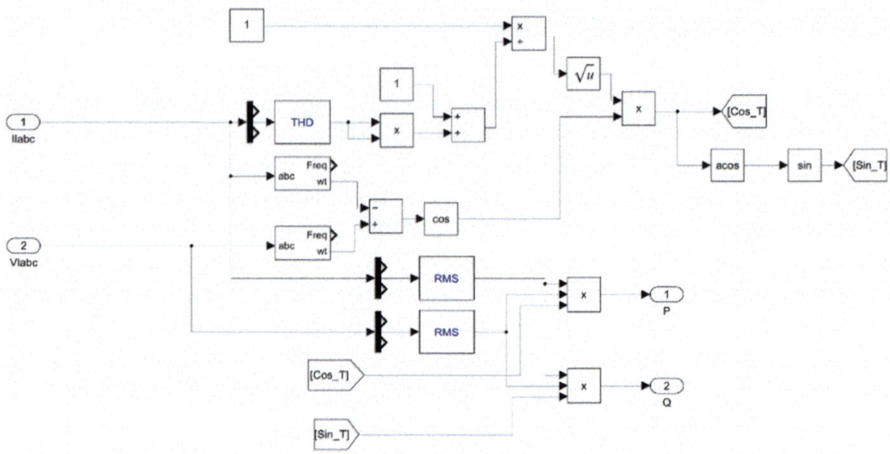

Fig. 16.6 Modified p-q theory active and reactive power simulation diagram

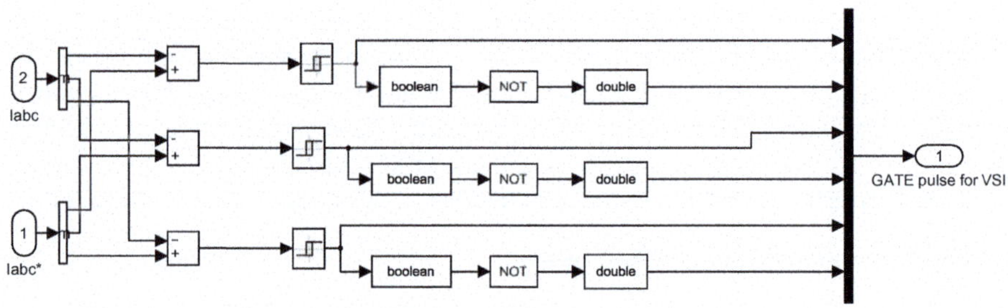

Fig. 16.7 Hysteresis current control simulation diagram

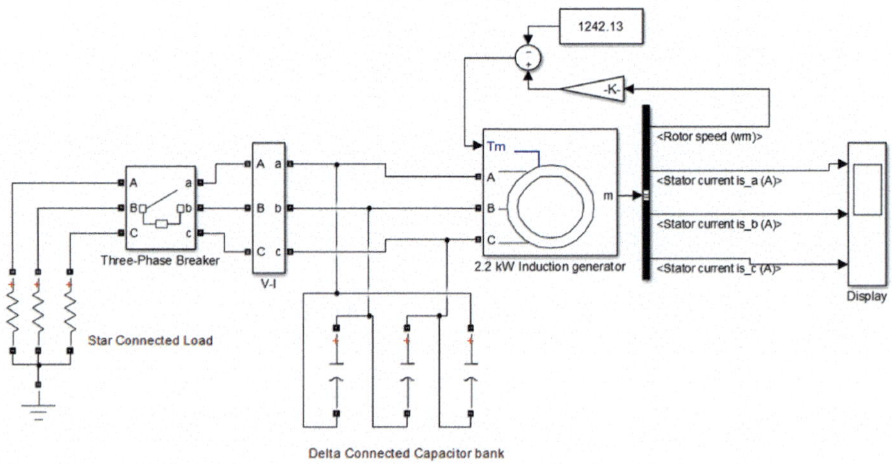

Fig. 16.8 SEIG Simulation diagram

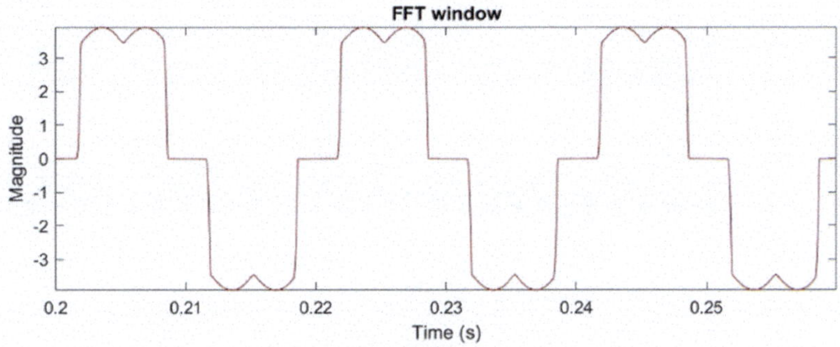

(a)

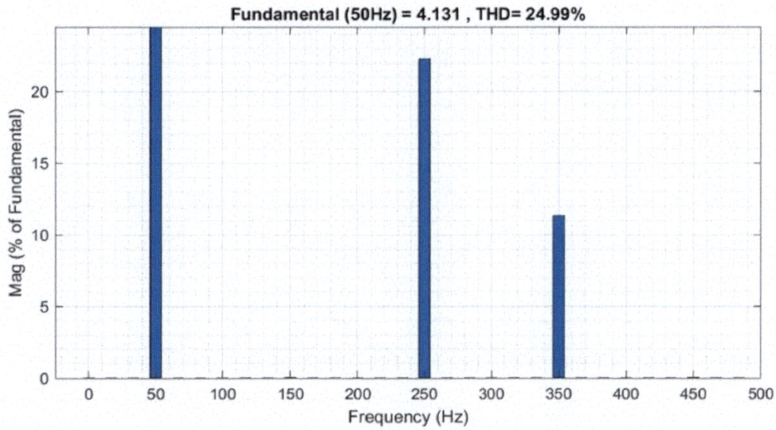

(b)

Fig. 16.9 THD before compensation of balanced non-linear load **a** FFT window, **b** THD%

capacitor voltage in Fig. 16.10. This peak voltage can be minimized using a limiter after the PI controller [12]. Figure 6.11a shows the sources voltage regulation after SAPF injection, (b) is the sources current profile before and after injection, and (c) is the results of power factor correction after compensation for balanced load, from 0.8026 to 0.992. Lagging power drawn by load has been compensated as shown in Fig. 16.12, source voltage lags compensation current and leads by the load current.

2 **Under unbalanced loaded conditions:**

Under an unbalanced non-linear load, THD is reduced to 3.27% from 16.47% shown in Figs. 16.14 and 16.17. In Fig. 16.15, (a) is the unbalanced source current, (b) is the compensating current, and (c) is the voltage across the DC capacitor. The SAPF is connected at 0.5 s, and a small ripple voltage is observed at DC voltage due to the nature of unbalanced load compensation. Figure 16.16 shows the improvement of the power factor to near unity.

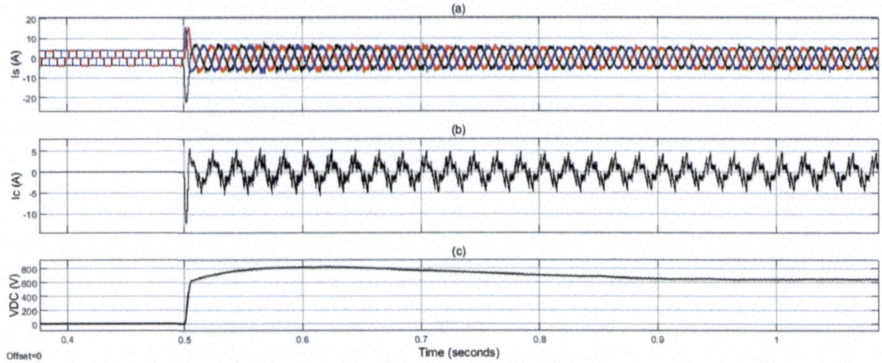

Fig. 16.10 Results of balanced non-linear load connected to SAPF **a** Source current, **b** Compensating current, **c** DC-link capacitor voltage

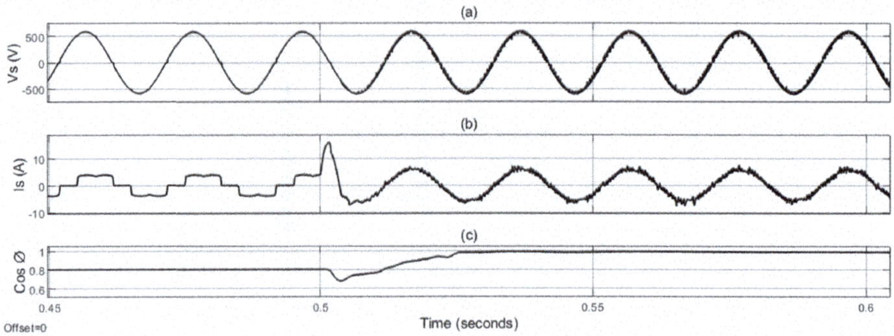

Fig. 16.11 Results of balanced non-linear load connected to SAPF **a** Source voltage, **b** Source current, **c** Power factor

The three-phase voltage buildup of 2.2 kW SEIG can be seen in Fig. 16.18. RMS voltage and frequency are given in Fig. 16.19. The building time is observed to be 1.2 s. The voltage is 415 V and the frequency of 50 Hz is obtained in simulation. As shown in the figures the terminal voltages build up to 1 s, during this time machine takes reactive power from capacitor bank, and rotor speed goes down and then picks up with rated speed, little variation may be seen in the frequency graph.

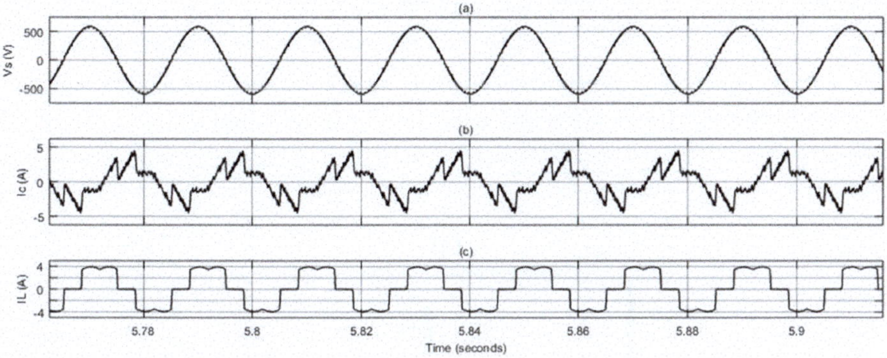

Fig. 16.12 Results of balanced non-linear load connected to SAPF **a** Source voltage, **b** Compensation current, **c** Load current

Fig. 16.13 THD after compensation of balanced non-linear load **a** FFT window, **b** THD%

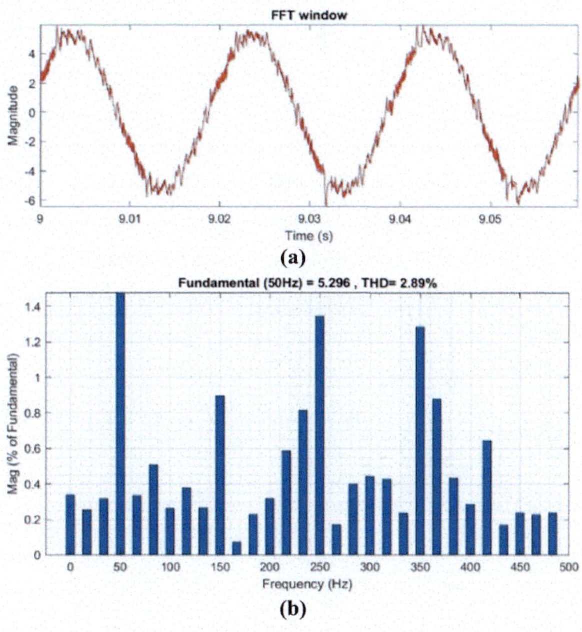

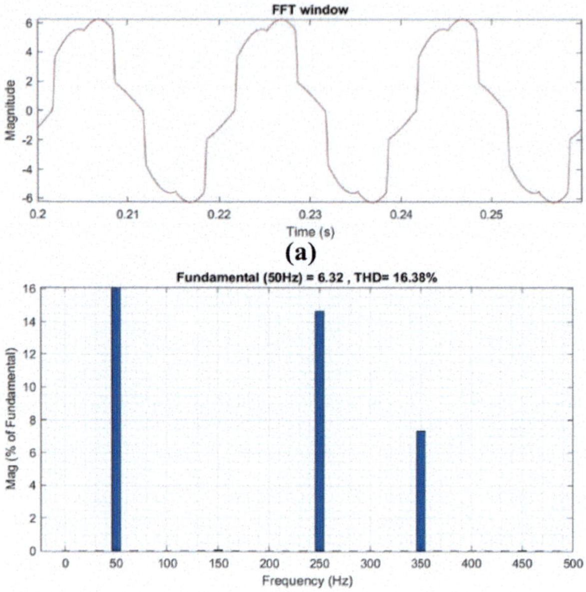

(b)

Fig. 16.14 THD before compensation of unbalanced non-linear load **a** FFT window, **b** THD%

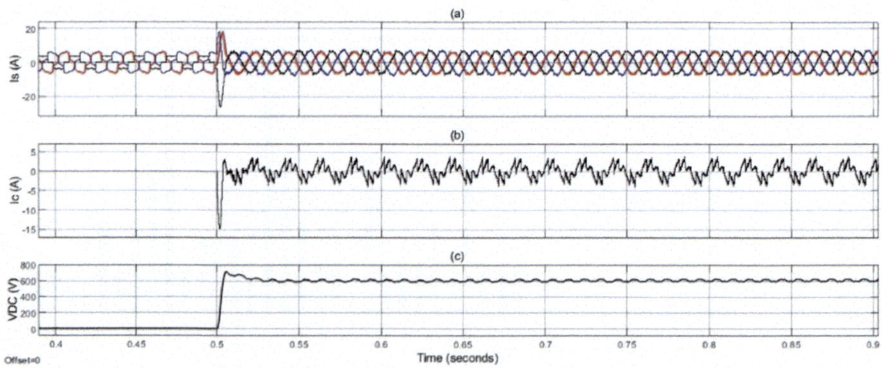

Fig. 16.15 Results of unbalanced non-linear load connected to SAPF **a** Source current, **b** Compensating current, **c** DC-link capacitor voltage

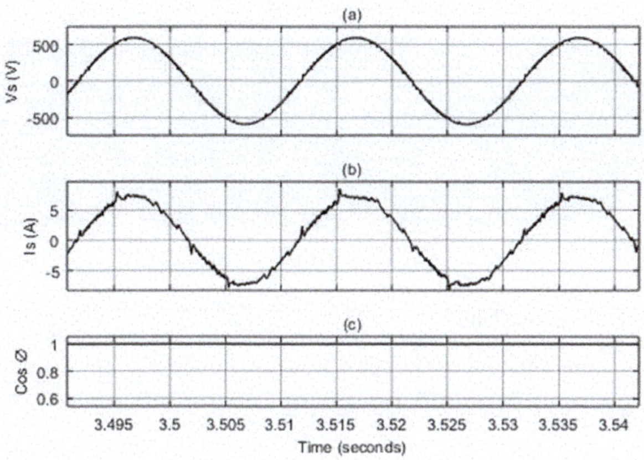

Fig. 16.16 Results of unbalanced non-linear load connected to SAPF **a** Source voltage, **b** Source current, **c** Power factor

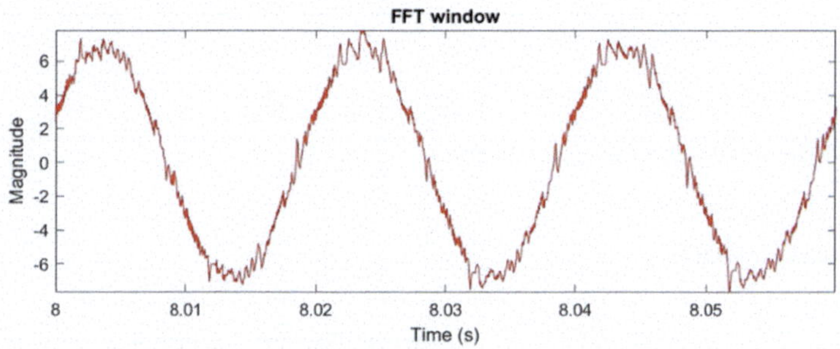

(a)

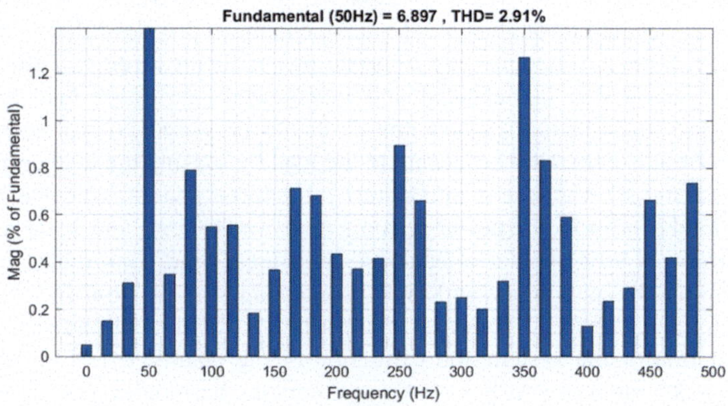

(b)

Fig. 16.17 THD after compensation of unbalanced non-linear load **a** FFT window, **b** THD%

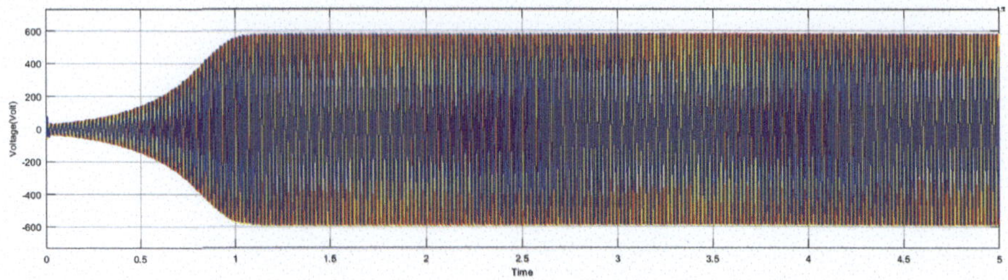

Fig. 16.18 Voltage buildup of SEIG

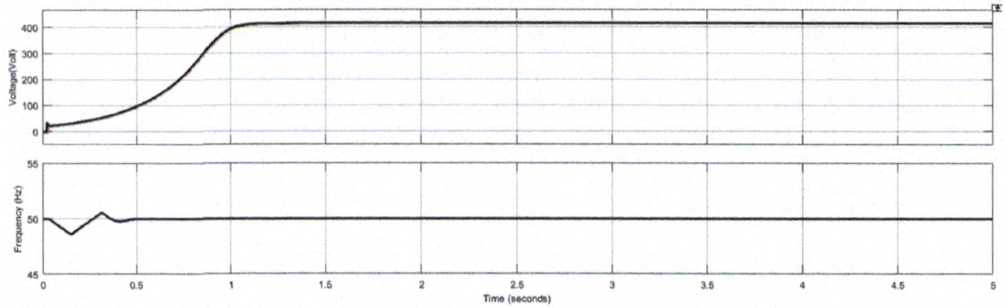

Fig. 16.19 RMS voltage and frequency of SEIG

16.6 Conclusion

Due to the advantages of SEIG, it has been chosen to generate electricity for hydropower plants. 2.2 kW SEIG is modeled in MATLAB, and the calculated parameters are verified through simulation. An enhanced p-q algorithm is presented to manage harmonics reduction and improve power factor through a shunt active power filter, applicable in balanced and unbalanced non-linear load scenarios. As load is the main factor for bad power, in the proposed control scheme, the reference current is generated according to the load by sensing the load current. The mathematical model was tested satisfactorily for three phases. Under a balanced non-linear load condition, the power factor is improved to 0.99, and THD is reduced to 2.89% from 24.99%. Under unbalanced non-linear load conditions, the power factor is improved to near unity, and THD is reduced to 2.91% from 16.38%. In this control technique, the THD obtained is according to IEEE 519 and IEC 1000–3. After simulation of the system and extraction of the results, it can be concluded that the control system of SAPF is capable of reducing harmonics, correcting power factor, and regulating the voltage of the DC-link capacitor. The proposed control scheme also gives independent reactive power control, which may be helpful in terminal voltage compensation in the future. As this SAPF can irradicate power quality issues, it should be tested in hardware by giving a supply from SEIG. Also, in hybrid renewable energy sources. The DC-link capacitors from SAPF can be replaced by solar and integrated solar with wind or hydropower to reduce the intermittency of power supply from renewable energy sources.

References

1. Hoon Y, Radzi MAM, Zainuri MAAM, Zawawi MAM (2019) Shunt active power filter: a review on phase synchronization control techniques. Electronics 8(7):791
2. Kumar R, Bansal H (2018) Shunt active power filter: current status of control techniques and its integration to renewable energy sources. Sustain Cities Soc 42:574–592
3. Kumaresan S, Habeebullah Sait H (2020) Design and control of shunt active power filter for power quality improvement of utility powered brushless DC motor drives. Automatika 61(3):507–521
4. Dai NY, Wong MC (2011) Design considerations of coupling inductance for active power filters. In: 6th IEEE Conference on Industrial Electronics and Applications, pp 1370–1375. IEEE
5. Imam AA, Kumar RS, Al-Turki YA (2020) Modeling and simulation of a PI controlled shunt active power filter for power quality enhancement based on PQ theory. Electronics 9(4):637
6. Hasim ASBA, Dardin SMFS, Ibrahim ZB, Azid ABA, Ishak AMB (2018) Comparative study; Different types of PWM control scheme in three-phase four-wire shunt active power filter (APF) topology. In 2018 international Conference on Electrical Engineering and Computer Science (ICECOS), pp 265-270. IEEE
7. Samanta J, Pudur, R (2022) Comparison of p-q and d-q Theory for Shunt Active Power Filter. In: IEEE International Power and Renewable Energy Conference (IPRECON), Kollam, India, pp 1–6. IEEE
8. Akagi H, Watanabe EH, Aredes M (2017) Instantaneous power theory and applications to power conditioning. John Wiley & Sons
9. Patel PJ, Patel RM, Patel V (2017) Implementation of FFT algorithm using DSP TMS320F28335 for shunt active power filter. J. Inst Eng Series B 98:321–327
10. Cividino L (1992) Power factor, harmonic distortion; causes, effects and considerations. In 1992, Fourteenth International Telecommunications Energy Conference-INTELEC'92, IEEE
11. Grady WM, Gilleskie RJ (1993) Harmonics and how they relate to power factor. In: Proceedings of the EPIR Power Quality Issues & Opportunities Conference, IEEE, pp 1–8
12. Wang Y, Xie YX, Liu X (2015) Analysis and design of DC-link voltage controller in shunt active power filter. J Power Electr 15(3):763–774
13. Chandra A, Singh B, Singh BN, Al-Haddad K (2000) An improved control algorithm of shunt active filter for voltage regulation, harmonic elimination, power-factor correction, and balancing of nonlinear loads. IEEE Trans Power Electr 15(3):495–507
14. Ullah I, Ashraf M (2019) Sliding mode control for performance improvement of shunt active power filter. SN Appl Sci 1(6):531
15. Ouchen S, Benbouzid M, Blaabjerg F, Betka A, Steinhart H (2021) Direct power control of shunt active power filter using space vector modulation based on supertwisting sliding mode control. IEEE J Emerg Select Topics Power Electr 9(3):3243–3253
16. Barbosa PG, Rolim LG, Watanabe EH, Hanitsch R (1998) Control strategy for grid-connected DC-AC converters with load power factor correction. In IEEE Proceedings-Generation, Transmission and Distribution, pp 487–492. IEEE
17. Murthy SS, Malik OP, Tandon AK (1982) Analysis of self-excited induction generators. In IET proceedings C (generation, transmission and distribution) 129(6), 260–265, IET Digital Library (1982)
18. Rathore, Umesh C, Sanjeev Singh (2014) Power quality control of SEIG based isolated pico hydro power plant feeding non-linear load. In 2014 IEEE 6TH India International Conference on Power Electronics (IICPE), pp 1–5. IEEE
19. Rana Kailash, Duli Chand Meena (2018) Self excited induction generator for isolated pico hydro station in remote areas. In 2018 2ND IEEE International Conference on Power Electronics, Intelligent Control and Energy Systems (ICPEICES), pp 821–826. IEEE
20. Rajak MK, Samanta J, Pudur R (2023) A hardware-based novel approach for parallel operation of two differently rated SEIGs. Resul Eng 17

21. Chakraborty S, Pudur R (2021) Supply of single-phase power for rural area using three-phase self-excited induction generator. In 2021 Asian Conference on Innovation in Technology (ASIANCON), pp 1–6. IEEE
22. ABB group, https://global.abb/group/en, last accessed 2023/07/10

Part V
Applications

Chapter 17
Distributed Power Generation Using a Pilot-Scale Downdraft Gasifier for Rural Application

Sweta Sharma, Ayush Garg, Ishita Sarkar, and Chanchal Loha

Abstract India is a significant consumer in the World Energy Market. Electricity in India is majorly produced by non-renewable sources causing huge environmental pollution which in turn affects climate changes. With the increase in energy demand, there is a genuine need to switch towards renewable sources to protect our planet. Rural areas being rich in agricultural products are the good source of biomass resources. Hence, biomass gasification is a promising technology that can play a significant role in addressing both energy needs and environmental concerns. The present study is aimed at the performance study and the techno-economic analysis of a pilot-scale 10 kW fixed-bed downdraft gasifier integrated with a gas engine for distributed power generation in rural areas utilizing agriculture waste such as rice husk, wood chips, and dry leaves along with that comparison with diesel fuel engine as an alternative due rising fuel prices. A maximum overall gasification efficiency of 22.99% has been achieved with the current operating conditions which is reasonable for biomass-to-power application. The levelized unit cost of electricity (LUCE) of the biomass gasification-based power generation unit is found as 15.21 INR which is lower than the LUCE of a similar capacity diesel generator-based power generation system.

Keywords Biomass gasification · Downdraft gasifier · Distributed power generation · Rural area

---

S. Sharma · A. Garg · I. Sarkar · C. Loha (✉)
Energy Research and Technology Group, CSIR-Central Mechanical Engineering Research Institute, Durgapur, West Bengal 713209, India
e-mail: chanchal.loha@gmail.com

Academy of Scientific and Innovative Research (AcSIR), Ghaziabad 201002, India

S. De et al. (eds.), *Challenges and Opportunities of Distributed Renewable Power*, Energy, Environment, and Sustainability, https://doi.org/10.1007/978-981-97-1406-3_17

17.1 Introduction

Electricity is a crucial factor in the development of any nation, impacting economic, social, and technological progress. It is important to note that India is the world's third largest energy-consuming country, it has doubled its energy use since 2000 [1]. In a large and geographically diverse country like India, relying solely on the extension of the national grid without considering decentralized power generation can present significant challenges and limitations. In centralized system, as the electricity travels long distance from the power generation source to the end-users, significant energy is lost during transmission. This can lead to inefficiencies and higher costs. Besides, India's energy resources, such as coal, natural gas, and renewable energy potential, are unevenly distributed across the country. Concentrating solely on the national grid extension may not tap into the full potential of decentralized renewable energy sources available in specific regions. By adopting a diversified and decentralized approach to power generation, India can better address the energy needs of its population, promote inclusive development, and ensure a more sustainable energy future. Rural areas in India are facing the challenge of meeting their increasing electricity demands. Therefore, exploring suitable power generation systems that are efficient, environmentally friendly, and economically viable becomes essential. Decentralized power generation utilizing renewable energy systems is increasing day by day which is viewed as an alternative to grid electricity and traditional fuels [2]. It is estimated that worldwide installed capacity of off-grid distributed generation system has grown by 365 MW in 2020 to reach 10.6 GW [3].

Small-scale generation systems, especially those based on renewable energy sources, can play a crucial role in rapidly increasing access to quality electricity services in under-electrified zones while also contributing to the de-carbonization of power supply. There are plenty of natural resources available in our country, but these resources are either untapped or they are inefficiently utilized in most of the cases. Among the different renewable energy sources, the most promising renewable energy source is biomass because it has a distinct advantage over the use of other renewable energy sources like wind power and solar power [4]. Unlike solar energy and wind energy, which are intermittent, biomass can provide energy for 24 h. Besides, biomass does not add net carbon dioxide (CO_2) to the atmosphere because it releases the same amount of carbon which is absorbed during its growth [5–7]. Also, the emission of sulfur dioxide (SO_2) is low because biomass contains a very low amount of sulfur (S). Biomass has indeed gained significant attention as a renewable energy resource in India, in line with the country's national energy policy objectives. The interest in biomass-based energy has been increasing due to the following reasons [4]:

1. Development of highly efficient conversion technologies and the increased production of biomass made the biomass-based energy competitive with the fossil fuel-based energy in many situations.
2. The environmental benefits of biomass energy over fossil fuel-based energy.
3. Concern over energy security and the diversity of energy supply.

4. Rural development and employment generation.
5. Restoration of degraded lands due to plantation of trees and possible increase of biodiversity.

It is estimated that around 32% of India's total primary energy is still derived from biomass and a significant rural population of our country still depends upon it for their primary energy needs [8]. As per the estimation of the International Institute for Applied Systems Analysis (IIASA) and the World Energy Council (WEC), the biomass consumption globally increased from 5.4 Gtoe in 1990 to 6.7–7.5 Gtoe in 2022. Biomass is abundantly available in agricultural-based countries like India. It is estimated that about 750 million metric tonnes of biomass (covering agricultural residues) is available per year out of which about 230 million metric tonnes per annum is surplus which can generate about 28 GW of energy [8]. Additionally, sugarcane bagasse from about 550 sugar mills running currently can produce about 5000 MW [9, 10]. Biomass available in India can be classified into different categories as given below in Table 17.1 [11–13].

A common technique for the utilization of biomass is the thermochemical conversion of biomass. Therefore, biomass gasification can be a potential technology for electricity production in rural India [14–16]. Biomass gasification involves the conversion of organic materials like agricultural residues, forestry waste, and energy crops into syngas, a combustible gas. Biomass gasification offers the advantage of utilizing renewable and carbon–neutral feedstocks, which can contribute to reducing greenhouse gas emissions compared to conventional fossil fuel-based power generation systems. Furthermore, the economic feasibility of power generation technologies is a critical factor for rural areas in India. The cost of installation, operation, and maintenance significantly influences the selection of a power generation system. Assessing the economic aspects of biomass gasification-based power generation in the context of rural India will provide valuable insights into their viability, levelized cost of electricity, and potential incentives or barriers for deployment. It is recommended in the literature that a small-scale biomass gasifier in the range of 10–200 kW is suitable to effectively utilize local biomass resources [13]. The gasification performance of different biomass materials is investigated by many researchers using fixed-bed biomass gasifiers. Gasifier performance of furniture wood and wood chips is studied in a small-scale downdraft fixed-bed gasifier to investigate the effect of varying equivalence ratio on the quality of the product gas and it is reported that 0.38 is the

Table 17.1 Classification of biomass

Wood	Wood obtained from forestry or waste from forest products such as kernel shells, wood chips, sawdust, etc.
Energy crops	These are the high-yield crops grown for energy application such as Silvergrass, Switchgrass, Napier grass, Jatropha, etc.
Agricultural residues	Residues of different crops such as rice husk, straw, sugarcane bagasse, corn husk, coconut shell, etc.
Waste	This includes food waste, industrial biomass waste, etc.

optimum equivalence ratio for the best performance of the downdraft gasifier as CO and CH_4 compositions and the calorific value reach to their peak [17]. The performance of an updraft gasifier is reported with wood, agricultural residue, and perennial grass as biomass materials [18]. The study reveals that different biomass generates producer gas with significantly calorific value, highest by wood chips. The gasification performance of wood biomass in a downdraft gasifier is reported by varying the operating parameters [19]. It is reported that the equivalence ratio of 0.3 and the moisture content <10% leads to the generation of syngas with higher value and lower tar content. Another study using a downdraft biomass gasifier shows that the temperature profile, the gas composition, the calorific value, and the pressure drop across the gasifier are very much sensitive to the gas flow rate. It is also reported that the high temperature in bed trends to better conversion of non-combustible components into combustible components in the resulting gas and thus improves the calorific value of product gas [20]. Studies on gasification of different types of biomass materials like wood chips, barks, olive pomace, and hazelnut shell are reported in a batch-scale fixed-bed downdraft gasifier with gas cleaning and flaring system [21]. The air–fuel ratio is adjusted to generate producer gas with higher heating value and less pollutant. Finally, the optimum system condition is identified (0.35 equivalence ratio) from the measurement of gas composition at which the higher heating value reaches maximum (5.5 MJ/Nm^3). Another study of biomass gasification in a pressurized staged fixed-bed gasifier connected with hot gas filtration and catalytic reforming of tars and hydrocarbon gases reports that the tar, an important factor for downstream application of producer gas, can be reduced to 1–12 g per m^3 of gas [22]. Others have simulated the biomass gasifier either for power generation system or combined heat and power generation system. Exergetic and economic analysis based on simulation of a compact cogeneration plant producing electric energy as well as hot and cold water from the wood gasifier shows that the overall efficiency of the system might reach 51.42% [23]. Performance of a commercial scale tri-generation system is reported by simulating with different biomass materials where the producer gas generated by the gasifier is used in an IC engine to produce electricity, the exhaust gas is used for generating hot water, and an adsorption cooling system is run with the excess heat [24]. From the literature survey, it can be observed that the experimental studies are mostly available for the performance of small-scale fixed-bed biomass gasifiers only. Some numerical studies are available considering integrated gasifier and power generation unit. However, there is a shortage of experimental studies conducted on pilot-scale biomass gasifiers integrated with power generation units. Besides, the comprehensive cost analysis of pilot-scale biomass gasifier-based power generation system is rare in the literature. Therefore, this study presents the performance of a 10 kW downdraft gasifier integrated with a gas engine for distributed power generation in rural areas utilizing agricultural waste such as rice husk, wood chips, and dry leaves. A techno-economic analysis is presented and the levelized unit cost of electricity (LUCE) of the biomass gasification-based power generation system is compared with a similar capacity diesel generator-based power generation system. This research is significant as it aims to investigate the advantages of biomass gasification and its potential for power generation in the context of rural

India because the environmental impact is a crucial concern in rural India, where preserving the environment and minimizing pollution is of utmost importance.

17.2 Materials and Methods

17.2.1 Biomass Properties

Three different biomass are used for the present investigation; rice husk, wood chips, and dry leaves biomass briquettes. Rice husk is the coating on grain of rice. It is also known as rice hull. Rice husk is collected from the local rice mill. Wood chips are formed by cutting or chipping larger pieces of wood from trees, branches, logging residues, roots, stumps, or wood waste into small- to medium-sized pieces. Wood chips are obtained from local sawmill, whereas dry leaves are collected from the residential colony of CSIR-Central Mechanical Engineering Research Institute, Durgapur, and converted into briquettes. The gasification performance of any biomass material depends upon its proximate analysis, ultimate analysis, and calorific value. Therefore, proximate analysis, ultimate analysis, and higher heating value (HHV) of these three biomass materials are carried out as per ASTM methods. ASTM E1756-08 method is used for determination of moisture content, ASTM E872-82 is used for the determination of volatile matter and ASTM E1755-01 is used for the determination of ash content. The ultimate analysis is done by an elemental analyzer (Elementar Analysensysteme GmbH, Vario ELV5.18.0 18). Table 17.2 shows the proximate analysis, the ultimate analysis, and the HHV of all three biomass materials.

From the proximate analysis, it can be observed that wood chips contain maximum amount of volatiles followed by dry leaves briquettes and rice husk, whereas the

Table 17.2 Proximate analysis, ultimate analysis, and HHV of biomass

	Rise husk	Wood chips	Dry leaves briquettes
Proximate analysis (wt %)			
Moisture	9.95	5.73	7.67
Volatiles	55.54	76.36	61.20
Fixed carbon	14.99	14.40	15.72
Ash	19.52	3.51	15.41
Ultimate analysis (wt %)			
C	49.07	60.91	45.37
H	3.79	4.58	5.80
O	46.42	33.65	47.60
N	0.63	0.86	1.02
S	0.09	0.00	0.21
HHV (MJ/kg)	14.07	17.40	15.12

fixed carbon content is almost similar for all three biomass materials. From ultimate analysis, it is evident that the carbon and hydrogen content is more and oxygen content is less in wood chips. The higher heating value of wood chips is maximum (17.40 MJ/kg) followed by dry leaves briquettes (15.12 MJ/kg) and the rice husk (14.07 MJ/kg). Therefore, it is clearly understood from the proximate analysis, the ultimate analysis, and the HHV values that the quality of wood chips biomass is superior to rice husk and dry leaves briquettes.

17.2.2 Experimental Setup

Figure 17.1 shows photograph of a 10 kW biomass gasifier-based power generation system. It is a small gasifier, movable, compact, and requires minimum maintenance. It consists of the main downdraft gasifier system, air feeding system, producer gas cleaning and cooling system, and 100% producer gas-based engine. The biomass gasification is performed in the downdraft fixed-bed gasification system. Figure 17.2 shows the different zones of a typical downdraft gasifier. Here, biomass is fed at the top and the air intake is from the sides some way down the fuel path where the reactor cross section is narrowed. Biomass moves from top to bottom and air moves concurrently with the biomass particles downward to the grate. While moving from top to bottom, biomass is first dried in the drying zone and then the volatile matter is released from the biomass in the pyrolysis zone. After the pyrolysis zone, biomass passes through the combustion zone where air is introduced with the help of an air supply pipe. In the combustion zone, char is combusted and heat is produced. Heat from the combustion zone is propagated to the drying and the pyrolysis zones mainly by radiation and partly by convection. Volatiles removed from biomass along with char pass from pyrolysis zone to the combustion zone. Then the remaining char and the combustion products from combustion zone pass through the reduction zone. So, not only the biomass but also the pyrolysis gas passes through the high-temperature combustion zone. This helps to crack the tars in the producer gas and the tar content in the product gas becomes very low. This is the main advantage of a downdraft gasifier which is crucial for power generation and niche thermal application [25].

17.2.3 Experimental Procedure

To start the gasifier first time, it is loaded with 02 kg of charcoal at the bottom of the gasifier above the perforated grate. Fresh biomass is loaded above charcoal up to top level and the gasifier lid is closed tightly. The charcoal is ignited using an auto-ignition system. Air is supplied at the throat of the gasifier through an air supply pipe connected with a blower, flow meter, and a flow control valve. Initially, air is supplied slowly, and the flow rate is increased once the charcoal is ignited. Air flow rate can be controlled with the help of the flow control valve. Heat from the

1) Gasifier, 2) Charcoal Filter, 3) Cooling Tower, 4) Condenser, 5) Biomass Filter, 6) Gas Engine, 7) Generator, 8) Control Panel

Fig. 17.1 Photograph of a 10 kW biomass gasifier-based power generation unit

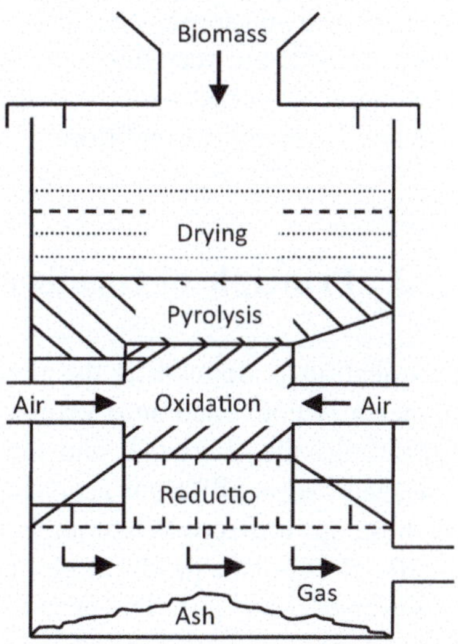

Fig. 17.2 Schematic of a downdraft gasifier

combustion zone ravels to the pyrolysis zone and then to the drying zone. In the drying zone, moisture is evaporated from biomass, and in the pyrolysis zone, volatiles are evaporated. These moisture-laden volatiles first pass through the combustion zone and then to reduction zone before leaving the gasifier. The grate is located at the bottom, below the reduction zone. Four vertical ribs are provided with the grate and it is connected with a motor through a shaft from the bottom. Once the charcoal is ignited, the motor of the grate is started which gives an agitation to the bed. The ribs connected with the grate also start rotating and break if any lumps are formed in the combustion zone during the operation of the gasifier. Below the reduction zone, it is the space where the char and ash are collected at the end of the reaction and only the syngas flows out of the gasifier. Initially, the syngas is passed through a by-pass line where the syngas is ignited to produce flares. The quality of the syngas is assessed by the color of the flare. Once, the colorless flame is produced, the flare valve is closed and the valve connected to the filter–engine line is opened. In order to introduce the produced gas to the engine, it should be free from tar and particulate matter to avoid any damage to the engine. So, the producer gas is passed through a gas cleaning and cooling system before supplied to the engine. There are different cleaning units—cyclone separator, dry charcoal filter, cooling tower, biomass filter, and fabric filter for cleaning the unwanted impurities efficiently. The producer gas coming out from the gasifier is passed through the cyclone separator first. It separates the particulate matters very efficiently from the produced gas which are >5 μm. Then, the producer gas is passed through a dry charcoal filter where small-size particles are arrested. After the charcoal filter, the producer gas is cooled down to about 50 °C with the help of the cooling tower. The purpose of the cooling tower is to condense the tar from the producer gas. Then the producer gas is passed through two biomass filters where the moisture and some particles are removed. Finally, the syngas is passed through a fabric filter where very fine particles are arrested before introducing into the engine. The engine is started with the help of a battery and it takes about 15–20 min to reach a steady fuel supply condition.

17.3 Principle of Gasification

Gasification is defined as the thermochemical conversion process in which solid biomass is converted into producer gas containing hydrogen, carbon monoxide, carbon dioxide, methane, and nitrogen depending upon the gasifying agent used. For gasification, different gasifying agents such as oxygen, air, steam, or a mixture of these are utilized. When air is used as gasifying agent, the process is called air gasification and the producer gas has a lower heating value due to the high percentage of nitrogen in the gas. The producer gas from air gasification can be used as fuel in boiler for heating purpose or in engine/gas turbine for power generation. When the gasifying agent used is steam or pure oxygen, it is called steam gasification or oxy-fuel gasification respectively. In steam gasification or pure oxygen gasification, the heating value of producer gas is higher which can be used for conversion to methanol and gasoline. The gasification process is carried out at elevated temperatures with a controlled amount of gasifying agent.

A series of steps takes place inside a gasifier reactor which contains: drying, pyrolysis, oxidation, and reduction reactions. The final producer gas obtained from gasification of biomass is a result of important endothermic and exothermic chemical reactions which are taking place inside the gasifier. The exothermic reactions in the combustion zone provide the heat to support the endothermic reactions at different zones. Gasification processes convert biomass into a mixture of combustible gases and the energy originally present in biomass is ideally transferred into the product gas mixture. Biomass gasification can convert 60%–90% of the energy originally present in biomass to the product gas mixture. The gasification process is generally divided into four basic zones; drying zone, pyrolysis zone, oxidation/combustion zone, and reduction zone. However, in most of the cases, these zones cannot be separated, they overlap.

Drying zone: The removal of moisture is the primary operation in the drying zone. Moisture content in biomass fuels ranges from 5 to 35%. Water is evaporated and transformed into steam at temperatures above 100 °C. In this zone, the biomass sample does not decompose in any way.

Pyrolysis zone: The thermal decomposition of biomass in the absence of oxygen is known as pyrolysis. The irreversible devolatilization reaction is the major reaction in this zone. The oxidation zone provides the necessary energy for the reaction, and the temperature ranges between 200 and 500 °C. After pyrolysis, solid biomass is converted into gaseous products (H_2, CO, CH_4, H_2O, and CO_2), tar, and char.

Oxidation/Combustion zone: This zone generates the energy required for the endothermic gasification process, which includes drying, pyrolysis, and reduction. The combustion takes place between 800 and 1200 °C. As seen in the reaction below, a heterogeneous reaction occurs between oxygen and solid carbonized fuel, resulting in carbon dioxide.

$$C + O_2 \rightarrow CO_2 \tag{17.1}$$

Hydrogen in fuel reacts with oxygen and transforms into steam. The reaction is expressed as follows:

$$H_2 + 1/2\,O_2 \rightarrow H_2O \tag{17.2}$$

Reduction zone: In the absence of oxygen, a number of high-temperature chemical processes occur in the reduction zone. The water gas reaction, the Boudouard reaction, the methanation reaction, and the water gas shift reaction are the principal reactions in this zone. The fuel in this zone is highly carbonized and all volatile materials are pushed out. The temperature range in this zone varies from 600 to 800 °C. These reactions mentioned above are given below:

$$\text{Water gas reaction: } C + H_2O \rightarrow CO + H_2 \tag{17.3}$$

$$\text{Boudouard reaction: } C + CO_2 \rightarrow 2CO \tag{17.4}$$

$$\text{Methanation reaction: } C + 2H_2 \rightarrow CH_4 \tag{17.5}$$

$$\text{Water gas shift reaction: } CO + H_2O \rightarrow CO_2 + H_2 \tag{17.6}$$

The producer gas produced from gasification process can be utilized in a number of processes such as to run gas engine or gas turbine for producing mechanical or electrical power, for heating and industrial processes, for converting into liquid fuels such as gasoline or diesel through additional chemical processes, and for producing chemical and other products.

Gasification process is carried out in a number gasification reactors, each with its own advantages and limitations. Some of the common types of reactors used in gasification process are fixed-bed gasifier, fluidized bed gasifier, and entrain bed gasifier. Fixed-bed gasifier is relatively simple in construction and suitable for small to medium-scale applications. Fixed-bed gasifier can be of three types such as updraft gasifier, downdraft gasifier, and cross-draft gasifier. Among these, downdraft gasifier is preferred for practical application in small to medium-scale biomass gasification projects because of its advantages as given below:

- Higher conversion efficiency: Downdraft gasifiers are known for their higher thermal efficiency compared to other gasifier types. This means they can convert a larger proportion of the feedstock's energy into producer gas, resulting in a more efficient gasification process.
- Lower tar content: Downdraft gasifiers typically produce producer gas with lower tar content compared to other gasifier designs. Tars are complex organic compounds that can cause issues in downstream processes and equipment, so reducing their presence in producer gas is beneficial.
- Fuel flexibility: Downdraft gasifiers can handle a wide range of feedstocks, including wood chips, agricultural residues, and other biomass materials. This versatility makes them suitable for various local biomass resources.
- Simple design and operation: Downdraft gasifiers have a relatively simple design, which makes them easier to construct and operate, especially in remote locations with limited technical resources.

17.4 Result and Discussion

17.4.1 Gasification Performance

Gasification of these three biomasses is performed in the pilot-scale biomass gasifier-based power generation system. Biomass is first gasified in the downdraft gasifier. Producer gas coming out from the gasifier is cleaned and the gas samples are collected before introducing it into the gas engine. Gas samples are collected using Tedlar gas sampling bags. Once the gasifier reaches the steady state of operation, three gas samples are collected at an interval of 15 min for each biomass. Gas samples are analyzed in a gas chromatography [Chemito, Thermo Fisher SCIENTIFIC, Model No: GC 1000]. In producer gas, H_2, CO, and CH_4 are the main combustible gas components. Therefore, the quality of producer gas depends upon the composition of these three gases. Figure 17.3 shows the concentration of three combustible gases in producer gas obtained from the gasification of rice husk, wood chips, and dry leaves briquettes. It can be seen that the producer gas contains 7.51, 6.20, and 6.85% H_2; 17.81, 15.52, and 16.32% CO; and 5.23, 4.61, and 3.86% CH_4, while wood chips, rice husk, and dry leaves briquettes are used as feedstock, respectively.

During the operation of the gasifier, the biomass feed rate and the power output are measured. Table 17.3 shows the biomass feed rate per hour and the corresponding power output from the engine. It can be observed that the biomass feed rate is 9 kg/h, 10 kg/h, and 12 kg/h and the corresponding power output is 10 kW, 8.5 kW, and 12 kW for wood chips, rice husk, and dry leaves briquettes, respectively. The overall gasification efficiency can be calculated from the power output, biomass feed rate, and the calorific value of biomass as given below:

$$\eta = \frac{Power\ Output}{Biomass\ Feed\ Rate \times LHV\ of\ biomass} \tag{17.7}$$

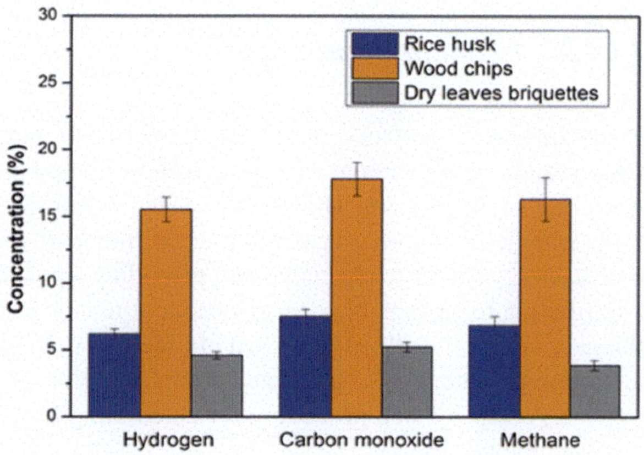

Fig. 17.3 Concentration of H_2, CO and CH_4 in producer gas from gasification of rice husk, wood chips and dry leaves briquettes

Table 17.3 Biomass feed rate and power output

Parameters	Rice husk	Wood chips	Dry leaves briquettes
Fuel feed rate (kg/h)	10	9	12
Power output (kW)	8.5	10.0	9.5
Power output per kg of biomass (kW/kg)	0.85	1.11	0.79

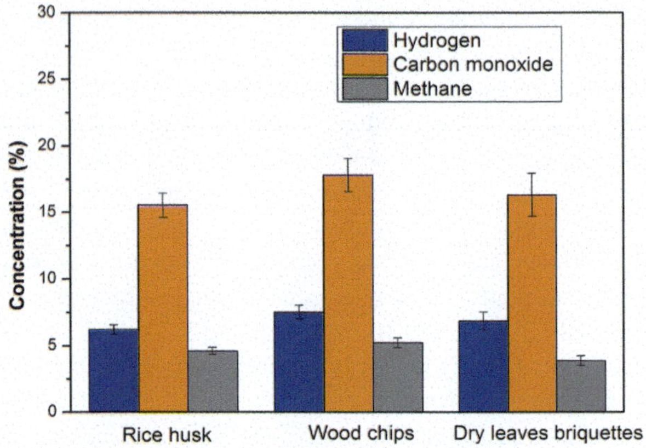

Fig. 17.4 Overall gasification efficiency obtained for rice husk, wood chips, and dry leaves briquettes

The overall gasification efficiency for three biomass materials is plotted in Fig. 17.4. It is observed that the overall efficiency varies from 18.85 to 22.99% and it is maximum for wood chips as expected. The overall gasification efficiency achieved from the present study is compared with some of the studies reported in the literature. It is found that the overall gasification efficiency reported for small-scale gasifier is in the range of 16–23% [26–28], whereas a little higher overall gasification efficiency of 25% is reported for a bigger size gasifier [29].

17.4.2 Economic Analysis

The main objective of an economic analysis of a biomass gasifier-based power generation system is to assess the financial feasibility and profitability of establishing and operating such a system. The economic analysis may help the stakeholders to determine whether the biomass gasifier plant is financially viable, competitive with alternative energy sources, and capable of providing a positive return on investment while contributing to sustainable energy production and waste management. Therefore, in this section, the Levelized Unit Cost of Electricity (LUCE) for biomass gasifier power generation system is calculated and compared with the conventional diesel generator system. A step-by-step procedure for calculating LUCE incorporating various costs is given below:

Power output by biomass gasifier:

Net electricity output per year (E_0) from a power generation system can be estimated from the equation given below:

$$E_0(kWh) = P \times (8760 \times CUF) \times (1 - a) \times (1 - l) \tag{17.8}$$

where P, CUF, a, and l are denoted as the rated power output of electricity generator, the capacity utilization factor, the fraction of generated power consumed by the auxiliaries, and the fraction of electrical losses in the local distribution network, respectively.

The economic analysis typically involves the capital cost, the operation and maintenance cost, and the fuel cost to calculate the LUCE. The detailed calculation procedure for calculating these costs is given below:

Capital cost:

The total capital cost (CP) of any power generation system can be calculated as per the equation given below:

$$\mathrm{CP} = \mathrm{CP_g} + \mathrm{CP_{eg}} + \mathrm{CP_{cw}} + \mathrm{CP_{dn}} \tag{17.9}$$

where $\mathrm{CP_g}$, $\mathrm{CP_{eg}}$, $\mathrm{CP_{cw}}$, and $\mathrm{CP_{dn}}$ are denoted as the capital cost of the gasifier, the engine–generator sets, the civil work, and the distribution network, respectively.

Total annualized cost:

The total annualized cost (AC) can be calculated from the annualized capital cost ($\mathrm{AC_C}$), the annualized operation and maintenance cost ($\mathrm{AC_{O\&M}}$), and the annualized fuel cost ($\mathrm{AC_F}$) as given below:

$$\mathrm{AC} = \mathrm{AC_C} + \mathrm{AC_{O\&M}} + \mathrm{AC_F} \tag{17.10}$$

The annualized capital cost ($\mathrm{AC_C}$) is calculated as given below:

$$AC_C = \left(C_g \times R_g\right) + \left(C_{eg} \times R_{eg}\right) + \left(C_{cw} \times R_{cw}\right) + \left(C_{dn} \times R_{dn}\right) \tag{17.11}$$

where $\mathrm{R_g}$, $\mathrm{R_{eg}}$, $\mathrm{R_{cw}}$, and $\mathrm{R_{dn}}$ represent the capital recovery factors for gasifier, engine–generator, civil works, and distribution network, respectively.

The recovery factor R can be calculated as given below:

$$R = \frac{d(1+d)^T}{(1+d)^{T-1}} \tag{17.12}$$

where d is the depreciation and T is the life of the system.

Similarly, the annualized operation and maintenance cost can be calculated as given below:

$$\begin{aligned} AC_{O\&M} = {} & \left(C_g \times m_g\right) + \left(C_{eg} \times m_{eg}\right) + \left(C_{cw} \times m_{cw}\right) + \left(C_{dn} \times m_{dn}\right) \\ & + (8760 \times CUF \times m_w \times n) \end{aligned} \tag{17.13}$$

where m_g, m_{eg}, m_{cw}, and m_{dn} represent the operation and maintenance cost factor as a fraction of the capital cost of the gasifier, the engine–generator, the civil works, and the distribution network, respectively. The manpower wage and the number of manpower required for operating the system are represented by m_w and n, respectively.

The annualized fuel cost can be calculated as follows:

$$AC_F = 8760 \times CUF \times \left(C_f \times S_f \times P\right) \tag{17.14}$$

where C_f and S_f represent the cost of fuel and the specific fuel consumption, respectively.

Therefore, the LUCE can be calculated as given below:

$$LUCE = \frac{AC}{E_0} \tag{17.15}$$

Table 17.4 shows the value of parameters used for calculating LUCE. All the costs are provided in terms of Indian Rupees (INR). LUCE of the present 10 kW biomass gasifier power generation system (BGPGS) is calculated based on the above equations. The LUCE of a diesel generator system (DGS) of same capacity is also calculated for comparison. It is found that the LUCE of the present BGPGS is 15.21 INR which is lower than the LUCE of the DGS, i.e., 16.38 INR. Although the initial investment requirement for setting up a 10 kW BGPGS is greater than that for a 10 kW DGS, the lower fuel cost in BGPGS results in a lower LUCE. Further, the price of crude oil in the international market and the diesel price in the Indian market keep on increasing. Due to this, the LUCE for DGS will keep on increasing for a long-term basis. Therefore, by considering the different factors, namely, capital cost, operation and maintenance cost, specific fuel consumption, and others, it can be concluded that the BGPPGS is the best option at present and it will be more attractive in the near future, particularly in rural areas where grid electricity is not available. A comparison of cost of electricity generation from biomass gasifier system reported in different literature is presented in Table 17.5.

17.5 Conclusion

A comprehensive study on the performance and economic viability of a 10 kW downdraft biomass gasifier-based power generation system is presented. The downdraft gasifier is integrated with a 100% producer-based gas engine for power generation. Gas cleaning and cooling systems are installed in between the gasifier and the gas engine. The gas cleaning system consists of a number of filters for cleaning the impurities from the syngas before introducing into the gas engine. The study evaluates the performance of the system with three different biomass materials; wood chip, rice husk, and dry leaves briquettes. Wood chip produces the best syngas composition with 7.51% H_2, 17.81% CO, and 5.23% CH_4 content compared to the other biomass materials. The overall gasification efficiency varies from 18.85% to 22.99%,

Table 17.4 Values of parameters for calculating LUCE [30–32]

Input parameters	Symbol	BGPGS	Diesel generator (DG)
Capital cost of gasifier	CP_g	3,91,000.00	NA
Capital cost of engine–generator	CP_{eg}	3,0,000.00	1,60,000.00
Capital cost of civil works	CP_{cw}	30,000.00	30,000.00
Capital cost of distribution network	CP_{dn}	60,000.00	60,000.00
The fraction of electrical losses in distribution network	l	0.10	0.10
Auxiliary power consumption	a	0.10	0.10
Capacity utilization factor	CUF	0.8	0.8
Recovery factor for civil work	R_{cw}	0.11	0.11
Recovery factor for distribution network	R_{dn}	0.11	0.11
Operation and maintenance cost factor as a fraction of the capital cost of gasifier	m_g	0.05	NA
Operation and maintenance cost factor as a fraction of the capital cost of the engine–generator	m_{eg}	0.1	0.1
Operation and maintenance cost factor as a fraction of the capital cost of civil work	m_{cw}	0.02	0.02
Operation and maintenance cost factor as a fraction of the capital cost of distribution network	m_{dn}	0.03	0.03
Manpower wages per hour	m_w	20	20
Number of manpower requirements	n	1	1
Cost of fuel (INR per kg for biomass and INR per L for diesel)	C_f	6	96
Specific fuel consumption (kg/kWh for biomass and L/kWh for diesel)	S_f	1.3	0.11

Table 17.5 Comparison of cost of electricity generation from biomass gasifier systems

Sl No	System type	Capacity	Cost per kWh	Refs.
01	Downdraft gasifier integrated with gas engine	10 kW	15.21 ₹	Present work
02	Dual fuel biomass gasifier power project	5–40 kW	12.46–22.55 ₹	Nauni et al. [30]
03	Downdraft gasifier integrated with an IC engine	10–100 kW	0.09–0.16 $	Susanto et al. [28]
04	A small-scale biomass gasification power plant in Brazil	100 kW	0.46 €	Fracaro et al. [29]

depending on the biomass material used. The maximum efficiency is achieved for the gasification with wood chips as biomass. These efficiency values are considered acceptable for biomass-to-power applications, indicating the feasibility of the technology. An economic analysis is conducted to determine the per unit cost of electricity generation from the biomass gasifier system and compare it to a diesel generator of the same capacity. While the initial setup cost of the biomass gasifier system is higher, the study finds that the levelized unit cost of electricity (LUCE) for the gasifier system is lower (INR 15.21) than that of the diesel generator (INR 16.38). This suggests that, despite the higher upfront investment, the biomass gasifier system offers a more cost-effective and sustainable long-term solution for electricity generation.

Moreover, the price of crude oil in the international market and the diesel price in the Indian market keep on increasing. Due to this, the LUCE for diesel generator system is expected to keep on increasing for a long-term basis. Therefore, it can be said that for distributed biomass supply, setting of pilot-scale biomass gasifier is more useful. It can provide employment opportunities to local communities, thereby improving their quality of life. Additionally, it offers a more sustainable and environmentally friendly energy solution compared to fossil fuels. Hence, the findings of the study can be valuable for policymakers, energy companies, and researchers. Policymakers can consider supporting the adoption of biomass gasification technologies in rural areas to address energy needs, reduce environmental impact, and create local economic opportunities.

Acknowledgements The authors express sincere gratitude to the Director, CSIR—Central Mechanical Engineering Research Institute, Durgapur for his continuous support and encouragement. Authors also acknowledge the valuable contributions of the technical and analytical staff of CSIR-CMERI for the construction, operation, and performing measurements of the pilot-scale biomass gasification-based power generation system.

References

1. Loha C, Chatterjee PK, Chattopadhyay H (2021) Performance of fluidized bed steam gasification of biomass – modeling and experiment. Energy Convers Manage 52:1583–1588
2. Loha C, Chattopadhyay H, Chatterjee PK (2014) Threedimensionalkinetic modelling offluidized bed biomass gasification. Chem Eng Sci 109:53–64
3. Pattanayak S, Loha C, Hauchhum L, Sailo L, Mishra L (2021) Experimental investigation on pyrolysis kinetics, reaction mechanisms and thermodynamic parameters of biomass and tar in N2 atmosphere. Sust Energy Technol Asses 48:101632
4. Pattanayak S, Hauchhum L, Loha C, Sailo L, Feasibility study of biomass gasification for power generation in Northeast India. https://doi.org/10.1007/s13399-021-01419-8
5. Lee U, Balu E, Chung JN (2013) An experimental evaluation of an integrated biomass gasification and power generation system for distributed power applications. Appl Energy 01:699–708
6. Daniel Martinez J, Mahkamov K, Andrade RV, Silva Lora EE (2012) Syngas production in downdraft biomass gasifiers and its application using internal combustion engines. Renew Energy 38:1–9

7. Henriksen U, Ahrenfeldt J, Jensen T, Gobel B, Bentzen J, Hindsgaul C (2006) The design, construction and operation of a 75 kW two-stage gasifier RID B-2579-2012. Energy 31:1542–1553
8. Zainal ZA, Rifau A, Quadir GA, Seetharamu KN (2002) Experimental investigation of a downdraft biomass gasifier. Biomass Bioenerg 23:283–289
9. Sharma AK (2009) Experimental study on 75 KW(th) downdraft (biomass) gasifier system. Renew Energy 34:1726–1733
10. Olgun H, Ozdogan S, Yinesor G (2011) Results with a bench scale downdraft biomass gasifier for agricultural and forestry residues. Biomass Bioenergy 35:572–580
11. Coronado CR, Yoshioka JT, Silveira JL (2011) Electricity, hot water and cold water production from biomass. Energetic and economical analysis of the compact system of cogeneration run with woodgas from a small downdraft gasifier. Renew Energy 36:1861–1868
12. Huang Y, Wang YD, Rezvani S, McIlveen-Wright DR, Anderson M, Hewitt NJ (2011) Biomass fuelled tri-generation system in selected buildings. Energy Convers Manage 52:2448–2454
13. IEA. India Energy Outlook (2021) World Energy Outlook Special Report, pp 1–250. www.iea.org
14. Buragohain B, Mahanta P, Vijayanand S, Moholkar VS (2010) Biomass gasification for decentralized power generation: the Indian perspective. Renew Sustain Energy Rev 2010(14):73–92
15. SinghJ GuS (2010) Biomass conversion to energy in India–a critique. Renew Sustain Energy Rev 2010(14):367–1378
16. MNRE. Bioenergy Overview. Ministry of New and Renewable Energy. Govt. of India. www.https://mnre.gov.in/bio-energy/current-status
17. Bisht AS, Thakur NS (2019) Small scale biomass gasificationplants for electricity generation in India: resources, installation, technical aspects, sustainability criteria & policy. Renew Energy Focus 28:112–126
18. Salem AM, Zaini IN, Paul MC, Yang W (2019) The evolution and formation of tar species in a downdraft gasifier: numerical modelling and experimental validation. Biomass Bioenerg 130:105377
19. James AM, Yuan W, Boyette MD, Wang D (2015) The effect of air flow rate and biomass type on the performance of an updraft biomass gasifier. Bio Res 10:3615–3624
20. Abhijeet P, Swagathnath G, Rangabhashiyam S, Asok Rajkumar M, Balasubramanian P (2020) Prediction of pyrolytic product composition and yield for various grass biomass feedstocks. Biomass Convers Biorefinery 10:663–674
21. Varma AK, Thakur LS, Shankar R, Mondal P (2019) Pyrolysis of wood sawdust: effects of process parameters on products yield and characterization of products. Waste Manag 89:224–235
22. Kurkela E, Kurkela M, Ilkka HI (2023) Pilot-scale development of pressurized fixed-bed gasification for synthesis gas production from biomass residues. Biomass Convers Bioref 13:6553–6574
23. Nadeem TB, Siddiqui M, Khalid M, Asif M (2023) Distributed energy systems: a review of classification, technologies, applications, and policies. Energ Strat Rev 48:101096
24. Chambon CL, Karia T, Sandwell P, Hallett JP (2020) Techno-economic assessment of biomass gasification-based mini-grids for productive energy applications: the case of rural India. Renew Energy 154:432–444
25. Uisung Lee (2014) Experimental and numerical evaluation of thermochemical conversion of biomass and municipal solid waste using a high temperature steam gasification
26. Situmorang YA, Zhao Z, Yoshida A, Abudula A, Guan G (2020) Small-scale biomass gasification systems for power generation (<200 kW class): a review. Renew Sustain Energy Rev 117:109486
27. Balat M, Ayar G (2005) Biomass energy in the world, use of biomass and potential trends. Energy Sources 27:931–940
28. Ding L, Yang M, Dong K, Vo DVN, Hungwe D, Jiahan YJ, Ryzhkov A, Kunio YK (2022) Mobile power generation system based on biomass gasification. Intern J Coal Sci Technol 9:34

29. IEA (2007) Biomass for power generation and CHP. IEA Energy Technology Essential 1–4. ETE03
30. Susanto H, Suria T, Pranolo SH (2018) Economic analysis of biomass gasification for generating electricity in rural areas in Indonesia. Mater Sci Eng 334:012012
31. Fracaro GPM, Souza SNM, Medeiros M, Formentini DF, Marques CA (2011) Economic feasibility of biomass gasification for small-scale electricity generation in Brazil. World Renew Energy Cong 295–302
32. Nouni MR, Mullick SC, Kandpal TC (2007) Biomass gasifier projects for decentralized power supply in India: a financial evaluation. Energy Policy 35:1373–1385

Part VI
Socio-economic Impact Studies

Chapter 18
Role of Solar Energy in the Development of the Indian Economy

Akash Dilip Kamble, Samar Das, Vijaya, Biraj Das, Urbashi Bordoloi, Puja Hazarika, and Pankaj Kalita

Abstract Solar energy-based power generation systems play a pivotal role in bolstering the Indian economy and contributing to India's energy security and independence. With reduced dependence on fossil fuel imports, solar power plants mitigate the risks associated with price volatility and supply disruptions in the conventional energy sector. By displacing fossil fuel-based electricity generation, solar power plants play a crucial role in mitigating climate change. India has implemented various solar-based power plants as part of its renewable energy initiatives. These power plants utilize different technologies and configurations to harness solar energy for electricity generation. Thus, the chapter explores the multifaceted role of solar energy by incorporating different solar-based technologies in developing the Indian economy. The chapter also enlightens their impact on job creation for economic growth, energy access in rural development as well as industrial development and technology transfer. Furthermore, government initiatives, along with supportive policies which possess significant importance in promoting solar energy adoption, attracting investments and driving the growth of the solar industry in India were also thoroughly covered because of their potential to bring transition in the sustainable development of the country.

Keywords Solar power plant economics · Solar economic growth · Solar energy adoption · Solar energy government initiatives

---

A. D. Kamble · S. Das · Vijaya · P. Hazarika · P. Kalita (✉)
School of Energy Science and Engineering, IIT, Guwahati, India
e-mail: pankajk@iitg.ac.in

B. Das · U. Bordoloi
School of Agro and Rural Technology, IIT, Guwahati, India

S. De et al. (eds.), *Challenges and Opportunities of Distributed Renewable Power*, Energy, Environment, and Sustainability, https://doi.org/10.1007/978-981-97-1406-3_18

18.1 Introduction

18.1.1 World Energy Scenario

In today's rapidly changing scenario of high fossil fuel prices, the gap between energy demand and energy supply has only been expanding. Countries are facing the serious threat of energy shortage due to surging energy consumption by the ever-increasing global population rate and industrialization. The global energy consumption scenario from the time period of 1978–2022 reports to be from 135 to 2414 TWh with a rise in energy consumption per capita from 17.7 to 21.03 MWh. With reference to economy, from the period of 1978 to 2022, the global gross domestic product's (GDP) annual average growth rate and GDP per capita annual average growth are reported to be 2.9 and 1.5%, respectively [1]. Global energy consumption, measured in exajoules (EJ), for various fuel sources from 2000 to 2020, based on data from BP's 2021 report, is given in Fig. 18.1. Additionally, the average annual growth rates for each fuel source during the period 2015 to 2020 are indicated within parentheses in Fig. 18.1 [2]. The major cause that led to the increase in annual global energy consumption is attributed to use of fossil fuels. As per the report of 2022, 81.8% of energy consumption still comes from sources powered by fossil fuel include coal, gas and oil [3]. Additionally, the fossil fuel exploitation possesses consequences on environmental degradation causing global warming, forest fires, droughts and storms to have doubled since 1980. The ten hottest years globally have been reported since 2004 and the timeframe 2015–2020 recorded the five hottest years worldwide [4]. Thus, of deep concerns to rising energy demands and consequences on environmental degradation, the focus of energy sources has been shifted to renewable energy. The share of renewable energy demand grew by an average of 4.7% each year in the period of 2010–2020. IEA reports electricity production alone from RES accounts for 29% in 2020, an increase from 3% in 2019 [5]. Of all the renewable energy sources available, solar PV and wind energy hold two third of renewable energy growth. Solar PV-generated power increased by 26% in 2022, contributing to a total of almost 1300TWh. This surge in power generation from PV surpassed power generation form wind for the first time, thus signifying solar PV as the largest renewable energy source for meeting the energy demand. The solar energy contribution is projected to escalate, reaching 25% of global renewable electricity generation by 2030. Similarly, wind power has also seen a remarkable 14% increase in electricity generation in the year 2022 and meeting the ambitious net zero emissions by 2050 requires reaching 7,400 TWh wind power by 2030. Biomass and bioenergy sectors have also made significant strides, including an impressive 8% boost in bioenergy power generation in 2020 [6]. Further, developments in the energy sector owing to substantial changes by emerging technologies indicate the next four decades to be distinct from past in terms of shift in global energy consumption pattern.

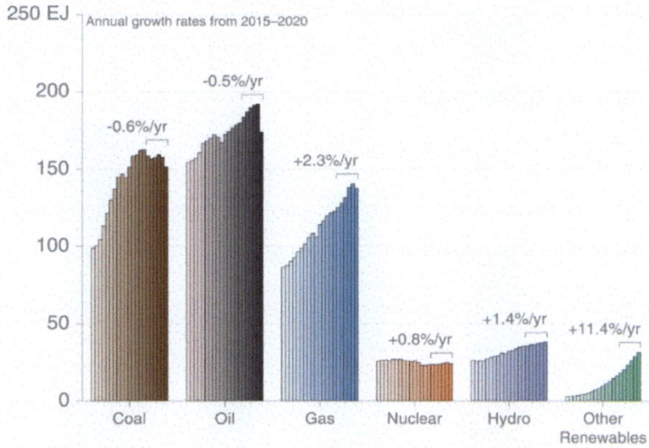

Fig. 18.1 Global energy consumption for various fuel sources from 2000 to 2020 [2]

18.1.2 Indian Energy Scenario

In alignment with the global trend, India has also transitioned towards utilization of renewable energy sources to meet the energy requirements. Back in 1974, electricity demand in India was hugely met by coal- and gas-powered power plants. Electricity demand met by an installed capacity of 16.7GW had a share of around 51% (more than half) by using coal as fuel of which, renewables accounted to nearly 0% [7]. However, the 1970 oil crisis served as the turning point for the Indian government to shift to RES. Subsequently, industrialization in 1991s also led to involvement of the private sector towards industrial development and growth of renewable energy sector [8]. With a current outstanding population of 1.36 billion and rapidly expanding economy, the high rise in energy demand meet now relies on utilization of several renewable energy sources without its sole dependence on fossil fuel-powered power plants. Currently, India has emerged to be the fourth largest producer of RE based on installed capacity, including large hydro [9]. Over the past 9 years, India has witnessed a remarkable 396% increase in its installed non-fossil fuel capacity, to 179.322 GW, accounting for approximately 43% of the country's total energy capacity as of July 2023 [10]. India holds a relatively lower position for per capita electricity consumption in the world at 1.29 MWh compared to China at 6.19 MWh [11]. This makes India implement more power generation capacity in the next two decades and match with current power capacity of the entire European Union. As a measure, multiple policies have been actively implemented in view of meeting energy requirements and reducing the effects of climate change. The Nationally Determined Contribution (NDC) has targeted to increase the power generation from RES to 40% by 2030 with a percentage holding at 16% in the year 2020. The commitment additionally involves reduction in greenhouse gas emission to 33–35% from the 2005 CO_2 emission levels. Significant progress in the addition of renewable energy sources to the power sector has been noticed since 2017 as shown in Fig. 18.2.

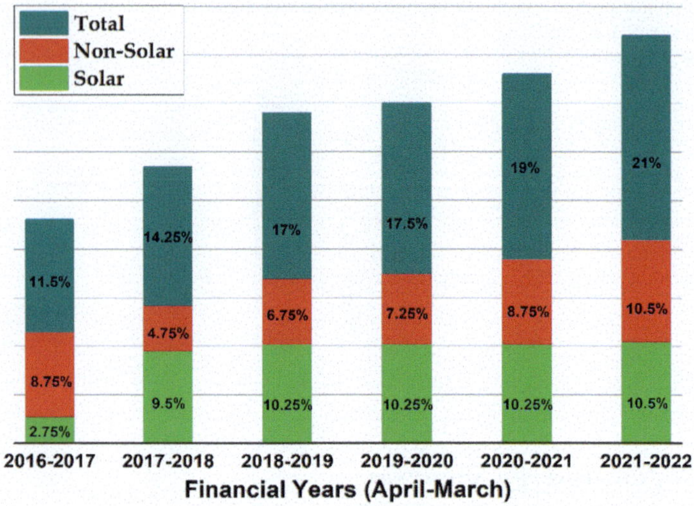

Fig. 18.2 Share of renewable energy in India's power consumption [12]

18.1.3 Solar Energy's Dominance Over Other Renewables

With immense resource capabilities, technological advancements, and cost efficiency, power generation from Solar Photovoltaic (PV) has expeditiously become the most economically viable solution. India at COP26 held at Glasgow, introduced the following schemes of—(i) Aims to achieve 500GW of renewable energy capacity by 2030. Renewable energy capacity as of 2019 holds at 100GW, (ii) Targets to meet 50% of its energy needs from RES by 2030, (iii) CO_2 emission reduction by 1bn tonnes by 2030, (iv) Reduce carbon intensity of the economy by 45% compared to 2005 levels by 2030 and (v) Reach the goal of net zero emissions by 2070 [13].

The solar sector has witnessed exponential expansion, driven by falling solar panel prices, technological advancements and a conducive policy environment, with an installed capacity of 67.07 GW as of July 2023. The large-scale solar parks (57) act as vital centres for large-scale solar power generation (39.28 GW), promoting energy efficiency and grid integration while reducing dependence on fossil fuels. Furthermore, the government's active promotion of rooftop solar installations complements these efforts, accelerating the expansion of solar capacity [14]. Moreover, private sector participation and falling solar technology costs have further bolstered the growth of solar energy [15]. This achievement showcases India's commitment to embracing solar energy as a major component of its renewable energy portfolio. Among the array of renewable energy alternatives, solar power consistently shines brightest in terms of its sustainable and potential attributes. It prevails as the prime alternative energy source to generate heat and electricity, ensuring no adverse effects on the environment.

Thus, this chapter aims to provide a comprehensive foundational perspective on solar energy and its role in shaping the future Indian economy. The chapter discusses solar photovoltaic and thermal technologies, their accessibility, existing status, key highlights from impactful research in the solar PV field, policies for advancement and the prospective trajectory of solar energy alternatives within India. The first segment of the chapter gives a brief about solar energy's dominance over other renewables and its increased potential in commercial market over the period of time. Further, the second and third segments of this book chapter is entitled in providing a brief overview of various solar technologies that include—photovoltaic systems, solar thermal technology, photovoltaic–thermal systems and solar concentrators along with their applications. These technologies harness the sun's radiant energy in distinct ways, with PV cells converting sunlight directly into electricity and solar concentrator systems using mirrors or lenses to focus sunlight for heat generation. The fourth segment narrates the pivotal role of government initiatives in propelling India's economy through solar energy. The fifth section consists of chronological advancements made in the field of solar power technology in India with a particular focus on installed capacity and upcoming prospects. The harnessed energy is then utilized across diverse sectors which underlines the versatility of solar power applications in residential, industrial and agricultural and space applications. This segment also enlightens on the maintenance of solar power plants along with technological innovations and the future prospects of Indian solar power development. Concluding this exposition, the sixth section includes the societal impacts of India's solar energy technological advancements in terms of solar awareness in rural areas, impact on job creation scenario and the overall CO_2 emission reduction.

18.2 Different Solar Technologies

Solar technologies can be predominantly categorized into Photovoltaics (PV), solar thermal, Photovoltaic–Thermal (PV/T) and concentrating collectors [16]. Photovoltaic systems are designed to transform solar energy into electricity, while solar thermal systems harness solar energy to produce heat. PV-T systems are a combination of photovoltaic and thermal systems. They are capable of converting solar energy into both electricity and heat. The concentrating collectors, on the other hand, efficiently capture solar irradiation and concentrate it to a focal point, generating substantial heat that can subsequently be converted into electricity.

18.2.1 Photovoltaic Systems

The photovoltaic systems mainly consist of silicon and germanium-based solar cells. It entails monocrystalline, polycrystalline as well as GaAs-based solar cells.

Currently, it holds a market share of 90% [17]. The PV technology can be further classified as silicon-based and non-silicon-based technologies [18].

18.2.1.1 Silicon-Based PV Technologies

The solar panels are generally made of layers of silicon doped with p-type (boron) and n-type (phosphorous) impurities [19]. The incidence of solar radiation on the panels results in release of electrons and the phenomenon is called photovoltaic effect. The flow of these electrons across a conductor results in electricity which is used for several on-grid and off-grid applications. Depending on the type of silicon used, PV cells can be further classified as monocrystalline, polycrystalline and amorphous or thin film as shown in Fig. 18.3 [18]. The monocrystalline cells are made out of a single crystal of silicon and are believed to offer the highest efficiency as compared to other types of PV cells. The efficiency of monocrystalline cells ranges from 15 to 20% [16]. However, the cost of monocrystalline PV systems is high owing to the complexity of the manufacturing process.

Multicrystalline or polycrystalline PV cells are developed by fine-layered ingots of silicon. The process of manufacturing multicrystalline cells is less complex and as a result, the system costs are low. The main limitation is however the low efficiency of these cells (12–15%). There have been several efforts to improve the efficiency of these cells and a recent reported efficiency is as high as 20%. The amorphous PV cells or thin-film cells are developed by spreading thin layer of amorphous silicon material on a surface [16]. The thin-film cells are flexible and can be bent in various shapes or directions as per the requirement. The manufacturing process is simple and the cost is lesser than monocrystalline PV. Thin-film PV systems are also less prone to temperature-dependent fluctuations in efficiency as compared to crystalline silicon-based PVs. The efficiency of these systems is generally in the range of 12–15% [20].

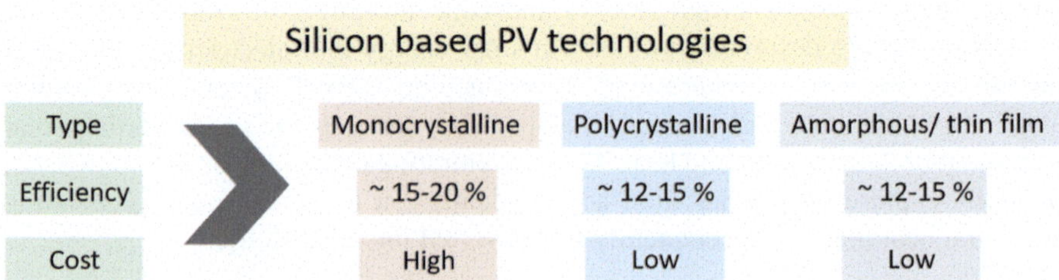

Fig. 18.3 Silicon-based PV technologies [18]

18.2.1.2 Non-silicon-Based PV Technologies

It comprises of materials such as Copper–Indium–Selenium (CIS), Copper–Indium–Gallium–Selenium (CIGS) and Cadmium Telluride (CdTe) [16, 18]. The manufacturing cost of these systems is lesser than crystalline silicon-based system and the efficiency is higher than amorphous silicon-based PV. While the market shares of CIS and CIGS are limited by the supply of indium, the toxic nature of Cd and the scarce availability of Te has limited the market share of CdTe [20].

18.2.1.3 Polymer-Based PV Technology

Further developments in PV technology aimed at improving the efficiency of cells by incorporating nanomaterials. As a result, upcoming solar cell generations consisting of dye-sensitized cells, quantum dot cells, organic and polymeric solar cells, perovskite cells and multijunction cells found a place under PV technology. The recent addition to solar cells also referred to as 'inorganics in organics' cells [21]. It consists of inorganic metal oxides and nanoparticles with organic materials such as carbon nanotubes, graphene, etc. These cells are flexible and cost lesser than cells in the previous generations [18]. To further enhance the efficiency of crystalline solar cells, manufacturers are shifting from aluminium back surface field (Al-BSF) cells to passivated emitter and rear cell (PERC) and silicon heterojunction (SHJ) cell technology. Another variation of PERC cell, Tunneling Oxide-Passivated Contact (TOPCon) is predicted to rapidly grow in the coming years owing to their higher efficiency [22].

18.2.1.4 Characteristics of PV Cells

The performance of a solar cell is determined by it's current–voltage (I–V) plot as shown in Fig. 18.4. The typical I–V characteristic plot consists of short-circuit current (I_{sc}), open-circuit voltage (V_{oc}), maximum current (I_{max}) and maximum voltage (V_{max}). The power (P) delivered by solar cell is plotted as a function of voltage. The delivered power is maximum when current and voltage correspond to I_{max} and V_{max}, respectively. A power electronic controller, called the maximum power point tracker (MPPT) is used for the purpose of tracking I_{max} and V_{max}. In addition to above-mentioned parameters, another parameter, namely fill factor (FF), is important in the context of PV cells as it determines the power delivered by the cell. Mathematically, it is expressed as the product of I_{sc} and V_{oc}. FF value lies between 0 and 1, 1 being ideal. All commercial cells have FF > 0.75 [21].

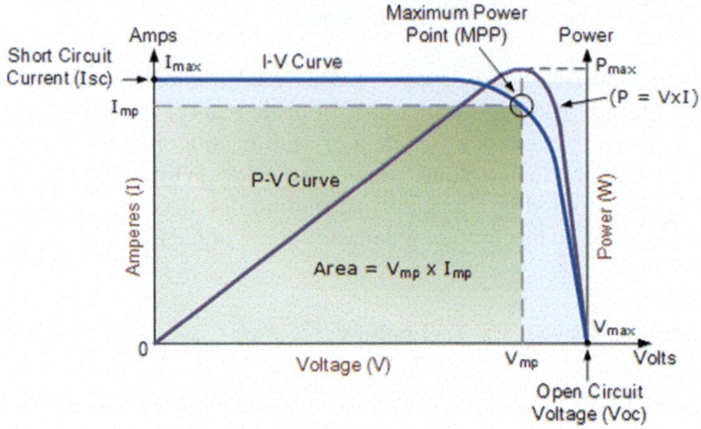

Fig. 18.4 Typical I–V and power characteristics of PV cells [23]

18.2.2 Solar Thermal Technology

Solar thermal panels use solar energy to heat up fluids such as water or air and thereby convert solar energy to thermal energy. The solar thermal collectors can be broadly categorized into (a) flat plate collectors and (b) evacuated tube collectors as shown in Fig. 18.5 [16].

The flat plate collectors (FPC) utilize flat plates inside a glass or plastic casing to absorb heat from the sun as shown in Fig. 18.6. The evacuated tube collectors (ETC),

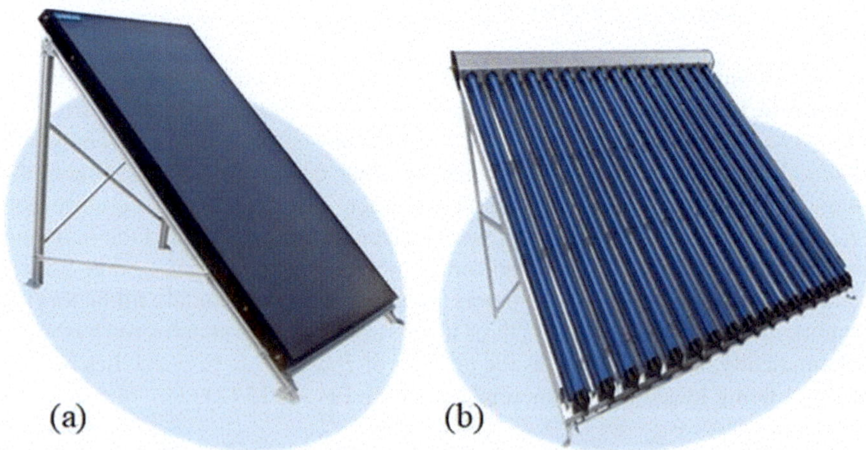

Fig. 18.5 Solar thermal technology: **a** Flat plate collectors, **b** Evacuated collectors [16]

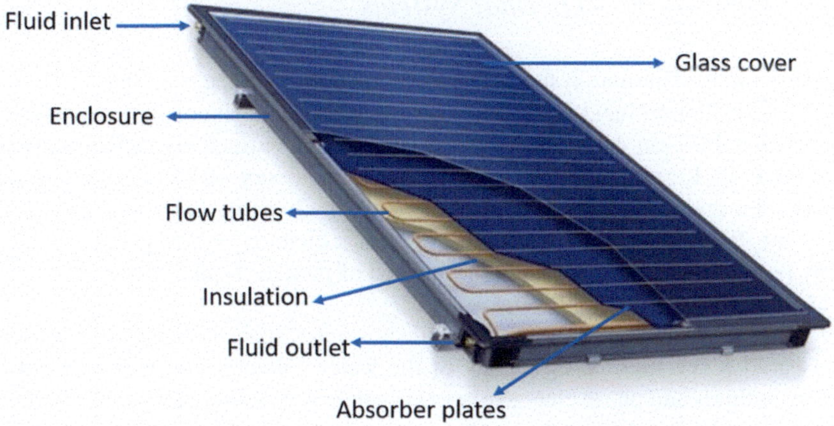

Fig. 18.6 Schematics of flat plate collector [24]

shown in Fig. 18.7, on the other hand, consist of tubes evacuated of air through which heat transfer medium (air or water) is passed.

The metallic absorbers in the evacuated tubes absorb heat and, in turn, heat up transfer fluid. Evacuated tube collectors are used for medium-temperature applications where the requirement of temperature is in the range of 60–80 °C [24].

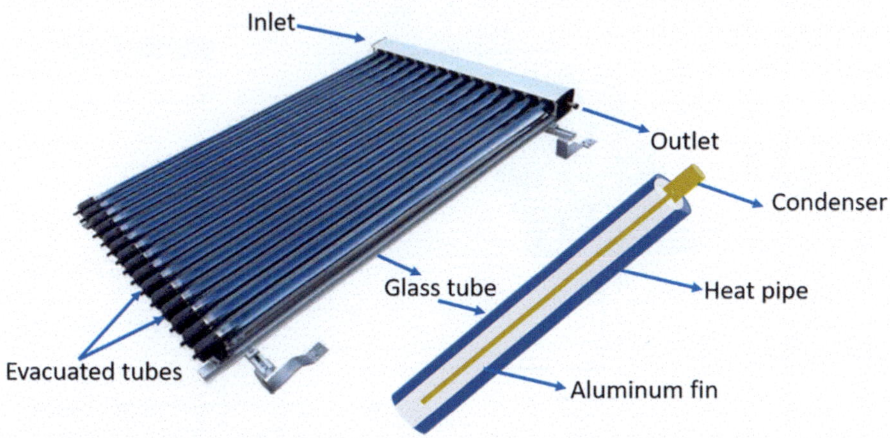

Fig. 18.7 Schematics of evacuated tube collector [24]

18.2.3 *Photovoltaic–Thermal Systems*

The Photovoltaic–Thermal system (PV-T) is developed on the idea that more than 50% of solar radiation incident on PV is converted to heat. Exposure to excess heat for prolonged duration damages solar cells and thereby affects the performance of PV panels [25]. Kern and Russel came up with the theory of PV-T, a system that converts solar energy to electricity as well as thermal energy simultaneously [26]. The hybrid PV-T systems improve the electrical efficiency of PV panels by up to 15% by removing excess heat from the surface of panels. Typically, air or water is used as transfer medium.

PV-T can be effectively utilized to provide hot water, space heating and electricity to households. The thermal efficiency of the water collector can be improved by adding nanofluids and phase change materials whereas the efficiency of air collectors can be increased by modifying the air channel by adding fins, thin films, porous media, etc. [25].

Depending on the type of application, PV-T systems can be designed as encapsulated, glazed or insulated type [24]. The PV-T system when integrated with buildings is referred to as building-integrated Photovoltaic–Thermal (BIPV-T) system. It is more efficient and occupies less space than PV or solar thermal alone [18, 24]. The general working and schematics of PV/T system are shown in Fig. 18.8 [27, 28].

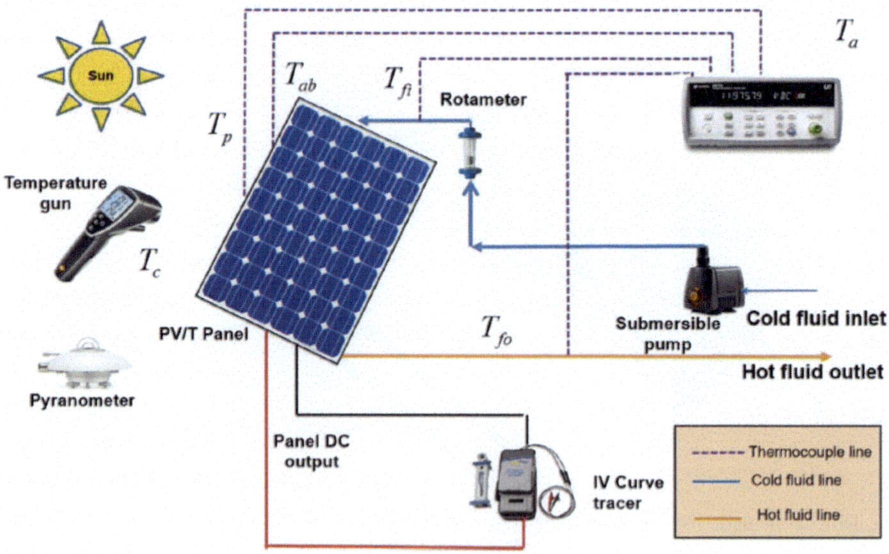

Fig. 18.8 Schematics of photovoltaic thermal system [28]

18.2.4 Solar Concentrators

Solar concentrators are promising technology for high-temperature applications. The concentrating solar collectors collect the solar radiation and focus it on a smaller area than that of the receiver using optical concentrators such as lens or mirrors [16, 29]. The concentrators can be further categorized into flat solar concentrators, compound solar concentrators, parabolic troughs, heliostat field collectors, Fresnel lenses and parabolic dish collectors [18]. Some of these are described below.

18.2.4.1 Parabolic Trough Concentrators

Parabolic trough concentrators consist of parabolic-shaped collectors which reflect the incident solar radiation onto central heat pipe or absorbing tube which acts as a receiver as shown in Fig. 18.9. The absorbed solar radiation heats up the fluid flowing through central heat pipe. A single-axis tracking system is used to track the movement of the sun and a temperature up to 400 °C can be achieved.

18.2.4.2 Linear Fresnel Collectors

Linear Fresnel collectors consist of long strips of mirrors placed close to ground with absorber tube a few metres above the mirrors as shown in Fig. 18.10. The mirrors reflect the sun rays onto absorber tube and heat up the transfer fluid flowing through the tube. To obtain higher temperature, parabolic dish collectors are used.

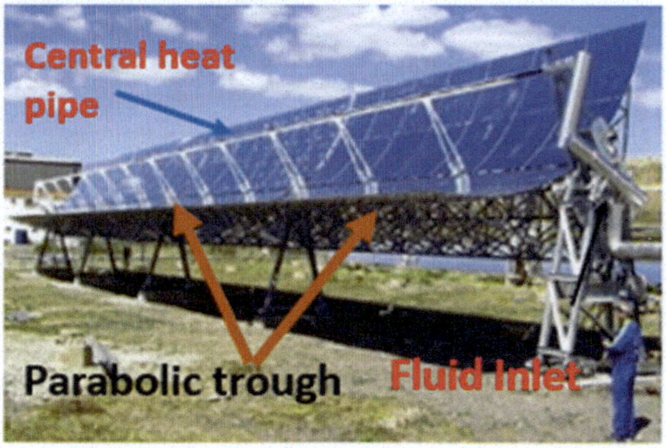

Fig. 18.9 Parabolic trough collector [30]

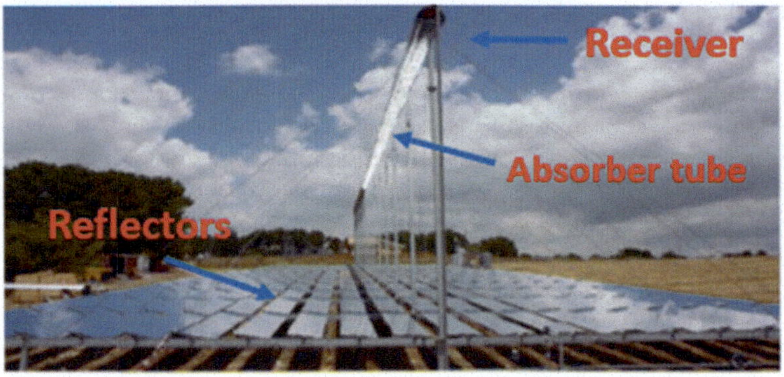

Fig. 18.10 Linear Fresnel collectors [31]

Fig. 18.11 Parabolic dish collector [32]

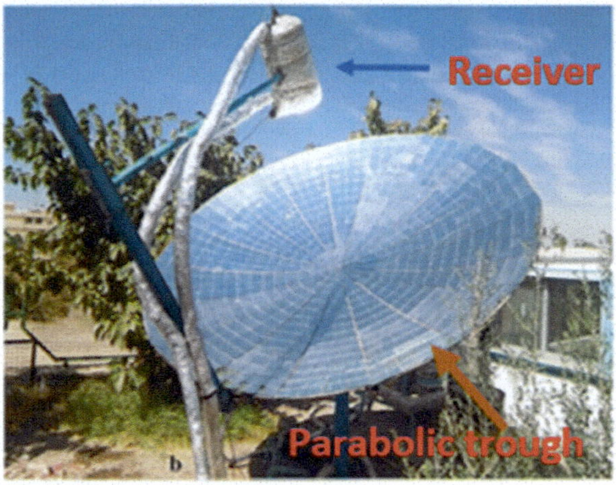

18.2.4.3 Parabolic Dish Collectors

Parabolic dish collectors consist of parabolic dish-shaped mirrors that concentrate the solar radiation towards the receiver placed on the focal point of the collector as shown in Fig. 18.11. The collectors are integrated with dual-axis tracking system for efficient tracking of the sun. The parabolic dish collectors can attain a temperature of up to 1000 °C. The advanced collectors are equipped with sterling engines for the conversion of heat to electricity.

18.2.4.4 Central Receiver System

A heliostat field consists of numerous mirrors which focus the sun's rays onto a central tower which houses a boiler as shown in Fig. 18.12. The concentrated energy

Fig. 18.12 Solar power tower with heliostats [33]

is used to generate steam which in turn is used to spin a turbine and thereby generate electricity.

The operating temperature of different types of solar collectors is summarized in Table 18.1.

Table 18.1 Operating temperature of different types of solar collectors [34]

Collector type	Operating temperature (°C)
Flat plate	30–80
Evacuated tube	50–200
Parabolic trough	60–300
Linear fresnel	60–250
Parabolic dish	100–500
Solar power tower	150–2000

18.3 Applications of Solar Technologies

Solar photovoltaic (PV) panels are used to convert sunlight directly into electricity. Solar power can be harnessed to generate electricity for residential, commercial and industrial purposes. Based on application, PV technologies can be classified into grid-connected or on-grid and off-grid systems, providing clean and renewable energy. An off-grid application of photovoltaic (PV) refers to a system that uses solar panels to generate electricity in a standalone manner, independent of the traditional power grid. Off-grid applications in India mainly comprise of street lighting, home lighting, solar lanterns, solar pumps and standalone power plants which are shown in the below mentioned Fig. 18.13.

On-grid application of PV refers to the use of photovoltaic systems that are connected to the utility grid. One of the key advantages of on-grid PV systems is that they enable two-way power flow. During times when the PV system generates more electricity than is being consumed locally, the excess power can be supplied to the grid. Conversely, when the PV system is not generating enough electricity to meet the demand, power can be drawn from the grid. On-grid PV systems can contribute to the stability of the electrical grid by diversifying the sources of electricity generation and reducing strain during peak demand periods which is shown in Fig. 18.14.

- **Controller protection interface in solar PV power plants**

The evolution of charge controllers in solar PV systems has significantly improved the efficiency and reliability of renewable energy generation. One notable advancement is the integration of Maximum Power Point Tracking (MPPT) technology which is shown in Fig. 18.15. MPPT controllers continuously monitor and adjust the operating conditions of solar panels to ensure they operate near their peak power output, regardless of changing environmental factors. This development has been widely recognized for its ability to enhance energy harvest and overall system performance.

The adoption of advanced multi-port DC–DC converters has brought substantial improvements to solar PV systems. These converters enable the concurrent management of both solar PV and battery sources, optimizing charging and discharging processes while efficiently utilizing solar-generated electricity. Research studies in the field highlight how these converters enhance system adaptability and grid integration.

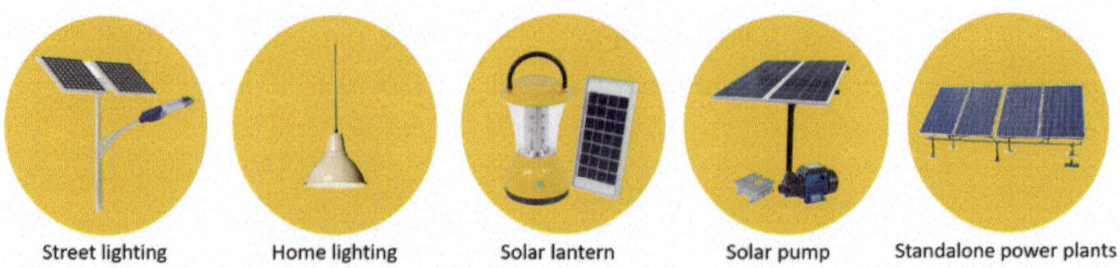

Fig. 18.13 Off-grid applications of PV [35–37]

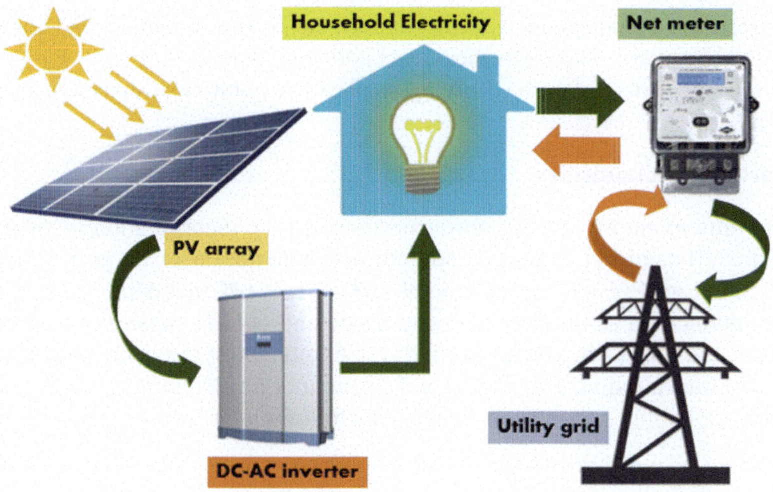

Fig. 18.14 On-grid fulfilling the domestic and grid power requirements

Fig. 18.15 Components of grid-connected solar PV power plants

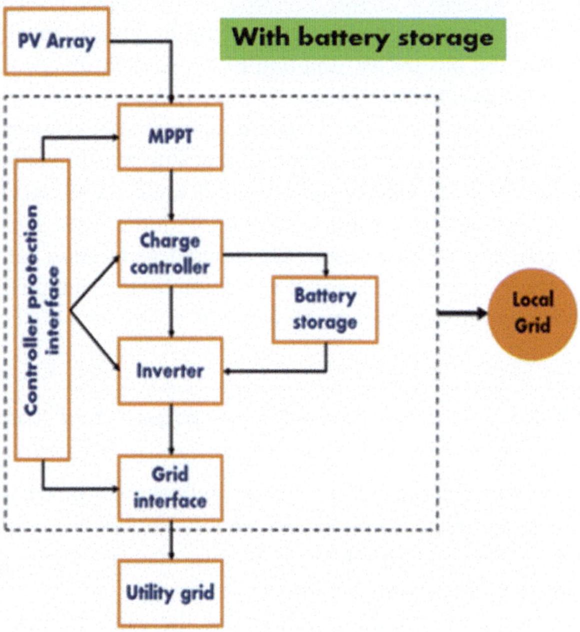

Advanced communication interfaces allow remote monitoring and control, while compatibility with various battery chemistries ensures tailored charging profiles. Integration with energy management systems enables coordinated control of multiple energy resources. Enhanced durability, load control features, automatic diagnostics and efficient heat management further contribute to the reliability and longevity of charge controllers. These advancements collectively empower solar PV systems to perform optimally, adapt to evolving technologies and meet diverse user needs [38, 39].

- **Smart grid technologies**

The intermittent nature of solar power necessitates the incorporation of smart grid system for efficient operation and smooth grid integration of on-grid solar technology. Smart grid system refers to an electricity network that deploys information and communication technology to create a safe and reliable network of interactive systems in real time [40]. The interactive system refers to systems involved in power generation, transmission, distribution and consumption [41]. Some of the features of smart grid technology adopted in India are as follows [42]:

- *Advanced meter infrastructure:* AMI refers to a modern system of utility meters (typically electricity, gas or water) that are equipped with advanced communication capabilities.
- *Peak load management:* Peak load management involves strategies and technologies to reduce or shift electricity consumption during peak demand periods.
- *Power quality management:* Power quality management focuses on ensuring that the electricity supplied to consumers meets specific voltage and frequency requirements.
- *Outage management:* Outage management systems help utilities detect, respond to and recover from power outages more efficiently.
- *Micro grids:* Microgrids are small-scale, localized energy systems that can operate independently or in conjunction with the main grid.
- *Distributed generation:* Distributed generation refers to the production of electricity from various small-scale sources distributed throughout the grid, as opposed to centralized power plants.

18.3.1 Solar-Powered Agriculture and Solar Water Pumping

Solar water pumps for irrigation and solar-powered agro-processing equipment are some examples of how solar technologies are being used in agriculture. They assist farmers in lessening their reliance on traditional energy sources and improving the effectiveness of agricultural operations.

Solar-powered water pumps use the sun's energy to move water for irrigation, livestock water intake and public water distribution. These systems are especially helpful in distant locations with inconsistent or inadequate grid electricity which is shown in Fig. 18.16 [37].

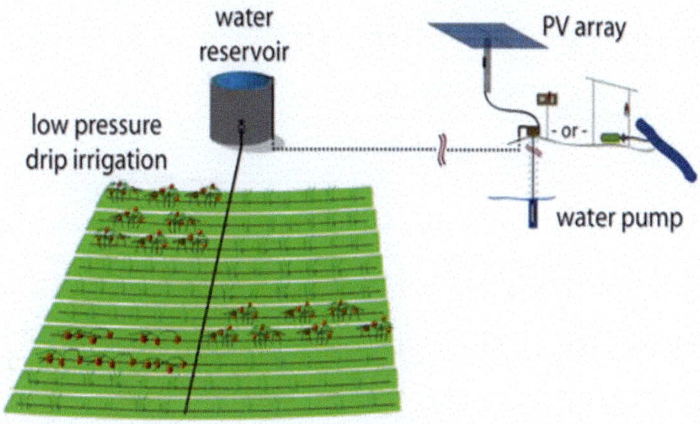

Fig. 18.16 Solar water pump for irrigation in agriculture [37]

18.3.2 Solar Desalination

Solar desalination systems use solar energy to convert seawater or brackish water into freshwater through the process of distillation or reverse osmosis with the use of solar radiation energy which is shown in Fig. 18.17. This technology helps address water scarcity issues in coastal regions [43].

Solar-Powered Water Treatment: Water treatment systems use solar technology, such as solar-powered ultraviolet (UV) disinfection machines. Solar energy is used by these systems to clean the water, making it suitable for drinking and other uses.

18.3.3 Solar-Powered Street Furniture

In order to provide charging stations for mobile devices, Wi-Fi access and lighting, solar energy is integrated into street furniture including benches, bus stops and shelters. These solar-powered features improve public areas and offer sustainable services. The solar-powered Japanese eco bench has a charging station and powers the street lighting systems which is shown in Fig. 18.18.

18.3.4 Solar Heating and Cooling

Solar thermal systems use sunlight to heat water or other fluids for a variety of purposes, including industrial processes, water heating and space heating. Solar air conditioning systems and other solar thermal technologies can be used for cooling.

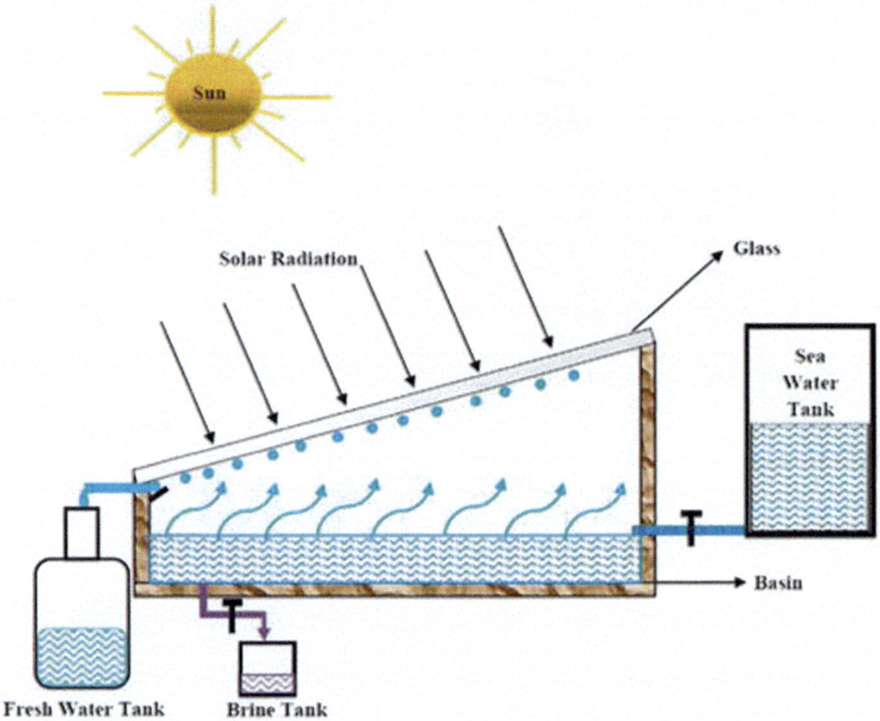

Fig. 18.17 Solar desalination [43]

Fig. 18.18 Eco benches with solar panels [44]

Fig. 18.19 Street lighting [45]

18.3.5 Solar Lighting

For outdoor illumination purposes, such as street lighting, garden lighting and walkway lighting, solar-powered LED lights are employed. As can be seen in Fig. 18.19, they include integrated photovoltaic panels that charge during the day and produce light at night.

18.3.6 Solar Charging Stations

Electric Vehicles (EVs) are refuelled with solar energy utilizing solar charging stations. These charging stations integrate solar panels with the necessary infrastructure to support zero-emissions travel which is shown in Fig. 18.20.

18.3.7 Solar-Powered Electronics

Integration of solar cells into mobile phones, smartwatches and other portable electronics can be powered by sunlight. This increases the device's battery life and lessens the dependency on conventional batteries.

Fig. 18.20 Solar charging stations [46]

18.3.8 Solar-Powered Remote Monitoring

Weather stations, environmental monitoring equipment and wireless communication systems in remote areas are just a few examples of solar-powered systems that are used for remote monitoring and communication.

18.3.9 Solar-Powered Satellites

Solar panels are essential components of satellites and space probes. They supply energy for the onboard systems that support communication, data gathering and research.

18.3.10 Solar-Powered Boats, Ships and Airports

Nowadays, onboard systems including navigational equipment, lighting and auxiliary power are all powered by solar panels, which are mounted on boats and ships which helps reduce fuel consumption leading to low emissions in marine transportation. Solar energy is being used more and more by airports to supply their energy requirements. To produce clean energy for lighting, heating, cooling and other operations, solar panels can be put on airport rooftops, parking garages and open spaces [47, 48].

18.3.11 Solar-Powered Educational Tools

Solar energy is used in educational settings to teach students about renewable energy and sustainability. Students can learn about solar power generation and experiment with building their own solar-powered systems using solar kits and instructional resources.

18.4 Government Initiatives and Policies for Boosting the Indian Economy

18.4.1 International Solar Alliance

Worldwide sustainable development at present is aided by energy, considered as one of the fundamental elements in its development. Currently, the worldwide nations need rapidly growing global energy coupled with a decline in the use of conventional non-renewable energy sources like petroleum and coal. The shift in particular from non-renewable to renewable energy sources is to reduce the effect of severe threat on climate change and degradation of environment. The source of renewable energy being the sunlight, tides, wind and geothermal heat offers significant and potential solutions, especially to developing nations to meet long-term energy demands while protecting the environment and promoting sustainability as well. To address these concerns, solar energy stands out as one of the cleanest energy sources by emitting minimal pollutants. Deployed and stored in several forms, expanding the usage of solar energy thus fosters the need to reduce dependence on fossil fuel usage.

With a rapid increase in population and a growing economy, India at this stage is struggling to meet the energy demands. While establishment of new power plants is itself a violation to exploitation of fossil fuels, it has thus become compelling to exploit the abundantly available renewable energy sources like solar energy that can support the growing population and economy. The involvement of government plays a significant role while employing and increasing the utilization of solar energy. Indian government with an aim to increase globalized energy security presented their proposal on the Paris Climate Summit in December 2015, wherein, countries seeking to leverage their solar resources utilization can make energy accessible in a coordinated approach. The collaborative platform was termed as International Solar alliance (ISA). Primary objective of ISA is to *'harmonize and aggregate demand for solar finance, solar technologies, innovation, R&D, and capacity building'*. Programme was launched on 30 November 2015 jointly by Narendra Modi, Prime Minister of India, and Francois Hollande, President of France during the Conference of Parties (COP 21) in Paris [49]. Leading the platform, the Indian Prime Minister emphasized on the clear vision to integrate solar energy into daily lives of individual along with extending the benefits to remote places. The prime goal of ISA is to attract investment of $1 trillion and develop 1000GW solar capacity by 2030 within the member states [50].

ISA endeavours to synchronize the solar energy requirements among 121 countries falling in the region between the Topic of Capricorn and the Topic of Cancer which has significant solar energy potential [51]. Developing countries located in the Southern Hemisphere although with maximum duration of sunshine hours cannot harness the benefit owing to challenges related with technology and funding. These countries are accountable for 36% of world GDP and 55% of world energy consumption. The '121 country coalition' Alliance formed thus seeks in addressing these gaps by convening all relevant organizations, industrial producers and stakeholders towards achieving the shared objectives. These countries with their shared ambition in view of making collective efforts for solar energy deployment and storage with better technologies have expressed their commitment in supporting India's proposal to establish the ISA also known as the International Agency for Solar Policy and Application (IASPA). The treat also extends its benefit to the non-tropical countries with an exemption of voting rights. Towards late 2018, the programme was extended to all the member countries of the United Nations (UN).

The progress of ISA to achieve energy access an energy security since COP21 has been summarized in few points below [51, 52]:

I. Since COP21, the treaty being joined by 60 countries has given it the International Organization status as of December 6, 2017.
II. The alliance has initiated five programmes that are dedicated to yield benefits in rural areas that need decentralized applications of solar energy, reducing the barrier between solar technologies and high cost involved, establishing mini-grids, decentralizing electrical output from PV rooftops, applicability of e-solar technologies that includes vehicles and roads. Additionally, the programme has also focused on creating lesser risked instruments that can ensure deployment of solar technologies on a larger scale.
III. So far, co-chaired by India and France, six gatherings have been conducted to discuss and make discussions regarding the activities of the International Solar Alliance.
IV. The 34 countries with membership of ISA have established 100 initial projects focusing on application of solar energy. These projects prioritize at harnessing solar energy potential across different locations.
V. Amendments has been made in meetings held in New Delhi and Paris to converge private sector, demand for solar energy and cost involved.

18.4.2 Advantage of India from International Solar Alliance

i. *Achieving the goal of SDG 7:* India stands to take advantage from its leadership in ISA while achieving the Sustainable Development Goal (SDG) 7—affordable and clean energy: *Ensure access to affordable, reliable, sustainable and modern energy for all.* Through collaborations with other member countries, India can gain benefit of technologies along with financial resources to expand its solar energy sector and thereby reduce dependence on fossil fuel use and carbon emissions.

ii. *Geopolitical significance:* The summit holds significant geopolitical denotations as for the first time an inter-governmental treaty will be headquartered in India. The alliance thus allows India to take the opportunity of a pivotal leadership role in the field of renewable and sustainable energy development and climate change. ISA in addition, through initiative of *'Make in India'* aims to position India as a technological hub by self-producing equipment like solar panels. The Indian government through this campaign intends to accelerate clean energy production within the country while creating employment opportunities and fostering economic growth.
iii. *Affordable finance at scale:* The proposals formed by the Indian think tank and Center for Energy, Environment and Water (CEEW) aim to reduce the risks involved in cost finance of solar projects through demand segregation [53, 54]. Demand segregation involves the stakeholders in investing in multiple projects, thereby reducing the risk of capital cost investment [55]. World Bank has incorporated the proposal with a modified version that might not meet the requirements of CEEW but nevertheless will reduce borrowing costs by the member countries. One of the successful solar PV rooftop proposals in Madhya Pradesh, India, that involved demand segregation was further launched in Ghana [55].
iv. *Advantage of possessing a large domestic market:* Average income of Indian consumers being lower in comparison to their Chinese counterparts, allow the actions to be taken for framework as both cost-effective and efficient. Pulling down the prices of certain products like solar appliances and air conditioners while maintaining their energy efficiency could benefit India as well as neighbouring countries and potentially have a global impact.
v. *Solar rooftop scale-up in India:* ISA's approach to solar rooftop scale-up was launched in India in March 2018. A total of 1105 MW rooftop solar demand was sanctioned for 13 ISA member countries that included India [56]. Indian approach for scaling up the solar energy sector encompasses both means of auctioning utility-scale projects and initiatives that enable PV rooftop and off-grid distribution of solar. India's endeavour to make avail the scale-up model domestically can be seen as an intellectual leadership under the framework of Andreson and Agarwala [49]. The model also seems to be appealing to other countries provided they are aware of it. Exploring the otherwise unexploited solar opportunities could also create employment opportunities directly or indirectly.
vi. *Scaling solar application for agricultural use:* A recent collaboration of ISA with Energy Efficiency Services Limited (EESL) headquartered in India includes creating bid for knowing the price with International Competitive Bid (ICB) and consultancy services with Project Management Consultancy (PMC) for solar water pumping systems (SWPS) [57]. The prices were discovered to be 47% lesser than the prevailing prices in South Africa after floatation of ICB.

A joint proposal laid by ISA-GCGI includes implementation of 1 million SWPs in Asia and the Asia Pacific countries. India under this zone will clearly benefit the agricultural sector through deployment of solar water pumps.

vii. *Collaborations and exchange of officials:* Exchange of experts in the field of solar energy within the member countries would allow to form collaborative initiatives for promoting energy access and energy security. The International Solar Alliance (ISA) actively engages in collaborations with various multilateral bodies, including the International Renewable Energy Agency (IRENA), Renewable Energy and Energy Efficiency Partnership (REEEP), International Energy Agency (IEA), Renewable Energy Policy Network for the twenty-first century (REN) and United Nations bodies [51]. Additionally, partnerships with bilateral organizations, corporates, industries and other stakeholders are sought to collectively contribute towards the shared objective of promoting the widespread adoption of solar energy in ISA member countries.

18.4.3 Indian Government Initiatives and Policies for Boosting Indian Economy

India possesses an extensive solar energy potential, with an annual influx of around 5,000 trillion kWh of energy across its land area. Most regions receive approximately 4–7 kWh/m^2/day. The utilization of solar photovoltaic power holds significant promise in India, offering considerable scalability. Moreover, solar energy allows decentralized power generation, facilitating swift capacity augmentation with short lead times. In terms of energy security, solar emerges as the most dependable source due to its abundant availability. Even a small portion of the total incident solar energy, if harnessed efficiently, can cater to the entire nation's power demands. The solar energy sector in India aligns with the government's sustainable growth objectives, progressively becoming a crucial component of the solution to fulfill the country's energy requisites and a fundamental contributor to ensuring energy security [58].

The solar potential of India, evaluated by the National Institute of Solar Energy, stands at approximately 748 GW, assuming that solar PV modules cover 3% of the country's unused land area. Solar energy holds a central role in India's strategy to address climate change, with the National Solar Mission being a prominent component of this effort.

A primary objective of this mission is to establish grid-connected solar power plants with a cumulative capacity of 100 GW by the year 2022. This objective aligns with India's Intended Nationally Determined Contributions (INDCs), which aim to achieve around 40% of the total installed electric power capacity from non-fossil fuel sources. Additionally, the INDCs target a reduction of 33–35% in the emission intensity of the GDP by 2030, compared to the levels recorded in 2005. The installed capacity of solar energy in India has increased from 2.63 GW in March 2014 to 49 GW in December 2021 which is shown in Fig. 18.21 [58].

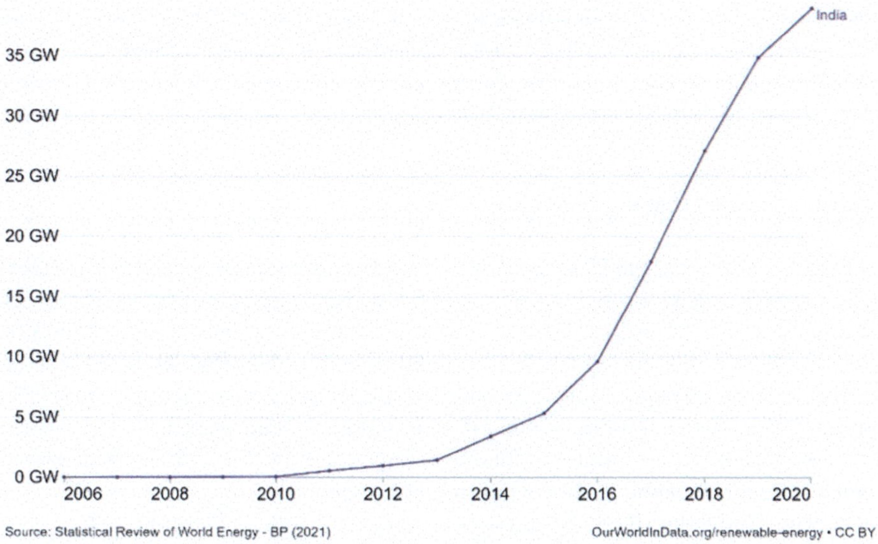

Fig. 18.21 Installed solar energy capacity in India: 2006–2020 [58]

18.4.3.1 Policy Support from the Government

By 2025, India's solar module manufacturing capacity is poised to quadruple compared to its 2021 levels. This surge is driven by robust demand, favourable policies, anticipated advancements in energy efficiency and improved price competitiveness, resulting in the commissioning of 30–35 GW of new module capacity [59]. The introduction of a basic customs duty (BCD) on imported solar cells (25%) and modules (40%) aims to bolster domestic manufacturing by enhancing the competitiveness of locally produced cells and modules. The government has also unveiled plans for dedicated corridors in Phase-II of the Green Energy Corridor (20 GW) to facilitate the transmission of renewable power. This initiative is designed to enable the seamless integration and evacuation of approximately 20 GW of Renewable Energy (RE) projects across seven states: Gujarat, Himachal Pradesh, Karnataka, Kerala, Rajasthan, Tamil Nadu and Uttar Pradesh [60].

18.4.3.2 Employment Generation

By achieving the goal of 500 GW of non-fossil electricity generation capacity by 2030 by installing 238 GW of solar and 101 GW of new wind capacity, India, has the potential to generate approximately 3.4 million jobs, encompassing both short-term and long-term opportunities [59]. These job prospects predominantly arise within the wind and on-grid solar energy sectors. The actual workforce required to seize these environmentally friendly job opportunities is projected to be around one million.

This indicates that the significant expansion of renewable energy infrastructure, as outlined in India's renewable energy targets, not only contributes to sustainable power generation but also plays a pivotal role in creating many employment opportunities across various skill levels and time frames.

18.4.3.3 Major Ongoing Schemes

The Pradhan Mantri Kisan Urja Suraksha Evam Utthan Mahabhiyan (PM KUSUM) is a pioneering endeavour that ranks among the world's largest initiatives. It focuses on furnishing over 3.5 million farmers with clean energy by transitioning their agricultural pumps to solar power. The fundamental objective of the PM-KUSUM scheme is to facilitate the incorporation of an additional solar capacity totalizing 30.80 GW. The scheme serves as a monumental effort to enhance energy security for farmers while concurrently promoting sustainable energy solutions through the widespread adoption of solar technology in the agricultural sector.

The Roof Top Solar (RTS) Programme commenced with its Phase-I on December 30, 2015, focusing on offering incentives and subsidies to residential, institutional and social sectors. Furthermore, achievement-linked incentives were extended to the government sector. Subsequently, in February 2019, the programme progressed to Phase-II with the ambitious objective of attaining a combined capacity of 40,000 MW by the year 2022 [61].

The solar parks development initiative aims to achieve a capacity of 40 GW, encompassing all States and Union Territories eligible for scheme benefits. These solar parks are being established through collaborations involving Central and State Government entities, as well as joint ventures between them, and private entrepreneurs. As of March 2020, the Inter-State Transmission System (ISTS) element, which includes 3200 km of transmission lines and 17,000 MVA substations, has been successfully finalized. Furthermore, the Intra-State Transmission System (InSTS) phase has been approved for eight states abundant in renewable energy resources: Tamil Nadu, Rajasthan, Karnataka, Andhra Pradesh, Maharashtra, Gujarat, Himachal Pradesh and Madhya Pradesh. This development aims to facilitate the evacuation of more than 20,000 MW of renewable power.

The government's vision entails the complete transformation of Andaman and Nicobar as well as Lakshadweep islands into sustainable energy hubs, relying entirely on renewable energy sources to fulfil their energy requirements. The Greening of Islands initiative strives to implement 52 MW of distributed grid-connected solar PV power projects as part of this effort. In each state of India, a designated city, which could be either the state capital or a prominent tourist destination, is undergoing transformation into a solar city. The primary focus of this initiative is to ensure that the entire electricity demand of the chosen city is completely satisfied through renewable energy sources, with a primary emphasis on solar energy. All solar and wind power projects commissioned until June 30, 2023, will have their Inter State Transmission System charges and losses exempted for inter-state power sales [61].

18.4.3.4 Major Initiatives and Achievements

India has introduced the Pradhan Mantri Kisan Urja Suraksha evam Utthaan Mahabhiyan (PM-KUSUM) Scheme, aiming to ensure energy and water security while also reducing reliance on diesel in the agricultural sector. This scheme generates additional income for farmers by enabling solar power generation. Its objective is to incorporate 30.8 GW of solar capacity, supported by central financial assistance exceeding ₹34,000 crore. As of December 31, 2021, over 77,000 standalone solar pumps, 25.25 MW of solar power plant capacity and more than 1026 pumps have been solarized under the individual pump solarization category.

For the promotion of large-scale grid-connected solar power projects, the 'Development of Solar Parks and Ultra Mega Solar Power Projects' initiative is currently in progress, aiming for a total capacity of 40 GW by March 2024. A cumulative capacity of around 9.2 GW has already been commissioned in these parks. Under the RTS Programme Phase-II, a goal of achieving 40 GW of installed solar rooftop capacity by December 2022 is being pursued. Financial support for up to 4 GW of residential solar rooftop capacity is provided, along with incentives for distribution companies to surpass previous-year achievements. So far, the country has seen the establishment of cumulative solar rooftop projects totalling 5.87 GW.

A scheme involving the setup of 12 GW grid-connected solar PV power projects by government entities, including Central Public Sector Undertakings, is in motion. The government has sanctioned approximately 8.2 GW of such projects. Till December 2021, more than 1.45 lakh solar street lights were installed, 9.14 lakh solar study lamps were distributed and roughly 2.5 MW of solar power packs were established according to State Nodal agencies' reports. As of December 31, 2021, around 4.25 GW of wind–solar hybrid capacity has been awarded. Among this, 0.2 GW is already operational, and another 1.2 GW of wind–solar hybrid projects are in various stages of bidding. The Indian Railways has set an ambitious target of achieving Net Zero Carbon Emission by 2030, primarily through the adoption of renewable energy sources. By December 2023, the Railways aims for complete electrification of its network [61].

18.5 Description of Installed Solar Power Plants in India

18.5.1 Chronological Development of Solar Power Plants in India

India's transition from smaller scale installations to large-scale utility projects, driven by breakthroughs in solar PV technology, can be seen in the chronological growth of its solar power plants. The emphasis now is on increasing effectiveness, lowering prices and combining solar energy with other renewable energy sources and energy storage. Future research and development initiatives are intended to advance solar technology and increase India's solar power capability. The development of solar power plants in India is described in the following sections.

18.5.1.1 Early Development (1990s–2000s)

Smaller scale installations, especially utilizing solar photovoltaic (PV) technology, were the primary focus of the original phase of solar power plant development in India. The capacity of the solar PV plants installed during this time period typically ranged from a few kilowatts to a few megawatts. Solar panels made of polycrystalline silicon were frequently used since they were affordable and easily accessible [20].

18.5.1.2 Jawaharlal Nehru National Solar Mission (JNNSM) Era (2010–2015)

An important turning point in India's development of solar energy was the launch of JNNSM in 2010. By encouraging large-scale solar power projects and lowering solar prices, the mission attempted to accelerate the use of solar energy. Solar PV technology maintained its dominance during this time, with cost- and efficiency-related advances. Thin-film solar technology, which turns sunlight into power using semiconductor materials, has also drawn interest for some uses.

18.5.1.3 Utility-Scale Solar Power Plants (2011–2015)

During this time, a number of utility-scale solar power facilities were put into operation, demonstrating India's desire to expand its solar energy potential. After opening in 2011, Gujarat Solar Park in Charanka grew to be one of the biggest solar parks in world. With a few demonstration projects utilizing parabolic trough and tower-based systems, concentrated solar power (CSP) technology began to receive attention.

18.5.1.4 Rapid Expansion and Technological Advancements (2015–2020)

During this time, India's solar power capacity rapidly increased because of policies that were favourable, competitive bidding and slipping solar module costs. Large-scale solar park developments include the Rewa Ultra Mega Solar Park in Madhya Pradesh and the Pavagada Solar Park in Karnataka. Solar panels made on monocrystalline silicon have grown in popularity because of their increased efficiency and attractive appearance. As a practical solution for locations with limited access to land, floating solar power plants where solar panels are installed on water bodies have begun to gain popularity.

18.5.1.5 Hybrid Solar Power Plants and Energy Storage (2018–Present)

In order to reduce unpredictable generation and improve grid stability, the combination of solar power plants with energy storage devices has gained popularity. Hybrid solar power plants have arisen to offer a more dependable and well-balanced renewable energy mix by combining solar with additional energy sources like wind or biomass. Due to its lower price and greater energy density, lithium-ion batteries have become the standard option for energy storage [62].

India has made significant advances in solar power production over the years as it works to harness renewable energy sources to meet its erupting energy needs and limit carbon emissions. Table 18.2 shows some key milestones in the development of solar power plants in India.

These milestones show how far India has come in developing solar energy, with a rising emphasis on innovation, funding and governmental assistance to accelerate the adoption of clean and sustainable energy sources.

18.5.2 Categorization of Developed Solar Power Plants Based on Different Solar Technologies in India

The diverse range of technologies utilized in solar power plants in India, including PV technology, Solar thermal technology, CSP technology, floating solar and hybrid systems. The choice of technology is influenced by various elements, including the availability of solar resources, the need for land, cost-effectiveness and project objectives. The various development in different categories is discussed in below-mentioned sections.

18.5.2.1 PV or Photovoltaic Power Plants

The developed solar PV or photovoltaic power plants in India are discussed in this section [65].

- *Kamuthi Solar Power Project, Tamil Nadu:* This project, commissioned in 2016, is located in Kamuthi and was the largest single-location solar power plant in the world at that time. It has a capacity of 648 MW and uses photovoltaic (PV) technology. The plant consists of over 2.5 million solar modules spread across 2,500 acres.
- *Pavagada Solar Park, Karnataka:* The Pavagada Solar Park, operational since 2017, is one of the largest solar parks in the world, covering an area of around 13,000 acres. It has an initial capacity of 600 MW, with plans for further expansion. The park primarily utilizes photovoltaic (PV) technology with the deployment of solar panels and inverters [66].

Table 18.2 Milestones in the development of solar power plants in India [63, 64]

Sl. No	Descriptions of solar power plants in India	Year	Power generation achieved
1	The first solar photovoltaic (PV) plant was installed in India in the state of Kerala with a capacity of 50 kilowatts (kW)	1992	50 (kW)
2	The Indian government launched the Jawaharlal Nehru National Solar Mission (JNNSM), which aimed to promote the deployment of solar energy across the country	2002	–
3	The first solar power plant under the JNNSM was commissioned in Gujarat. The Charanka Solar Park had an initial capacity of 5 MW	2009	5 (MW)
4	The first utility-scale solar power plant, the Shirdi Solar Power Plant, was commissioned in Maharashtra with a capacity of 3.4 MW	2010	3.4 (MW)
5	The Gujarat Solar Park in Charanka became the largest solar power plant in India at the time with a total capacity of 214 MW	2011	214 (MW)
6	The Kamuthi Solar Power Project in Tamil Nadu became the largest single-location solar power plant in the world with a capacity of 648 MW	2012	648 (MW)
7	The Rewa Ultra Mega Solar Park in Madhya Pradesh was commissioned with a capacity of 750 MW. It became one of the largest solar parks in the world at that time	2015	750 (MW)
8	The Pavagada Solar Park in Karnataka, spanning across 13,000 acres, started operations with an initial capacity of 600 MW	2017	600 (MW)
9	India's cumulative installed solar capacity crossed the 20 GW mark	2018	20 (GW)
10	The largest solar power plant in India, the Bhadla Solar Park in Rajasthan, reached a capacity of 2.25 GW	2019	2.25 (GW)
11	India's cumulative installed solar capacity crossed the 35 GW mark	2020	35 (GW)
12	The Government of India announced a target of achieving 500 GW of renewable energy capacity by 2030, including a substantial share from solar power	2021	500 (GW) ***(Set to achieve)***

- *Bhadla Solar Park, Rajasthan:* The Bhadla Solar Park, located in Rajasthan, has become the largest solar power plant in India with a capacity of 2.25 GW. It primarily utilizes photovoltaic (PV) technology, with solar panels covering a vast area. The park benefits from the region's high solar irradiance and favourable weather conditions for solar power generation.
- *Kurnool Ultra Mega Solar Park, Andhra Pradesh:* This massive solar park has a total capacity of 1 GW and utilizes PV technology. It is one of the largest solar parks in India.
- *Rewa Ultra Mega Solar Park, Madhya Pradesh:* With a capacity of 750 MW, this solar park is one of the largest single-site solar projects in India, and it relies on PV panels.
- *Kamuthi Solar Power Project, Tamil Nadu:* Located in Kamuthi, Tamil Nadu, this is one of the largest single-location solar power plants in the world, with a capacity of 648 MW. It is primarily PV-based.
- *Shakti Sthala Solar Park, Karnataka:* Shakti Sthala Solar Park in Karnataka has a total capacity of 2 GW, making it one of the largest solar parks in India, and it utilizes PV technology.
- *Pavagada Solar Park, Karnataka:* Pavagada Solar Park in Karnataka is one of the largest solar parks globally, with an initial capacity of 600 MW, and it relies on PV panels.
- *Dhursar Solar Park, Rajasthan:* Dhursar Solar Park in Rajasthan has a capacity of 1 GW and is predominantly based on PV technology.
- *Ananthapuramu Solar Park, Andhra Pradesh:* This solar park has a total capacity of 1.5 GW and utilizes PV technology to generate electricity.
- *Nokh Solar Park, Rajasthan:* Nokh Solar Park has a capacity of 925 MW and relies on PV panels for solar power generation.
- *Charanka Solar Park, Gujarat:* Charanka Solar Park in Gujarat, one of the earliest solar parks in India, has multiple solar power plants with a cumulative capacity of over 790 MW, primarily based on PV technology.

18.5.2.2 PV-T or Photovoltaic–Thermal Power Plants

PV-T (Photovoltaic–Thermal) technology is the new emerging solar technology. So, solar power plants operating specifically on PV-T (Photovoltaic–Thermal) technology in India were relatively less common compared to conventional PV (Photovoltaic) solar power plants. However, some examples of PV-T technology existed with installed PV power plant are shown below.

- *Cochin International Airport PV-T System, Kerala:* Cochin International Airport is known for its sustainability initiatives, and it has implemented a PV-T system on its premises. This PV-T system includes both solar PV panels and solar thermal collectors which simultaneously generating electricity and collecting thermal energy [67].

- *Solar Rooftop Installations:* Rooftop solar installations across India use various PV technologies, primarily crystalline silicon, to simultaneously generating electricity and collecting thermal energy at the consumer level. These are often connected to the grid or used for captive consumption.

18.5.2.3 CSP or Concentrated Solar Power Plants

The developed CSP or concentrated solar power plants in India discussed in this section [68, 69].

- *Rewa Ultra Mega Solar Park, Madhya Pradesh:* The Rewa Solar Park, commissioned in 2018, is one of the largest solar parks in India, with a capacity of 750 MW. It incorporates three solar power plants within the park. The project features a combination of both PV and Concentrated Solar Power (CSP) technologies. It includes large-scale deployment of solar panels as well as the use of solar tracking systems to optimize sunlight capture.
- *NTPC Dadri Solar Power Plant, Uttar Pradesh:* The NTPC Dadri Solar Power Plant, operational since 2021, is located in Uttar Pradesh. It is an innovative solar project that combines both PV and thermal technologies. The plant integrates a concentrating solar power (CSP) system with a 1.1 MW solar thermal field alongside a 15 MW PV field. This hybrid approach enables the plant to generate electricity from both solar radiation and heat.
- *Dhirubhai Ambani Solar Park (formerly known as Rajasthan Sun Technique Energy Pvt. Ltd.), Rajasthan:* This project is one of the larger CSP installations in India and uses parabolic trough collectors to concentrate sunlight and generate steam for electricity production., Capacity: 125 MW, Technology: Parabolic trough, Location: Pokhran, Rajasthan.
- *Gujarat Solar One (GSL1), Gujarat:* GSL1 is a demonstration project for CSP technology in India. It uses parabolic trough collectors to generate steam and produce electricity, Capacity: 2.5 MW, Technology: Parabolic trough, Location: Gujarat.
- *Reliance Power's CSP Project, Rajasthan:* Reliance Power had announced plans for a 100 MW CSP project in Rajasthan. However, project statuses can change, so it's advisable to verify the current status of this project, Proposed Capacity: 100 MW, Technology: CSP (specific technology details not provided), Location: Rajasthan.
- *National Solar Thermal Power Test Facility (NSTPTF), Rajasthan:* NSTPTF is a government facility established by the Ministry of New and Renewable Energy (MNRE). It is dedicated to research, development and testing related to solar thermal and CSP technologies. While not a commercial power plant, it plays a significant role in advancing CSP technology in India, Location: Jodhpur, Rajasthan.

- *Adani Group's CSP Projects:* The Adani Group has expressed interest in developing CSP projects in India. Specific details and project statuses may vary, so it's advisable to check for the latest updates on their CSP projects.

The solar power industry in India is projected to continue to innovate as a result of ongoing improvements in solar technology, including more efficient panels and energy storage options.

18.5.3 Technological Innovations and Future Prospects

India is still researching cutting-edge solar technology to boost their effectiveness and affordability. Research is being done to determine the commercial viability of perovskite solar cells, which have the potential for better conversion efficiencies and cheaper production costs. In some areas, the use of concentrated photovoltaic (CPV) systems which use mirrors or lenses to focus sunlight onto high-efficiency solar cells is being studied [70, 71].

One of the most appealing waste heat recovery technologies is the thermoelectric generator (TEG), which has a number of special benefits including zero emissions, solid-state operation, quiet operation, the absence of moving parts, high reliability and maintenance-free operation [72]. A thermoelectric generator is a solid-state device that uses the Seebeck effect to produce energy directly from heat [73]. Thermoelectric generators' poor conversion efficiency, however, has limited their ability to be used widely. Due to its enormous potential, research on hybrid photovoltaic-thermoelectric generation (PV-TEG) has recently accelerated more quickly than that of other hybrid photovoltaic systems. The thermoelectric uses the infrared area of the sun spectrum, whereas photovoltaic uses the visible and ultraviolet regions. Therefore, combining photovoltaic and thermoelectric systems will result in a more effective and extensive use of the solar spectrum [74].

The use of heat pipe-based PV-Ts is increasing for improvement in overall efficiency of photovoltaic system. Additionally, different types of predictive and optimization techniques were emerging with the help of artificial intelligence and machine learning is the new breakthrough in Indian solar energy research [75].

18.5.3.1 Solar Power Plants Based on Advanced Technology

- *Sakri Solar Plant, Maharashtra:* The Sakri Solar Plant, commissioned in 2020, is a floating solar power plant located on the reservoir of the Gangapur Dam in Maharashtra. It has a capacity of 1 MW and is built using floating PV technology. The solar panels are mounted on floating platforms, which not only generate electricity but also help in reducing water evaporation from the reservoir.

- *Omkareshwar Floating Solar Park, Madhya Pradesh:* A floating photovoltaic (FPV) solar power plant is proposed. It will cover a portion of the Omkareshwar reservoir with solar panels mounted on floating platforms. With a capacity of 600 MW, it will showcases India's utilization of FPV technology to maximize energy generation while conserving land and water resources [76].
- *Dholera Solar Park, Gujarat:* Dholera Solar Park is envisioned as a smart city and is equipped with advanced solar infrastructure. It aims to incorporate IoT (Internet of Things) and AI (Artificial Intelligence) technologies for optimal solar energy management. The park is designed for a total capacity of 5 GW when fully developed [77].

These examples demonstrate India's commitment to adopting advanced technologies and innovative solutions to harness solar energy more efficiently and sustainably. As the renewable energy sector continues to grow, we can expect further advancements and adoption of cutting-edge technologies in India's solar power projects.

18.5.4 Maintenance of Solar Power Plants

To guarantee long-term performance, increase energy output and safeguard the system's investment, maintenance is essential for solar power plants. Solar panels and related equipment can last longer with proper maintenance procedures. Here are some important factors of maintaining solar power plants [78–80]:

- *Regular Cleaning:* The solar panels must be kept clear of dirt, dust, leaves and other debris. More sunlight can be absorbed by clean panels, which will produce more electricity. With a gentle brush or sponge and a moderate detergent, the panels can be cleaned. Avoid using anything abrasive that might scratch the glass. Consider cleaning in accordance with your area and the level of pollution or dust in the air.
- *Visual Inspections:* Conduct visual inspections of the solar panels and mounting structure to check for any physical damage or signs of wear and tear. Look for loose connections, damaged wiring or any other visible issues. If you notice any problems, contact a professional for repairs.
- *Monitoring:* Many solar PV systems come with monitoring software that allows you to track the system's performance. Check the monitoring data frequently to make sure the system is producing the predicted amount of electricity, and keep watching out for any sudden drops in performance which could indicate a problem.
- *Electrical Checks:* Turn off the system and perform routine electrical checks to ensure all connections are secure. Check for any loose wires, corrosion or signs of overheating at the junction boxes and inverters.

- *Inverter Maintenance:* Inverters are a critical component of a solar PV system. Check the inverter's display for error messages or unusual behaviour. Inverter maintenance should only be performed by qualified technicians. If you suspect an issue with the inverter, contact a professional.
- *Warranty and Documentation:* Keep all warranty documents, installation records and maintenance logs handy. Ensure you are aware of the warranty terms for your solar panels, inverters and other components.
- *Safety Precautions*: Before performing any maintenance, always turn off the solar PV system and follow proper safety procedures. Regular maintenance and monitoring will help maximize the lifespan of your solar PV system and ensure it continues to generate clean, renewable energy efficiently. If you're unsure about any aspect of maintenance, it's best to consult with a professional solar technician.
- *Temperature and Climate Considerations:* Extreme temperatures can affect the performance of solar panels. Check for any signs of overheating or temperature-related damage. In regions with harsh winters, ensure that snow is removed from the panels to allow sunlight to reach them.
- *Dust and Pollution:* If you live in an area with high levels of air pollution or dust, more frequent cleaning may be necessary to prevent a build-up that can reduce energy production.
- *Sealing and Weatherproofing:* Ensure that all roof penetrations and panel seals are intact and properly sealed. Leaks can lead to moisture damage and reduce the efficiency of the system.
- *Battery Maintenance (if applicable):* If your solar PV system includes energy storage batteries, follow the manufacturer's guidelines for maintenance and ensure they are kept in a cool, dry place.
- *Protection from Hail and Storms:* In areas prone to hail or severe storms, consider investing in protective measures such as hail guards or storm shut-off features that can safeguard the system during adverse weather conditions.
- *Upgrading and System Expansion:* As technology advances, you may consider upgrading components of your solar PV system, such as replacing old inverters or adding more panels. Consult with a professional for any system modifications.
- *Training and Education:* Educate yourself and any household members about the basics of solar PV system operation and safety. This knowledge can be valuable in identifying potential issues.

The upkeep requirements for solar power plants might vary based on the technology chosen, the size of the plant, the environment and other considerations. Many operators of solar power plants also decide to hire professional maintenance services to guarantee that all maintenance tasks are completed accurately and on time. Maximizing energy production and safeguarding the long-term investment in solar energy are both benefits of proper maintenance.

18.6 Societal and Climatic Impact of Solar Energy Adoption in India

18.6.1 Rural Development Schemes Through Solar Awareness Programmes

India, as a country with almost 65% population staying in rural areas, is highly benefited from the evolution of solar energy-based applications. The impact of solar energy can be seen both in society and household levels. Under the off-grid solar PV programme of the Ministry of New and Renewable Energy, Government of India, solar home lighting systems in rural households, solar street lights, solar pumps in fields, solar lanterns and solar study lamps have been installed across the country. Solar pumps are beneficial for small and marginal farmers for irrigation purposes. The use of solar irrigation systems and other solar-assisted post-harvesting machines gives reliable production of crops and value addition of cultivated products, which helps to enhance the economic condition of small-scale farmers. Apart from agriculture, the development of different cottage industries can be observed with the interventions of solar energy. The implementation of solar cooking in rural households reduces the hard labour of the ladies collecting forest wood. Solar cooking replaces the unhygienic smoky kitchen, which minimizes health-related issues. Solar study lamps have been supplied to rural students to provide clean and affordable lighting solutions for their study hours. The un-electrified households relying on kerosene lamps are being covered and benefitting.

The success of the policy 'Power for All' by the Government of India depends on the active participation of all the stakeholders in the development, progress and maintenance of different renewable-based energy schemes. Social awareness is the key booster for the uniform and speedy spread of solar energy-based installations in a populous and developing country like India. It is reported that 'lack of awareness' is the main cause under the social barriers to the smooth implementation of different solar energy schemes, especially in rural areas [81]. Reluctance has been shown by people to install solar PV systems as people are not aware of the benefits of the new technology over the conventional technology. The solar energy sector faces significant challenges due to general public disinterest and disengagement. Many households are hesitant to invest in solar energy due to concerns about its reliability. Additionally, the lack of awareness and knowledge in rural communities about solar energy and its environmental benefits contributes to slow adoption. Moreover, the perception of high maintenance further deters consumers from embracing this sustainable technology. Another serious issue related to the execution of grid-connected solar projects is the conflict of interest between agricultural land and land required for setting up PV panels. The central and state government, with the coordination of the local authorities, should identify suitable land allocation strategies so that the conflict can be resolved [82]. The rooftop solar energy harnessing technology is the way forward for sustainable power generation for both urban and rural societies. Off-grid solar power presents a sustainable energy solution enabling villages in India to overcome the limitations of rural electrification projects and the power sector. Though different schemes have already been executed throughout the country to harness solar power, the lack of awareness of domestic solar energy utilization is still prevailing. The rural people are familiar with but not educated. It has been reported significant concerns about the involvement of private enterprises and the existing rural–urban disparities in the implementation of this technology [83]. Therefore, it is the responsibility of the government to assign some agency to create not only awareness but also to provide technical support during the installation and maintenance of solar projects. Creating a comprehensive and easily accessible information hub that documents real-world experiences related to the operation, maintenance and installation of solar energy technologies would be beneficial.

Under the Ministry of New and Renewable Energy department of the Government of India, a programme named 'Information and public awareness program' is launched to create awareness among the people through different modes of media like electronic, social, print, student exhibitions, seminars, workshops and outdoor publicity. The programme also facilitates a human resource development programme to empower young working professionals to work towards the efficient management of solar energy implementation on field levels. The programme also provides financial support to the organization for organizing different workshops and seminars, which gives a platform for the stockholders to share their inputs. Also, different private organizations and non-government organizations have been actively involved in creating awareness of the existing schemes and benefits provided by the government. The awareness programmes include campaigning, technical training, advertisement of different schemes and subsidies provided by the government etc. More such targeted awareness programmes focusing on the reliability of solar energy and its advantages should be available in each rural micro societies of the villages of the country for a rapid and smooth adoption of solar energy schemes.

18.6.2 Job Creation and Economic Growth by Solar Market

The government of India, through MNRE, has consistently introduced policies aimed at fostering the growth of domestic manufacturing for solar energy equipment in the nation. MNRE has initiated Production Linked Incentive (PLI) scheme to boost high efficiency solar PV module production with a substantial budget of INR 24,000 crore. The Indian government also implemented the 'Public Procurement order' (preference to make in India) to prioritize the purchase and utilization of domestically manufactured solar PV modules and inverters by government entities. Some measures like Basic Customs Duty (BCD) on imported solar modules and discontinuation of Customs Duty Concession Certificates for solar PV power project etc. underscore the commitment of the government to promote the use of domestically produced components [84].

The ministry is also providing financial assistance for grid-connected rooftop solar PV projects in different sectors. As an example, assistance of up to 40% is given for capacities up to 3kW_p for the residential sector, up to 20% for capacities between 3 and 10kW_p and up to 20% for housing societies with a maximum capacity of 500 kW_p [84].

As shown in Fig. 18.22. The India solar energy market was valued at USD 38 billion in 2022 and is forecasted to surge to around USD 238 billion by 2030, with a projected annual growth rate of about 40% between 2023 to 2032. The renewable energy market of India ranks fourth globally, with substantial growth in solar power. The remarkable growth has placed India at the forefront of renewable energy on the EY Renewable Energy Country Attractive Index 2021, ranking third [85].

As shown in the Fig. 18.23. India achieved a record low solar tariff of INR 1.99 per kWh in December 2020 but has an average 22% increase in winning tariffs since then in utility scale solar tenders [86]. This rise is attributed to factors such as increased project costs, driven by surging commodity prices, the impact of basic custom duty, approved list of models and manufacturers, higher demand for solar power and increased insurance premiums.

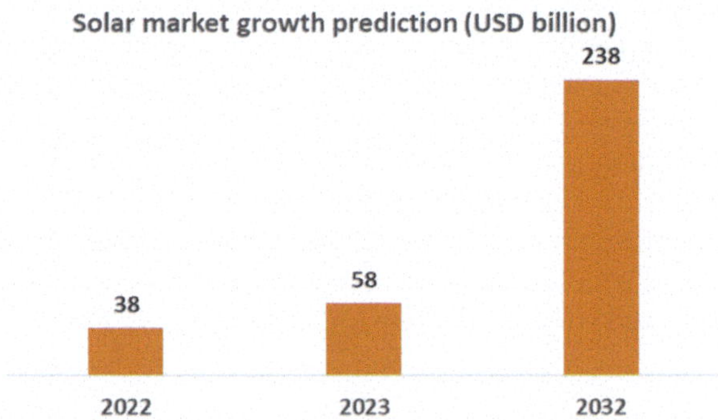

Fig. 18.22 Solar market growth prediction in India [85]

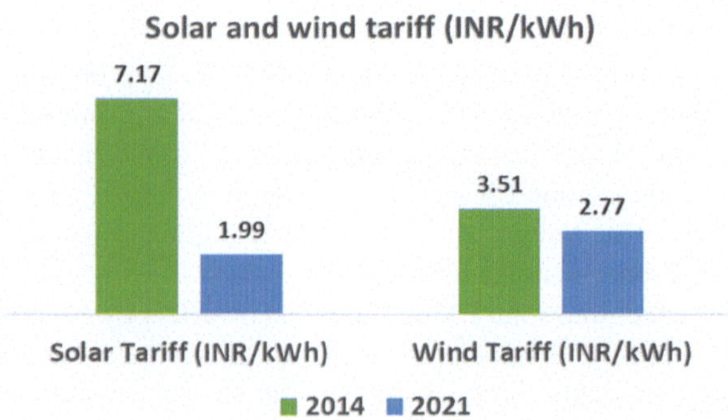

Fig. 18.23 The change of solar and wind tariff in India over the years [86]

A study conducted by The Energy and Resource Institute (TERI), projects that the cost of solar electricity in India could potentially decrease to as low as INR 2.30 per kWh. Furthermore, with the widespread adoption of tracking technology, there is the potential for even cheaper solar costs, possibly reaching as low as INR 1.99 per kWh, by increasing the capacity utilization factor of new solar plants beyond current levels [87].

The use of energy storage systems is helping overcome the intermittent nature of solar power, making solar power a viable option. Declining costs, government tax incentives and competitive pricing with fossil fuels like coal are driving the adoption of solar energy in India. The nation's potential for importing low-cost solar panels from China enhances the growth of solar power market in India.

Solar energy is found to be the largest growing sector in terms of job creation with respect of Indian economy. According to the international organization 'Power for All', in 2021, the decentralized renewable energy sector was able to support 80,000 employees in India, and it was expected to do more in the coming years. Solar energy projects are the new sources of business opportunities, revenue and employment generation and thus boost the rural economic conditions. As solar PV plants require large amounts of land areas, landowners from rural areas are getting extra revenue from land leases. The solar energy sector has the potential to create jobs both at the technical and administrative levels. The small-scale solar installation near to the consumption source is a major benefit for creating job opportunities in remote places. Both large- and small-scale solar power plants require skilled workers in the development and maintenance stage. The work involves electrical and civil works during installation, daily operation and technical and maintenance works. Some unskilled work, like module cleaning and security maintenance of the PV plant, has also required some labor and thus provides rural employment.

According to a study by International Renewable Energy Agency, 7% of the world's renewable sector jobs have been created in India itself. The report analysed that the on-grid solar system of India has an estimated 1,37,000 jobs, and off-grid systems have 80,600 jobs in the country. The policymakers are expecting that more livelihood opportunities will be coming with the per megawatt installation of solar energy in the nation. Proper vocational, mechanical and technical training should be arranged so that skilled labourers and professionals can take the job opportunity in the rural areas. Continuous upskilling of the workers is necessary as the technology gets improved from time to time. The role of the Ministry of Panchayat Raj is very crucial to bridge the gap between the government bodies and the village people. It is also expected that the rural self-help groups will promote the highest number of participations from the women population in the coming days. As solar power generation mainly focuses on installation in rural areas, the migration of workers from rural places to urban areas can also be reversed with proper planning and implementation.

18.6.3 Greenhouse Effect and Carbon Reduction Through Solar Power in India

The escalating demand for energy worldwide is primarily due to increasing population, economic growth and urbanization. India being the most populous country across the globe makes them the highest consumers of convention energy, leading to greenhouse emission. 65% of energy demand in India is met through coal-based power plants, which are the highest greenhouse gas emitters. These power plants emit chemicals like carbon-dioxide (CO_2), nitrogen oxides (NO_2) and sulphur oxides (SO_2) that possesses serious threat to human health as well as causing environmental degradation. Quantity of CO_2 emitted per kWh for coal powered electricity is taken as 0.96 kg approximately. This value turns up as 2 kg CO_2/kWh when taken into consideration the distribution and transmission losses which accounts to 20 and 40% respectively. The increasing trend of population growth rate will further cause expansion of industrialization and rising energy demand. In view of meeting the energy requirements and mitigating carbon emission, Indian government has augmented the utilization of solar power through several solar projects across the country. Contribution of solar power has been considered owing to India's availability of good solar radiation throughout the year as 300 days and lesser CO_2 emission. Comparison of a lifecycle CO_2 emission from various energy sources are presented in Fig. 18.24 [7].

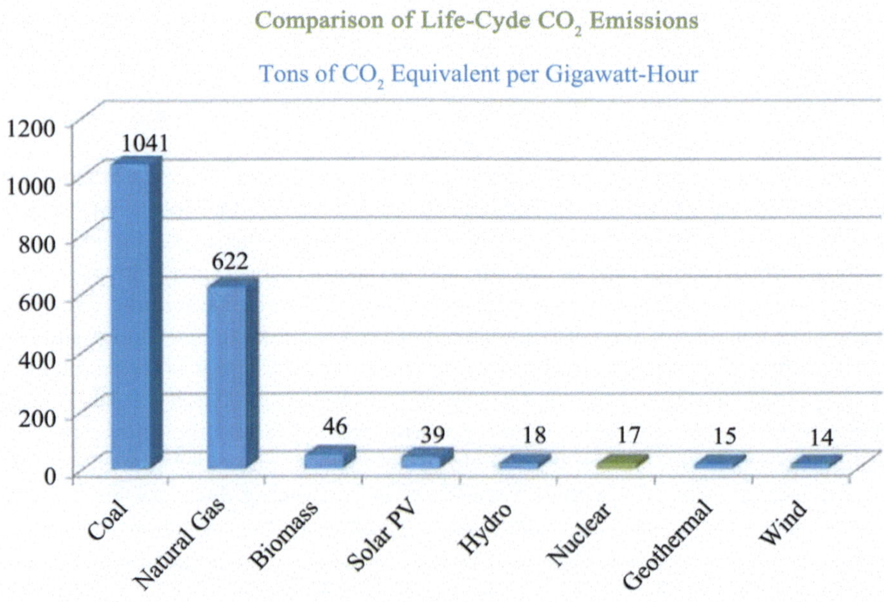

Fig. 18.24 Comparison of lifecycle CO_2 emissions [7]

18.6.3.1 *Estimated Solar Installations to Meet the CO_2 Equivalent Gap of India*

India's CO_2 equivalent gap is around 9.3 Gigatons, of which power sector is expected to contribute 3.28 Gigatons. With abundant potential of solar radiation levels, the Renewable Energy Corporation (REC) estimates that solar energy sector can be accountable for contributing 40% of necessary reduction in emission amounting to 1.31 Gigatons. These reductions can be achieved with solar applications higher by 374GW than the current estimated solar energy installations by 2025, i.e. 2.5 times higher the current estimated solar installations. This is in assumptions with 15% compound annual growth rate (CAGR), 900g/kWh of CO_2 emission and solar radiation yield of 1440 kWh/kWp [52]. Several solar installation initiatives have been taken by Indian administration as a measure of remedy to reduce the emission rate like utility scale Solar installations and rooftop-based solar installations.

18.6.3.2 Role of Indian Government to Reduce Carbon Emission Through Solar Energy Applications

The National Determined Contribution (NDC) as per its communication with United Nations Framework Convention on Climate Change (UNFCC) on October, 2015 aimed to reduce greenhouse emissions by 33–35% of country's GDP compared to 2005 levels by 2030. Indian Prime Minister recently on India's 26th session of Conference of Parties (COP26) held in Glasgow, United Kingdom raised the emission percentage reduction to 45% of GDP by 2030. During the period of 2005–2016, the emission intensity of GDP has been reduced by 24%. The updated NDC of COP26 further quotes *'To put forward and further propagate a healthy and sustainable way of living based on traditions and values of conservation and moderation, including through a mass movement for 'LIFE'– 'Lifestyle for Environment' as a key to combating climate change'*. LIFE visions to provide a lifestyle for humans by not harming the planet. A document titled 'India's long-term Low Carbon Development Strategy' was also submitted to UNFCC in November, 2022 with an aim to reach the net-zero carbon emission goal of India by 2070 [88, 89].

Updated NDC of COP26 additionally represents India's framework to achieve a cleaner energy transition during the period 2021–2030. One of the key schemes that has been laid down to combat climate change includes National Solar Mission (NSM) under National Plan on Climate Change (NAPCC) to promote India's energy security and solar energy access. NSM is responsible for setting up large solar PV plants and aims to generate 40% of power by use of renewable energy sources by 2030. NSM launched initially by ex-PM Manmohan Singh as Jawaharlal Nehru National Solar Mission (JNNSM) in the year 2010 to develop 20GW of power by 2022. Coming to year 2015, Current Indian PM Narendra Modi raised the demand to meet 100 GW by 2022. The previous goal of 20 GW was achieved by 2018, 4 years prior to 2022. Several other solar schemes have been introduced in the year 2022 that includes- Pradhan Mantri Kisan Urja Suraksha evam Utthaan Mahabhiyan (PMKUSUM) for utilization of solar energy in agricultural sector and remote villages and Atal Jyoti Yojana (AJAY) for installation of solar street lights [90]. The goal of introducing these schemes in state and central level was to meet the energy demand while also reduce the greenhouse gas emissions obtained otherwise from the use of fossil fuel powered power plants.

As reported by MNRE, overall renewal energy (RE) capacity of India was 104GW dated November 2021. Out of which solar energy contributed the maximum with 47% followed by 38% from wind derived power, 10% from biomass related sector and 5% from hydro power projects. Among the states, considering utility scale solar installations highest was accounted for Rajasthan with capacity of 3615 MW followed by Gujarat and Uttar Pradesh with 1538 MW and 674 MW respectively. Gujarat however had the maximum rooftop-based solar PV installations with capacity of 660 MW trailed by Maharashtra, Haryana and Rajasthan with 597MW, 266MW and 225MW respectively [52]. The installations accounts to period ranging from January-November 2021. Every mission taken by the administration is in hope to shift the power generation to a greener source that emits less greenhouse gases compared to fossil fuels. The initiatives taken by government are all in line with the NDC's objective to reduce the CO_2 emission rate by 45% from 2005 levels.

18.7 Conclusion

Solar power systems that harness energy from the sun are instrumental in fortifying India's economy while enhancing its energy security and self-sufficiency. As a part of its renewable energy endeavours, India has established a range of solar-powered facilities leading to a target of net zero emission by 2070 at COP26 in 2021. India's cumulative installed solar capacity crossed the 35 GW mark in 2020. In 2019, the Bhadla Solar Park in Rajasthan reached a capacity of 2.25 GW making it the largest solar power plant in India.

Moreover, the exploration extends to encompass government actions and policies like the launch of Jawaharlal Nehru National Solar Mission (JNNSM) in 2010, given their paramount significance in advancing solar energy uptake, attracting investments and propelling the upward trajectory of India's solar sector. These initiatives are of profound importance as they also focused on reducing country's carbon footprints. With abundant potential of solar radiation levels, the Renewable Energy Corporation (REC) estimates that solar energy sector can be accountable for contributing 40% of necessary reduction in emission amounting to 1.31 Gigatons from power sectors contribution of 3.28 Gigatons.

The distributed renewable power sector was able to support approximately 80,000 employees in India, and it was expected to do more in the coming years sorting the unemployment issue of Indian youths which fuels economic expansion. It also underscores their cost-effectiveness, role in providing energy access to rural communities and their pivotal contribution to industrial progress and knowledge dissemination.

Further, The Government of India announced a target of achieving 500 GW of renewable energy capacity by 2030, including a substantial share from solar power. These installations employ diverse technologies and setups to capture solar energy for generating electricity to foster the growth of the Indian economy. Thus, by expanding its utilization of abundant solar potential through innovative policies, India has committed to achieve its both the goals of sustainable and economic development along with mitigating climate change towards a greener world for generations to come.

Acknowledgements This work is a part of the start-up project (**Grant number:** ASTEC/S&T/1802/1/2019-20/1574-1588) awarded to **Dr. Pankaj Kalita**, Associate Professor, **School of Energy Science and Engineering, Indian Institute of Technology Guwahati, Assam, India.** The financial support extended by IIT Guwahati is gratefully acknowledged.

References

1. GDP per capita growth (annual %). https://data.worldbank.org/indicator/NY.GDP.PCAP.KD.ZG?end=2022&start=1978&view=chart. Accessed 8 Aug 2023
2. Jackson RB (2022) Global fossil carbon emissions rebound near pre- COVID-19 levels OPEN ACCESS
3. Ritchie H, Roser M, Rosado P Energy. In: 2022. https://ourworldindata.org/energy. Accessed 8 Aug 2023
4. Holechek JL, Geli HME, Sawalhah MN (2022) A global assessment: can renewable energy replace fossil fuels by 2050? pp 1–22
5. (2020) Global Energy Review 2020. In: Glob. Energy Rev. 2020. https://www.iea.org/reports/global-energy-review-2021/renewables. Accessed 8 Aug 2023
6. Renewable Energy Agency I (2023) Global geothermal market and technology assessment 2 about Irena. In: 2023 www.irena.org. Accessed 8 Aug 2023
7. Khanna MK, Malik S, Kumar H (2023) Indian solar panel initiatives in reducing carbon dioxide emissions, pp 191–203. https://doi.org/10.4236/epe.2023.154009
8. Raina G, Sinha S (2019) Outlook on the Indian scenario of solar energy strategies: policies and challenges. Energy Strateg Rev 24:331–341. https://doi.org/10.1016/j.esr.2019.04.005
9. Ministry of Micro S & ME (2022) Year End Review- 2022. In: Press Relase Punjab Gov. https://pib.gov.in/PressReleasePage.aspx?PRID=1886709. Accessed 8 Aug 2023
10. Invest India (2023) Renewable Energy. https://www.investindia.gov.in/sector/renewable-energy. Accessed 8 Aug 2023
11. Singh A (2010) Per Capita Electricity Consumption. In: 2021. https://ourworldindata.org/grapher/per-capita-electricity-consumption?tab=chart&country=~ETH%0Ahttps://ourworldindata.org/grapher/per-capita-electricity-consumption. Accessed 8 Aug 2023
12. International Energy Agency I (2021) Annual power sector capacity additions in India. https://www.iea.org/data-and-statistics/charts/annual-power-sector-capacity-additions-in-india-2010-2019. Accessed 8 Aug 2023
13. Das A, Saini V, Parikh K, Parikh J, Ghosh P, Tot M (2023) Pathways to net zero emissions for the Indian power sector. Energy Strateg Rev 45:101042. https://doi.org/10.1016/j.esr.2022.101042
14. Ministry of Commerce & Industry Initiatives taken by the government to boost manufacturing
15. Thapar S (2022) Centralized vs decentralized solar: a comparison study (India). Renew Energy 194:687–704. https://doi.org/10.1016/j.renene.2022.05.117

16. Ahmad L, Khordehgah N, Malinauskaite J, Jouhara H (2020) Recent advances and applications of solar photovoltaics and thermal technologies. Energy 207. https://doi.org/10.1016/j.energy.2020.118254
17. Fazal MA, Rubaiee S (2023) Progress of PV cell technology: Feasibility of building materials, cost, performance, and stability. Sol Energy 258:203–219. https://doi.org/10.1016/j.solener.2023.04.066
18. Diwania S, Agrawal S, Siddiqui AS, Singh S (2020) Photovoltaic–thermal (PV/T) technology: a comprehensive review on applications and its advancement. Intern J Energy Environ Eng 11(1):33–54. https://doi.org
19. Wilson GM, Al-Jassim M, Metzger WK, Glunz SW, Verlinden P, Xiong G, Mansfield LM, Stanbery BJ, Zhu K, Yan Y, Berry JJ, Ptak AJ, Dimroth F, Kayes BM, Tamboli AC, Peibst R, Catchpole K, Reese MO, Kling Pastuszak J, Węgierek P (2022) Photovoltaic cell generations and current research directions for their development. Materials (Basel) 15. https://doi.org/10.3390/ma15165542
20. Green MA (2019) Photovoltaic technology and visions for the future. In Progress in Energy (Vol. 1, Issue 1). Institute of Physics. https://doi.org/10.1088/2516-1083/ab0fa8
21. Suman SP, Goyal P (2020) Evolution of PV technology from conventional to nano-materials. Mater Today Proc 28:1593–1597. https://doi.org/10.1016/j.matpr.2020.04.846
22. Shiradkar N, Arya R, Chaubal A, Deshmukh K, Ghosh P, Kottantharayil A, Kumar S, Vasi J (2022) Recent developments in solar manufacturing in India. Sol Compass 1:100009. https://doi.org/10.1016/j.solcom.2022.100009
23. IV relation changing load resistor of a solar cell. https://electronics.stackexchange.com/questions/289250/iv-relation-changing-load-resistor-of-a-solar-cell. Accessed 8 Aug 2023
24. Evangelisti L, De Lieto Vollaro R, Asdrubali F (2019) Latest advances on solar thermal collectors: a comprehensive review. In Renewable and Sustainable Energy Reviews (Vol. 114). Elsevier Ltd. https://doi.org/10.1016/j.rser.2019.109318
25. Herrando M, Markides CN (2016) Hybrid PV and solar-thermal systems for domestic heat and power provision in the UK: techno-economic considerations. Appl Energy 161:512–532. https://doi.org/10.1016/j.apenergy.2015.09.025
26. Rockendorf G, Sillmann R, Podlowski L, Litzenburger B (1999) PV-hybrid and thermoelectric collectors (Vol. 67). www.elsevier.com/locate/solener
27. Das D, Kamble AD, Kalita P (2022) Performance investigation of transparent photovoltaic-thermal collector with horizontal oscillating and rectangular spiral flow patterns. Int J Energy Res 46:239–253. https://doi.org/10.1002/er.6298
28. Das D, Bordoloi U, Kamble AD, Muigai HH, Pai RK, Kalita P (2021) Performance investigation of a rectangular spiral flow PV/T collector with a novel form-stable composite material. Appl Therm Eng 182:116035. https://doi.org/10.1016/j.applthermaleng.2020.116035
29. Herrando M, Wang K, Huang G, Otanicar T, Mousa OB, Agathokleous RA, Ding Y, Kalogirou S, Ekins-Daukes N, Taylor RA, Markides CN (2023) A review of solar hybrid photovoltaic-thermal (PV-T) collectors and systems. In Progres
30. Kalogirou SA (2012) A detailed thermal model of a parabolic trough collector receiver. Energy 48:298–306. https://doi.org/10.1016/j.energy.2012.06.023
31. Zhu G, Wendelin T, Wagner MJ, Kutscher C (2014) History, current state, and future of linear Fresnel concentrating solar collectors. Sol Energy 103:639–652. https://doi.org/10.1016/j.solener.2013.05.021
32. Karimi R, Gheinani TT, Madadi Avargani V (2018) A detailed mathematical model for thermal performance analysis of a cylindrical cavity receiver in a solar parabolic dish collector system. Renew Energy 125:768–782. https://doi.org/10.1016/j.renene.2018.03.015
33. Solar tower power. https://commons.wikimedia.org/wiki/File:CSIRO_ScienceImage_2141_Solar_Array_at_the_CSIRO_Energy_Centre.jpg. Accessed 8 Aug 2023
34. Savsani M, Motwani K, Patel J (2017) Medium temperature application of parabolic trough collector for industrial process heating-review. 6th International Conference on Recent Trends in Engineering, Science and Management, pp 1171–1181. http://ijates.com/images/short_pdf/1484056368_N303_IJATES.pdf

35. Solar Street Lighting Design. https://www.flyinglighting.com/street-light/led-street-light/best-selling-street-lamp-72w-led-street-light.html. Accessed 8 Aug 2023
36. Swamy KN, Kondamudi R, Vijayalakshmi P, Jaldhani V, Suchandranath BM, Kiran TV, Srikanth B, Subhakar RI, Sailaja N, Surekha K, Neeraja CN, Subba Rao LV (2015) Afr J Agric Res 10:4364–4369. https://doi.org/10.5897/AJAR2015.10323
37. Bs V, Sw S (2015) Solar photovoltaic water pumping system for irrigation: a review. African J Agric Res 10:2267–2273. https://doi.org/10.5897/ajar2015.9879
38. Energy Management Systems: Maximizing Energy Savings. https://www.energy.gov/scep/energy-management-systems-maximizing-energy-savings-text-version. Accessed 8 Aug 2023
39. Albarakati AJ, Boujoudar Y, Azeroual M, Eliysaouy L, Kotb H, Aljarbouh A, Khalid Alkahtani H, Mostafa SM, Tassaddiq A, Pupkov A (2022) Microgrid energy management and monitoring systems: A comprehensive review. Front Energy Res 10:1–18. https://doi.org/10.3389/fenrg.2022.1097858
40. Lamnatou C, Chemisana D, Cristofari C (2022) Smart grids and smart technologies in relation to photovoltaics, storage systems, buildings and the environment. Renew Energy 185:1376–1391. https://doi.org/10.1016/j.renene.2021.11.019
41. Qays MO, Ahmad I, Abu-Siada A, Hossain ML, Yasmin F (2023) Key communication technologies, applications, protocols and future guides for IoT-assisted smart grid systems: A review. Energy Rep 9:2440–2452. https://doi.org/10.1016/j.egyr.2023.01.085
42. SM technologies, MoP, GoI. https://www.nsgm.gov.in/. Accessed 8 Aug 2023
43. Alnaimat F, Klausner J, Mathew B (2018) 'Solar desalination', Desalination and water treatment. InTech. https://doi.org/10.5772/intechopen.76981
44. Robert B, Brown EB (2004) Solar_park_bench. https://commons.wikimedia.org/wiki/File:Solar_park_bench_at_Iowa_State_Fair_grounds.gk.jpg. Accessed 8 Jul 2023
45. Nakanishi S, Matsumura K, Mori H, Ishii K, Ogawa K (2005) Solar street light. In: Shapu Giho/Sharp Tech. J. https://commons.wikimedia.org/wiki/File:Solar_Street_Light_Balatan.JPG. Accessed 8 Jul 2023
46. Solar-charging-station. https://commons.wikimedia.org/wiki/File:Baka-charging-station.jpg. Accessed 8 Aug 2023
47. KSINC launches solar-powered tourist vessel Sooryamshu. https://www.thehindu.com/news/cities/Kochi/ksinc-launches-solar-powered-tourist-vessel-sooryamshu/article66699921.ece. Accessed 8 Aug 2023
48. Solar-Powered Airports in the World. https://www.solarfeeds.com/mag/solar-powered-airports-in-the-world/. Accessed 8 Aug 2023
49. Rajawat K (2019) International solar alliance: India's potential in clean energy, pp 9–11
50. ISA annual report 2021. https://isaorgwebsite.blob.core.windows.net/media/isa/media/pdf/governancepdfs/isa-annual-report-2021.pdf. Accessed 8 Aug 2023
51. Shidore S, Busby JW (2019) One more try: the international solar alliance and India's search for geopolitical in fluence, 26. https://doi.org/10.1016/j.esr.2019.100385
52. Schulze A (REC), Lea H (2016) Climate change: closing the COP21 Gap by Going Solar. https://www.recgroup.com/sites/default/files/documents/rec_cop21_study_final_web.pdf. Accessed 8 Aug 2023
53. TWI CCT (2017) Common risk mitigation mechanism feasibility study. Common Risk Mitig Mech Feasibility Study
54. Sivaram V, Shrimali G, Reicher D (2015) How India's audacious solar ambitions could make or break its climate commitments
55. Busby JW (2011) Aids drugs for all: social movements and market transformations. AIDS Drugs All Soc Movements Mark Transform 1–322. https://doi.org/10.1017/CBO9781139565967
56. Annual report (2022). https://isolaralliance.org/uploads/docs/52e970d74272c9079053d966043b80.pdf. Accessed 8 Aug 2023
57. ISA (2021) Scaling solar application for agricultural use. https://isolaralliance.org/work/scaling-solar-application-agricultural-use. Accessed 8 Aug 2023
58. MNRE. https://mnre.gov.in/annual-report/. Accessed 8 May 2023

59. Budget 2022: State of solar energy sector. https://www.investindia.gov.in/team-india-blogs/budget-2022-state-solar-energy-sector. Accessed 8 Aug 2023
60. Committee S, Energy ON (2010) Ministry of new and renewable energy. http://mnre.gov.in/schemes/new-technologies/tidal-energy/. Accessed 8 Aug 2023
61. MNRE (2021) Ministry of new and renewable energy. Govern India: Annual Report 2020–2021(53):1689–1699
62. Potential locations for solar, wind hybrid projects in India. https://www.pv-magazine.com/2023/05/01/potential-locations-for-solar-wind-hybrid-projects-in-india/. Accessed 8 Aug 2023
63. Prakash O, Ahmad A, Kumar A, Chatterjee R, Sharma S, Alayi R, Monfared H (2022) The compressive study of energy security prospects in India through solar power. Int J Low-Carbon Technol 17:962–979. https://doi.org/10.1093/ijlct/ctac075
64. India's Solar Parks Achieve Milestones. https://solarquarter.com/2023/07/21/indias-solar-parks-achieve-milestones-11-parks-completed-with-8521-mw-capacity-7-parks-partially-finished/. Accessed 8 Aug 2023
65. Shah K (2020) India is home to the world's largest utility-scale solar installations, pp 1–14
66. Pavagada-solar-park. https://www.nsenergybusiness.com/projects/pavagada-solar-park/. Accessed 8 Aug 2023
67. Sukumaran S, Sudhakar K (2017) Fully solar powered airport: A case study of Cochin International airport. J Air Transp Manag 62:176–188. https://doi.org/10.1016/j.jairtraman.2017.04.004
68. MNRE (2019) Annual report ministry of new and renewable energy
69. Tulika Tandon (2021) List of solar power plants / solar parks in India. In: https://www.jagranjosh.Com/. https://www.jagranjosh.com/general-knowledge/list-of-solar-power-plants-in-india-1607686203-1. Accessed 8 Aug 2023
70. Technologies revolutionizing solar energy. https://www.pv-magazine-india.com/2022/06/07/technologies-revolutionizing-solar-energy/. Accessed 8 Aug 2023
71. Kamath HG, Ekins-Daukes NJ, Araki K, Ramasesha SK (2019) The potential for concentrator photovoltaics: a feasibility study in India. Prog Photovoltaics Res Appl 27:316–327. https://doi.org/10.1002/pip.3099
72. Shittu S, Li G, Zhao X, Ma X, Akhlaghi YG, Ayodele E (2019) Optimized high performance thermoelectric generator with combined segmented and asymmetrical legs under pulsed heat input power. J Power Sources 428:53–66. https://doi.org/10.1016/j.jpowsour.2019.04.099
73. Rowe DM (1999) Thermoelectrics, an environmentally-friendly source of electrical power. Renew Energy 16:1251–1256. https://doi.org/10.1016/s0960-1481(98)00512-6
74. Li G, Chen X, Jin Y (2017) Analysis of the primary constraint conditions of an efficient photovoltaic-thermoelectric hybrid system. Energies 10:1–12. https://doi.org/10.3390/en10010020
75. Deka MJ, Kalita P, Das D, Kamble AD, Jyoti Bora B, Sharma P, Jyoti Medhi B (2023) An approach towards building robust neural networks models using multilayer perceptron through experimentation on different photovoltaic thermal systems. Energy Convers Manag 292. https://doi.org/10.1016/j.enconman.2023.117395
76. Floating solar farms. https://www.power-technology.com/features/worlds-biggest-floating-solar-farms/?cf-view. Accessed 8 Aug 2023
77. Dholera Solar Park, Gujarat. https://www.power-technology.com/projects/dholera-solar-park-gujarat/. Accessed 8 Aug 2023
78. Cleaning-solar-panels. https://blog.ecoflow.com/us/cleaning-solar-panels/. Accessed 8 Aug 2023
79. Monitoring-platforms-solar-photovoltaic-system. https://www.energy.gov/femp/monitoring-platforms-solar-photovoltaic-systems. Accessed 8 Aug 2023
80. Solar-guides/what-do-solar-inverter-error-codes-mean/. https://instylesolar.com/solar-guides/what-do-solar-inverter-error-codes-mean/. Accessed 8 Aug 2023
81. Punia Sindhu S, Nehra V, Luthra S (2016) Recognition and prioritization of challenges in growth of solar energy using analytical hierarchy process: Indian outlook. Energy 100:332–348. https://doi.org/10.1016/j.energy.2016.01.091

82. Yenneti K (2016) The grid-connected solar energy in India: structures and challenges. Energy Strateg Rev 11–12:41–51. https://doi.org/10.1016/j.esr.2016.06.002
83. Rathore N, Panwar NL (2022) Outline of solar energy in India: advancements, policies, barriers, socio-economic aspects and impacts of COVID on solar industries. Int J Ambient Energy 43:7630–7642. https://doi.org/10.1080/01430750.2022.2075925
84. Singh SRK (2023) A number of incentives being provided by Government for implementation of major renewable energy programmes - Union Power & NRE Minister. https://pib.gov.in/PressReleasePage.aspx?PRID=1911485. Accessed 8 Aug 2023
85. India-solar-energy-market. https://www.custommarketinsights.com/report/india-solar-energy-market/. Accessed 8 Aug 2023
86. Duty C (2022) Solar tariffs to rise by ~21% in the next 12 months impact of module price volatility and basic customs duty, pp 1–18
87. Solar power cost. https://energy.economictimes.indiatimes.com/news/renewable/solar-power-cost-will-fall-to-rs-1-9-per-unit-in-india-by-2030-teri-study/67972162. Accessed 8 Aug 2023
88. Government of India (2015) India's updated first nationally determined contribution under Paris agreement. https://unfccc.int/sites/default/files/NDC/2022-08/India Updated First Nationally Determined Contrib.pdf. Accessed 8 Aug 2023
89. Aathira Perinchery, The Wire (2022) India updates its nationally determined contribution to tackle climate change. In: Wire. https://thewire.in/environment/india-updates-its-nationally-determined-contribution-to-tackle-climate-change. Accessed 8 Aug 2023
90. India's clean energy transition is rapidly underway, benefiting the entire world. https://www.iea.org/commentaries/india-s-clean-energy-transition-is-rapidly-underway-benefiting-the-entire-world. Accessed 8 Aug 2023

The manufacturer's authorised representative in the EU is Springer Nature Customer Service Centre GmbH, Europaplatz 3, 69115 Heidelberg, Germany. If you have any concerns regarding our products, please contact ProductSafety@springernature.com

Printed and bound by CPI Group (UK) Ltd, Croydon, CR0 4YY
23/09/2026
02221471-0008